Benchmark Papers in Electrical Engineering and Computer Science Series

Editor: John B. Thomas—Princeton University

A selection of published volumes in this series

SYSTEM SENSITIVITY ANALYSIS / *José B. Cruz, Jr.*
RANDOM PROCESSES: Multiplicity Theory and Canonical Decompositions / *Anthony Ephremides and John B. Thomas*
ALGEBRAIC CODING THEORY: History and Development / *Ian F. Blake*
TWO-DIMENSIONAL DIGITAL SIGNAL PROCESSING / *Robert M. Gray and Lee D. Davisson*
COMPUTER ARITHMETIC / *Earl E. Swartzlander, Jr.*
ARRAY PROCESSING: Applications to Radar / *Simon Haykin*
NONPARAMETRIC DETECTION: Theory and Applications / *Saleem A. Kassam and John B. Thomas*
DECONVOLUTION OF SEISMIC DATA / *V. K. Arya and J. K. Aggarwal*
REPRODUCING KERNEL HILBERT SPACES: Applications in Statistical Signal Processing / *Howard L. Weinert*
DISTRIBUTED PARAMETER SYSTEMS THEORY, PART I: Control / *Peter Stavroulakis*
DISTRIBUTED PARAMETER SYSTEMS THEORY, PART II: Estimation / *Peter Stavroulakis*
QUANTIZATION / *Peter F. Swaszek*

QUANTIZATION

Edited by
PETER F. SWASZEK
University of Rhode Island

A Hutchinson Ross Benchmark® Book

VAN NOSTRAND REINHOLD COMPANY
New York

Benchmark Papers in Electrical Engineering and Computer Science Series,
Volume 29
Library of Congress Catalog Card Number: 84-23449
ISBN: 0-442-28124-2

Manufactured in the United States of America.

Published by Van Nostrand Reinhold Company Inc.
135 West 50th Street
New York, New York 10020

Van Nostrand Reinhold Company Limited
Molly Millars Lane
Wokingham, Berkshire RG11 2PY, England

Van Nostrand Reinhold
480 Latrobe Street
Melbourne, Victoria 3000, Australia

Macmillan of Canada
Division of Gage Publishing Limited
164 Commander Boulevard
Agincourt, Ontario MIS 3C7, Canada

15 14 13 12 11 10 9 8 7 6 5 4 3 2 1

Library of Congress Cataloging in Publication Data
Main entry under title:
Quantization.
(Benchmark papers in electrical engineering and computer science; v. 29)
"A Hutchinson Ross Benchmark book."
Includes indexes.
1. Analog-to-digital converters. 2. Signal processing—digital techniques.
I. Swaszek, Peter F. II. Series.
TK7887.6.Q36 1985 001.64'42 84-23449
ISBN 0-442-28124-2

CONTENTS

PART III: VECTOR SIGNAL QUANTIZERS

PART IV: ROBUST SIGNAL QUANTIZERS

SERIES EDITOR'S FOREWORD

This Benchmark Series in Electrical Engineering and Computer Science is aimed at sifting, organizing, and making readily accessible to the reader the vast literature that has accumulated. Although the series is not intended as a complete substitute for a study of this literature, it will serve at least three major critical purposes. In the first place, it provides a practical point of entry into a given area of research. Each volume offers an expert's selection of the critical papers on a given topic as well as his views on its structure, development, and present status. In the second place, the series provides a convenient and time-saving means for study in areas related to but not contiguous with one's principal interests. Last, but by no means least, the series allows the collection, in a particularly compact and convenient form, of the major works on which present research activities and interests are based.

Each volume in the series has been collected, organized, and edited by an authority in the area to which it pertains. In order to present a unified view of the area, the volume editor has prepared an introduction to the subject, has included his comments on each article, and has provided a subject index to facilitate access to the papers.

We believe that this series will provide a manageable working library of the most important technical articles in electrical engineering and computer science. We hope that it will be equally valuable to students, teachers, and researchers.

This volume, *Quantization*, has been edited by Peter F. Swaszek of the University of Rhode Island. It contains thirty-seven papers on signal quantization and its applications. Although considerable attention has been given to the classical contributions, most of the papers reflect recent advances in quantizer design, particularly as it relates to vector and robust signal quantization.

JOHN B. THOMAS

PREFACE

Quantization is the process of transforming a discrete-time, continuously valued random process into a discrete-time, discretely valued random process. Typically this "analog-to-digital conversion" of data is employed before digital processing of the data occurs, "digital processing" being defined to include such diverse operations as arithmetic computation (with finite wordlengths), digital communications (pulse code modulation–type schemes), and computer memory storage. Since quantization is an irreversible process, mapping a continuum of values into a finite set of values, it introduces distortion, typically unwanted, onto the original data. The focus of the published research on quantization has been on the design, analysis, and application of schemes that accomplish this data simplification with minimum distortion (equivalently, with highest fidelity). In a typical application, A/D (analog/digital) conversion is applied to a large set of random inputs. Since the distortion is different for each input value, a useful measure of the quality of the quantization is a statistical average of the distortion for each input. In fact, a measure common to many of the papers in this volume is mean-square error, the power in the difference signal between the input and the output of the quantizer.

As a subset of data-compression techniques, quantization has been considered in the technical literature for many years. A reader interested in signal quantization who attempted a search of the open literature would find hundreds of articles on various topics related to quantizer design, use, and performance. The articles range from evaluations of the distortion for a particular source and quantizer pair (which has been done for a variety of sources and distortion measures) through theoretical investigations of the best achievable performance to discussions of the complexity and convergence of iterative design algorithms. The goal of this volume is to partition this vast literature into several subtopics, providing selections from the available literature on each topic with editorial comments describing the papers' contributions to and contextual position in that area. To that end, this volume is divided into six parts:

Part I: Overview of Quantization
Part II: Scalar Signal Quantizers
Part III: Vector Signal Quantizers

Part IV: Robust Signal Quantizers
Part V: Suboptimum Vector Signal Quantizers
Part VI: Applications of Quantizers

Of these six parts, the first and last contain only a few articles; the aim of this volume is to present in detail the four major areas, Parts II through V. Part I provides an introduction to the quantization problem and its solutions. The applications section, Part VI, provides continuity for the reader to past and present uses of the techniques described in the previous sections.

Scalar quantizers, historically the most commonly employed form of quantizer because of the ease of their implementation and analysis, are discussed in Part II. In this section are collected several of the older, commonly referenced classics on scalar quantization. Vector quantizers are the subject of Part III. By releasing the constraint of memoryless (scalar) devices, additional gain in performance can be obtained. The papers in this section discuss the attainable performance gain and describe methods for designing and implementing these quantizers. Part IV contains investigations of the robustness of quantizers to changes in the input source's statistics. Fixed source statistics are commonly assumed in the papers of Parts II and III. In practical problems such as speech or image quantization, the source statistics are time varying (non-stationary), and the above assumption is a poor one. The topics under this general heading of robust signal quantizers include performance and design rules under source/quantizer mismatch and design techniques for incompletely defined source situations. Finally, the papers in Part V consider suboptimum vector quantizers. The vector quantizers discussed in Part III are typically difficult to design and implement, requiring a large number of mathematical operations. Suboptimum vector quantizers are defined as those schemes that outperform scalar quantizers but with design and implementations less complex than those of optimum vector quantizers.

The approach employed in the selection of papers for this volume is to concentrate on the subject areas of Parts III, IV, and V, these being more recent areas of research interest, while also providing easy access in Part II to a few of the classic references on scalar quantization. As a further guide to the literature, the editorial comments to each major subsection include bibliographies with references in the text to articles on similar material.

This volume was assembled with full knowledge of the other reprint volumes available on quantization; it updates and supplements them while also providing a different view of the problem and its solutions. In particular, those readers interested in early works on scalar quantization and other data-compression techniques should consult another volume in the Benchmark series, *Data Compression* (L. D. Davisson, and R. M. Gray, eds., Dowden, Hutchinson, & Ross, Inc., 1976). Those interested in applications of scalar quantizers to speech and image data should consult *Waveform Quantization and Coding* (N. Jayant, ed., IEEE Press, 1976). In addition, a special issue of *IEEE Transactions on Information Theory* published in March 1982 was dedicated to recent research in quantization. Several of the papers reprinted here are contained in that issue.

Finally, I would like to acknowledge support for this project from the National Science Foundation under NEFRI Grant ECS-8206237 and from a Physio-Control Corporation Career Development Award while I was at the University of Washington. Also, special thanks go to the series editor, John B. Thomas, for suggesting the production of this volume.

PETER F. SWASZEK

CONTENTS BY AUTHOR

INTRODUCTION

Quantization, which is defined as the mapping of data into a coarser, more easily handled form, is of ubiquitous need in today's information processing systems. Information-bearing signals are typically modeled as analog in nature while most processing of the signals is digital. The abundance of digital processing results from the increased availability and reliability of computers and other digital equipment. As examples, consider the following: Digital communications systems encode samples of an input signal (commonly speech or image data) into a bit stream of finite length—finite because of channel capacity or bandwidth constraints. The mathematical operations of digital signal processing techniques, including digital holography, FFT computation, speaker recognition, image enhancement, pattern recognition, and digital filtering, are implemented on digital computers using finite-length binary representations for the data values. For these applications a simple and efficient method of transforming continuously valued signals into discretely valued signals is both desirable and necessary to ensure good system performance. As an added note, it is sometimes useful to be able to map discrete or finely digitized data into a coarser form. The definition of the quantization process will include this situation also.

Typically, information-bearing signals exist on a continuum of time. For the discussions reprinted here, a time sampling of the signal is assumed to have already occurred and the information-theoretic model of a discrete-time, continuously valued random process will be employed for the input source to the quantizer. The input space will be the k-dimensional Euclidean space $\mathcal{R}^k$ with the input data a k-dimensional random vector $\mathbf{x}$. Occasionally, the input signal will fall naturally into a group of length k-that is, positional coordinates, time-sequential samples, or a state vector. A natural grouping is not necessary, however; a length k input vector is defined here to be any ordered set of k input variables, whether they occur concurrently, sequentially, or both. The source is modeled statistically by its multivariate probability density function $p(\mathbf{x})$, often assuming the exact form of $p(\mathbf{x})$ to be known and to be continuous on $\mathcal{R}^k$. Relaxation of these assumptions has also been considered in the literature (see the editorial comments in Parts III and IV).

With these assumptions, the problem of interest is to decide in some way how to round off a vector observation to one of a set of N values. For the above-defined source, an N-level quantizer is defined to be a mapping of $\mathfrak{R}^k$ onto N distinct points in $\mathfrak{R}^k$. The mapping is noninvertible in the sense that many points in the original space are mapped to a particular output point. A k-dimensional, N-level vector quantizer can be characterized by the set $\{S_i, \mathbf{y}_i; i = 1, 2, \ldots N\}$, where the S_i are the N quantization regions (a partitioning of $\mathfrak{R}^k$) and the $\mathbf{y}_i$ are the N output points associated with these regions. Letting $Q(\mathbf{x})$ represent the result of quantizing an input vector $\mathbf{x}$, the operation of the quantizer is to map every point in S_i into $\mathbf{y}_i$:

$$Q(\mathbf{x}) = \mathbf{y}_i \text{ if } \mathbf{x} \in S_i.$$

Clearly, for a nonambiguous mapping, the S_i should be disjoint regions that cover $\mathfrak{R}^k$.

Another characterization of the quantizer, especially convenient for scalar ($k = 1$) quantizers, is the compander model. For this characterization a quantizer is modeled as a series connection of three elements: a compressor, g, mapping the range of the input onto the unit hypercube, an N-level uniform quantizer on the hypercube, and an expander, g^{-1}, mapping the hypercube back into the range of the input. The term *compander* results from combining the words "compressor" and "expander." In the scalar case, the three elements are of particularly simple form: the compressor and expander are memoryless nonlinearities mapping the real line to the unit interval and vice versa, and the uniform quantizer is an equal-width interval quantizer on [0,1] (region widths all equal to $1/N$). It is easily argued that any N-level scalar quantizer can be implemented in this fashion by appropriate choice of the compressor, g; hence, the second characterization for a scalar quantizer is the number of levels, N, and the compressor function, g. This model can also be used for vector quantizers (see Papers 11 and 13) even though the compressor, g (now a vector nonlinearity), and the uniform quantizer on the hypercube are somewhat more difficult to select and implement (see the editorial comments in Parts III and IV).

In many applications of quantizers, the system's goal is the eventual reconstruction of the signal or of some well-defined function of the signal. For this case the quantizer representation of the input is expected to be, in some sense, close to the original input. To evaluate the performance of our quantization scheme a measure of this closeness between the quantizer's input and output is required. Unfortunately, choosing such a measure is, in general, a difficult engineering problem. Often a statistical measure of performance, useful if the quantizer is employed to digitize a large number of input samples, is chosen. Letting $E\{\cdot\}$ represent the expectation operation over the source density, a common form of performance measure is the expectation of some function e of the quantizer's input $\mathbf{x}$ and output $Q(\mathbf{x})$:

$$E\{\, e[\mathbf{x}, Q(\mathbf{x})]\, \}.$$

In the majority of the articles reprinted in this volume, the functional e is the square of the Euclidean distance between the input and the output, $|\mathbf{x} - Q(\mathbf{x})|^2$, the performance criterion being the mean squared error (MSE):

$$\mathrm{MSE} = E\{[\mathbf{x} - Q(\mathbf{x})]^T[\mathbf{x} - Q(\mathbf{x})]\} = \int_{\mathcal{R}^k} [\mathbf{x} - Q(\mathbf{x})]^T [\mathbf{x} - Q(\mathbf{x})]\, p(\mathbf{x})\, d\mathbf{x}.$$

The design goal of minimization of the MSE has traditionally been employed in the engineering literature. If $Q(\mathbf{x}) - \mathbf{x}$ is imagined to be a random noise vector added to the input by the quantizer, then the MSE is equivalent to the second moment of the noise variable or "noise power," an intuitive measure for communications engineers. Furthermore, the MSE criterion often yields tractable results. Intuition about the quantization problem and its solution can be developed by studying the MSE case. Many of the results presented herein for MSE also generalize directly to mean rth error. Instances of other measures considered in the literature are referenced in the editorial comments.

The quantization problem, then, is to analyze the performance of various quantization schemes and, more important, to design efficient (high-performance) schemes. Often constraints exist upon the possible quantizer characterization (scalar schemes, fixed entropy of the resulting discrete output, fixed region shapes for the S_i, or a specific functional form of the compressor, g) or upon the statistical environment in which it must operate (a variety of source densities or for a nonstationary source). Thus, depending upon the application, the minimization problem can have a variety of solutions. For a further introduction to the various aspects and terminology of the quantization problem, the reader is referred to Paper 1 and to the editorial comments for the six parts.

The first published reference to the quantization problem is Sheppard's (1896) study of uniform scalar quantizers. Interest in scalar quantizers grew with the advent of pulse code modulation techniques (Cutler, 1952, 1955), several studies appearing in the *Bell System Technical Journal* on uniform and logarithmic companders (Papers 2 and 3) and in the literature on optimum quantizer performance (Panter and Dite, 1951). The frequency of published material on scalar quantizers increased dramatically after Lloyd's and Max's classic papers on optimum scalar quantizers (Papers 4 and 5). Of late, the articles on optimum scalar quantizers appearing in the literature often deal with the complexity and use of the design algorithms, extensions of the available MSE results to other, more general, distortion measures, or applications of quantizers to problems beyond minimum MSE signal representation/reconstruction. Also of concern are techniques for designing quantizers that are insensitive to changes in the source statistics; so-called robust signal quantizers. Several of the recent papers on this subject appear in Part IV.

The improvement possible with vector quantization was initially considered by Schutzenberger (1958) with refinement by Zador (1963) in his often-referenced thesis. Unfortunately, Zador's results demonstrated only how much performance gain was available, not how to obtain it. The majority of results in vector quantization have appeared within the last eight years and are considered in Part III. The thrust of these articles has been in several directions: extensions of Zador's performance bounds to other distortion criteria, advances in the design, performance, and implementation of uniform vector quantizers on the hypercube, companding in *k*-dimensions, and an extension of Lloyd's Method I quantizer design algorithm to the vector case. Along with the design algorithm, several examples of optimum vector quantizers have appeared for small *N* and *k* (for low-bit-rate applications such as speech or image coding). Another area that has developed within the last six years is suboptimum vector quantizers, as discussed in Part V. The goal of these schemes is to obtain most of the performance gain available from vector quantization without the design and implementation complexity inherent with optimum vector quantizers. To date, suboptimum vector quantizers have been considered only for spherically symmetric sources, in particular the multivariate Gaussian source.

As is to be expected in any volume of fixed size, several important areas and papers must be neglected. In this volume, sequential quantizers, including differential quantizers and delta modulators, are not covered. The literature on these schemes is equal in size to that on the types of quantization schemes considered herein. Also, examples of the application of quantization schemes to current data reduction or signal representation problems are limited to only four papers. For both of these areas, the interested reader is referred to the literature and the volume edited by Jayant (1976). Finally, the editor apologizes for any omissions of specific papers but hopes that the included selections provide a useful introduction to the field of quantization.

REFERENCES

Cutler, U.S. Patent Office, #2605361, July 29, 1952, and #2724740, Nov. 20, 1955.

Jayant, N., ed., *Waveform Quantization and Coding,* IEEE Press, New York, 611p.

Panter, P. F., and W. Dite, 1951, Quantization Distortion in Pulse Count Modulation with Non-linear Spacing of Levels, *IRE Proc.* **39**(1):44–48.

Schutzenberger, M. P., 1958, On the Quantization of Finite Dimensional Messages, *Inform. Control* **1**(5):153–158.

Sheppard, W. F., 1896, On the Calculation of the Most Probable Values of Frequency Constants for Data Arranged according to Equidistant Divisions of a Scale, *London Math. Soc. Proc.* **24**(pt.2):353–380.

Zador, P., 1963, *Development and Evaluation of Procedures for Quantizing Multivariate Distributions,* dissertation, Department of Statistics, Stanford University, 64p.

Part I

OVERVIEW OF QUANTIZATION

Editor's Comments on Paper 1

1 **GERSHO**
Principles of Quantization

The first paper in this volume, and the only paper in this section, presents an overview of the quantization problem. A well-written tutorial on the subject, this paper has also appeared in an earlier version (Gersho, 1977) and in an IEEE reprint volume (Lawrence, LoCicero, and Milstein, 1983). Gersho introduces many of the topics considered in further detail in other papers in this volume, including the characterization of scalar (zero-memory) quantizers by breakpoints and output values, quantization noise versus overload distortion with a discussion of mean squared error as a design performance criterion, the performance of uniform scalar quantizers, the companding model for nonuniform scalar quantizers, robust scalar quantization by A-law and μ-law companding, optimum scalar quantization employing centroids and nearest-neighbor regions, optimal companders, and a short introduction to block (vector) quantization including entropy coding and rate distortion bounds. From the above list, it is clear that this expository article is an excellent introduction to the quantization literature; the discussions are simple and to the point, and technical details are referenced to the literature.

REFERENCES

Gersho, A., 1977, Quantization, *IEEE Comm. Mag.* **15**(9):16–29.
Lawrence, V. B., J. L. LoCicero, and L. B. Milstein, 1983, *Tutorials in Modern Communications,* Computer Science Press, Rockville, Md., 348p.

Reprinted from *IEEE Trans. Circuits Syst.* **CAS-25** (7):427–436 (1978)

Principles of Quantization

ALLEN GERSHO, SENIOR MEMBER, IEEE

Abstract—**Quantization is the process of replacing analog samples with approximate values taken from a finite set of allowed values. The approximate values corresponding to a sequence of analog samples can then be specified by a digital signal for transmission, storage, or other digital processing. In this expository paper, the basic ideas of uniform quantization, companding, robustness to input power level, and optimal quantization are reviewed and explained. The performance of various schemes are compared using the ratio of signal power to mean-square quantizing noise as a criterion. Entropy coding and the ultimate theoretical bound on block quantizer performance are also compared with the simpler zero-memory quantizer.**

I. Introduction

THE PROCESSING and transmission of digital signals is rapidly approaching a dominant role in communication systems. Nevertheless, the physical origin of many information-bearing signals (speech, image, telemetry, seismic, etc.) is intrinsically analog and continuous-time in nature. Therefore, an effective interface between the analog and digital worlds is of crucial importance in modern signal-processing. Very often the quality of analog-to-digital (A/D) conversion is the critical limiting factor in overall system performance. A clear understanding of quantization, the essential mechanism of A/D conversion, is needed to answer such questions as how many bits per second (or bits per sample) are really needed, or how much distortion (quantizing noise) is inevitable for a given bit rate.

A/D conversion may be viewed as made up of four operations: prefiltering, sampling, quantizing, and coding. In this paper, we focus on quantization and specifically on "zero-memory" quantization.

Quantization begins with the availability of analog samples. Each sample may in general take on any of a continuum of amplitude values ranging from $-\infty$ to $+\infty$. The quantizer replaces each of these sample values with an output value which is an approximation to the original amplitude. The key feature is that each output value is one of a *finite* set of real numbers. Hence, a symbol from a finite alphabet can be used to represent and identify the particular output value that occurs. A different n-bit binary word can be associated with each output value if the set of output values contains no more then 2^n members. With this procedure, a sequence of analog samples can be transformed into a sequence of binary words suitable for storage, transmission, or some other form of digital signal processing. A receiver having the table of output values (sometimes called "quanta" or "quantum levels") associated with the set of binary words, can then reconstruct an approximation to the original sequence of samples. Hence, with some appropriate form of interpolation, a continuous waveform can be created which approximates the waveform originally applied to the A/D system. The reconstruction process is called digital-to-analog (D/A) conversion.

The simplest and most common form of quantizer is the zero-memory quantizer. In this case, the output value is determined by the quantizer only from one corresponding input sample, independent of the values taken on by earlier (or later) analog samples applied to the quantizer input. More sophisticated (but less well understood theoretically) is the block quantizer which looks at a group or "block" of input samples simultaneously and produces a block of output values, chosen from a finite set of possible output blocks, approximating the corresponding input samples. In general, for a given number of bits per sample representing the output values, a better quality approximation can be achieved by block quantization. Of theoretical interest is the limiting case where the block length approaches infinity. Studying this limiting situation provides information about the ultimate quality of approximation achievable for a given bit rate. Another class of quantizers which could be described as sequential quantizers includes such well-known digitization schemes as delta modulation, differential PCM, and other adaptive versions. A sequential quantizer stores some information about the previous samples and generates the present quantized output using both the current input and the stored information. In this paper, we shall focus primarily on zero-memory quantization. Quantization with memory will be discussed only for the purpose of examining how much can be gained through the use of memory.

II. Zero-Memory Quantization

A zero-memory N-point quantizer Q may be defined by specifying a set of $N+1$ decision levels $x_0, x_1, \cdots, x_N$ and a set of N output points $y_1, y_2, \cdots, y_N$. When the value x of an input sample lies in the ith quantizing interval, namely

$$R_i = \{x_{i-1} < x < x_i\}$$

the quantizer produces the output value y_i. Since y_i is used to approximate samples contained in the interval R_i, y_i is itself chosen to be some value in the interval R_i. The end

Manuscript received November 10, 1977; revised January 30, 1978.
The author is with Bell Laboratories, Murray Hill, NJ 07974. This work was done while the author was visiting the Department of System Science, University of California, Los Angeles, CA.

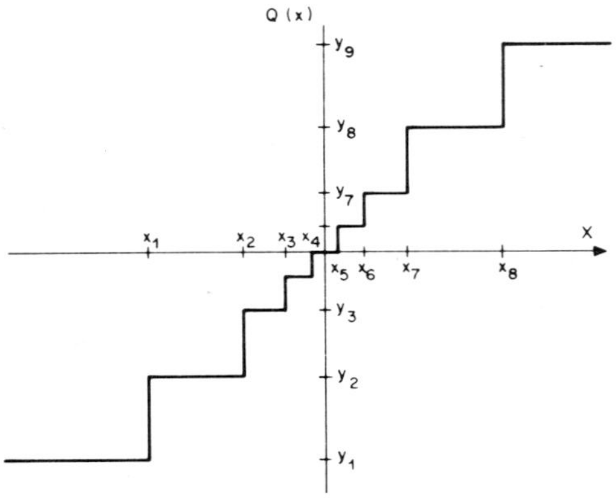

Fig. 1. Input–output characteristic of a midtread quantizer with $N=9$.

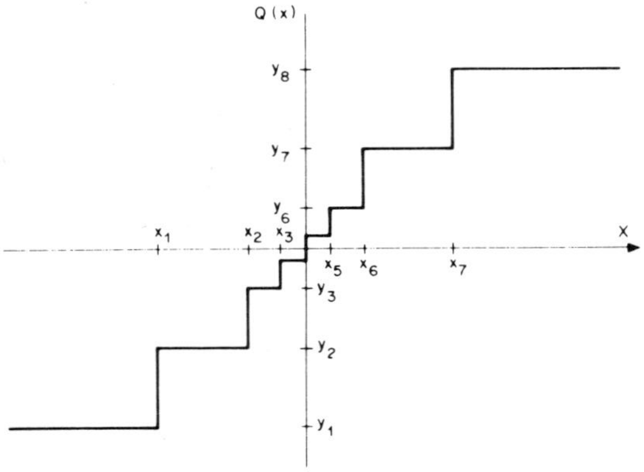

Fig. 2. Input–output characteristic of a midriser quantizer with $N=8$.

levels x_0 and x_N are chosen equal to the smallest and largest values, respectively, that the input samples may have. Usually, the sample values are unbounded, which we henceforth assume, so that $x_0=-\infty$ and $x_N=+\infty$. The N output points always have finite values. If $N=2^n$, a unique n-bit binary word can be associated with each output point, yielding an "n-bit quantizer."

The input–output characteristic $Q(x)$ of the quantizer has the staircase form shown in Figs. 1 and 2. The midtread characteristic produces zero output for input samples that are in the neighborhood of zero. The midriser characteristic has a decision level located at zero. A quantizer is simply a memoryless nonlinearity whose characteristic may be viewed as a staircase approximation to the "identity" operation $y=x$.

When the input sample is located in the end regions R_1 or R_N the quantizer is said to be overloaded. All other quantizing intervals R_i are finite in size.

Fundamental to an analytical study of quantization is the recognition that the input samples must be regarded as random in character. The input samples are not known in advance and thus can be regarded as an information-bearing signal. Quantization is actually a mechanism whereby information is thrown away, keeping only as much as is really needed to allow reconstruction of the original signal

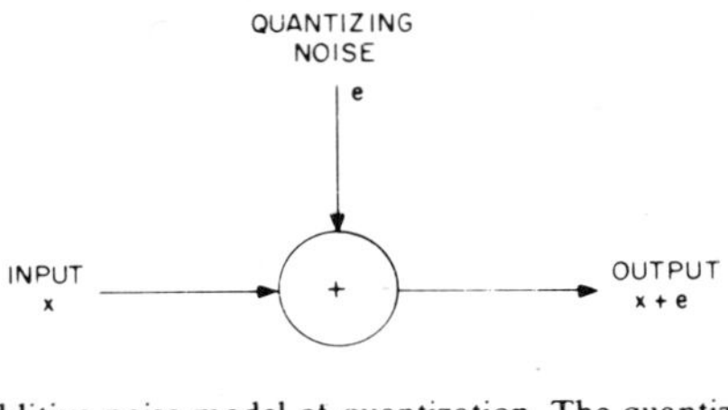

Fig. 3. Additive noise model of quantization. The quantizing noise e is often approximated as being independent of the input samples when the number of levels is large.

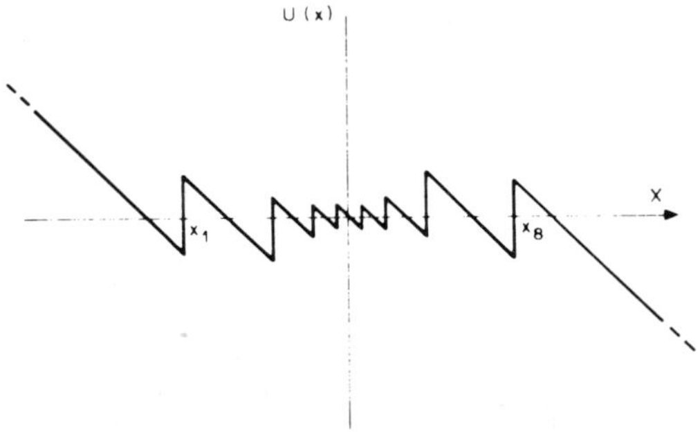

Fig. 4. Quantizing error as a function of input sample value for the quantizer of Fig. 1.

to within a desired accuracy as measured by some fidelity criterion. We define $p(x)$ as the first-order probability density function (PDF) of each input sample to the quantizer. Assume for convenience that the mean value of the input samples is zero and that $p(x)$ has even symmetry about zero. The zero-mean assumption implies that any dc bias has been removed. The symmetry assumption is satisfied by most common density functions including the Gaussian (normal) density. With the symmetry assumption, the quantizer characteristic $Q(x)$ is ordinarily chosen to have odd symmetry.

The quantization process can be modeled as the addition of a random noise component $e=Q(x)-x$ to the input sample as indicated in Fig. 3. Unlike the usual signal-plus-noise models in communication theory, here the noise is actually dependent on the signal amplitude. The quantization noise may be regarded as the response when the input sample is applied to the nonlinear characteristic

$$U(x)=Q(x)-x$$

shown in Fig. 4. When the input sample lies within the interval $x_1<x<x_{N-1}$, the output noise is described as granular noise and is bounded in magnitude. When the input lies outside this interval, the output is described as overload noise and the amplitude is unbounded. It is often convenient to artificially model quantization noise as the sum of granularity and overload noise as if they were two separate noise sources.

An effectively designed quantizer should be "matched" to the particular input density, to the extent that this density is known to the designer. In particular, for a fixed number N of levels, the choice of overload levels x_1 and x_{N-1} controls a tradeoff between the relative amounts of granularity and overload noise.

In modeling quantization error as an additive noise source as in Fig. 3, it is often convenient to treat the noise as having a flat spectral density and as being uncorrelated with the input samples. This idea was used by Widrow [1] for uniformly spaced quantization levels. More generally, it may be shown that the quantizing noise is approximately white (i.e., successive noise samples are uncorrelated) and uncorrelated with the input process if (1) successive input samples are only moderately correlated, (2) the number of output points N is large, and (3) the output points are very close to the midpoints of the corresponding quantization intervals. For a more precise treatment of the spectrum of quantizing noise, see Bennett [2].

III. Performance Measures

Since the quantization error is modeled as a random variable, a measure of the performance of a quantizer must be based on a statistical average of some function of the error. Most common is the mean-square distortion measure D defined by the usual expectation of the square of $U(x)$ above.

$$D=\int_{-\infty}^{\infty}[Q(x)-x]^2 p(x)\,dx. \tag{1}$$

This quantity can be used to measure the degradation introduced by the quantizer for a fixed input PDF $p(x)$. Frequently, it is more useful to describe the quantizer's performance by the "signal-to-noise ratio (SNR)," often defined as

$$\mathrm{SNR}=10\log_{10}(\sigma^2/D) \tag{2}$$

where σ^2 is the variance of the input samples. Other error criteria have also been considered in the study of quantization, such as the average of the kth power of the error magnitude. Frequently, the performance measure adopted is a subjective choice and psychological studies are used to determine preferred quantization schemes among a set of schemes considered. Another approach is to consider the quality of approximation of a segment of the reconstructed waveform to the original waveform. Mean-square distortion may be viewed as a special case of this approach where the performance measure is the sum of the squared errors for all sampling instants of the waveform segment. However, this measure does not distinguish between different approximations having the same total squared error. For example, it might be subjectively preferable to have a very high squared error at one isolated sampling instant rather than to have moderately high squared errors at several adjacent sampling instants. Hence, a more sophisticated distortion measure might be more meaningful than the usual mean-square distortion criterion.

In most applications of quantization, the number of levels N is very large, so that a sufficiently high SNR is obtained. A useful formula for mean-squared error can then be used. Equation (1) can be written in the form

$$D=\sum_{i=1}^{N}\int_{x_{i-1}}^{x_i}(y_i-x)^2 p(x)\,dx \tag{3}$$

by breaking up the region of integration into the separate intervals R_i and noting that $Q(x)=y_i$ when x is in R_i. For large N, each interval R_i can be made quite small (with the exception of the overload intervals R_1 and R_N which are unbounded). Then it is reasonable to approximate the probability density $p(x)$ as being constant within the interval R_i. On setting $p(x)\cong p(y_i)$ when x is in R_i and approximating $p(x)\cong 0$ for x in the overload regions, the integral for each term of the sum (3) is readily found, and we get

$$D=\frac{1}{12}\sum_{i=2}^{N-1}p(y_i)\Delta_i^3 \tag{4}$$

where $\Delta_i=x_i-x_{i-1}$, the length of interval R_i. This approximate formula is based on the assumption that, for N large, a sufficient number of quantizing levels are available for both the granularity and overload noise to be very small. Equation (4) implies that the overload points x_0 and x_N are chosen so that overload noise is negligible compared to granular noise. Equation (4) will be used later to derive an integral formula for distortion.

Of frequent interest is the special case of uniform quantization where the decision levels are equally spaced so that the intervals R_i are of constant length, i.e., $\Delta_i=\Delta$, sometimes called the step size of the quantizer. In this case, the staircase quantizer characteristic of Fig. 1 has equal width and equal height steps. The expression for mean-square error simplifies to

$$D=\frac{\Delta^2}{12}\sum_{i=2}^{N-1}p(y_i)\Delta.$$

But

$$\sum p(y_i)\Delta\approx\int p(s)\,ds=1$$

so that

$$D\approx\frac{\Delta^2}{12}. \tag{5}$$

Thus the mean-square distortion of a uniform quantizer grows as the square of the step size. This is perhaps the most often used result concerning quantization. This expression may be obtained directly by regarding the granularity noise as a uniformly distributed random variable over the interval $-\Delta/2$ to $+\Delta/2$ and neglecting overload noise.

A symmetric uniform quantizer if fully described by specifying the number of levels and either the step size Δ or the overload level V, where $V=x_N=-x_0$. To avoid significant overload distortion the overload level is chosen to be a suitable multiple $y=V/\sigma$ called the loading factor, of the rms signal level σ. A common choice is the so-called four-sigma loading where $y=4$. Then the step size is $\Delta=8\sigma/(N-2)$ since the total amplitude range of the quantizing intervals is 8σ and there are $N-2$ levels in

$N=2^n$

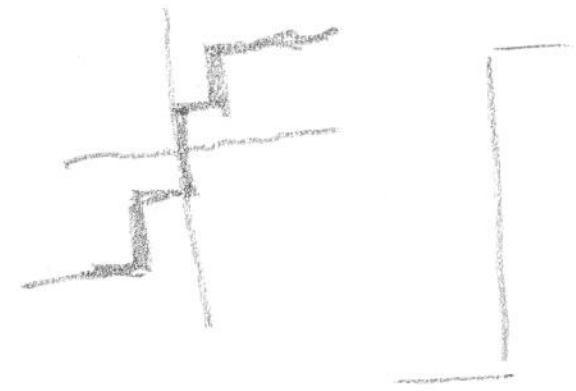

$V=4\sigma$

$y \to$ loading factor.
$\sigma \to$ signal standard deviation:

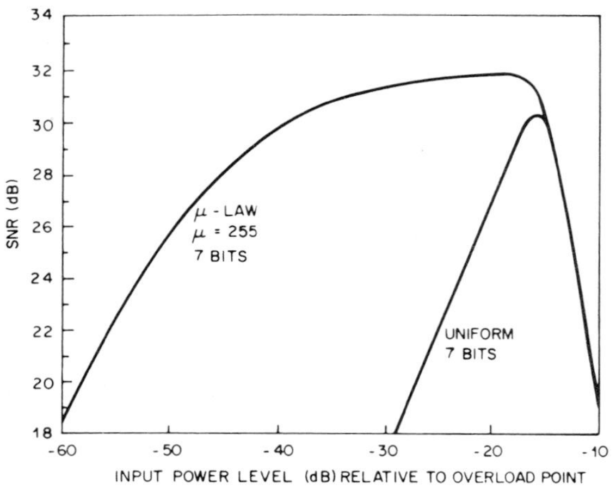

Fig. 5. Dependence of SNR on the input power level for uniform and μ-law quantizers both having 7 bits of quantization (128 levels). For a minimum acceptable quality of 25 dB, it can be seen that the μ-law quantizer has a dynamic range of about 40 dB while the uniform quantizer has a range of about 10 dB. The curves may also be used to show how SNR depends on the choice of overload point when the input power level is fixed. Curves are based on a Laplacian input PDF.

that range. Then, for an n-bit quantizer with $N=2^n$ and $N \gg 2$, we find using (2) and (5) that

$$\mathrm{SNR}=6n-7.3. \tag{6}$$

This linear increase of SNR with the number of bits of quantization was noted by Oliver, Pierce, and Shannon [3] in 1948. Note that changing the loading factor modifies the constant term 7.3 but does not alter the rate of increase of SNR with n. (The rate is actually $20\log_{10}2 \approx$ 6.02 dB per bit.)

Varying the loading factor for a particular input power level σ^2 is equivalent to varying the input power level for a fixed loading factor. In Fig. 5, the dependence of SNR on input power level is sketched for a uniform quantizer with $N=128$. The curve takes into account the effect of overload noise which rapidly becomes dominant as the signal level reaches a critical value. The curve is based on the assumption of a Laplacian PDF

$$p(x)=\frac{1}{\lambda}e^{-2|x|/\lambda}$$

which is occasionally used to approximate the PDF of speech. In this case, it may be seen that the best performance is achieved when the loading factor of 6.1 (15.7 dB) is used. If the input power level deviates a few decibels from the anticipated value (used in designing the quantizer), a substantial drop in SNR will result.

IV. Companding

Uniform quantization is not, in general, the most effective way to achieve good performance. For a given number of quantizing intervals, taking into account the input probability density, nonuniform spacing of the decision levels can yield lower quantizing noise and less sensitivity to variations in input signal statistics. An effective technique for studying nonuniform quantization, used by Bennett [2], is to model the quantizer as a memoryless nonlinearity $F(x)$, the "compressor," followed by a uniform quantizer as shown in Fig. 6. The nonlinearity spreads out low-amplitude sample values over a larger range while shrinking the higher amplitude values into a smaller region. The effect is to allocate more quantizer levels to the lower amplitudes, which generally have higher probability, and fewer levels to the less frequently occurring higher amplitudes. The compressed signal is uniformly quantized and the output values are then applied to the inverse nonlinearity $F^{-1}(x)$, producing an approximation to the signal originally applied to the compressor. The overall scheme in Fig. 6 is called companding, a term combining the words "compressing" and "expanding."

Fig. 6. Companding model of nonuniform quantization.

The characteristic $F(x)$ is a monotonically increasing function having odd symmetry, ranging from values $-V$ to $+V$, and with $F(V)=V$ and $F(0)=0$. This nonlinear operation, being monotonic, is completely invertible. That is, an input sample x applied to the compressor produces the response value $F(x)$; the original value x could be recovered by applying the value $y=F(x)$ to the inverse nonlinearity, the "expander" $F^{-1}(y)$, and obtaining x again. Thus there is no loss of information due to the nonlinear operation itself. The uniform quantizer is chosen to have $N-2$ ($\approx N$) intervals, not including overload regions, so that $\Delta=2V/N$. The combined effect of the compressor and the uniform quantizer is equivalent to the operation of a particular nonuniform quantizer whose decision levels and output points are determined by the shape of the compressor. Every possible nonuniform quantizer can be modeled in this way by a suitable choice of the function $F(x)$. Fig. 7 shows how the nonuniform quantizer decision levels are related to the uniform quantizer levels.

An important approximate formula for mean-square error in nonuniform quantizers can be derived based on the preceding model of nonuniform quantization. For large N, we approximate the curve of $F(y)$ in the ith quantizing interval by a straight-line segment with slope $F'(y_i)$, the derivative of $F(y)$ evaluated at y_i, where y_i is the output point of the equivalent nonuniform quantizer. Then

$$F'(y_i)\Delta_i=F(x_i)-F(x_{i-1})=2V/N$$

so that, defining the slope of the compressor curve

$$g(y) \triangleq F'(y)$$

we have

$$\Delta_i=\frac{2V}{Ng(y_i)}. \tag{7}$$

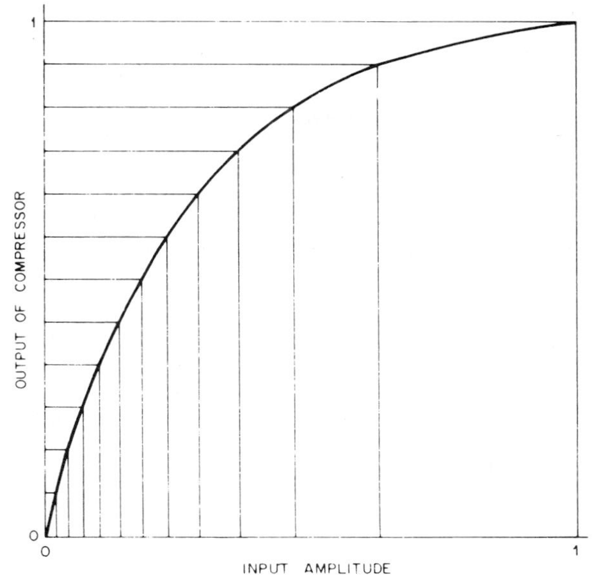

Fig. 7. Compressor mapping of decision levels (shown for positive amplitudes only).

Now, applying (4) yields

$$D=\frac{V^2}{3N^2}\int_{-V}^{V}\frac{p(s)}{[g(s)]^2}\,ds. \tag{8}$$

This formula, due to Bennett [2], accurately specifies the granular quantizing noise when N is large. Given a proposed compressor characteristic $F(x)$ and choice of overload point V, (8) can be used to evaluate the resulting granular distortion. By adding the overload noise, the tail components of (3), formulas for SNR can be derived, leading to curves such as those of Figs. 5 and 9. The formula is also of analytic value for optimizing the compressor characteristic. (See Section VI.)

For speech signals as well as many other analog sources, lower amplitude values occur with higher probability than the higher amplitude values, so that it would be reasonable to have quantizer levels more densely packed in the low signal region. For very low signal levels, the relevant step sizes will be approximately uniform with size

$$\Delta_0=\frac{2V}{Ng(0)}.$$

The improvement in performance of the nonuniform quantizer for low signal level inputs over the uniform quantizer is then determined by the ratio

$$c_A=\frac{\Delta}{\Delta_0}=g(0)$$

which is called the companding advantage. This quantity is frequently used in comparing different compressor characteristics. Increasing the companding advantage concentrates more levels in the low-amplitude region and improves the SNR for weak signal inputs. At the same time, a higher companding advantage means fewer levels in the high-amplitude region, tending to reduce the SNR for strong signal inputs.

V. Robust Quantization

In certain applications, notably in speech transmission, the same quantizer must accommodate signals with widely varying power levels. The use of "robust" quantizers, which are relatively insensitive to changes in the probability density of the input samples, has become of great practical importance.

To obtain robust performance, the SNR of the quantizer should ideally be independent of the particular PDF of the input signal. If the slope of the compressor curve is chosen to be

$$g(x)=\frac{V}{b|x|} \tag{9}$$

then (8) reduces to

$$D=\frac{b^2}{3N^2}\sigma^2$$

so that the SNR σ^2/D reduces to the constant $3N^2/b^2$, which is in fact independent of $p(x)$. Integrating (9) gives

$$F(x)=V+c\log(x/V) \tag{10}$$

for $x>0$, where c is a constant. This result shows that such a logarithmic compressor curve would give the desired robust performance. Of course, (8) neglects overload noise so that the SNR will not remain constant but will begin to drop when the input power level becomes large enough. Also, the compressor curve (10) is not in fact realizable, since $F(0)$ is not finite. To circumvent the latter difficulty, a modified compressor curve is used which behaves well for small values of x and retains the logarithmic behavior elsewhere.

A compressor curve widely used for speech digitization is the μ-law curve (see Fig. 8) given by

$$F(x)=V\frac{\log(1+\mu x/V)}{\log(1+\mu)} \tag{11}$$

for $x>0$. As always, $F(x)$ is an odd function so that $F(x)=-F(-x)$ for negative x. This characteristic was first described in the literature by Holzwarth [5], studied extensively by Smith [11], and reportedly was used by Bennett as early as 1944 in unpublished work. For $\mu\gg 1$ and $\mu x\gg V$, $F(x)$ approximates the form (10). From (8), the mean-square granular quantizing noise can be calculated, leading to the result

$$D/\sigma^2=\frac{[\log(1+\mu)]^2}{3N^2}\left\{1+\frac{2\alpha y}{\mu}+\left(\frac{y}{\mu}\right)^2\right\} \tag{12}$$

where α is the ratio of mean absolute value to rms value of the input samples, and y is the loading factor defined earlier. The effects of different choices of μ (corresponding to different companding advantages) has been examined by Smith [12]. Typical values are 100 for 7-bit and 255 for 8-bit speech quantizers. PCM systems in the

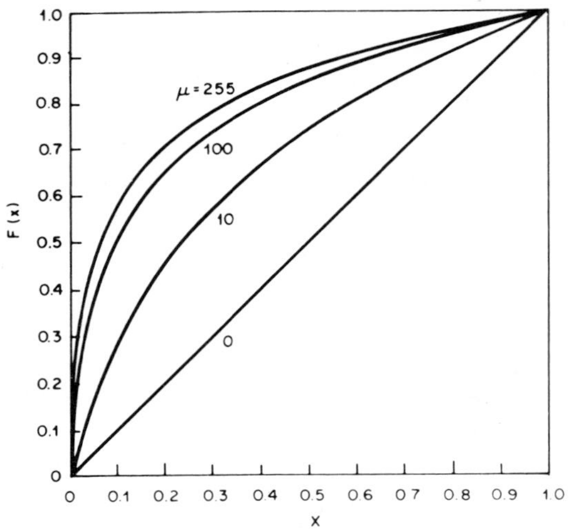

Fig. 8. μ-law compressor curves.

United States, Canada, and Japan use μ-law companding.

Another robust logarithmic characteristic due to Cattermole [6] is A-law companding where

$$F(x)=\begin{cases} \dfrac{Ax}{1+\log A}, & 0 \leqslant x \leqslant V/A \\ \dfrac{V+V\log(Ax/V)}{1+\log A}, & V/A \leqslant x \leqslant V. \end{cases} \tag{13}$$

A typical value of A is 87.6 for a 7-bit speech quantizer. The A-law characteristic is used in European PCM telephone systems. Both A law and μ law have the desired robust quality and can achieve more or less the same performance.

To illustrate the advantage of a robust quantizer, Fig. 5 shows curves of SNR versus input signal power level for both uniform and μ-law quantizers when the number of levels is 128. For a wide range of power levels, a high SNR of the μ-law quantizer is maintained while the SNR of the uniform quantizer drops rapidly with diminishing power levels. In order to achieve the same quality over a significant dynamic range, an 11-bit uniform quantizer must be used. Thus a saving of 4 bits per sample is achieved by using nonuniform quantization.

In practice, companders are now designed as piece-wise linear approximations to a desired characteristic. These "segmented" companding laws are conveniently implemented with digital circuitry. For 8-bit μ-law quantization of speech, the coded binary word has one bit to identify the sign of the analog sample, 3 bits to specify the segment number, and 4 bits to identify the particular level from among the 16 levels on that segment. A major advantage of this approach is the ease with which digital signal processing can be performed on the quantized signal.

VI. Optimum Quantization

For applications where one particular probability density function is known to describe adequately the distribution of input samples to be quantized, it is natural to seek the best possible quantizer characteristic for that density. Two approaches have been taken to this problem. One uses the assumption that N is large and leads to explicit solutions; the other is valid for any N, and leads to algorithmic procedures for finding the optimum decision levels and output points. We begin with the latter approach.

In a little-known Polish article, Lukaszewicz and Steinhaus [7] in 1955 found necessary conditions for optimality of a set of decision levels and output points for both the mean-square and the mean-absolute error criterion. (In the latter case $[Q(x)-x]^2$ is replaced by $|Q(x)-x|$ in (1).) Independently, Lloyd [4] in 1957, using the mean-square error criterion, found necessary conditions for optimality and an effective algorithm for computing the optimal solution. In 1960, Max [8] independently formulated the necessary conditions for optimality for a kth absolute mean error criterion (including $k=2$), and rediscovered the same algorithm used by Lloyd. In addition, Max examined the optimization of the step size for uniform quantization. Max also tabulated the optimum quantizer levels for the Gaussian distribution for various values of N.

For the mean-square error criterion, with some fixed value of N, the necessary conditions for optimality on the values of $x_1, x_2, \cdots, x_{N-1}$ and $y_1, y_2, \cdots, y_N$ are found simply by setting derivatives of D as given in (3) with respect to each of these parameters to zero. The resulting conditions are as follows.

(1) Each output level of y_i must be the centroid or center of mass of the interval R_i with respect to the input density $p(x)$. In other words, y_i is the conditional mean value of the input random variable x, given that x is in the region R_i.

(2) Each decision level x_i must be halfway between the two adjacent output points.

These conditions do not give the optimum values explicitly, since the value of the output point y_i for an interval R_i depends on the value of the decision levels x_{i-1} and x_i defining R_i, and the decision levels x_i depend on the output levels y_i and y_{i+1}. However, these conditions are used in the Lloyd–Max algorithm (see Max [8]) for computing iteratively a set of parameters that simultaneously satisfy both conditions. Using the Lloyd–Max algorithm, Paez and Glisson [9] tabulated the optimum quantizer parameters for the Laplacian and a particular form of the gamma density.

Lloyd also observed that the conditions, while necessary, are not sufficient conditions for a minimum. In fact, he gave a counterexample of a probability density function and an associated quantizer that satisfies the conditions and is not optimal. Fleischer [9] obtained sufficient conditions which, if satisfied, will guarantee that the quantizer is in fact optimal. In particular, he showed that, if the input density $p(x)$ satisfies the property that

$$\frac{d^2}{dx^2}\{\log p(x)\}<0 \qquad (14)$$

for all x, in other words, if $\log p(x)$ is concave, then only one quantizer exists which satisfies the Lloyd–Max conditions (1) and (2), and that quantizer is indeed optimal. It should be noted that the converse is not true, so that it is possible to have a density $p(x)$ not satisfying (14) and yet a unique optimal quantizer may exist. Nonetheless, condition (14) holds for the Gaussian density as well as for many other common densities. Hence, the tabulated quantizer parameters given by Max for the Gaussian density are in fact unique and optimal.

An alternate approach to the search for optimal quantizers begins with the use of Bennett's formula (8), which is based on the assumption that N is large. Minimization of (8) over the class of all curves of compressor slope $g(x)$ having a fixed area under the curve $g(x)$ yields the result that the optimum level density function is proportional to the cube root of the PDF

$$g^*(s)=c_1[p(s)]^{1/3}.$$

By integrating $g^*(s)$, one obtains the compressor characteristic

$$F^*(s)=c_1\int_0^s[p(\alpha)]^{1/3}d\alpha, \qquad \text{for } s>0 \qquad (15)$$

where c_1 is the constant chosen so that $F(V)=V$. Equation (15) was first obtained by Panter and Dite [11] in a classic and often overlooked paper. Their approach started with (4) and did not make use of Bennett's formula. Direct minimization of (8) was first examined by Smith [12]. Roe [13], while unaware of the works of Panter and Dite and Smith, derived a formula for the optimal decision levels that is equivalent to the result (15) but does not use the companding model. Algazi [14] used the companding model to obtain results on optimal quantizers for a general class of error criteria. His results include (15).

Finally, we note that (15) determines the optimum quantizer for a given choice of overload point. A separate one-dimensional minimization of D can be used to obtain the best overload point (see [14]). From Fig. 4, it is evident that, once the compressor curve is known, the decision levels and output points are readily obtained by a mapping of the uniform quantizer parameters. Computation of the minimum mean-square error obtained with this approach leads to values in good agreement with Max's tabulations (for the Gaussian PDF), even for values of N as small as 6. For $N=6$, the individual decision levels are within 3 percent of the correct values (see Roe [12]). Naturally, as N increases, the discrepancy approaches zero, since (15) is based on the assumption that N is large.

An example of optimum quantization studied by Smith

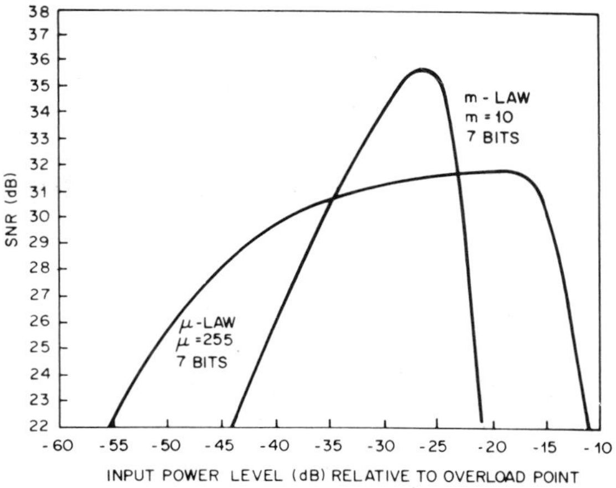

Fig. 9. Dependence of SNR on input power level for μ-law and m-law quantizers when the input PDF is Laplacian and the number of levels is 128 (7 bits). The m-law quantizer is optimal for the Laplacian PDF with input power level 26.5 dB. At this power level, there is a 4-dB improvement in SNR over the suboptimum but more robust μ-law quantizer.

[11] is based on the Laplacian PDF. The optimum compressor according to (15) has the form

$$F(x)=\frac{V(1-e^{-mx})}{1-e^{-mV}}, \qquad \text{for } x>0$$

giving rise to the "m-law" quantizer. Fig. 9 shows the dependence of SNR on input power level for the robust μ-law quantizer with $\mu=255$ and for the optimum m-law quantizer with $m=10$ when the input density is Laplacian. Comparison of SNR performance of μ-law and m-law quantizers shows that, for $\mu=255$ and $m=10$, the m-law curve has a 5-dB advantage at the power level for which it is designed. The μ-law quantizer maintains its reasonably high SNR over a broad range of power levels, while the m-law quantizer becomes inferior for input power levels about 10 dB below or 5 dB above the designed value. Comparing the m-law SNR curve in Fig. 9 with the uniform quantizer SNR curve in Fig. 5 shows that there is less than a 6-dB improvement in using the optimum quantizer rather than the uniform quantizer. In some applications, this gain might not justify the extra cost of implementing a specially designed nonuniform quantizer as opposed to the simpler uniform quantizer.

A convenient and general way to describe the performance of optimal quantizers is based on the application of the optimal compressor slope, $g^*(y)$, to the Bennett formula (8) for mean-square distortion. The result is that the minimum granular distortion for optimal N-point quantization and for large N is given by

$$D=\left\{\frac{1}{12N^2}\left[\int_{-v}^{v}[p_0(x)]^{1/3}dx\right]^3\right\}\sigma^2 \qquad (16)$$

where $p_0(x)$ denotes the input density normalized to have unit variance. This formula, first derived by Panter and

Dite [10], is useful for estimating the number of quantization levels needed for a desired performance (i.e, SNR specification). The integral in square brackets, H, depends only on the shape of the input density function and not on the actual power level. It can be seen from (16) that the SNR for an optimal quantizer has the form $10\log_{10} N^2 - C$ where C is a constant determined by $p_0(x)$ using (16). Letting $n = \log_2 N$ gives the result

$$\mathrm{SNR} = 6n - C$$

where $C = 10\log_{10}(H^3/12)$. The 6-dB-per-bit improvement in SNR is the same as for the nonoptimum uniform quantizer where the SNR is given by (6). However, the value of C obtained from (16) is as small as possible since (16) gives the minimum granular distortion attainable for any zero-memory quantizer. For the Gaussian density, C is found to be 4.35 dB by approximating (16) using $V = \infty$.

VII. Quantization with Memory

In block quantization, more commonly considered for image digitization rather than speech, a block of k input samples $(x_1, x_2, \cdots, x_k) = \boldsymbol{x}$ (which may be regarded as a vector in k dimensions) is simultaneously quantized, producing an output "point" or vector $(y_{i1}, y_{i2}, \cdots, y_{ik}) = \boldsymbol{y}_i$ approximating $\boldsymbol{x}$. Thus the output y_{ij} is an approximation to x_j for each $j = 1, 2, \cdots, k$. An N-point quantizer selects one of N output "points" $\boldsymbol{y}_1, \boldsymbol{y}_2, \cdots, \boldsymbol{y}_N$ to approximate $\boldsymbol{x}$. Unlike zero-memory quantization, the value y_{ij} depends not only on the corrresponding input sample x_j but also on the values of all other samples x_i in the block. Even if the input samples are statistically independent, an advantage can be gained by quantizing a block at a time rather than one sample at a time. A convenient measure of the distortion of the block quantizer is

$$D = \frac{1}{k} \sum_{i=1}^{k} \overline{e_i^2}$$

where $\overline{e_i^2}$ is the mean-square error in the ith sample. The performance of block quantization could be compared with zero-memory quantization by examining how the bit rate or average number of bits per sample, $B = (\log_2 N)/k$, depends on D, the distortion per sample. Clearly, as the block length k increases, the minimum bit rate needed for a given distortion will decrease. In the limit as $k \to \infty$, the minimum bit rate B approaches a limiting value R depending on D. The function $R(D)$ is the rate-distortion function due to Shannon (who defined it in a different way). For certain classes of input process $\{x_i\}$, explicit solutions for $R(D)$ have been found, and for many other cases, upper and lower bounds are available. For a treatment of rate-distortion theory, see Berger [15].

One simple technique for reducing the bit rate without the full complexity of block quantization is by entropy coding the successive output symbols of a zero-memory quantizer. The output of an N-point zero-memory quantizer is one of N different symbols $y_1, y_2, \cdots, y_N$, each having a corresponding probability $p_1, p_2, \cdots, p_N$ of occurring. Instead of transmitting $\log_2 N$ bits per sample (or the next largest integer if $\log_2 N$ is not an integer) to identify each output sample, variable-length coding such as the Huffman code can be used. Such a code assigns a word with a high number of bits to a low probability symbol and a short word for a high probability symbol. The resulting average number of bits per sample attainable approaches or equals the entropy of the quantizer output

$$H = -\sum_{1}^{N} p_i \log_2 p_i.$$

This scheme requires buffering in order to produce a steady output bit stream.

In general, optimal quantizers do not result in equal probabilities for the output symbols, in which case H is always smaller than $\log_2 N$. For example, a 16-point optimal quantizer for Gaussian samples produces output symbols with entropy 4.73 bits (from Max [8]) compared to the 5 bits per symbol needed for equal-length coding. Once entropy coding is to be used, the preceding optimization theory is no longer relevant. It is more appropriate to find a compressor curve which leads to minimum mean-square error for a constraint on output entropy rather than on number of output points. This leads to the surprising result that the uniform quantizer is nearly optimal (see Gish and Pierce [16]).

Finally, another class of quantizers with memory are the sequential quantizers such as delta modulation, differential PCM, and the various adaptive versions of these schemes. In essence, all of these schemes take advantage of correlation in the successive input samples by using a feedback loop around the quantizer. However, this is a subject for a separate paper.

VIII. Quantizer Performance

From a user's viewpoint, the performance of a quantizer is determined by how many bits per sample are needed to digitize a given analog source so that it can be reproduced with a prescribed maximum amount of distortion (or minimum SNR). Alternatively, the performance is determined by how high a SNR can be achieved for a prescribed average bit rate, B measured in bits per sample. We take the later approach here and survey some key results on achievable quantizer performance. For convenience, we focus only on the case of input samples with a Gaussian PDF and the mean-square distortion measure. The issue of robustness is not considered in this discussion.

From rate-distortion theory it is known that, in the limit as the block length approaches infinity, block quantization of a Gaussian source with statistically independent samples can achieve the bit rate

$$B = \tfrac{1}{2}\log_2(\sigma^2/D)$$

where D is the average mean-square distortion per sample (and $D < \sigma^2$). Converting to SNR, then, gives the result

$$\mathrm{SNR}_1 = 6B \qquad (17)$$

which is also a lower bound on attainable SNR for any realizable quantization scheme regardless of the input PDF, as long as the samples are independent. If the source samples are correlated, a higher SNR can always be achieved (see Berger [15]).

If a zero-memory uniform quantizer is used with entropy coding of the output symbols, an efficient quantization scheme is achieved. By optimizing the overload point, Goblick and Holsinger [17] found that H^*, the highest output entropy attainable with uniform quantizing, satisfies the equation

$$H^* = \tfrac{1}{4} + \tfrac{1}{2}\log_2(\sigma^2/D).$$

With entropy coding, H^* may be taken as the attainable bit rate B, so that solving for SNR gives

$$\text{SNR}_2 = 6B - 1.50. \tag{18}$$

Hence, the uniform quantizer with entropy coding achieves an SNR only 1.5 dB below the very best attainable performance with block quantization.

If the best nonuniform quantizer for minimizing distortion is combined with entropy coding, taking the bit rate as the output entropy gives the result

$$\text{SNR}_3 = 6B - 2.45 \tag{19}$$

where (19) was empirically found to fit the data tabulated by Max for quantizers with more than eight levels. Clearly, the nonuniform quantizer is inferior when entropy coding is being used.

Of course, entropy coding adds a significant amount of complexity to the implementation of a quantizer. Without entropy coding, we have seen that the highest SNR achievable with nonuniform quantization is given using (16) by

$$\text{SNR}_4 = 6B - 4.35. \tag{20}$$

Recall that (16) is based on the assumption of large N and it neglects overload noise. The exact SNR values for N between 2 and 36 can be obtained from Max's tables. It turns out that (20) is about 3.6 percent too small for $N=12$ and becomes progressively more accurate as N increases (see Fig. 10).

Finally, the simplest quantization scheme, using a uniform quantizer without entropy coding, gives the least favorable performance. Using a loading factor of 4 and neglecting overload noise led to the SNR formula (6)

$$\text{SNR}_5 = 6B - 7.3. \tag{21}$$

However, the optimum loading factor depends on the number of levels used and the effect of overload distortion. Goblick and Holsinger [17] fitted the curve

$$B = 0.125 + 0.6\log_2(\sigma^2/D)$$

to Max's tabulated data for uniform quantizers with optimized loading factor. Converting this expression to a SNR formula gives

$$\text{SNR}_6 = 5B - 0.63. \tag{22}$$

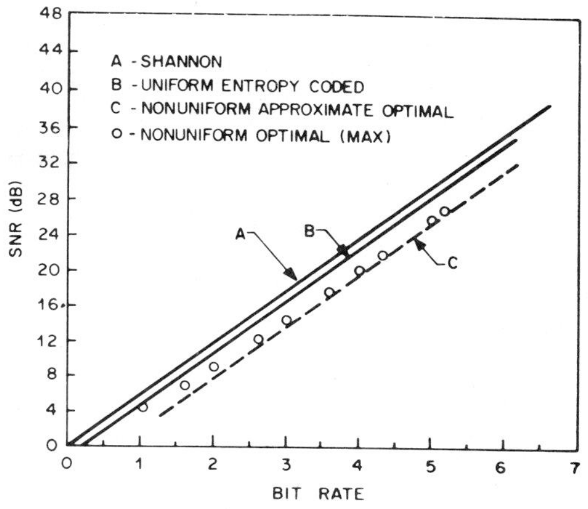

Fig. 10. Quantizer performance in SNR as a function of the average number of bits per sample needed to encode an analog source. Curve A is the best theoretically attainable performance for a Gaussian source with independent samples. It is also a lower bound for any source with independent samples. Curve B is the performance achieved for Gaussian samples with a uniform quantizer followed by entropy encoding. Curve C, based on (16), is asymptotically the optimal performance obtainable for a large number of levels, with nonuniform quantization of Gaussian samples without entropy coding. Circled points are based on Max's tabulated values for optimal nonuniform quantization of Gaussian samples without entropy encoding.

Since Max's tables go up to $N=36$, it is not known how accurate (22) is for $B>5.2$. For $B<6.7$, (22) gives higher SNR values than (21), which shows that four-sigma loading is not an optimal choice for a Gaussian PDF.

Summarizing, we have seen that uniform quantizing followed by entropy coding can achieve SNR values within 1.5 dB of the best performance theoretically attainable with any quantization scheme whatever. For an additional 3-dB penalty in SNR, an optimum nonuniform quantizer without the complexity of entropy coding can be used. Simplest of all, the uniform quantizer can achieve an SNR within 7 dB or so of the best performance theoretically attainable.

It should be emphasized that this modest difference in performance between the simplest and most complex quantization schemes is based on the assumption that the input samples are statistically independent. Zero-memory quantization can be grossly inadequate when there is substantial correlation between successive input samples. However, the utility of zero-memory quantizers does not end when the input is correlated. In such situations, the zero-memory quantizer is still used as a component part of more sophisticated quantization schemes. Sequential quantization schemes all use a zero-memory quantizer of one form or another imbedded in a feedback loop. Also, block quantization schemes generally attempt to transform the vector of input samples into a new vector with independent components. These components are then individually quantized with a zero-memory quantizer. Indeed, the basic zero-memory quantizer plays a ubiquitous role in the digital coding of analog sources.

References

[1] B. Widrow, "A study of rough amplitude quantization by means of Nyquist sampling theory," *IRE Trans. Circuit Theory*, vol. CT-3, pp. 266–276, 1956.

[2] W. R. Bennett, "Spectrum of quantized signals," *Bell Syst. Tech. J.*, vol. 27, pp. 446–472, July 1948.

[3] B. M. Oliver, J. R. Pierce, and C. E. Shannon, "The philosophy of PCM," *Proc. IRE*, vol. 36, pp. 1324–1331, 1948.

[4] S. P. Lloyd, "Least-squares quantization in PCM," unpublished memorandum, Bell Laboratories, 1957 (copies available from the author).

[5] H. Holzwarth, "PCM and its implementation by logarithmic quantization" (in German), *Arch. Elek. Übertragung*, vol. 3, pp. 277–285, 1949.

[6] K. W. Cattermole, *Principles of Pulse Code Modulation*. London, England: Elsevier, 1969.

[7] J. Lukaszewicz and H. Steinhaus, "On measuring by comparison" (in Polish), *Zastos. Mat.*, vol. 2, pp. 225–231, 1955.

[8] J. Max, "Quantizing for minimum distortion," *IRE Trans. Inform. Theory*, vol. IT-6, pp. 7–12, 1960.

[9] M. D. Paez and T. H. Glisson, "Minimum mean-squared-error quantization in speech PCM and DPCM systems," *IEEE Trans. Commun.*, vol. COM-20, pp. 225–230, Apr. 1972.

[10] P. Fleischer, "Sufficient conditions for achieving minimum distortion in a quantizer," *IEEE Int. Conv. Rec.*, pp. 104–111, 1964.

[11] P. F. Panter and W. Dite, "Quantizing distortion in pulse-count modulation with nonuniform spacing of levels," *Proc. IRE*, vol. 39, pp. 44–48, 1951.

[12] B. Smith, "Instantaneous companding of quantized signals," *Bell Syst. Tech. J.*, vol. 27, pp. 446–472, 1948.

[13] G. M. Roe, "Quantizing for minimum distortion," *IEEE Trans. Inform. Theory*, vol. IT-10, pp. 384–385, 1964.

[14] V. R. Algazi, "Useful approximation to optimum quantization," *IEEE Trans. Commun. Technol.*, vol. COM-14, pp. 297–301, 1966.

[15] T. Berger, *Rate Distortion Theory*. Englewood Cliffs, NJ: Prentice-Hall, 1971.

[16] H. Gish and J. N. Pierce, "Asymptotically efficient quantization," *IEEE Trans. Inform. Theory*, vol. IT-14, pp. 676–681, 1968.

[17] T. J. Goblick and J. L. Holsinger, "Analog source digitization: A comparison of theory and practice," *IEEE Trans. Inform. Theory*, vol. IT-13, pp. 323–326, Apr. 1967.

Part II

SCALAR SIGNAL QUANTIZERS

Editor's Comments on Papers 2 Through 9

The scalar quantizer is by far the most commonly employed and simplest to design, implement, and analyze. In comparison to the general k-dimensional quantizer described in the introduction, the scalar (one-dimensional) case has the advantage that the only permissible region whose union covers the real line (or a portion of it) is the interval, each region being identified by its two endpoints. This description is much simpler than attempting to specify a region in a k-dimensional space (e.g., specifying the hyperplanes and vertices that bound a polytope in k dimensions). With this realization, an N-level scalar quantizer is uniquely specified by $2N-1$ variables: the $N-1$ interval endpoints (the other two being minus and plus infinity, respectively) and the N output values. Being memoryless devices, scalar quantizers are often visualized as staircaselike nonlinearities, each tread representing a quantization interval. Examples of scalar quantizer designs for the MSE criterion can be found in

Adams and Giesler (1978), Goblich and Holsinger (1967), Jones (1977), and Pearlman and Senge (1979).

This section presents eight selections from the vast literature on scalar quantization; the first six are historical references, and the other two provide further insight into the design and performance of scalar quantizers. The first two selections, by Bennett and by Smith, are edited versions of these often-referenced classics from the *Bell System Technical Journal.* Both papers deal with the quantizing of signals for pulse code modulation systems. In Paper 2 Bennett explores properties of the quantization noise, in particular noting that for uniform quantization with stepsize Δ, the effect of the quantization is similar to the addition of a signal-independent "quantization noise" and that the noise power (MSE) is the ubiquitous $\Delta^2/12$. For a further discussion of this noise and its correlation properties, see a later paper by Widrow (1956). Also introduced in Paper 2 is the technique of companding for the implementation of nonuniform scalar quantizers. The MSE of the compander scheme is presented through "Bennett's integral," a function of the source density and the compressor slope (for necessary conditions upon the validity of this result, see Bucklew 1984). Smith (Paper 3) further advances the compander model for nonuniform scalar quantizers, presenting voluminous results on logarithmic companders deriving the form of the optimal compander and from this compander, producing an expression for the minimum MSE equivalent to that of Panter and Dite (1951).

Papers 4 and 5 by Lloyd (the rewrite of his 1959 paper) and by Max, are the original discussions of optimum scalar quantization for the finite N case. Both argue that for a fixed set of quantization intervals, the minimum MSE output assignment is each interval's centroid (the weighting function being the source density). Conversely, for fixed output values, the best interval selection has breakpoints spaced halfway between adjacent outputs. For neither fixed, this pair of conditions provides a set of $2N-1$ simultaneous equations, necessary for a minimum of MSE. Both Lloyd and Max also include a trial-and-error technique for solving this set of simultaneous equations. In essence, this technique starts with a guess of one of the $2N-1$ quantizer parameters. The other $2N-2$ parameters are found from the application of all but one of the simultaneous equations, the last being used to measure the quality of the original guess. If the last equation is satisfied within some small threshold, the quantizer design is complete; if not, the initial parameter guess is modified.

In Paper 4 Lloyd considers only the mean square error criterion, presenting the necessary conditions of centroids and nearest-neighbor regions. In addition, he includes a large N performance analysis and another design algorithm (commonly called Lloyd's Method I) that applies the two sets of necessary conditions iteratively to converge to a local minimum of MSE. This algorithm sequentially holds either the outputs or the regions fixed, redefining the other set from the remaining necessary conditions. Every application of these conditions reduces the MSE, converging to a local minimum. In comparison to the other design algorithm, this technique relieves the difficulty of making a good initial guess. (See Kieffer, 1982 for further details on the convergence of this technique.) Currently, this iterative design algorithm is being used (in a vector version) for vector quantizer design (see Papers 18 and 19). In Paper 5, Max derives the necessary conditions of centroids and nearest-neighbor regions but for a more general error functional (resulting in nearest-neighbor regions as long as the error functional is increasing in the quantization error and generalized centroids). He also tabulates performance results (MSE and entropy) of optimum nonuniform and uniform quantizers for the Gaussian source with $N \leq 36$.

The other two early papers included in this section are those of Fleischer and Bruce from the 1964 *IEEE International Convention Record*. In Papers 4 and 5 Max and Lloyd establish *necessary* conditions upon the quantizer's parameters (output values and breakpoints) for a local minimum of MSE. In Paper 6 Fleischer establishes a *sufficient* condition for centroidal outputs and nearest neighbor regions to provide a global minimum of MSE (see also Kieffer, 1983, and Trushkin, 1982 for discussions with other performance criteria). Fleischer's sufficient condition is based upon the logarithm of the source's density function being a concave function. The discussion concerns the scalar case only (no such result exists for the vector case) and demonstrates that the Gaussian source, in particular, has a globally optimum quantizer for all values of N. In Paper 7 Bruce extends the work of Max and Lloyd on design algorithms for optimum scalar quantizers by presenting a dynamic programming algorithm suitable for any design criterion (see also Sharma, 1978).

Paper 8 by Bucklew and Gallagher is a much more recent contribution on the iterative design algorithm discussed by both Max and Lloyd (trial-and-error method). In particular, Bucklew and Gallagher were concerned with the starting values for the iteration, presenting a comparison of two starting techniques. Another discussion along similar lines is that of Lu and Wise (1983).

The final paper in this section discusses uniform scalar quantizers. Uniform quantizers are of constant interest since they are simple to implement, do not distort the relative values (arithmetic operations upon the data are still simple to perform), and are typically insensitive to variation in the source's statistical characterization. Bucklew and Gallagher, in Paper 9, present several interesting properties of uniform quantizers, including a discussion of when Bennett's approximation of $\Delta^2/12$ holds. Uniform quantizers have also been shown to yield asymptotically (large N) the smallest entropy of all quantizers for a given level of performance (see Gish and Pierce, 1968, or Gray and Gray, 1977).

REFERENCES

Adams, W. C., Jr., and C. E. Giesler, 1978, Quantization Characteristics for Signals Having Laplacian Amplitude Probability Density Functions, *IEEE Trans. Commun.*, **COM-26**(8):1295–1297.

Bucklew, J. A., 1984, Two Results on the Asymptotic Performance of Quantizers, *IEEE Trans. Inform. Theory* **IT-30**(2):341–348.

Goblich, T. J., and J. L. Holsinger, 1967, Analog Source Digitization: a Comparison of Theory and Practice, *IEEE Trans. Inform. Theory,* **IT-13**(2):323–326.

Gray, R. M., and A. H. Gray, Jr., 1977, Asymptotically Optimal Quantizers, *IEEE Trans.* Inform. Theory **IT-23**(1):143–144.
Jones, H. W., Jr., 1977, *Minimum Distortion Quantizers,* NASA Tech. Note TN-D-8384.
Kieffer, J. C., 1982, Exponential Rate of Convergence for Lloyd's Method 1, *IEEE Trans. Inform. Theory* **IT-28**(2):205–210.
Kieffer, J. C., 1983, Uniqueness of Locally Optimal Quantizer for Log-Concave Density and Convex Error Weighting Function, *IEEE Trans. Inform. Theory* **IT-29**(1):42–47.
Lu, F. S., and G. L. Wise, 1983, A Further Investigation of the Lloyd-Max Algorithm for Quantizer Design, *Allerton Conf. Commun. Control, Comput. Proc.* pp. 481–490.
Panter, P. F., and W. Dite, 1951, Quantization Distortion in Pulse Count Modulation with Non-linear Spacing of Levels, *IRE Proc.* **39**(1):44–48.
Pearlman, W. A., and G. H. Senge, 1979, Optimal Quantization of the Rayleigh Probability Density, *IEEE Trans. Commun.* **COM-27**(1):101–112.
Sharma, D., 1978, Design of Absolutely Optimal Quantizers for a Wide Class of Distortion Measures, *IEEE Trans. Inform. Theory* **IT-24**(6):693–702, Comments on and Reply, 1982, **IT-28**(3):555.
Trushkin, A. V., 1982, Sufficient Conditions for Uniqueness of a Locally Optimal Quantizer for a Class of Convex Error Weighting Functions, *IEEE Trans. Inform. Theory* **IT-28**(3):187–198.
Widrow, B., 1956, A Study of Rough Amplitude Quantization by Means of Nyquist Sampling Theory, *IRE Trans. Circuit Theory* **CT-3**(4):266–276.

BIBLIOGRAPHY

Algazi, V., 1966, Useful Approximations to Optimal Quantization, *IEEE Trans. Commun. Tech.* **COM-14**(6):297–301.
Berger, T., 1972, Optimum Quantizers and Permutation Codes, *IEEE Trans. Inform. Theory* **IT-18**(6):759–765.
Berger, T., 1982, Minimum Entropy Quantizers and Permutation Codes, *IEEE Trans. Inform. Theory* **IT-28**(2):149–157.
Bruce, J. D., 1965, *Optimum Quantization,* MIT Research Laboratory of Electronics Technical Report 429.
Bucklew, J. A., and N. C. Gallagher Jr., 1979, A Note on Optimal Quantization, *IEEE Trans. Inform. Theory* **IT-25**(3):365–366.
Cambanis, S., and N. L. Gerr, 1983, A Simple Class of Asymptotically Optimal Quantizers, *IEEE Trans. Inform. Theory* **IT-29**(5):664–676.
Elias, P., 1970, Bounds on the Performance of Optimum Quantizers, *IEEE Trans. Inform. Theory* **IT-16**(3):172–184.
Farvardin, N., and J. W. Modestino, 1984, Optimum Quantizer Performance for a Class of Non-Gaussian Memoryless Sources, *IEEE Trans. Inform. Theory* **IT-30**(3):485–497.
Garey, M., D. S. Johnson, and H. S. Witsenhausen, 1982, The Complexity of the Generalized Lloyd-Max Problem, *IEEE Trans. Inform. Theory* **IT-28**(3):255–256.
Gish, H., and J. N. Pierce, 1968, Asymptotically Efficient Quantizing, *IEEE Trans. Inform. Theory* **IT-14**(5):676–683.
Habibi, A., 1975, A Note on the Performance of Memoryless Quantizers, *National Telecomm. Conf. Rec.* **2:**16–21.
Kassam, S. A., 1978, Quantization Based on the Mean Absolute Error Criteria, *IEEE Trans. Commun.* **COM-26**(2):267–270.
Mauersberger, W., 1981, An Analytic Function Describing the Error Performance of Optimum Quantizers, *IEEE Trans. Inform. Theory* **IT-27**(4):519–521.
Messerschmitt, D. G., 1971, Quantizing for Maximum Output Entropy, *IEEE Trans. Inform. Theory* **IT-17**(5):612.

Netravali, A. N., and R. Saigal, 1976, Optimum Quantizer Design Using a Fixed-Point Algorithm, *Bell System Tech. J.* **55**(11):1423–1435.

O'Neal, J. B., Jr., 1967, A Bound on Signal-to-Quantizing Noise ratios for Digital Encoding Systems, *IEEE Proc.* **55**(3):287–292.

Roe, G. M., 1964, Quantizing for Minimum Distortion, *IEEE Trans. Inform. Theory* **IT-10**(4):384–385.

Schuchman, L., 1964, Dither Signals and their Effects on Quantizing Noise, *IEEE Trans. Commun. Tech.* **Com-12**(12):162–165.

Sripad, A. B., and D. L. Snyder, 1977, A Necessary and Sufficient Condition for Quantization Errors to Be Uniform and White, *IEEE Trans. Acoust., Speech, and Signal Process.* **ASSP-25**(5):442–448.

Widrow, B., 1960, Statistical Analysis of Amplitude Quantized Sampled-Data Systems, *AIEE App. & Ind. Trans.* **79**(pt. II):555–568.

Williams, G., 1967, Quantizing for Minimum Error with Particular Reference to Speech, *IEE Electronic Letters* **3:**134–135.

Wood, R. C., 1969, On Optimal Quantization, *IEEE Trans. Inform. Theory* **IT-5**(2):248–252.

2

Reprinted with permission from pages 446–452, 455–459, and 471–472 of *Bell System Tech. J.* **27**(3):446–472 (1948)

Spectra of Quantized Signals

By W. R. BENNETT

1. Discussion of Problem and Results Presented

SIGNALS which are quantized both in time of occurrence and in magnitude are in fact quite old in the communications art. Printing telegraph is an outstanding example. Here, time is divided into equal divisions, and the number of magnitudes to be distinguished in any one interval is usually no more than two, corresponding to the closed or open positions of a sending switch. It is only in recent years, however, that the development of high speed electronic devices has progressed sufficiently to enable quantizing techniques to be applied to rapidly changing signals such as produced by speech, music, or television. Quantizing of time, or time division, has found application as a means of multiplexing telephone channels.[1] The method consists of connecting the different channels to the line in sequence by fast moving switches synchronized at the transmitting and receiving ends. In this way a transmission medium capable of handling a much wider band of frequencies than required for one telephone channel can be used simultaneously by a group of channels without mutual interference. The plan is the same as that used in multiplex telegraphy. The difference is that ordinary rotating machinery suffices at the relatively low speeds employed by the latter, while the high speeds needed for time division multiplex telephony can be realized only by practically inertialess electron streams. Also the widths of frequency band required for multiplex telephony are enormously greater than needed for the telegraph, and in fact have become technically feasible only with the development of wide-band radio and cable transmission systems. As far as any one channel is concerned the result is the same as in telegraphy, namely that signals are received at discrete or quantized times. In the limiting case when many channels are sent the speech voltage from one channel is practically constant during the brief switch closure and, in effect, we can send only one magnitude for each contact or quantum of time. The more familiar word "sampling" will be used here interchangeably with the rather formidable term "quantizing of time".

Quantizing the magnitude of speech signals is a fairly recent innovation. Here we do not permit a selection from a continuous range of magnitudes but only certain discrete ones. This means that the original speech signal is to be replaced by a wave constructed of quantized values selected on a minimum error basis from the discrete set available. Clearly if we assign the quantum values with sufficiently close spacing we may make the quantized wave indistinguishable by the ear from the original. The purpose of quantization of magnitudes is to suppress the effects of interference in the transmission medium. By the use of precise receiving instruments we can restore the received quanta without any effect from superposed interference provided the interference does not exceed half the difference between adjacent steps.

By combining quantization of magnitude and time, we make it possible to code the speech signals, since transmission now consists of sending one of a discrete set of magnitudes for each distinct time interval.[2,3,4,5,6,7] The maximum advantage over interference is obtained by expressing each discrete signal magnitude in binary notation in which the only symbols used are 0 and 1. The number which is written as 4 in decimal notation is then represented by 100, 8 by 1000, 16 by 10,000; etc. In general, if we have N digit positions in the binary system, we can construct 2^N different numbers. If we need no more than 2^N different discrete magnitudes for speech transmission, complete information can be sent by a sequence of N on-or-off pulses during each sampling interval. Actually a total of $2^N!$ different coding plans (sets of one-to-one correspondences between signal magnitudes and on-or-off sequences) is possible. The straightforward binary number system is taken as a representative example convenient for either theoretical discussion or practical instrumentation. We assume that absence of a pulse represents the symbol 0 and presence of a pulse represents the symbol 1. The receiver then need only distinguish between two conditions: no transmitted signal and full strength transmitted signal. By spacing the repeaters at intervals such that interference does not reach half the full strength signal at the receiver, we can transmit the signal an indefinitely great distance without any increment in distortion over that originally introduced by the quantizing itself. The latter can be made negligible by using a sufficient number of steps.

To determine the number of quantized steps required to transmit specific signals, we require a knowledge of the relation between distortion and step size. This problem is the subject of the present paper.* We divide the problem into two parts: (1) quantizing the magnitude only and (2) combined quantizing of magnitude and time. The first part can be treated by a simple model: the "staircase transducer", which is a device having the instantaneous ouput vs. input curve shown by Fig. 1. Signals impressed on the stair-

* Other features of the quantizing and coding theory are discussed in forthcoming papers by Messrs. C. E. Shannon, J. R. Pierce, and B. M. Oliver.

case transducer are sorted into voltage slices (the treads of the staircase), and all signals within plus or minus half a step of the midvalue of a slice are replaced in the output by the midvalue. The corresponding output when the input is a smoothly varying function of time is illustrated in Fig. 2. The output remains constant while the input signal remains within the boundaries of a tread and changes abruptly by one full step when the signal crosses the boundary. It is not within the scope of the present paper to discuss the internal mechanism of a staircase transducer, which may have many different physical embodiments. We are concerned rather with the distortion produced by such a device when operating perfectly.

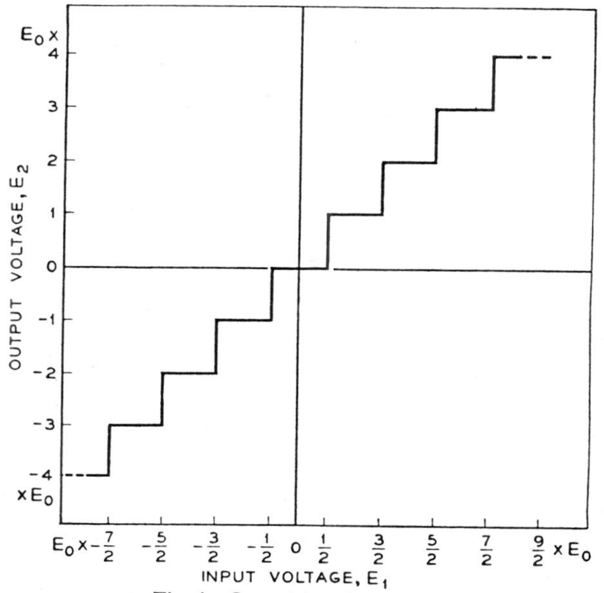

Fig. 1—Quantizing characteristic.

The distortion or error consists of the difference between the input and output signals. The maximum instantaneous value of distortion is half of one step, and the total range of variation is from minus half a step to plus half a step. The error as a function of input signal voltage is plotted in Fig. 3 and a typical variation with time is indicated in Fig. 2. If there is a large number of small steps, the error signal resembles a series of straight lines with varying slopes, but nearly always extending over the vertical interval between minus and plus half a step. The exceptional cases occur when the signal goes through a maximum or minimum within a step. The limiting condition of closely spaced steps enables us to derive quite simply an approximate value for the mean square error, which will later be shown to be sufficiently accurate in most cases of practical importance. This approximation consists of calculating the mean square value of a straight line going from minus half a step to plus half a step with arbitrary slope. If

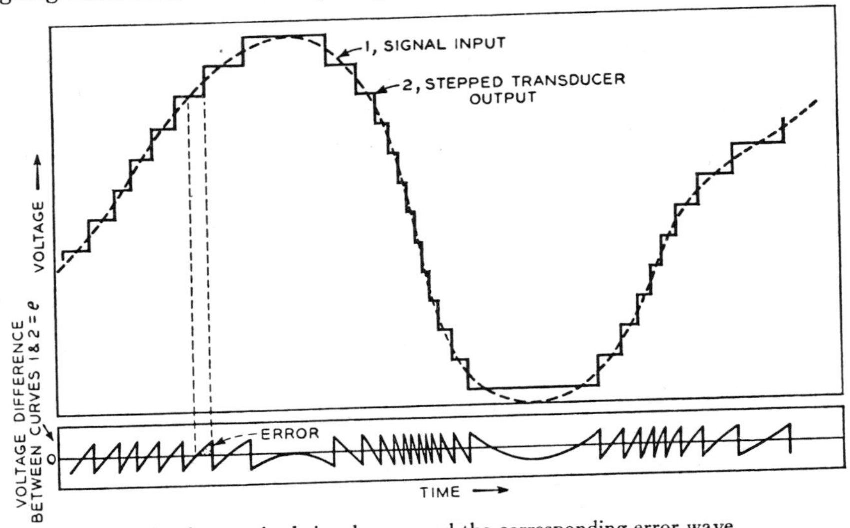

Fig. 2—A quantized signal wave and the corresponding error wave.

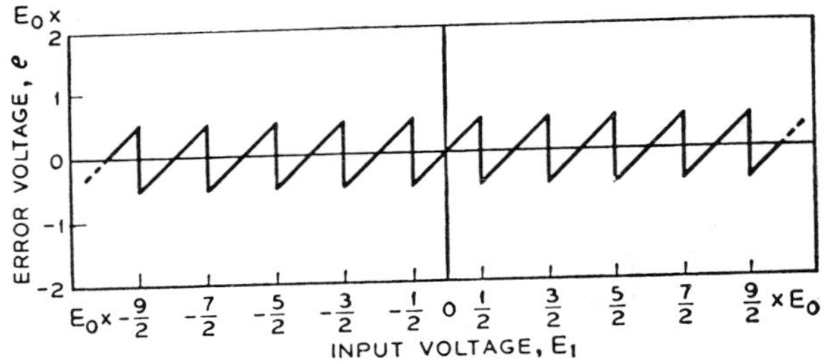

Fig. 3—Characteristic of the errors in quantizing.

E_0 is the voltage corresponding to one step, and s is the slope, the equation of the typical line is:

$$\epsilon = st, \qquad -\frac{E_0}{2s} < t < \frac{E_0}{2s} \tag{1.0}$$

where ϵ is the error voltage and t is the time referred to the midpoint as origin. Then the mean square error is

$$\overline{\epsilon^2} = \frac{s}{E_0}\int_{-E_0/2s}^{E_0/2s} \epsilon^2\,dt = \frac{E_0^2}{12}, \tag{1.1}$$

or one twelfth the square of the step size.

Not all the distortion falls within the signal band. The distortion may be considered to result from a modulation process consisting of the application of the component frequencies of the original signal to the non-linear staircase characteristic. High order modulation products may have frequencies quite remote from those in the original signal and these can be excluded by a filter passing only the signal band. It becomes of importance, therefore, to calculate the spectrum of the error wave. This we shall do in the next section for a generalized signal using the method of correlation, which is based on the fact that the power spectrum of a wave is the Fourier cosine transform of the correlation function. The result is then applied to a particular kind of signal, namely one having energy uniformly distributed throughout a definite frequency band and with the phases of the components randomly distributed. This is a particularly convenient type of signal because it in effect averages over a large number of possible discrete frequency components within the band. Single or double-frequency signal waves are awkward for analytical purposes because of the ragged nature of the spectra produced. The amplitudes of particular harmonics or cross-products of discrete frequency components are found to oscillate violently with magnitude of input. The use of a large number of input components smooths out the irregularities.

The type of spectra obtained is shown in Fig. 4. Anticipating binary coding, we have shown results in terms of the number of binary digits used. The number of different magnitudes available are 16, 32, 64, 128, and 256 for $N = 4, 5, 6, 7$ and 8 digits, respectively. Here a word of explanation is needed with respect to the placing of the scale of quantized voltages. A signal with a continuous distribution of components along the frequency scale is theoretically capable of assuming indefinitely great values of instantaneous voltage at infrequent instants of time. An actual quantizer (staircase transducer) has a finite overload value which must not be exceeded and hence can have only a finite number of steps. This difficulty is resolved here by the experimentally observed fact that thermal noise, which has the type of spectrum we have assumed for our signal, has never been observed to exceed appreciably a voltage four times its root-mean-square value. Hence we have placed the root-mean-square value of the input signal at one-fourth the overload input to the staircase. This fixes the relation between step size and the total number of steps. In the actual calculation the number of steps is taken as infinite; the effect of the assumed additional steps beyond 2^N is negligible because of the rarity of excursion into this range.

The curves of Fig. 4 are drawn for the case in which the signal band starts at zero frequency. The original signal band width is represented by one unit on the horizontal scale. The relatively wide spread of the distortion spectrum is clearly shown. As the number of digits (or steps) is increased

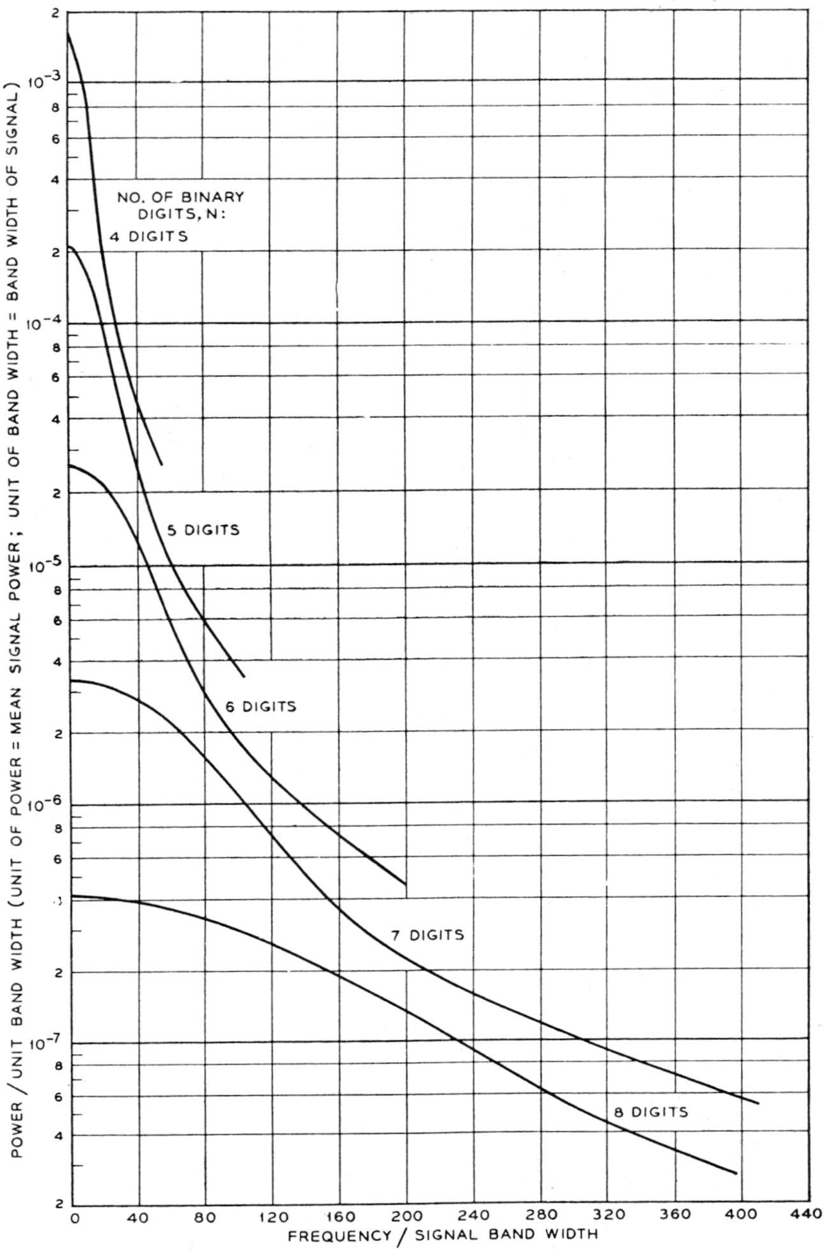

Fig. 4—Spectrum of distortion from quantizing the magnitudes of a random noise wave. Full load on the quantizer is reached by peaks 12 db above the r.m.s. value of input.

the spectrum becomes flatter over a wider range, but with a smaller maximum density. The area under each curve represents the total mean power in the corresponding error wave and is found to agree quite accurately with the approximate result of Eq. (1.1). The distortion power falling in the signal band is represented by the area included under the curve from zero to unit abscissa.

[*Editor's Note:* Material has been omitted at this point.]

Distortion caused by quantizing errors produces much the same sort of effects as an independent source of noise. The reason for this is that the spectrum of the distortion in the receiving filter output is practically independent of that of the signal over a wide range of signal magnitudes. Even when the signal is weak so that only a few quantizing steps are operated, there is usually enough residual noise on actual systems to determine the quantizing noise and mask the relation between it and the signal. Eq. (1.1) yields a simple rule enabling one to estimate the magnitude of the quantizing noise with respect to a full load sine wave test tone. Let the full load test tone have peak voltage E; its mean square value is then $E^2/2$. The total range of the quantizer must be $2E$ because the test signal swings between $-E$ and $+E$. The ratio $2E/E_0 = r$ is a convenient one to use in specifying the quantizing; it is the ratio of the total voltage range to the range occupied by one step. The ratio of mean square signal to mean square quantizing noise voltage is

$$\frac{E^2/2}{E_0^2/12} = \frac{6E^2}{4E^2/r^2} = \frac{3r^2}{2} \tag{1.3}$$

Actual systems fail to reproduce the full band $f_s/2$ because of the finite frequency range needed for transition from pass-band to cutoff. If we introduce a factor κ to represent the ratio of equivalent rectangular noise band to $f_s/2$, the actual received noise power is multiplied by κ. Then the signal-to-noise ratio in db for a full load test tone is

$$D = 10 \log_{10} \frac{3r^2}{2\kappa} \text{ db} \tag{1.4}$$

In practical applications the value of κ is about 3/4 which gives the convenient rule:

$$D = 20 \log_{10} r + 3 \text{ db} \tag{1.5}$$

In other words, we add 3 db to the ratio expressed in db of peak-to-peak quantizing range to the range occupied by one step. For various numbers of binary digits the values of D are:

TABLE I

Number of Digits	D
3	21
4	27
5	33
6	39
7	45
8	51

From Table I we can make a quick estimate of the number of digits required for a particular signal transmission system provided that we have some idea of the required signal-to-noise ratio for a full load test tone. The latter ratio may be expressed in terms of the full load test tone which the system is required to handle and the maximum permissible unweighted noise power at the same level point. Since quantizing noise is uniformly distributed throughout the signal band, its interfering effect on speech or other program material is probably similar to that of thermal noise with the same mean power. Requirements given in terms of noise meter readings must be corrected by the proper weighting factor before applying the table. If the signal transmitted is itself a multiplex signal with channels allotted on a frequency division basis, the noise power falling in each channel is the same fraction of the total noise power as the band width occupied by the signal is of the total band width of the system.

We have thus far considered only the case in which the quantized steps are equal. In actual systems designed for transmission of speech it is found advantageous to taper the steps in such a way that finer divisions are available for weak signals. For a given number of total steps this means that coarser quantization applies near the peaks of large signals, but the larger absolute errors are tolerable here because they are small relative to the bigger signal values. Tapered quantizing is equivalent to inserting complementary nonlinear transducers in the signal branch before and after the quantizer. In the usual case, the transducer ahead of the quantizer is of the "compressing" type in which the loss increases as the signal increases. If the full load signal just covers all the linear quantizing steps, a weak signal gets a bigger share of the steps than it would if the transducer were linear. The transducer after the quantizer must be of the "expanding" type which gives decreased loss to the large signals to make the overall combination linear.

On the basis of the theory so far discussed, we can say that the error spectrum out of the linear quantizer is virtually the same whether or not the signal input is compressed. The operation of the expandor then magnifies the errors produced when the signal is large. When weak signals are applied, the mean square error is given by Eq. (1.1), as before, but when the signal is increased an increment in noise occurs. The mean square value of noise voltage under load may be computed from the probability density of the signal values and the output-vs-input characteristic of the expandor, or its inverse, the compressor. A first order approximation, valid when the steps are not too far apart, replaces (1.1) by:

$$\overline{\epsilon^2} = \frac{E_0^2}{12} \int_{Q_2}^{Q_1} \frac{p_1(E_1)\, dE_1}{[F'(E_1)]^2} \tag{1.6}$$

where Q_1 and Q_2 are the minimum and maximum values of the input signal voltage E_1, $p_1(E_1)$ is the probability density function of the input voltage, and $F'(E_1)$ is the slope of $F(E_1)$, the compression characteristic.

Some experimental results obtained with a laboratory model of a quantizer are given in Figs. 6–9. Figs. 6–7 show measurements on the third harmonic associated with 6-digit quantizing. As mentioned before, the amplitude of any one harmonic oscillates with load. The calculated curves shown were obtained by straightforward Fourier analysis. In the measurements it was convenient to spot only the successive nulls and peaks.

In Fig. 6 the bias was set to correspond to the stair-case curve of Fig. 1, while in Fig. 7 the origin is moved to the point $(E_0/2, E_0/2)$, i.e., to the middle of a riser instead of a tread. The peaks of ratio of harmonic to fundamental decrease steadily as the amplitude of the signal is increased to full load, which is just opposite to the usual behavior of a communication system. It is difficult to extrapolate experience with other systems to specify quality in terms of this type of harmonic distortion.

Figure 8 shows measurements of the total distortion power falling in the signal band when the signal is itself a flat band of thermal noise. The technique of making such measurements has been described in earlier articles.[9,10] Measurements are shown for quantizing with both equal and tapered steps. The particular taper used is indicated by the expandor characteristic of Fig. 9. The compression curve is found by interchanging horizontal and vertical scales. The measurements were made on a quantizer with 32, 64, and 128 steps, and a sampling rate of 8,000 cycles per sec-

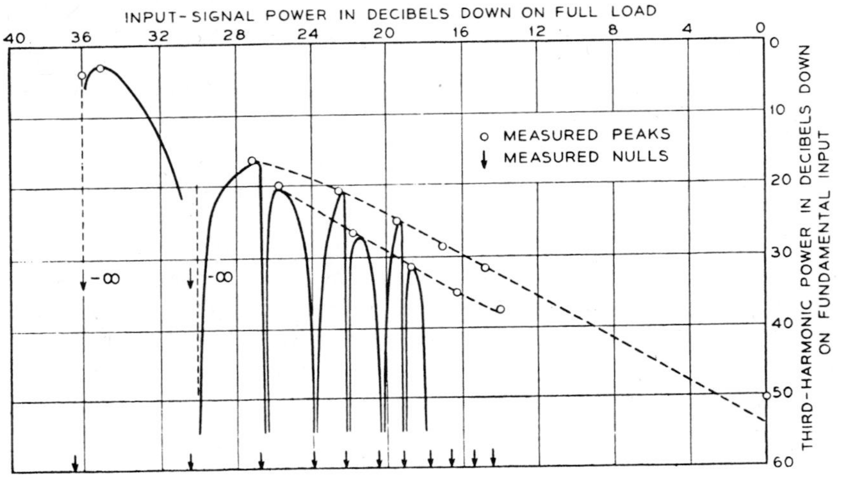

Fig. 6—Third harmonic in 64-step quantized output with bias at mid tread. The smooth curves represent computed values.

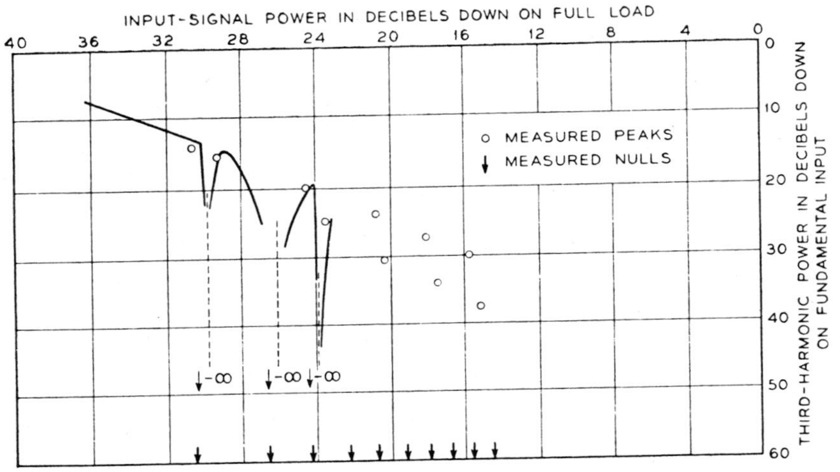

Fig. 7—Third harmonic in 64-step quantized output with bias at mid-riser. The smooth curves represent computed values.

ond. The applied signal was confined to a range below 4,000 cycles per second. With equal steps the distortion power is practically independent of load as shown by the db-for-db straight lines. With tapered steps, the distortion is less for weak signals, and only slightly greater for large signals.

The vertical line designated "full load random noise input" represents the value of noise signal power at which peaks begin to exceed the quantizing

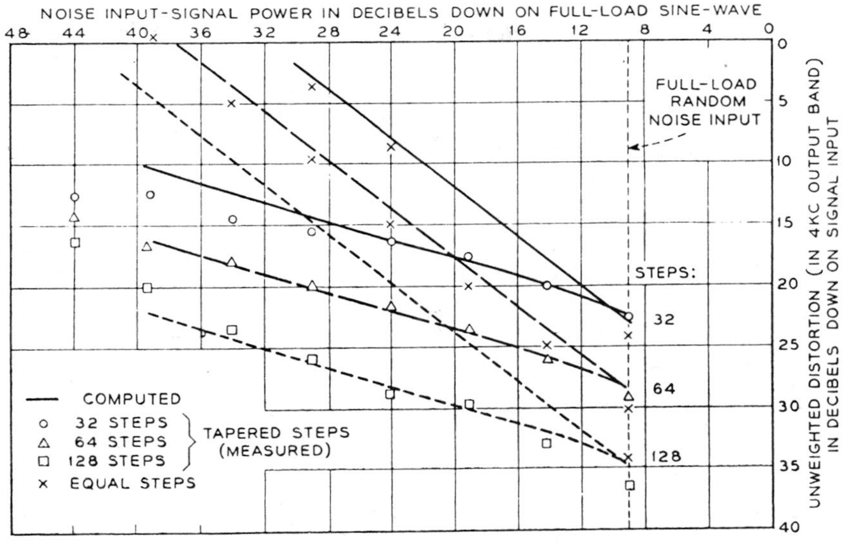

Fig. 8—Total distortion in signal band from quantizing with equal and tapered steps.

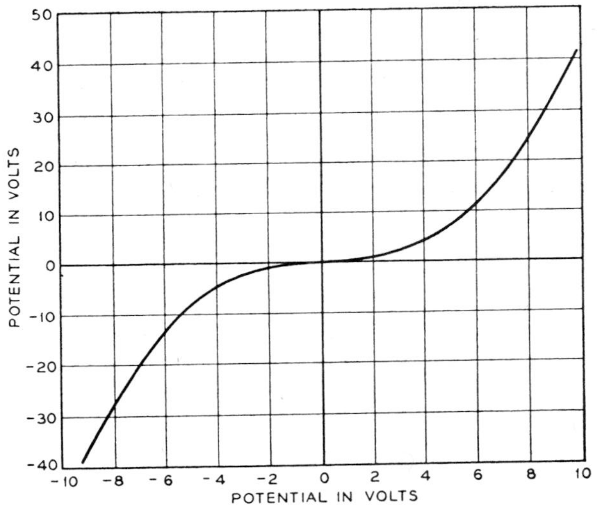

Fig. 9—Expanding characteristic applied to noise in tapered steps of Fig. (8).

range. This occurs when the rms value of input is 9 db below the rms value of the sine wave which fully loads the quantizer.

[*Editor's Note:* Material has been omitted at this point.]

REFERENCES

1. W. R. Bennett, Time Division Multiplex Systems, *Bell Sys. Tech. Jour.*, Vol. 18, pp. 1–31; Jan. 1939.
2. H. S. Black, Pulse Code Modulation, *Bell Lab. Record*, Vol. 25, pp. 265–269; July, 1947.
3. W. M. Goodall, Telephony by Pulse Code Modulation, *Bell Sys. Tech. Jour.*, Vol. 26, pp. 395–409; July, 1947.
4. D. D. Grieg, Pulse Count Modulation System, *Tele-Tech.*, Vol. 6, pp. 48–50, 98; Sept. 1947; also *Elect. Comm.*, Vol. 24, pp. 287–296; Sept. 1947.
5. A. G. Clavier, P. F. Panter, and D. D. Grieg, PCM Distortion Analysis, *Elec. Engg.*, Vol. 66, pp. 1110–1122; Nov. 1947.
6. H. S. Black and J. O. Edson, PCM Equipment, *Elec. Engg.*, Vol. 66, pp. 1123–1125; Nov. 1947.
7. L. A. Meacham and E. Peterson, An Experimental Pulse Code Modulation System of Toll Quality, *Bell Sys. Tech. Jour.*, Vol. 27, pp. 1–43; Jan., 1948.
8. H. Nyquist, Certain Topics in Telegraph Transmission Theory, *A. I. E. E. Trans.*, pp. 617–644; April, 1928.
9. E. Peterson, Gas Tube Noise Generator for Circuit Testing, *Bell Lab. Record*, Vol. 18, pp. 81–83; Nov. 1939.
10. W. R. Bennett, Cross-Modulation in Multichannel Amplifiers, *Bell Sys. Tech. Jl.*, Vol. 19, pp. 587–610; Oct. 1940.
11. N. Wiener, Generalized Harmonic Analysis, *Acta. Math.*, Vol. 55, pp. 177–258; 1930.
12. S. O. Rice, Mathematical Analysis of Random Noise, *Bell Sys. Tech. Jour.*, Vol. 24, p. 50; Jan. 1945.
13. R. Courant and D. Hilbert, Methoden der Mathematischen Physik, Vol. 1, p. 64; Berlin, 1931.
14. W. R. Bennett, New Results in the Calculation of Modulation Products, *Bell Sys. Tech. Jour.* Vol. 12, pp. 228–243; April 1933.
15. G. A. Campbell, The Practical Application of the Fourier Integral, *Bell Sys. Tech. Jour.*, Vol. 7, pp. 639–707; Oct. 1928.
16. S. C. Kleene, Analysis of Lengthening of Modulated Repetitive Pulses, *Proc. I. R. E.*, Vol. 35, pp. 1049–1053; 1947.

3

Reprinted with permission from pages 653, 656–661, 663–676, 698–699, and 708–709 of *Bell System Tech. J.* **36**(3):653–709 (1957)

Instantaneous Companding of Quantized Signals

By BERNARD SMITH

Instantaneous companding may be used to improve the quantized approximation of a signal by producing effectively nonuniform quantization. A revision, extension, and reinterpretation of the analysis of Panter and Dite permits the calculation of the quantizing error power as a function of the degree of companding, the number of quantizing steps, the signal volume, the size of the "equivalent dc component" in the signal input to the compressor, and the statistical distribution of amplitudes in the signal. It appears, from Bennett's spectral analysis, that the total quantizing error power so calculated may properly be studied without attention to the detailed composition of the error spectrum, provided the signal is complex (such as speech or noise) and is sampled at the minimum information-theoretic rate.

These calculations lead to the formulation of an effective process for choosing the proper combination of the number of digits per code group and companding characteristic for quantized speech communication systems. An illustrative application is made to the planning of a hypothetical PCM system, employing a common channel compandor on a time division multiplex basis. This reveals that the calculated companding improvement, for the weakest signals to be encountered in such a system, is equivalent to the addition of about 4 to 6 digits per code group, i.e., to an increase in the number of uniform quantizing steps by a factor between $2^4 = 16$ and $2^6 = 64$.

Comparison with the results of related theoretical and experimental studies is also provided.

[*Editor's Note:* Material has been omitted at this point.]

B. Quantizing Impairment in PCM Systems

From the foregoing it is clear that quantization (i.e., the representation of a bounded continuum of values by a finite number of discrete magnitudes), permits the encoded, and therefore essentially noise-free transmission of approximate, rather than exact values of sampled amplitudes. In fact, *the deliberate error imparted to the signal by quantization is the significant source of PCM signal impairment.*[1-5] Adequate limitation of this quantization error is therefore of prime importance in the application of PCM to communication systems.

A number of methods of reducing quantizing error suggest themselves on a purely qualitative and intuitive basis. For example, one may obtain a finer-grained approximation by providing more, and therefore smaller, quantizing steps for a given range of amplitudes. Alternatively, one may provide a more complete description of the signal by increasing the sampling rate beyond the minimum information-theoretic value already assumed.†

It is also possible to vary the size of the quantizing steps (without adding to their number) so as to provide smaller steps for weaker signals.

Whereas the first two techniques result in an increase of bandwidth and system complexity, the third requires only a modest increase in instrumentation without any increase in bandwidth.* This investigation is therefore devoted to the study of nonuniform step size as a means of reducing quantizing impairment.

C. *Physical Implications of Nonuniform Quantization*

1. Quantizing Error as a Function of Step Size

Quantizing impairment may profitably be expressed in terms of the total mean square error voltage since the ratio of the mean square signal voltage to this quantity is equal to the signal-to-quantizing error power ratio.

In evaluating the mean square error voltage, we begin by considering a complex signal, such as speech at constant volume, whose pulse samples yield an amplitude distribution corresponding to the appropriate probability density. These pulse samples may be expected to fall within, or "excite", all the steps assigned to the signal's peak-to-peak voltage range. It will be assumed that, for quality telephony, the steps will be sufficiently small, and therefore numerous, to justify the assumption that the probability density is effectively constant within each step, although it may be expected to vary from step to step. Thus the continuous curve representing the probability density as a function of instantaneous amplitude is to be replaced by a suitable histogram.

If the midstep voltage is assigned to all amplitudes falling in a particular quantizing interval, the absolute value of the error in any pulse sample will be limited to values between zero and half the size of the step in question; when combined with the assumed approximation of a uniform probability density within each step, this choice minimizes the mean square error introduced at each level.[5] Summation of the latter quantity over all levels then yields the result that the total mean square quantizing error voltage is equal to one-twelfth the weighted average of the square of the size of the voltage steps traversed (i.e., excited) by the input signal. The direct consideration of the physical meaning of this result (which, as (6) below, will constitute the basis of all subsequent calculations) will now be shown to provide a simple qualitative description of the implications of nonuniform quantization.

* We refer to bandwidth in the transmission medium as determined by the pulse repetition rate, which, in the time division multiplex applications envisioned herein, is given by the product: (sampling rate) × (number of digits or pulses per sample) × (number of channels).

2. Properties of the Mean Square Excited Step Size

Fig. 1(a) shows the range of input voltages, between the values $+V$ and $-V$ divided into N equal quantizing steps (i.e., uniformly quantized); Fig. 1(b) depicts the same range divided into N tapered steps, corresponding to nonuniform quantization.

Consider a complex signal, such as speech, whose distribution of instantaneous amplitudes at constant volume results in the excitation (symmetrically about the zero level) of the steps in the moderately large interval $X-X'$. The quantizing error power will be shown to be proportional to the (weighted) average of the square of the excited step size. For uniform quantization, it is clear, from Fig. 1(a), that this average is a constant, independent of the statistical properties of the signal. For a nonuniformly quantized signal, [Fig. 1(b)], the mean square excited step size is reduced by the division of the identical interval $X-X'$ into more steps, most of which are smaller than those shown in Fig. 1(a). Appreciation of the full extent to which the quantizing error power may so be reduced requires the added recognition that the few larger quantizing steps in the range $X-X'$, corresponding to excitation by comparatively rare speech peaks, are far less significant in their contribution to the weighted average than the small steps in the vicinity of the origin, due to the nature of the probability density of speech at constant volume.[18]

It is also clear that weaker signals, corresponding to a contraction of the interval $X-X'$, enjoy the greatest potential tapering advantage since their excitation may be confined to steps which are all appreciably smaller than those in Fig. 1(a). However, if the interval $X-X'$ were to

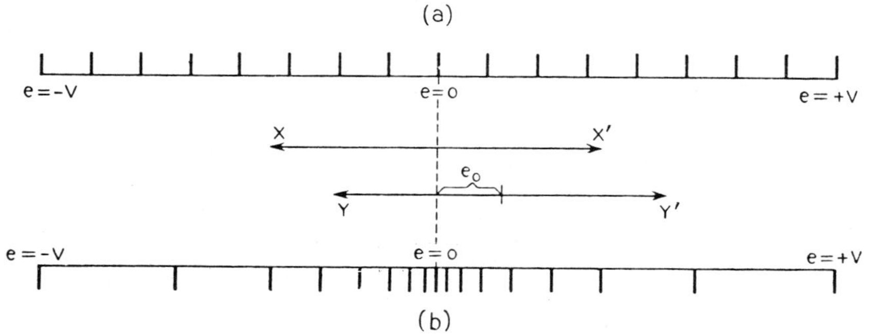

Fig. 1 — (a) Distribution of steps of equal size corresponding to direct, uniform quantization; (b) nonuniform quantization of this range into the same number of steps. The function of the instantaneous compandor is to provide such nonuniform quantization in the manner illustrated in Fig. 2.

increase in size and approach the full range, $+V$ to $-V$, (to accommodate stronger signals), the excitation of extremely large steps might result in an rms step size exceeding the uniform size shown in Fig. 1(a).

Fig. 1 also indicates that signals (including unwanted noise) too weak to excite even the first quantizing step (and therefore absolutely incapable of transmission) when uniformly quantized, may successfully be transmitted as a result of the excitation of a few steps following nonuniform quantization.

Although the assumption that the average value of the signal is zero is quite proper for speech, subsequent discussion will disclose the possibility that the quiescent value of the signal, as it appears at the input to the quantizing equipment, may not always coincide with the exact center of the voltage range depicted in Fig. 1. This effect may formally be described in terms of the addition of an equivalent dc bias to the speech input at the quantizer. As shown in Fig. 1, the addition of such a dc component, e_0, to the signal which previously excited the band of steps labeled $X-X'$, transforms $X-X'$ into an array of equal extent $Y-Y'$, centered about $e = e_0$ instead of $e = 0$. This causes the excitation of some larger steps, in Fig. 1(b), as well as the assignment of greatest weight[18] to the steps in the vicinity of $e = e_0$, which are larger than those near $e = 0$; the net result is an increase in the rms excited step size, and the quantizing error power. This effect will depend on the comparative size of e_0 and the signal as well as on the degree of step size variation. In particular, Fig. 1(a) indicates that the presence of e_0 does not affect the rms excited step size under conditions of uniform quantization.

It is clear from the foregoing that the effect of nonuniform quantization of PCM signals will vary greatly with the strength of the signal; greatest improvement is to be expected for weak signals, whereas an actual impairment may be experienced by strong signals. The range of signal volumes is therefore of prime importance in the choice of the proper distribution of step sizes.

D. *Nonuniform Quantization Through Uniform Quantization of a Compressed Signal*

Nonuniform quantization is logically equivalent to uniform quantization of a "compressed version" of the original input signal. When applied directly, tapered quantization provides an acceptably high ratio of sample amplitude to sample error for weak pulses, by decreasing the errors (i.e., the step sizes) assigned to small amplitudes. Signal compression achieves the same goal by increasing weak pulse amplitudes without altering the step size.

The instantaneous compressor envisioned herein is, in essence, a non-linear pulse amplifier which modifies the distribution of pulse amplitudes in the input PAM signal by preferential amplification of weak samples. A satisfactory compression characteristic will have the general shape shown in Fig. 2. Thus the amplification factor, (v/e), varies from a large value for small inputs to unity for the largest amplitude (V) to be accommodated, so that the distribution of pulse sizes may be modified without changing the total voltage range. Fig. 2 also illustrates how uniform quantization of the compressor output produces a tapered array of input steps similar to those already considered in connection with Fig. 1(b).

A complementary device, the expandor, employs a characteristic inverse to that of the compressor to restore the proper (quantized) distribution of pulse amplitudes after transmission and decoding. Taken together, the compressor and expandor constitute a compandor.

The resolution of tapered quantization into the sequential application of compression, uniform quantization, and expansion is operationally convenient,[6] as well as logically sound. Since there is a one-to-one correspondence between step size allocations and compression characteristic

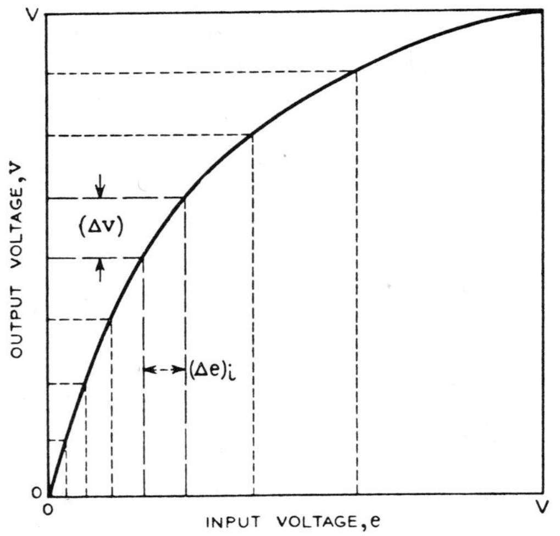

Fig. 2 — Curve illustrating the general shape of a suitable instantaneous compression characteristic. All continuous, single-valued curves connecting the origin to the point (V, V) and rising from the origin with a slope greater than one, i.e., $(dv/de)_{e=0} > 1$, are potential compression characteristics. The symmetrical negative portion $[v(e) = -v(-e)]$ is not shown. The production of a tapered array of input steps $(\Delta e)_i$ by uniform quantization of the output into steps of (equal) size Δv, is also represented.

curves, the central problem of choosing the proper distribution of step sizes will be discussed in terms of the choice of the appropriate compression characteristic; the reduction of quantizing error, corresponding to nonuniform quantization without change in the total number of steps, will be termed companding improvement.

[*Editor's Note:* Material has been omitted at this point.]

F. *Applicability of the Present Analysis*

Before we proceed to a detailed analysis, it is important to emphasize certain restrictive conditions required for the meaningful application of the results to be derived.

1. Signal Spectrum

A signal with a sufficiently complex spectrum, such as speech, is required to justify consideration of the total quantizing error power without regard to the detailed composition of the error spectrum. Although it is known that quantization of simple signals (e.g., sinusoids) results in discrete harmonics and modulation products deserving of individual attention,[8, 9] Bennett has shown that the error spectrum for complex signals is sufficiently noise-like to justify analysis on a total power basis.[2, 12]

2. Sampling Rate

The consistent comparison of signal power with the total quantizing error power, rather than with the fraction of the latter quantity appropriate to the signal band, might at first appear to impose serious limitations on the present analysis. Furthermore, the role of sampling has not been discussed explicitly. It is therefore important to note that the justification for this treatment, in the situation of actual interest, has also been given by Bennett.[2] We need only add the standard hypothesis[1-6] that the sampling rate chosen for a practical system would equal the minimum acceptable rate (slightly in excess of twice the top signal frequency[3]) in order to invoke Bennett's results, which tell us that, for this sampling rate, the quantizing error power in the signal band and the total quantizing error power are identical.[2] Thus, sampling at the minimum rate is assumed throughout.

3. Number of Quantizing Steps

As already remarked, the present results are based on the assumption that N is not small, inasmuch as we assume a probability density which, although varying from step to step, remains effectively constant within each quantizing step; indeed the step sizes will be treated as differential quantities.

Experimental evidence[6, 7, 10] (as well as the analysis to follow) argues against the consideration of fewer than five digits (i.e., $2^5 = 32$ quantizing steps) for high quality transmission of speech. Numerical estimates indicate that the present approximation should be reasonable for five or more digits per code group. These estimates are confirmed by the consistency of actual measurements of quantizing error power with calculations based on the same approximation (see Fig. 8 of reference 2 for 5, 6, and 7 digit data obtained with an input signal consisting of thermal noise instead of speech).

Further indication of the adequacy of this approximation is provided by the knowledge that Sheppard's corrections (see Section II-B) appear adequate even when (Δe) is not very small, for a probability density which (as is the case for speech[18]) approaches zero together with its derivatives at both ends of the (voltage) range under consideration.[24]

Therefore, we are not presently concerned with the limitations imposed by this approximation.

4. Subjective Effects Beyond the Scope of the Present Analysis

We shall have occasion to study graphs depicting the signal to quantizing error power ratio as a function of signal power. Although these curves, and the equations they represent, will always be of interest for the case where even the weakest signal greatly exceeds the corresponding error power, there exists the possibility of rash extrapolation to the region where this inequality is reversed. Unfortunately, such extrapolation may have little or no meaning.* This is particularly clear when one considers that signals incapable of exciting at least the first quantizing step, in the absence of companding, will be absolutely incapable

* This is implicit in the deduction of Equation (6).

of transmission. Under these circumstances, companding may actually *resuscitate* a signal; the mathematical description of resuscitation (as anything short of infinite improvement) is clearly beyond the scope of the present analysis.

At the other extreme, it is probable that there exists a limit of error power suppression beyond which listeners will fail to recognize any further improvement. Our analysis will not be useful in describing this region of subjective saturation. Furthermore, it is possible that the subjective improvement afforded a listener by adding to the number of quantizing steps, or companding, may depend on the initial and final states, even before subjective saturation is reached. For example, it is entirely possible that the change from 5 to 6 digits per code group may provide a degree of improvement which appears different to the listener from that corresponding to the increase from 6 to 7 digits, although the present mathematical treatment does not recognize such a distinction.

II. EVALUATION OF MEAN SQUARE QUANTIZATION ERROR (σ)

A.* *Generalization of the Analysis of Panter and Dite*

The mean square error voltage, σ_j, associated with the quantization of voltages assigned to the jth voltage interval, e_j, is adopted as the significant measure of the error introduced by quantization. If e_j is to represent any voltage, e, in the range

$$Q_j = \left[e_j - \frac{(\Delta e)_j}{2}\right] \leqq e \leqq \left[e_j + \frac{(\Delta e)_j}{2}\right] = R_j \tag{1}$$

then

$$\sigma_j = \int_{Q_j}^{R_j} (e - e_j)^2 P(e)\, de \tag{2}$$

where $(e - e_j)$ is the voltage error imparted to the sample amplitude by quantization and $P(e)$ is the probability density of the signal. The location of e_j at the center of the voltage range assigned to this level minimizes σ_j since we shall assume an effectively constant value of $P(e)$ within the confines of a single step.

If the value of $P(e)$ is approximated by the constant value $P(e_j)$ appropriate to e_j in (2), it follows that

$$\sigma_j = (\Delta e)_j^3 P(e_j)/12 \tag{3}$$

* This passage contains mathematical details which may be omitted, in a first reading, without loss of continuity.

The total mean square voltage error, σ, is equal to the sum of the mean square quantizing errors introduced at each level, so that,

$$\sigma = \sum_j \sigma_j = \tfrac{1}{12}\sum_j P(e_j)(\Delta e)_j^3 \tag{4a}$$

$$= \tfrac{1}{12}\sum_j (\Delta e)_j^2[P(e_j)(\Delta e)_j] \tag{4b}$$

which may be rewritten as

$$\sigma \cong \tfrac{1}{12}\sum_j (\Delta e)_j^2 p_j \tag{4c}$$

since the discrete probability appropriate to the jth step is given by

$$p_j = \int_{Q_j}^{R_j} P(e)\,de \cong [P(e_j)(\Delta e)_j] \tag{5}$$

Hence,

$$\sigma \cong \tfrac{1}{12}[(\Delta e)^2]_{\text{AV}} = \overline{(\Delta e)^2}/12 \tag{6}$$

Thus, the total mean square error voltage is equal to one-twelfth the average of the square of the input voltage step size when the steps are sufficiently small (and therefore numerous) to justify the approximations employed in the deduction of (6). In applying (6), it is important to note that (4) implies that this is a *weighted* average over the steps traversed (or "excited") by the signal.

In the special case of uniform step size, substitution of $(\Delta e)_j = (\Delta e) =$ const. reduces (6) to the simple form

$$\sigma_0 = [\sigma]_{\Delta e = \text{const}} = (\Delta e)^2/12 \tag{7}$$

Equations (6) and (7) provide the basis for the qualitative interpretation of quantizing error power which has already been discussed in connection with Fig. 1.

The deduction of (6) from (4a) is implicit in the work of Panter and Dite.[5] The absence of an explicit formulation of (6) therein* results from the direct application of the equivalent of (4b) to a specific problem involving a particular algebraic expression for $(\Delta e)_j$.

A prior, equivalent derivation of (7), based on a graphical representation of $(e - e_j)$ as a sawtooth error function for uniform quantization has been given by Bennett.[2] Although this derivation bypassed (6), Bennett has also analyzed compressed signals by means of an expression

* The present notation has been chosen to resemble that of Reference 5 in order to facilitate direct comparison by the reader.

[(1.6) of Reference 2] which is equivalent to (6), when the average is expressed as an integral over a continuous probability distribution and (Δe) is replaced by $(de/dv)(\Delta v)$, with $(\Delta v) = \text{const}$. This form of (6) is the point of departure for the calculation in the Appendix.

B. *Operational Significance of* σ

Manipulation of (2) may be shown to result in the expression

$$\sigma = \sum_j \sigma_j = \sum_j e_j^2 p_j - \int e^2 P(e)\, de,$$

which is the difference between the mean square signal voltages following and preceding quantization. Hence σ is proportional to the difference between the quantized and unquantized signal powers. Since σ is intrinsically positive, the quantizing error power is *added* to the signal by quantization and is, in principle, measurable as the difference between two wattmeter readings.

In addition to providing an operational interpretation of the quantizing error power, the rewritten expression for σ reveals the equivalence of σ to the "Sheppard correction" to the grouped second-moment in statistics,[24-27] where calculations are facilitated by grouping (i.e., uniform quantization) of numerical data. The reader who is interested in a more elaborate deduction of (7) from the Euler-Maclaurin summation formula, as well as discussions of the validity of (7), may therefore consult the statistical literature.

III. CHOICE OF COMPRESSION CHARACTERISTIC

A. *Restriction to Logarithmic Compression*

We shall consider the properties of the logarithmic type of compression characteristic,* defined by the equations†

$$v = \frac{V \log\,[1 + (\mu e/V)]}{\log\,(1 + \mu)}, \qquad \text{for} \qquad 0 \leqq e \leqq V \tag{8a}$$

and

$$v = \frac{-V \log\,[1 - (\mu e/V)]}{\log\,(1 + \mu)}, \qquad \text{for} \qquad -V \leqq e \leqq 0 \tag{8b}$$

* The author first encountered this characteristic in the work of Panter and Dite[5] and the references thereto cited by C. P. Villars in an unpublished memorandum. He has since learned that such characteristics had been considered by W. R. Bennett as early as 1944 (unpublished), as well as by Holzwarth[16] in 1949.

† Unless otherwise specified, natural logarithms will be used throughout.

In (8), v represents the output voltage corresponding to an input signal voltage e, and μ is a dimensionless parameter which determines the degree of compression.

Typical compression characteristics, corresponding to various choices of the compression parameter, μ, in (8a), are shown in Fig. 3. The logarithmic replot of Fig. 4 provides an expanded picture of small amplitude behavior, as well as evidence of the probable realizability of such characteristics.

Although restriction of attention to (8) may at first appear to impose serious limitations on the generality of the analysis, this impression is not confirmed by more careful scrutiny of the problem.

Perusal of Fig. 3 indicates that (8) generates a considerable variety of curves which meet the general requirements already enunciated in connection with Fig. 2. Thus, the constant factor, $V/\log(1 + \mu)$, has been chosen to satisfy the condition

$$[v]_{e=V} = V \tag{9}$$

Evidence of the significance of the μ-characteristics may be derived

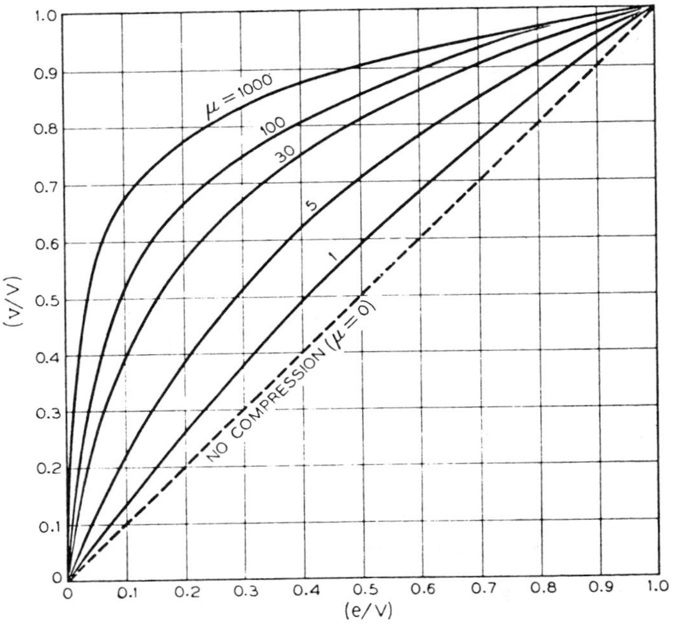

Fig. 3 — Typical logarithmic compression characteristics determined by equation (8a). The symmetrical negative portions, corresponding to equation (8b), are not shown.

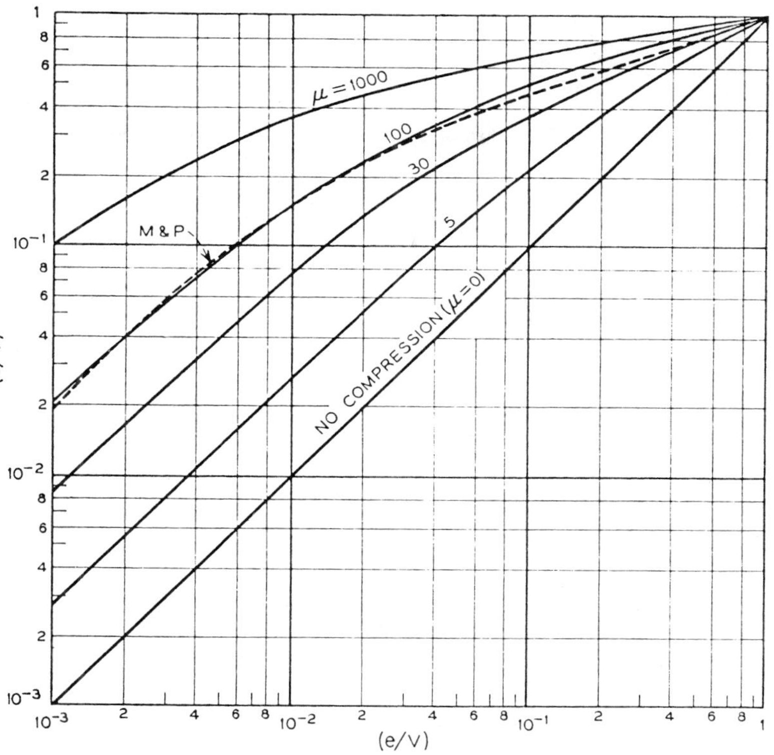

Fig. 4 — Logarithmic replot of compression curves shown in Fig. 3, to indicate detailed behavior for weak samples. The characteristic employed in the experiments of Meacham and Peterson[6] (M & P) is also shown. Similarity between this characteristic and the $\mu = 100$ curve testifies to the probable realizability of these logarithmic characteristics.

from consideration of the ratio of step size to corresponding pulse amplitude, $(\Delta e/e)$, since this quantity is a measure of the maximum fractional quantizing error imposed on individual samples. Hence the relation,

$$(e/\Delta e) = [N/2 \log (1 + \mu)](1 + V/\mu e)^{-1}$$

[which follows from (12a)] has been plotted, for $\mu = 10$, 100, and 1000, in Fig. 5. These curves reflect the fact that the sample to step size ratio reduces to the asymptotic forms:

$$(e/\Delta e) \rightarrow N/2 \log (1 + \mu) = \text{const} \qquad \text{for} \qquad (e/V) \gg \mu^{-1}$$

and

$$(e/\Delta e) \rightarrow [N/2 \log (1 + \mu)](\mu e/V) \qquad \text{for} \qquad (e/V) \ll \mu^{-1}$$

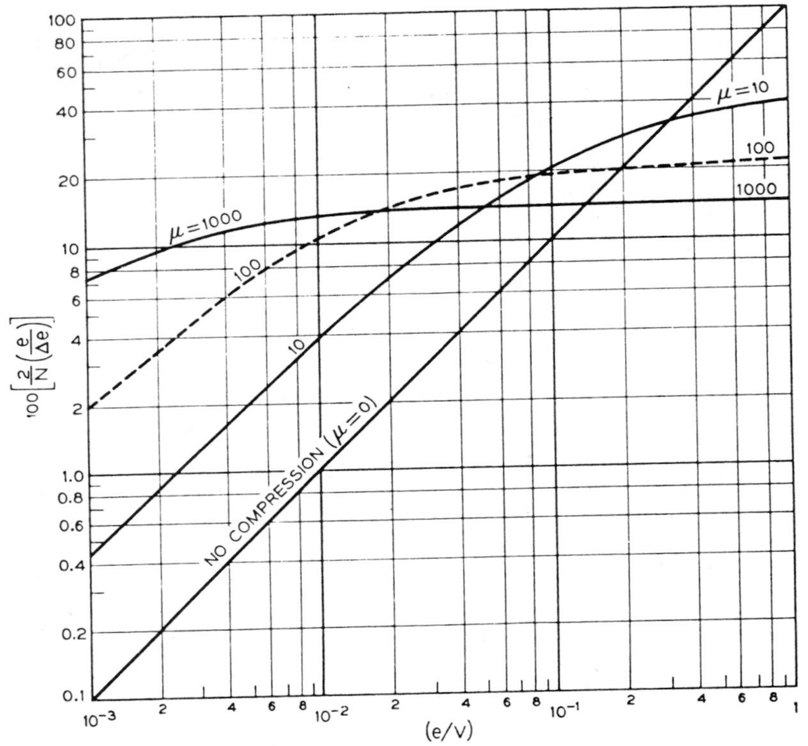

Fig. 5 — Pulse sample to step size ratios, as a function of relative sample amplitude, for various degrees of logarithmic companding (i.e., values of μ). The factor $(2/N)$ in the ordinate permits the curves to be drawn without reference to the total number of quantizing steps (N); the factor (100) is included to permit the ordinates directly to convey the proper order of magnitude for $(e/\Delta e)$, since present interest will be found to center about values of N for which $100(2/N) \sim 1$. As noted in the text, the ordinates, which constitute an index of the precision of quantization, approach constancy for $(e/V) \gg \mu^{-1}$, and vary linearly with abscissa for $(e/V) \ll \mu^{-1}$.

The essentially logarithmic behavior ($e/\Delta e \cong$ const) for large pulse amplitudes is intuitively desirable since it implies an approach to the equitable reproduction of the entire distribution of amplitudes in a specified signal. Although existing experimental evidence indicates that the small amplitudes are not only most numerous,[18] but also most significant for the *intelligibility*[21-23] of speech at constant volume, the absence of comparable evidence on the properties of *naturalness* makes it plausible to consider only those compression characteristics which give promise of providing the same, acceptably small, upper limit on the fractional quantizing error for pulse samples of all sizes.

For sufficiently small input pulses, $(e/\Delta e)$ becomes proportional to e, as a result of the linearity of the logarithmic function in (8) for small arguments. In view of our professed preference for logarithmic behavior, with $(e/\Delta e) \cong$ const., it is important to emphasize that the transition to linearity is not peculiar to (8), but is rather an example of the linearity to be expected of any suitably behaved (i.e., continuous, single-valued, with $(dv/de)_{e=0} > 1$) odd compression function, $v(e)$, capable of power series expansion, in the vicinity of the origin. In (8) this transition to linearity takes place where (e/V) is comparable to μ^{-1}. The extension of the region where $(e/\Delta e) \cong$ const. to lower and lower pulse amplitudes requires an increase in μ, and a concomitant reduction of the $(e/\Delta e)$ ratio for strong pulses.

Further evidence of the significance of the parameter μ may be deduced by evaluating the ratio of the largest to the smallest step size from the asymptotic expressions for $(e/\Delta e)$. Thus we find

$$\frac{(\Delta e)_{e=V}}{(\Delta e)_{e=0}} \rightarrow \mu \qquad \text{for} \qquad \mu \gg 1$$

which is a special form of the more general relation[5]

$$\frac{(\Delta e)_{e=V}}{(\Delta e)_{e=0}} = \frac{(dv/de)_{e=0}}{(dv/de)_{e=V}} = \mu + 1$$

which follows from our standard approximation of

$$(de/dv) \cong (\Delta v/\Delta e)$$

with $\Delta v =$ const.

B. *Comparison with Other Compandors*

An upper bound for companding improvement, which permits the quantitative evaluation of the penalty incurred (if any) through the restriction to logarithmic companding, is established in the Appendix. Comparison of the results to be derived from (8) with this upper bound will reveal that nonlogarithmic characteristics, which provide somewhat more companding improvement at certain volumes, are apt to prove too specialized for the common application to a broad volume range envisioned herein. The μ-characteristics do not suffer from this deficiency since the equitable treatment of large samples, which we have hitherto associated with an "intuitive naturalness conjecture," will be seen to tend to equalize the treatment of all signal volumes.

Finally, it will develop that (8), when applied to (6), has the added merit of calculational simplicity.

IV.* THE CALCULATION OF QUANTIZING ERROR

A. *Logarithmic Companding in the Absence of "DC Bias"*

As previously noted, we consider the effect of *uniformly* quantizing a *compressed* signal. If we designate the uniform output voltage step size by (Δv), then

$$(\Delta v) = \frac{2V}{N} \tag{10}$$

since the full voltage range between $-V$ and $+V$, of extent $2V$, is to be divided into N equal steps. For a number of levels, N, which is sufficiently large to justify the substitution of the differentials dv and de for the step sizes Δv and Δe, differentiation of (8a) yields

$$\frac{(\Delta v)}{V} = k\left[\frac{1}{1 + (\mu e/V)}\right]\frac{\mu(\Delta e)}{V} \tag{11}$$

where $k = 1/\log(1 + \mu)$.

Combining (10) with (11) and the counterpart of the latter in the domain of (8b), we find

$$\Delta e = \alpha(V + \mu e) \qquad \text{for} \qquad 0 \leqq e \leqq V \tag{12a}$$

and

$$\Delta e = \alpha(V - \mu e) \qquad \text{for} \qquad -V \leqq e \leqq 0 \tag{12b}$$

where

$$\alpha = 2\log(1 + \mu)/\mu N \tag{13}$$

Substitution of (12) into (6) yields

$$\sigma = (\alpha^2/12)[V^2 + \mu^2\overline{e^2} + 2\mu V\,\overline{|e|}] \tag{14}$$

where the quantity $\overline{|e|}$ is introduced by the difference in sign in (12a) and (12b). For ordinary compandor applications, we may write

$$\overline{|e|} = 2\int_0^V eP(e)\,de \tag{15}$$

since the symmetry of the input signal provides that $P(-e) = P(e)$ and $\bar{e} = 0$.

---

* This passage contains mathematical details which may be omitted, in a first reading, without loss of continuity.

It is convenient to define the quantization error voltage ratio,

$$D = \frac{\text{RMS Error Voltage}}{\text{RMS Input Signal Voltage}} = (\sigma/\overline{e^2})^{\frac{1}{2}} \tag{16}$$

which takes the form

$$D = \log(1 + \mu)[1 + (C/\mu)^2 + 2AC/\mu]^{\frac{1}{2}}/\sqrt{3}N \tag{17}$$

when we define the quantities

$$A = \overline{|e|}/\sqrt{\overline{e^2}} = \frac{\text{Average Absolute Input Signal Voltage}}{\text{RMS Input Signal Voltage}} \tag{18}$$

and

$$C = V/\sqrt{\overline{e^2}} = \frac{\text{Compressor Overload Voltage}}{\text{RMS Input Signal Voltage}} \tag{19}$$

The simple linear proportionality of Δe to $(V \pm \mu e)$ results from the properties of the logarithmic function in differentiation. Other, seemingly more simple compression equations, when differentiated, yield much more complicated and unwieldy expressions for Δe. The value of this simplicity is evident in the absence, from (14), of moments of e higher than the second.

If we set $A = 0$, (17) reduces to one deduced by Panter and Dite[5]; their analysis erroneously associated A with $\bar{e} = 0$ rather than with $\overline{|e|}$, as a result of their tacit assumption that (12a) and (12b) are identical. They also imposed the restriction of considering only that class of input signals having peak values coincident with the compandor overload voltage, by defining V as the peak value of the signal in specifying C. The definition of C in terms of the independent properties of both signal $(\overline{e^2})$ and compandor (V) is then converted into one based solely on the properties of the signal. This interpretation leads to conclusions quite different from those to be presented here.

B. *Logarithmic Companding in the Presence of "DC Bias"*

It has heretofore been assumed that the input signal is symmetrically disposed about the zero voltage level since it may be expected that $\bar{e} = 0$ for speech. Although this is a standard assumption, subsequent discussion will disclose that it is probable; in actual practice, for the average value of the input signal to be introduced at a point other than the origin of the compression characteristic. In terms of Fig. 1, the signal is presented to the array of quantizing steps with its quiescent value displaced by an amount e_0 from the center of the voltage interval ($-V$ to $+V$).

Such an effect, regardless of its origin, may formally be described by considering the composite input voltage

$$E = e + e_0 \tag{20}$$

where e is the previously considered symmetrical speech signal and e_0 is the superimposed constant voltage.

Substitution of E for e in (8) and (12) yields

$$\sigma_E = (\alpha^2/12)[V^2 + \mu^2\overline{E^2} + 2\mu V \overline{|E|}] \tag{21}$$

where the subscript E is introduced to distinguish this result from (14). Note that the value $[e]_{E=0} = -e_0$ now separates the domain of applicability of (8a) and (12a) from that of (8b) and (12b), so that (15) is replaced by

$$\overline{|E|} = \int_{-V}^{-e_0} (-E)P(e)\, de + \int_{-e_0}^{V} EP(e)\, de \tag{22}$$

which reduces to

$$\overline{|E|} = \overline{|e|} + 2e_0 \int_0^{e_0} P(e)\, de - 2\int_0^{e_0} eP(e)\, de \tag{23}$$

Since $\bar{e} = 0$, and $e_0 = $ const., we also find

$$\overline{E^2} = \overline{e^2} + e_0^2 \tag{24}$$

C. *Application to Speech as Represented by a Negative Exponential Distribution of Amplitudes*

It is necessary to assume an explicit function for $P(e)$ in (15) and (23) before applying the general results which have thus far been deduced. We shall assume, as a simple but adequate first approximation, that the distribution of amplitudes in speech at constant volume[18] may be represented by

$$P(e) = G \exp(-\lambda e) \qquad \text{for} \qquad e \geqq 0 \tag{25}$$

where $P(-e) = P(e)$, $G = \lambda/2$, and $\lambda^2 = 2/\overline{e^2}$. The values of G and λ

follow from the standard relations

$$\int_{-\infty}^{\infty} P(e)\, de = 1$$

and

$$\int_{-\infty}^{\infty} e^2 P(e)\, de = \overline{e^2}$$

When applied to (15) and (18), with the upper limit in (15) replaced by ∞ with negligible error, (25) implies that

$$A = \overline{|e|}/\sqrt{\overline{e^2}} = 1/\sqrt{2} = 0.707 \tag{26}$$

Hence, (17) will be replaced, for numerical calculations, by the relation

$$\sqrt{3}ND = \log(1 + \mu)[1 + (C/\mu)^2 + \sqrt{2}C/\mu]^{\frac{1}{2}} \tag{27}$$

The corresponding substitution of (25) into (23) yields, for the case of $e_0 \neq 0$,

$$\overline{|E|} = e_0 + (\overline{e^2}/2)^{\frac{1}{2}} \exp(-\sqrt{2}C/B) \tag{28}$$

where we have introduced the "bias parameter,"

$$B = V/e_0 \tag{29}$$

When (28) is combined with (13), (21), and (24), we find, after some algebraic manipulation, that

$$\begin{aligned}\sqrt{3}ND_E &= \log(1 + \mu)\\ &\cdot[1 + (C/\mu)^2(1 + \mu/B)^2 + (\sqrt{2}C/\mu)\exp(-\sqrt{2}C/B)]^{\frac{1}{2}}\end{aligned} \tag{30}$$

where $D_E^2 = (\sigma_E/\overline{e^2})$. It is to be noted that D_E has been defined in terms of the ratio of σ_E to $\overline{e^2}$ rather than $\overline{E^2}$, so that

$$D_E^2 = \frac{\text{Mean Square Error Voltage}}{\text{Mean Square Speech Voltage}} \tag{31a}$$

$$= \frac{\text{Average Error Power}}{\text{Average Speech Power}} \tag{31b}$$

Examination of (30) reveals that it has the required property of reducing to (27) for $e_0 = 0$, i.e., for $B \rightarrow \infty$. Furthermore (27) and (30) indicate that $D_E \geqq D$ so that the addition of a dc component increases the quantizing error power when companding is used. The existence of such an impairment may easily be understood in terms of the physical interpretation of (6), as discussed in connection with Fig. 1.

Equations (27) and (30) also reveal that the penalty inflicted by a finite e_0 is largely determined by the ratio (μ/B). If $(\mu/B) \ll 1$, the presence of e_0 will be unimportant. At the other extreme, if $(\mu/B) \gg 1$, $(1 + \mu/B)^2 \to (\mu/B)^2$ and

$$\sqrt{3}ND_E \to \log (1 + \mu) \cdot [1 + (C/B)^2 + (\sqrt{2}C/\mu) \exp (-\sqrt{2}C/B)]^{\frac{1}{2}} \tag{32}$$

which proves to be relatively insensitive to changes in μ for the values of μ, C and B considered herein. In this case B largely usurps the algebraic role previously assigned to μ in (27).

D. *Uniform Quantization*: $\mu = 0$

The mean square quantization voltage error in the absence of companding, corresponding to direct, uniform quantization of the input signal, follows immediately from (7) and (10) since $\Delta v = \Delta e$ under these conditions. Thus

$$\sigma_0 = (\Delta v)^2/12 = V^2/3N^2$$

whence

$$D_0 = (\sigma_0/\overline{e^2})^{\frac{1}{2}} = C/\sqrt{3}N \tag{33}$$

This inverse proportionality of D_0 and N is well known.[2, 3, 5]

Equation (33) may also be deduced by letting μ approach zero in the expressions for D and D_E, since (8) implies that v approaches e as μ approaches zero. The fact that $D_0 = (D_E)_{\mu \to 0}$ reveals that, in the absence of companding, the addition of e_0 does not change the quantizing error power. This conclusion was anticipated in the discussion of Fig. 1.

[*Editor's Note:* Material has been omitted at this point.]

APPENDIX

THE MINIMIZATION OF QUANTIZING ERROR POWER

In spite of the demonstrated utility of the μ-characteristics, one cannot avoid speculating about the possibility of achieving substantially more companding improvement by using a characteristic which differs from (8). We shall therefore outline a study of the actual minimization of quantizing error power without regard to the relative treatment of various amplitudes in the signal. The results will confirm that a significant reduction of the quantizing error power beyond that attainable with logarithmic companding is self-defeating — for it not only imposes the risk of diminished naturalness, but also implies a compandor too "volume-selective" for the applications envisioned herein.

1. The Variational Problem and Its Formal Solution

Equation (6) may be expressed in the form

$$\sigma = \frac{2V^2}{3N^2} \int_0^V (dv/de)^{-2} P(e)\, de \qquad \text{(A-1)}$$

where $P(e)$ has been assumed to be an even function. The function, $v(e)$, which will minimize (A-1), subject to the usual boundary conditions at $e = 0$ and $e = V$, may be obtained by solving the Euler differential equation of the variational problem.[28] For (A-1), this takes the form

$$(dv/de) = KP^{1/3} \qquad \text{(A-2)}$$

where the constant K is given by

$$K = V \Big/ \int_0^V P^{1/3}\, de \qquad \text{(A-3)}$$

Hence the minimum quantizing error is given by

$$\sigma_{\text{MIN}} = 2\left[\int_0^V P^{1/3}\, de\right]^3 \Big/ 3N^2 \qquad \text{(A-4)*}$$

[*Editor's Note:* Material has been omitted at this point.]

---

* An alternate derivation of (A-2) and (A-4), has been given by Panter and Dite,[5] who also acknowledge a prior and different deduction by P. R. Aigrain. Upon reading a preliminary version of the present manuscript, B. McMillan called my attention to S. P. Lloyd's related, but unpublished work, which proved to contain still another derivation. I am grateful to Dr. Lloyd for access to this material.

REFERENCES

1. H. S. Black, Modulation Theory, Van Nostrand, N. Y., 1953.
2. W. R. Bennett, Spectra of Quantized Signals, B.S.T.J., **27**, pp. 446–472, July, 1948.
3. Oliver, Pierce and Shannon, The Philosophy of PCM, Proc. I.R.E., **36**, pp. 1324–1331, Nov., 1948.
4. Clavier, Panter and Dite, Signal-to-Noise-Ratio Improvement in a PCM System, Proc. I.R.E., **37**, pp. 355–359, April, 1949.
5. P. F. Panter and W. Dite, Quantization Distortion in Pulse-Count Modulation with Nonuniform Spacing of Levels, Proc. I.R.E., **39**, pp. 44–48, Jan., 1951.
6. L. A. Meacham and E. Peterson, An Experimental Multichannel Pulse Code Modulation System of Toll Quality, B.S.T.J., **27**, pp. 1–43, Jan., 1948.
7. H. S. Black and J. O. Edson, Pulse Code Modulation, Trans. A.I.E.E., **66**, pp. 895–899, 1947.
8. Clavier, Panter and Grieg, Distortion in a Pulse Count Modulation System, Elec. Eng., **66**, pp. 1110–1122, 1947.
9. J. P. Schouten and H. W. F. Van' T. Groenewout, Analysis of Distortion in Pulse Code Modulation Systems, Applied Scientific Research, **2B**, pp. 277–290, 1952.
10. W. M. Goodall, Telephony by Pulse Code Modulation, B.S.T.J., **26**, pp. 395–409, July, 1947.
11. C. B. Feldman and W. R. Bennett, Bandwidth and Transmission Performance, B.S.T.J., **28**, pp. 490–595, July, 1949.
12. W. R. Bennett, Sources and Properties of Electrical Noise, Elec. Eng., **73**, pp. 1001–1008, Nov., 1954.
13. C. Villars, Etude sur la Modulation par Impulsions Codées, Bulletin Technique PTT, pp. 449–472, 1954.
14. J. Boisvieux, Le Multiplex à 16 Voies à Modulation Codée de la C.F.T.H., L'Onde Electrique, **34**, pp. 363–371, Apr., 1954.
15. E. Kettel, Der Störabstand bei der Nachrichtenübertragung durch Codemodulation, Archiv der Elektrischen Übertragung, **3**, pp. 161–164, Jan., 1949.
16. H. Holzwarth, Pulsecodemodulation und ihre Verzerrungen bei logarithmischer Amplitudenquantelung, Archiv der Elektrischen Übertragung, **3**, pp. 277–285, Jan., 1949.
17. Herreng, Blondé and Dureau, Système de Transmission Téléphonique Multiplex à Modulation par Impulsions Codées, Cables & Transmission, **9**, pp. 144–160, April, 1955.
18. W. B. Davenport, Jr., An Experimental Study of Speech-Wave Probability Distributions, J. Acous. Soc. Amer., **24**, pp. 390–399, July, 1952.
19. S. B. Wright, Amplitude Range Control, B.S.T.J., **17**, pp. 520–538, Oct., 1938.
20. Carter, Dickieson and Mitchell, Application of Compandors to Telephone Circuits, Trans. A.I.E.E., **65**, pp. 1079–1086, Dec., 1946.
21. J. C. R. Licklider and I. Pollack, Effects of Differentiation, Integration and Infinite Peak Clipping upon the Intelligibility of Speech, J. Acous. Soc. Amer., **20**, pp. 42–51, Jan., 1948.
22. D. W. Martin, Uniform Speech-Peak Clipping in a Uniform Signal to Noise Spectrum Ratio, J. Acous. Soc. Amer., **22**, pp. 614–621, Sept., 1950.
23. J. C. R. Licklider, The Intelligibility of Amplitude-Dichotomized, Time-Quantized Speech Waves, J. Acous. Soc. Amer., **22**, pp. 820–823, Nov., 1950.
24. H. Cramér, Mathematical Methods of Statistics, Princeton Univ. Press, Princeton, N. J., 1946, see pp. 359–363.
25. T. C. Fry, Probability and its Engineering Uses, Van Nostrand, N. Y., 1928, see pp. 310–312.
26. A. C. Aitken, Statistical Mathematics, Interscience Pub. Inc., N. Y., 3d Ed., 1944, see pp. 44–47.
27. W. F. Sheppard, On the Calculation of the Most Probable Values of Frequency-Constants for Data Arranged According to Equidistant Divisions of a Scale, Proc. London Math. Soc., **29**, pp. 353–380, 1898.
28. R. Courant and D. Hilbert, Methods of Mathematical Physics, Vol. 1, Interscience Pub. Inc., N. Y., English Ed., 1953, see Chapt. IV, especially pp. 184–187, and p. 206.
29. E. T. Whittaker, A Treatise on the Analytical Dynamics of Particles and Rigid Bodies, Cambridge Univ. Press, 4th Ed., 1937, see p. 54.

4

Reprinted from *IEEE Trans. Inform. Theory* **IT-28**(2):129–137 (1982)

Least Squares Quantization in PCM

STUART P. LLOYD

Abstract—It has long been realized that in pulse-code modulation (PCM), with a given ensemble of signals to handle, the quantum values should be spaced more closely in the voltage regions where the signal amplitude is more likely to fall. It has been shown by Panter and Dite that, in the limit as the number of quanta becomes infinite, the asymptotic fractional density of quanta per unit voltage should vary as the one-third power of the probability density per unit voltage of signal amplitudes. In this paper the corresponding result for any finite number of quanta is derived; that is, necessary conditions are found that the quanta and associated quantization intervals of an optimum finite quantization scheme must satisfy. The optimization criterion used is that the average quantization noise power be a minimum. It is shown that the result obtained here goes over into the Panter and Dite result as the number of quanta become large. The optimum quantization schemes for 2^b quanta, $b = 1, 2, \cdots, 7$, are given numerically for Gaussian and for Laplacian distribution of signal amplitudes.

I. Introduction

The basic ideas in the pulse-code modulation (PCM) system [1], [2, ch. 19] are the Shannon–Nyquist sampling theorem and the notion of quantizing the sample values.

The sampling theorem asserts that a signal voltage $s(t)$, $-\infty < t < \infty$, containing only frequencies less than W cycles/s can be recovered from a sequence of its sample values according to

$$s(t) = \sum_{j=-\infty}^{\infty} s(t_j)K(t - t_j), \qquad -\infty < t < \infty, \quad (1)$$

where $s(t_j)$ is the value of s at the jth sampling instant

$$t_j = \frac{j}{2W}, \qquad -\infty < j < \infty,$$

and where

$$K(t) = \frac{\sin 2\pi Wt}{2\pi Wt}, \qquad -\infty < t < \infty, \quad (2)$$

is a "sin t/t" pulse of the appropriate width.

The pulse-amplitude modulation (PAM) system [2, ch. 16] is based on the sampling theorem alone. One sends over the system channel, instead of the signal values $s(t)$ for all times t, only a sequence

$$\cdots, s(t_{-1}), s(t_0), s(t_1), \cdots \quad (3)$$

of samples of the signal. The (idealized) receiver constructs the pulses $K(t - t_j)$ and adds them together with the received amplitudes $s(t_j)$, as in (1), to produce an exact reproduction of the original band-limited signal s.

PCM is a modification of this. Instead of sending the exact sample values (3), one partitions the voltage range of the signal into a finite number of subsets and transmits to the receiver only the information as to which subset a sample happens to fall in. Built into the receiver there is a source of fixed representative voltages—"quanta"—one for each of the subsets. When the receiver is informed that a certain sample fell in a certain subset, it uses its quantum for that subset as an approximation to the true sample value and constructs a band-limited signal based on these approximate sample values.

We define the *noise signal* as the difference between the receiver-output signal and the original signal and the *noise power* as the average square of the noise signal. The problem we consider is the following: given the number of quanta and certain statistical properties of the signal, determine the subsets and quanta that are best in minimizing the noise power.

II. Quantization

Let us formulate the quantization process more explicitly. A quantization scheme consists of a class of sets $\{Q_1, Q_2, \cdots, Q_\nu\}$ and a set of quanta $\{q_1, q_2, \cdots, q_\nu\}$. The $\{Q_\alpha\}$ are any ν disjoint subsets of the voltage axis which, taken together, cover the entire voltage axis. The $\{q_\alpha\}$ are any ν finite voltage values. The number ν of quanta is to be regarded throughout as a fixed finite preassigned number.

We associate with a partition $\{Q_\alpha\}$ a label function $\gamma(x)$, $-\infty < x < \infty$, defined for all (real) voltages x by

$$\begin{aligned} \gamma(x) &= 1 \quad \text{if} \quad x \quad \text{lies in} \quad Q_1, \\ \gamma(x) &= 2 \quad \text{if} \quad x \quad \text{lies in} \quad Q_2, \\ &\ \ \vdots \\ \gamma(x) &= \nu \quad \text{if} \quad x \quad \text{lies in} \quad Q_\nu. \end{aligned} \quad (4)$$

If $s(t_j)$ is the jth sample of the signal s, as in Section I, then we denote by a_j the label of the set that this sample falls in:

$$a_j = \gamma(s(t_j)), \qquad -\infty < j < \infty.$$

In PCM the signal sent over the channel is (in some code or another) the sequence of labels

$$\cdots, a_{-1}, a_0, a_1, \cdots, \quad (5)$$

each a_j being one of the integers $\{1, 2, \cdots, \nu\}$. The technology of this transmission does not concern us, except that

Manuscript received May 1, 1981. The material in this paper was presented in part at the Institute of Mathematical Statistics Meeting, Atlantic City, NJ, September 10–13, 1957.
The author is with Bell Laboratories, Whippany Road, Whippany, NJ 07981.

we assume that such a sequence can be delivered to the receiver without error.

The receiver uses the fixed voltage q_α as an approximation to all sample voltages in Q_α, $\alpha = 1, 2, \cdots, \nu$. That is, the receiver, being given the value of a_j in the sequence (5), proceeds as if the jth sample of s had value q_{a_j} and produces the receiver-output signal

$$r(t) = \sum_{j=-\infty}^{\infty} q_{a_j} K(t - t_j), \qquad -\infty < t < \infty.$$

To put it another way, the system mutilates an actual sample voltage value x to the quantized value $y(x)$ given by

$$y(x) = q_{\gamma(x)}, \qquad -\infty < x < \infty, \tag{6}$$

and we may express the receiver output in terms of this as

$$r(t) = \sum_{j=-\infty}^{\infty} y(s(t_j)) K(t - t_j), \qquad -\infty < t < \infty. \tag{7}$$

Hence the noise signal, defined as

$$n(t) = r(t) - s(t), \qquad -\infty < t < \infty,$$

is given by

$$n(t) = \sum_{j=-\infty}^{\infty} z(s(t_j)) K(t - t_j), \qquad -\infty < t < \infty, \tag{8}$$

where

$$z(x) = y(x) - x, \qquad -\infty < x < \infty, \tag{9}$$

may be regarded as the quantization error added to a sample which has voltage value x.

Note that we assume that the receiver uses the nonrealizable pulses (2). If other pulses are used (e.g., step functions or other realizable pulses) there will be sampling noise, in general, even without quantization [3]. Our noise (8) is due strictly to quantization.

Finally we must emphasize that we assume that the $\{Q_\alpha\}$ and $\{q_\alpha\}$ are constant in time. In deltamodulation and its refinements the $\{Q_\alpha\}$ and $\{q_\alpha\}$ change from sampling instant to sampling instant, depending on the past behavior of the signal being handled. Such systems are very difficult to treat theoretically.

III. Noise Power

Instead of working with a particular band-limited signal, we assume that there is given a probabilistic family of such signals. That is, the s of the preceding sections and hence the various signals derived from it are to be regarded as stochastic processes [4]. We denote the underlying probability measure by $P\{\cdot\}$ and averages with respect to this measure (expectations) by $E\{\cdot\}$.

We use the following results of the probabilistic treatment. We assume that the s process is stationary, so that the cumulative probability distribution function of a sample,

$$F(x) = P\{s(t) \le x\}, \qquad -\infty < x < \infty,$$

is independent of t, $-\infty < t < \infty$, as indicated by the notation. Then the average power of the s process, assumed to be finite, is constant in time:

$$S = E\{s^2(t)\} = \int_{-\infty}^{\infty} x^2\, dF(x), \qquad -\infty < t < \infty. \tag{10}$$

Moreover, the r and n processes have this same property; the average receiver-output power R is given by

$$R = E\{r^2(t)\} = \int_{-\infty}^{\infty} y^2(x)\, dF(x), \qquad -\infty < t < \infty, \tag{11}$$

where $y(x)$ is defined in (6), and the noise power N is

$$N = E\{n^2(t)\} = \int_{-\infty}^{\infty} z^2(x)\, dF(x), \qquad -\infty < t < \infty, \tag{12}$$

with $z(x)$ as in (9). (Detailed proofs of these statements, together with further assumptions used, are given in Appendix A.) The stochastic process problem is thus reduced to a problem in a single real variable: choose the $\{Q_\alpha\}$ and $\{q_\alpha\}$ so that the rightmost integral in (12) is as small as possible.

IV. The Best Quanta

We consider first the problem of minimizing N with respect to the quanta $\{q_\alpha\}$ when the $\{Q_\alpha\}$ are fixed preassigned sets.

The dF integral in (12) may be written more explicitly as

$$N = \sum_{\alpha=1}^{\nu} \int_{Q_\alpha} (q_\alpha - x)^2\, dF(x). \tag{13}$$

(The sets $\{Q_\alpha\}$ must be measurable $[dF]$ if (11)–(13) are to have meaning, and we assume always that this is the case.) If we regard the given F as describing the distribution of unit probability "mass" on the voltage axis [5, p. 57], then (13) expresses N as the total "moment of inertia" of the sets $\{Q_\alpha\}$ around the respective points $\{q_\alpha\}$. It is a classical result that such a moment assumes its minimum value when each $\{q_\alpha\}$ is the center of mass of the corresponding $\{Q_\alpha\}$ (see, e.g., [5, p. 175]). That is,

$$q_\alpha = \frac{\int_{Q_\alpha} x\, dF(x)}{\int_{Q_\alpha} dF(x)}, \qquad \alpha = 1, 2, \cdots, \nu, \tag{14}$$

are the uniquely determined best quanta to use with a given partition $\{Q_\alpha\}$.

To avoid the continual mention of trivial cases we assume always that F is increasing at least by $\nu + 1$ points, so that the quantization noise does not vanish. Then none of the denominators in (14) will vanish, at least in an optimum scheme. For if Q_α has vanishing mass it can be combined with some set Q_β of nonvanishing mass (discarding q_α) to give a scheme with $\nu - 1$ quanta and the same noise. Then one of the sets of this scheme can be divided into two sets and new quanta assigned to give a scheme with ν quanta and noise less than in the original scheme. (We omit the details.)

If the expression on the right in (14) is substituted for q_α in (13), there results

$$N = S - \sum_{\alpha=1}^{\nu} q_\alpha^2 \int_{Q_\alpha} dF(x),$$

where the $\{q_\alpha\}$ here are the optimum ones of (14). The sum on the right is the receiver-output power from (11). Hence when the $\{q_\alpha\}$ are centers of mass of the $\{Q_\alpha\}$, optimum or not, then $S = R + N$, which implies that the noise is orthogonal to the receiver output. One expects this in a least squares approximation, of course.

V. The Best Partition

Now we find the best sets $\{Q_\alpha\}$ to use with a fixed preassigned set of quanta $\{q_\alpha\}$. The considerations of this section are independent of those of the preceding section. In particular, the best $\{Q_\alpha\}$ for given $\{q_\alpha\}$ may not have the $\{q_\alpha\}$ as their centers of mass.

We assume that the given $\{q_\alpha\}$ are distinct since it will never happen in an optimum scheme that $q_\alpha = q_\beta$ for some $\alpha \neq \beta$. For if $q_\alpha = q_\beta$, then Q_α and Q_β are effectively one set $Q_\alpha \cup Q_\beta$ as far as the noise is concerned (13), and this set can be redivided into two sets and these two sets can be given distinct quantum values in such a way as to reduce the noise. (We omit the details.)

Consider the probability mass in a small interval around voltage value x. According to (13) any of this mass which is assigned to q_α (i.e., which lies in Q_α) will contribute to the noise at rate $(q_\alpha - x)^2$ per unit mass. To minimize the noise, then, any mass in the neighborhood of x should be assigned to a q_α for which $(q_\alpha - x)^2$ is the smallest of the numbers $(q_1 - x)^2, (q_2 - x)^2, \cdots, (q_\nu - x)^2$. In other words,

$$Q_\alpha \supset \left\{x : (q_\alpha - x)^2 < (q_\beta - x)^2 \text{ for all } \beta \neq \alpha\right\}, \quad \alpha = 1,, \cdots, \nu,$$

modulo sets of measure zero $[dF]$.[1] This simplifies to

$$Q_\alpha \supset \left\{x : (q_\beta - q_\alpha)\left(x - \tfrac{1}{2}(q_\alpha + q_\beta)\right) < 0 \text{ for all } \beta \neq \alpha\right\}, \quad \alpha = 1, 2, \cdots, \nu. \tag{15}$$

It is straightforward that the best $\{Q_\alpha\}$ are determined by (15) as the intervals whose endpoints bisect the segments between successive $\{q_\alpha\}$, except that the assignment of the endpoints is not determined. To make matters definite we let the $\{Q_\alpha\}$ be left-open and right-closed, so that the best partition to use with the given quanta is

$$\begin{aligned} Q_1 &= \{x : -\infty < x \le x_1\} \\ Q_2 &= \{x : x_1 < x \le x_2\} \\ &\vdots \\ Q_{\nu-1} &= \{x : x_{\nu-2} < x \le x_{\nu-1}\} \\ Q_\nu &= \{x : x_{\nu-1} < x < \infty\}, \end{aligned} \tag{16}$$

where the endpoints $\{x_\alpha\}$ are given

$$\begin{aligned} x_1 &= \tfrac{1}{2}(q_1 + q_2) \\ x_2 &= \tfrac{1}{2}(q_2 + q_3) \\ &\vdots \\ x_{\nu-1} &= \tfrac{1}{2}(q_{\nu-1} + q_\nu). \end{aligned} \tag{17}$$

We have assumed, as we shall hereafter, that the indexing is such that $q_1 < q_2 < \cdots < q_\nu$.

VI. Quantization Procedures

From Sections IV and V we know that we may confine our attention to quantization schemes defined by $2\nu - 1$ numbers

$$q_1 < x_1 < q_2 < x_2 < \cdots < q_{\nu-1} < x_{\nu-1} < q_\nu, \tag{18}$$

where the $\{x_\alpha\}$ are the endpoints of the intervals $\{Q_\alpha\}$, as in (16), and the $\{q_\alpha\}$ are the corresponding quanta. We will regard such a set of numbers as the Cartesian coordinates of a point

$$\rho = (q_1, x_1, \cdots, q_\nu)$$

in $(2\nu - 1)$-dimensional Euclidean space $E_{2\nu-1}$. The noise as a function of ρ has the form

$$N(\rho) = \int_{-\infty}^{x_1} (q_1 - x)^2 \, dF(x) + \int_{x_1}^{x_2} (q_2 - x)^2 \, dF(x) + \cdots + \int_{x_{\nu-1}}^{\infty} (q_\nu - x)^2 \, dF(x). \tag{19}$$

In an optimum scheme the $\{q_\alpha\}$ will be centers of mass of the corresponding $\{Q_\alpha\}$, (14), and the $\{x_\alpha\}$ will lie midway between adjacent $\{q_\alpha\}$, (17). From the derivations these conditions are sufficient that $N(\rho)$ be a minimum with respect to variations in each coordinate separately and hence are necessary conditions at a minimum of $N(\rho)$. As it turns out, however, they are not sufficient conditions for a minimum of $N(\rho)$. Points at which (14) and (17) are satisfied, which we term *stationary points*, while never local maxima, may be saddle points of $N(\rho)$. Moreover, among the stationary points there may be several local minima, only one of which is the sought absolute minimum of $N(\rho)$. These complications are discussed further in Appendix B. The author has not been able to determine sufficient conditions for an absolute minimum.

The derivations suggest one trial-and-error method for finding stationary points. A trial point $\rho^{(1)}$ in $E_{2\nu-1}$ is

[1] If $C(x)$ is a condition on x, then $\{x: C(x)\}$ denotes the set of all x which satisfy $C(x)$.

chosen as follows. The endpoints

$$-\infty < x_1^{(1)} < x_2^{(1)} < \cdots < x_{\nu-1}^{(1)} < \infty$$

are chosen arbitrarily except that each of the resulting $\{Q_\alpha^{(1)}\}$ should have nonvanishing mass. Then the centers of mass of these sets are taken as the first trial quanta $\{q_\alpha^{(1)}\}$.

These values will not satisfy the midpoint conditions (17), in general, so that the second trial point $\rho^{(2)}$ is taken to be

$$q_\alpha^{(2)} = q_\alpha^{(1)}, \qquad \alpha = 1, 2, \cdots, \nu$$

$$x_\alpha^{(2)} = \tfrac{1}{2}\left(q_\alpha^{(2)} + q_{\alpha+1}^{(2)}\right), \qquad \alpha = 1, 2, \cdots, \nu - 1,$$

with appropriate modifications if any of the resulting $\{Q_\alpha^{(2)}\}$ have vanishing mass. This step does not increase the noise, in view of the discussion in Section V; that is, $N(\rho^{(2)}) \leq N(\rho^{(1)})$.

The new $\{q_\alpha^{(2)}\}$, centers of mass (c.m.) of the old $\{Q_\alpha^{(1)}\}$, will not be centers of mass of the new $\{Q_\alpha^{(2)}\}$, in general; trial point $\rho^{(3)}$ is determined by

$$x_\alpha^{(3)} = x_\alpha^{(2)}, \qquad \alpha = 1, 2, \cdots, \nu - 1,$$

$$q_\alpha^{(3)} = \left(\text{c.m. of } Q_\alpha^{(3)}\right), \qquad \alpha = 1, 2, \cdots, \nu.$$

For the resulting noise we have $N(\rho^{(3)}) \leq N(\rho^{(2)})$.

We continue in this way, imposing conditions (14) and (17) alternately. There results a sequence of trial points

$$\rho^{(1)}, \rho^{(2)}, \cdots \tag{20}$$

such that

$$N(\rho^{(1)}) \geq N(\rho^{(2)}) \geq \cdots.$$

The noise is nonnegative, so that $\lim_m N(\rho^{(m)})$ will exist, and we might hope that the sequence (20) had as a limit a local minimum of $N(\rho)$.

If the sequence (20) has no limit points then some of the $\{x_\alpha^{(m)}\}$ must become infinite with m; this corresponds to quantizing into fewer than ν quanta. Since we have assumed that F increases at least by $\nu + 1$ points there will be quantizing schemes with ν quanta for which the resulting noise is less than the optimum noise for $\nu - 1$ quanta, obviously. If $\rho^{(J)}$ is such a scheme then (20) will have limit points, using the property that $N(\rho^{(m)})$ is a decreasing sequence.[2]

Suppose $\rho^{(\infty)}$ is such a limit point. If each of the coordinate values $\{x_\alpha^{(\infty)}\}$ of $\rho^{(\infty)}$ is a continuity point of F then it is easy to see that the coordinates of $\rho^{(\infty)}$ will satisfy both (14) and (17). In particular, if $N(\rho)$ has a unique stationary point ρ_0 (which is the minimum sought), then the sequence (20), unless it diverges, will converge to ρ_0.

Note, by the way, that at a local minimum of $N(\rho)$ the numbers $\{x_\alpha\}$ are necessarily continuity points of F. Suppose to the contrary that there is a nonvanishing amount of mass concentrated at one of the endpoints $\{x_\alpha\}$, and that the adjacent sets Q_α and $Q_{\alpha+1}$ are as in (16), so that the mass at x_α belongs to Q_α. The centers of mass q_α and $q_{\alpha+1}$ will lie equidistant from x_α (17), and from (19) the noise will not change if we reassign the mass at x_α to $Q_{\alpha+1}$, retaining the given $\{q_\alpha\}$ as quanta. But q_α and $q_{\alpha+1}$ are definitely not centers of mass of the corresponding modified sets, and the noise will strictly decrease as q_α and $q_{\alpha+1}$ are moved to the new centers of mass. Thus the given configuration is not a local minimum, contrary to assumption. From this result and (19) we see that $N(\rho)$ is continuous in a neighborhood of a local minimum. We have proved also that there is no essential loss of generality in assuming the form (16) for the $\{Q_\alpha\}$.

We refer to the above trial-and-error method as Method I. Another trial-and-error method is the following one, Method II. To simplify the discussion we assume for the moment that F is continuous and nowhere constant. We choose a trial value q_1 satisfying

$$q_1 < \int_{-\infty}^{\infty} x\, dF(x).$$

The condition that q_1 be the center of mass of Q_1 determines x_1 as the unique solution of

$$q_1 = \frac{\int_{-\infty}^{x_1} x\, dF(x)}{\int_{-\infty}^{x_1} dF(x)}.$$

The quantities q_1 and x_1 now being known, the first of conditions (17) determines q_2 as

$$q_2 = 2x_1 - q_1.$$

If this q_2 lies to the right of the center of mass of the interval (x_1, ∞) then the trial chain terminates, and we start over again with a different trial value q_1. Otherwise, x_1 and q_2 being known, the second of conditions (14):

$$q_2 = \frac{\int_{x_1}^{x_2} x\, dF(x)}{\int_{x_1}^{x_2} dF(x)}$$

serves to determine x_2 uniquely. Now the second of conditions (17) gives

$$q_3 = 2x_2 - q_2.$$

We continue in this way, obtaining successively $q_1, x_1, \cdots, q_{\nu-1}, x_{\nu-1}, q_\nu$; the last step is the determination of q_ν according to

$$q_\nu = 2x_{\nu-1} - q_{\nu-1}. \tag{21}$$

However in this procedure we have not used the last of conditions (14):

$$q_\nu = \frac{\int_{x_{\nu-1}}^{\infty} x\, dF(x)}{\int_{x_{\nu-1}}^{\infty} dF(x)}. \tag{22}$$

and the q_ν obtained from (21) will not satisfy (22) in general. The discrepancy between the right members of (21) and (22) will vary continuously with the starting value q_1, and the method consists of running through such chains

[2]It seems likely that this condition $N(\rho^{(1)}) \leq$ (optimum noise for $\nu - 1$ quanta) is stronger than necessary for the nondivergence of (20).

using various starting values until the discrepancy is reduced to zero.

This method is applicable to more general F, with some obvious modifications. When F has intervals of constancy the $\{x_\alpha\}$ may not be uniquely determined by conditions (14), and a trial chain may involve several arbitrary parameters besides q_1. Discontinuities of F will cause no real trouble, since we know that the $\{x_\alpha\}$ of an optimum scheme are continuity points of F; a trial chain that does not have this property is discarded. We note that Method II may be used to locate all stationary points of $N(\rho)$.

VII. Examples

In all of the examples we now consider, the distribution of sample values is absolutely continuous with a sample probability density $f = F'$, which is an even function. If $N(\rho)$ has a unique stationary point, which we assume to be the case in the examples treated, then the optimum $\{q_\alpha\}$ and $\{x_\alpha\}$ will clearly be symmetrically distributed around the origin. In applications we are usually interested in having an even number of quanta, $\nu = 2\mu$; so we renumber the positive endpoints and quanta according to

$$0 = x_0 < q_1 < x_1 < \cdots < q_{\mu-1} < x_{\mu-1} < q_\mu; \quad (23)$$

the endpoints and quanta for the negative half-axis are the negatives of these.

We normalize to unit signal power $S = 1$. The $\{q_\alpha\}$ and $\{x_\alpha\}$ for other values of S are to be obtained by multiplying the numbers in the tables by $\sqrt{S}$.

The simplest case is the uniform distribution:

$$f(x) = \frac{1}{2\sqrt{3}}, \qquad -\sqrt{3} \le x \le \sqrt{3}$$

$$= 0, \qquad \sqrt{3} < |x| < \infty.$$

Method II of the preceding section shows that $N(\rho)$ in this case has a unique stationary point, which is necessarily an absolute minimum. The optimum scheme is the usual one with ν equal intervals of width $1/(2\nu\sqrt{3})$ each; the quanta being the midpoints of these intervals. The minimum value of the noise is the familiar $N = 1/\nu^2$.

Another case of possible interest is the Gaussian:

$$f(x) = \frac{e^{-1/2x^2}}{\sqrt{2\pi}}, \qquad -\infty < x < \infty.$$

The optimum schemes for $\nu = 2^b$, $b = 1, 2, \cdots, 7$, are given in Tables I–VII,[3] respectively. The corresponding noise values appear in Table VIII together with the quantities $\nu^2 N$ and νx_1. The behavior of these latter with increasing ν hint at the existence of asymptotic properties; we examine this question in the next section.

[3] Since some of the tables were never completed, those tables although mentioned in text are not included in this paper.

TABLE I
Gaussian, $\nu = 2$

α	q_α	x_α
1	0.7979	∞

TABLE II
Gaussian, $\nu = 4$

α	q_α	x_α
1	0.4528	0.9816
2	1.5104	∞

TABLE III
Gaussian, $\nu = 8$

α	q_α	x_α
1	0.2451	0.5006
2	0.7560	1.0500
3	1.3439	1.7480
4	2.1520	∞

TABLE IV
Gaussian, $\nu = 16$

α	q_α	x_α
1	0.1284	0.2582
2	0.3880	0.5224
3	0.6568	0.7996
4	0.9423	1.0993
5	1.2562	1.4371
6	1.6181	1.8435
7	2.0690	2.4008
8	2.7326	∞

TABLE VIII
Gaussian; Optimum Noise for Various Values of ν

ν	N	$\nu^2 N$	νx_1
2	0.3634	1.452	
4	0.1175	1.880	3.93
8	3.455×10^{-2}	2.205	4.00
16	9.500×10^{-3}	2.430	4.13
32			
64			
128			
(∞)	(0)	(2.72)	(4.34)

For speech signals a distribution which has been found useful empirically is the Laplacian:[4]

$$f(x) = \frac{e^{-|x|\sqrt{2}}}{\sqrt{2}}, \qquad -\infty < x < \infty.$$

The optimum quantizing schemes for this distribution for $\nu = 2^b$, $b = 1, 2, \cdots, 7$, are given in Tables IX–XV, respectively. The corresponding N, $\nu^2 N$, and νx_1 values are given in Table XVI; again, we notice certain regularities.

VIII. Asymptotic Properties

Let us assume that the distribution F is absolutely continuous with density function $f = F'$, which is itself dif-

[4] The author is indebted to V. Vyssotsky of the Acoustics Research Group for this information (private communication).

ferentiable, and that for each ν there is a unique optimum quantization scheme. We revert to our original numbering (18).

Let the quantities $\{h_\alpha\}$ be defined by

$$h_\alpha = x_\alpha - q_\alpha = q_{\alpha+1} - x_\alpha, \qquad \alpha = 1, 2, \cdots, \nu - 1,$$

so that, for $\alpha = 2, 3, \cdots, \nu - 1$, Q_α consists of an interval of length h_α to the right of q_α together with an interval of length $h_{\alpha-1}$ to the left of q_α. We have already imposed the optimizing conditions (17) in the very definition of the $\{h_\alpha\}$. The center of mass conditions (14) (except for the first and last) may be written as

$$\int_{q_\alpha - h_{\alpha-1}}^{q_\alpha + h_\alpha} (x - q_\alpha) f(x)\, dx = 0, \qquad \alpha = 2, 3, \cdots, \nu - 1.$$

If we expand f here in Taylor's series around q_α, the integration gives

$$\tfrac{1}{2}\left(h_\alpha^2 - h_{\alpha-1}^2\right) f(q_\alpha) + \tfrac{1}{3}\left(h_\alpha^3 + h_{\alpha-1}^3\right) f'(q_\alpha) = o\left(h_\alpha^3\right) + o\left(h_{\alpha-1}^3\right), \qquad \alpha = 2, 3, \cdots, \nu - 1. \quad (24)$$

The numbers in Tables VIII and XVI suggest the existence of an asymptotic fractional density of quanta. Accordingly, we define the function $g_\nu(x)$, $-\infty < x < \infty$, by

$$\begin{aligned} g_\nu(x) &= 0, && -\infty < x \le q_1 \\ &= \frac{1}{2\nu h_\alpha}, && q_\alpha < x \le q_{\alpha+1}, \quad \alpha = 1, 2, \cdots, \nu - 1, \\ &= 0, && q_\nu < x < \infty. \end{aligned} \quad (25)$$

The definition is arranged so that (for given ν) the sets $Q_2, Q_3, \cdots, Q_{\nu-1}$ subtend equal areas of $1/\nu$ each under the graph of $g_\nu(x)$ versus x. We will proceed as if a limiting density,

$$g(x) = \lim_{\nu \to \infty} g_\nu(x), \qquad -\infty < x < \infty,$$

existed.

We wish to express g in terms of the given sample density function f. To do this we will use conditions (24), together with the following further assumptions. We assume that g has a derivative, and we assume that for given x and k the difference $\epsilon_\nu(x) = g_\nu(x) - g(x)$, $-\infty < x < \infty$, has the property[5]

$$\epsilon_\nu\left(x + \frac{k}{\nu}\right) - \epsilon_\nu(x) = o\left(\frac{1}{\nu}\right).$$

In (24), then, we may approximate $h_\alpha - h_{\alpha-1}$ by

$$\begin{aligned} h_\alpha - h_{\alpha-1} &= \frac{1}{2\nu}\left[\frac{1}{g_\nu(q_\alpha + h_\alpha)} - \frac{1}{g_\nu(q_\alpha - h_{\alpha-1})}\right] \\ &= -\frac{g'(q_\alpha)}{2\nu^2 g^3(q_\alpha)} + o\left(\frac{1}{\nu^2}\right), \qquad \alpha = 2, 3, \cdots, \nu - 1, \end{aligned}$$

[5] The notation $u(\nu) = o(v(\nu))$ means in our case $\lim_{\nu\to\infty} u(\nu)/v(\nu) = 0$.

TABLE XVII
APPROXIMATE LAST ENDPOINT FROM ASYMPTOTIC FORMULA

ν	Gaussian b_ν	Laplacian b_ν
2	0	
4	1.168	
8	1.992	
16	2.657	
32		
64		
128		

and we find that the left-hand member of (24) is indeed $o(h^3) = o(1/\nu^3)$ provided that

$$\frac{g'(x)}{g(x)} = \frac{f'(x)}{3f(x)}, \qquad -\infty < x < \infty. \quad (26)$$

The normalized solution of (26) is

$$g(x) = \frac{f^{1/3}(x)}{\int_{-\infty}^{\infty} f^{1/3}(x')\, dx'}, \qquad -\infty < x < \infty, \quad (27)$$

provided that the integral in the denominator exists.

The noise power becomes

$$\begin{aligned} N &= \frac{1}{12\nu^2}\int_{-\infty}^{\infty} \frac{f(x)}{g^2(x)}\, dx + o\left(\frac{1}{\nu^2}\right) \\ &= \frac{1}{12\nu^2}\left[\int_{-\infty}^{\infty} f^{1/3}(x)\, dx\right]^3 + o\left(\frac{1}{\nu^2}\right), \end{aligned} \quad (28)$$

neglecting the contributions from the end quanta.[6] For the Gaussian example then, the numbers $\nu^2 N$ of Table VIII should have a limit easily evaluated from (28) as $\nu^2 N \to \pi\sqrt{3}/2 (\approx 2.72)$, and in the Laplacian case, Table XVI, we find $\nu^2 N \to 9/2$.

The quantities denoted by νx_1 in Tables VIII and XVI should have the limiting value

$$\lim_{\substack{\nu \to \infty \\ q_\alpha \to 0}} \nu(h_{\alpha-1} + h_\alpha) = \frac{1}{g(0)},$$

comparing with (25). In the Gaussian example we find $1/g(0) = \sqrt{6\pi} (\approx 4.34)$, and for the Laplacian: $1/g(0) = 3\sqrt{2} (\approx 4.24)$.

For large values of ν the sets $\{Q_\alpha\}$ should subtend approximately equal areas of $1/\nu$ each under the graph of $g(x)$ versus x, so that the number b_ν defined by

$$\frac{1}{\nu} = \int_{b_\nu}^{\infty} g(x)\, dx$$

might be expected to be near the rightmost division point. Comparing Table XVII with Tables I–VII and IX–XVI we see that the approximation is surprisingly good, at least in the examples considered.

[6] Other derivations of (27)–(28) are given in [6] and [7].

ACKNOWLEDGMENT

The numerical results presented in the tables are due to Miss M. C. Gray and her assistants in the Numerical Analysis and Digital Processes Group; the programming of Method I for the IBM-650 electronic computer was done by Miss C. A. Conn.

After substantial progress had been made on the work described here there appeared in [11] a review of a paper by J. Lukaszewicz and H. Steinhaus on optimum go/no-go gauge sets. The present author has not been able to obtain a copy of this paper, but it seems likely that these authors have treated a problem similar or identical to the one discussed in Sections IV–VI. M. P. Schützenberger in [12] examines the quantization problem in the case where $\nu = 2$ and where F increases at 3 or 4 points.

APPENDIX A

Suppose $s(t)$, $-\infty < t < \infty$, is a continuous parameter stochastic process, real, separable, measurable, stationary, and of finite power:

$$S = E\{s^2(t)\} = \int_{-\infty}^{\infty} x^2\, dF(x) < \infty, \qquad -\infty < t < \infty,$$

(where F is the first-order distribution of the process, Section III). Then s has a spectral representation

$$s(t) = \int_{-\infty}^{\infty} e^{2\pi i\lambda t} d\xi(\lambda), \qquad -\infty < t < \infty, \tag{29}$$

where the spectral process $\xi(\lambda)$, $-\infty < \lambda < \infty$, has orthogonal increments [4, p. 527]. To say that s is band-limited to the frequency band $-W \le \lambda \le W$ is to say that the ξ process has vanishing increments outside of this band with probability one, and (29) becomes

$$s(t) = \int_{-W-0}^{W+0} e^{2\pi i\lambda t} d\xi(\lambda), \qquad -\infty < t < \infty. \tag{30}$$

Since we are particularly concerned with the behavior of ξ at the band edges, we rewrite (30) as

$$s(t) = \int_{-W+0}^{W-0} e^{2\pi i\lambda t} d\xi(\lambda) + 2\delta_1 \cos 2\pi Wt - 2\delta_2 \sin 2\pi Wt,$$

$$-\infty < t < \infty,$$

where the real random variables δ_1 and δ_2 describe the jumps of ξ at the band edges:

$$\xi(\pm W + 0) - \xi(\pm W - 0) = \delta_1 \pm i\delta_2.$$

For fixed t, the function $e^{2\pi i\lambda t}$, $-W \le \lambda \le W$, has Fourier coefficients

$$\begin{aligned} c_j &= \frac{1}{2W}\int_{-W}^{W} e^{-2\pi i j\lambda/(2W)} e^{2\pi i\lambda t} d\lambda \\ &= \frac{\sin 2\pi W(t - j/(2W))}{2\pi W(t - j/(2W))} \\ &= K(t - t_j), \qquad -\infty < j < \infty, \end{aligned}$$

in the notation of Section I. This function is of bounded variation, so that the partial sums

$$S_l(\lambda) = \sum_{j=-l}^{l} e^{2\pi i j\lambda/(2W)} K(t - t_j), \qquad -W \le \lambda \le W,$$

converge boundedly to

$$\begin{aligned} S(\lambda) &= \lim_{l\to\infty} S_l(\lambda) \\ &= e^{2\pi i\lambda t}, \qquad -W < \lambda < W, \\ &= \cos^{2\pi Wt}, \qquad \lambda = \pm W, \end{aligned}$$

from [8]. Hence, using the representation (30) for the samples, the sampling series

$$\hat{s}(t) = \sum_{j=-\infty}^{\infty} s(t_j) K(t - t_j) \tag{31}$$

converges (in stochastic mean square) to

$$\begin{aligned} \hat{s}(t) &= \underset{l\to\infty}{\text{l.i.m.}} \int_{-W-0}^{W+0} S_l(\lambda)\, d\xi(\lambda) \\ &= \int_{-W+0}^{W-0} e^{2\pi i\lambda t} d\xi(\lambda) + 2\delta_1 \cos 2\pi Wt \\ &= s(t) + 2\delta_2 \sin 2\pi Wt, \qquad -\infty < t < \infty, \end{aligned} \tag{32}$$

from [4, p. 429]. (The corresponding result for deterministic functions is given in [9].) Since the orthogonal increments property of ξ requires $E\{\delta_1\delta_2\} = 0$ together with

$$E = \{\delta_1^2\} = E\{\delta_2^2\} = \tfrac{1}{2}E\{|\xi(\pm W + 0) - \xi(\pm W - 0)|^2\},$$

we see from (32) that the sampling series (31) represents s with probability 1, if and only if, the ξ process has no power concentrated at the band edges,

$$E\{|\xi(\pm W + 0) - \xi(\pm W - 0)|^2\} = 0.$$

(Other proofs of this result appear in [3] and in [10].)

Let s be as above and suppose $\varphi(x)$, $-\infty < x < \infty$, is a Baire function. Then the random variables

$$\cdots, \varphi(s(t_{-1})), \varphi(s(t_0)), \varphi(s(t_1)), \cdots \tag{33}$$

constitute a stationary discrete-parameter stochastic process. If the number

$$\Phi = E\{\varphi^2(s(t_j))\} = \int_{-\infty}^{\infty} \varphi^2(x)\, dF(x), \qquad -\infty < j < \infty,$$

is finite then the process (33) admits a spectral representation

$$\varphi(s(t_j)) = \int_{-W}^{W} e^{2\pi i j\lambda/(2W)} d\eta(\lambda), \qquad -\infty < j < \infty,$$

where the η process has orthogonal increments ([4, p. 481], with a change of scale).

A certain continuous parameter stochastic process $\theta(t)$, $-\infty < t < \infty$, may be defined in terms of the η process by

$$\theta(t) = \int_{-W}^{W} e^{2\pi i\lambda t} d\eta(\lambda), \qquad -\infty < t < \infty.$$

This process is stationary in the wide sense and has the given process (33) as its samples, clearly

$$\theta(t_j) = \varphi(s(t_j)), \qquad -\infty < j < \infty.$$

Moreover,

$$E\{\theta^2(t)\} = \int_{-W}^{W} E\{|d\eta(\lambda)|^2\} = \int_{-\infty}^{\infty} \varphi^2(x) dF(x),$$

$$-\infty < t < \infty.$$

The θ process is represented by the sampling series

$$\theta(t) = \sum_{j=-\infty}^{\infty} \varphi(s(t_j)) K(t - t_j), \qquad -\infty < t < \infty, \tag{34}$$

if and only if, the spectral process η has no power concentrated at the band edges. The arguments are identical to those given above for the s process itself.

The r and n processes of Sections II and III are of the form just described, since the functions $y(x)$, (6), and $z(x)$, (9), will differ from certain Baire functions only on sets of measure zero $[dF]$ when we assume, as we do, that the sets $\{Q_\alpha\}$ are measureable $[dF]$.

The well-known mean-erogodic property,

$$\underset{m''-m'\to\infty}{\text{l.i.m.}} \sum_{j=m'}^{m''} \frac{(-1)^j\varphi(s(t_j))}{m''-m'+1}$$
$$= \eta(W+0) - \eta(W-0) + \eta(-W+0) - \eta(-W-0)$$
$$= 2\,\mathrm{Re}[\eta(\pm W+0) - \eta(\pm W-0)],$$

([4, p. 491]) shows that the requirement that η have no power at the band edges is equivalent to the condition

$$\lim_{m''-m'\to\infty} E\left\{\left|\sum_{j=m'}^{m''} \frac{(-1)^j\varphi(s(t_j))}{m''-m'+1}\right|^2\right\} = 0.$$

Finally we note that if the ξ process has a discrete component at frequencies $\pm\lambda_0$, then depending on the form of φ, the derived η process is likely to have discrete components at all of the harmonic frequencies $\pm m\lambda_0$ (modulo $2W$) of λ_0, $m = 1, 2, \cdots$. In particular, if λ_0 is rational then the η process may have a discrete component at the band edges, a possibility which must be excluded if (34) is to hold.

Appendix B

A simple example shows that the conditions (14) and (17) are not sufficient for an absolute minimum of N. Suppose F is absolutely continuous, with a density $f = F'$ as shown in Fig. 1, where $c_1(b_2 - b_1) + c_2(b_4 - b_3) = 1$. If $\nu > 1$ quanta are desired, let $\nu_1, \nu_2 > 0$ be any integers such that $\nu_1 + \nu_2 = \nu$, and divide the interval (b_1, b_2) into ν_1 equal intervals and (b_3, b_4) into ν_2 equal intervals; let the quanta be the midpoints of these ν intervals. If we suppose that $b_2 < \frac{1}{2}(b_1 + b_3)$ and $b_3 > \frac{1}{2}(b_2 + b_4)$ then the division point which separates the right-hand $\{Q_\alpha\}$ in (b_1, b_2) and the left-hand $\{Q_\alpha\}$ in (b_3, b_4) will lie in the interval (b_2, b_3), so that the conditions (14) and (17) will be satisfied. Thus we have $\nu - 1$ distinct local minima of N. (If c_1, respectively c_2, is small enough there may even be another minimum, corresponding to $\nu_1 = 0$, respectively $\nu_2 = 0$.) Which of these is the true minimum depends on the values of the parameters. (Explicitly, the noise has the value

$$N = \frac{c_1(b_2 - b_1)^3}{12\nu_1^2} + \frac{c_2(b_4 - b_3)^3}{12\nu_2^2},$$

and the ν_2/ν_1 for which this is a minimum is given by

$$\frac{\nu_2/(b_4 - b_3)}{\nu_1/(b_2 - b_1)} = \left(\frac{c_2}{c_1}\right)^{1/3},$$

agreeing with (27).)

The following interesting example is due to J. L. Kelly, Jr., of the Visual Research Group. Let the density f be as in Fig. 2, with $c_2 > c_1$, $c_1 + c_2 = 1$. The signal power is $S = 1/3$, independently of c_1 and c_2. Suppose $\nu = 2$. One configuration for which conditions (14) and (17) are satisfied is $q_1 = -\frac{1}{2}$, $x_1 = 0$, $q_2 = \frac{1}{2}$, clearly, and the resulting noise in $N = 1/12$. When $c_2 > 3c_1$; however, there is another solution—it is the one which in the limit $c_1 = 0$, $c_2 = 1$ goes into the scheme where $(0, 1)$ is divided

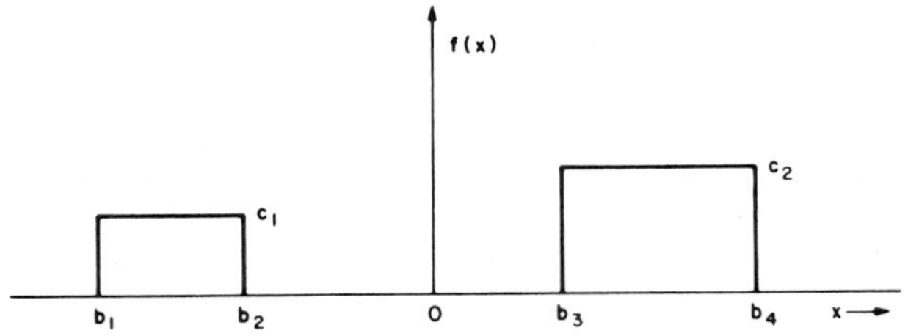

Fig. 1. The density $f(x)$ vanishes outside of the intervals (b_1, b_2), (b_3, b_4).

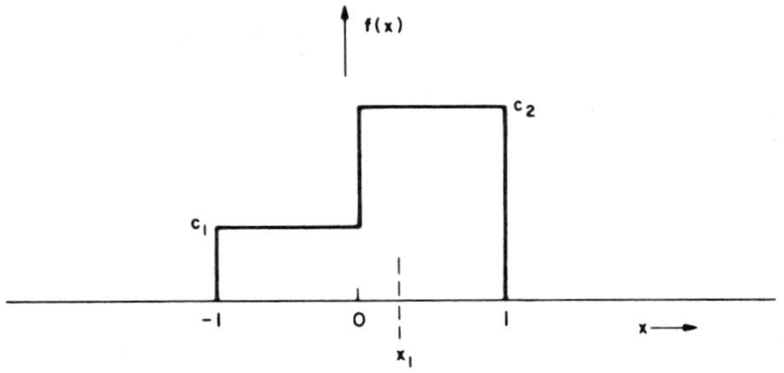

Fig. 2. The density $f(x)$ vanishes outside of the interval $(-1, 1)$.

into two equal parts and $(-1, 0)$ is ignored. The parameters for this configuration work out to be

$$q_1 = \frac{c_2 - 9c_1}{4c_2},$$
$$x_1 = \frac{c_2 - 3c_1}{2c_2},$$
$$q_2 = \frac{3(c_2 - c_1)}{4c_2},$$
$$N = \frac{1}{12} - \frac{(c_2 - 3c_1)^3}{16c_2^2}.$$

Hence if this configuration exists then it is better than the one first mentioned. We note, by the way, that method I of Section VI will converge to the $(-1/2, 0, 1/2)$ configuration, if the starting value $x_1^{(1)}$ is negative; if $c_2 > 3c_1$, however, this configuration is only a saddle point of N.

Author's Note 1981:

This is nearly a verbatim reproduction of a draft manuscript, which was circulated for comments at Bell Laboratories; the Mathematical Research Department log date is July 31, 1957. I wish to thank the editors for their invitation to publish this antique *samizdat* in the present issue.

The main reason the paper was not submitted for publication previously was that the numerical calculations were never completed. The Gaussian $\nu = 32$ case was done on the IBM 650 card programmable calculator; the Laplacian cases were done only for $\nu = 2$. Some time later the 650 was replaced by an IBM 704 electronic computer, but no quantizing program was written for it.

I was not satisfied with not having conditions for a unique minimum but would have published the paper without this. Later, P. E. Fleischer of Bell Laboratories gave a neat sufficient condition in his paper [13].

In the examples of Appendix B, the direct current can be removed by changing the origin; the noise is not affected. The results of the paper are valid for the uncentered processes used.

I was aware when I wrote the paper that the methods for quantizing a real random variable extend to other loss functions. In the least squares case the process quantizing noise is just the noise per sample; the generalization is more complicated and was omitted.

References

[1] B. M. Oliver, J. R. Pierce, and C. E. Shannon, "The philosophy of PMC," *Proc. I.R.E.*, vol. 36, pp. 1324–1331, 1948.

[2] H. S. Black, *Modulation Theory*. Princeton, NJ: Van Nostrand, 1953.

[3] S. P. Lloyd and B. McMillan, "Linear least squares filtering and prediction of sampled signals," in *Proc. Symp. on Modern Network Synthesis*, vol. 5. Brooklyn, NY: Polytechnic Institute of Brooklyn, 1956, pp. 221–247.

[4] J. L. Doob, *Stochastic Processes*. New York: Wiley, 1953.

[5] H. Cramér, *Mathematical Methods of Statistics*. Princeton, NJ: Princeton University, 1951.

[6] P. F. Panter and W. Dite, "Quantization distortion in pulse-count modulation with nonuniform spacing of levels," *Proc. I.R.E.*, vol. 39, pp. 44–48, 1951.

[7] B. Smith, "Instantaneous companding of quantized signals," *Bell Syst. Tech. J.*, vol. 36, pp. 653–709, 1957.

[8] A. Zygmund, *Trigonometrical Series*. New York: Dover, 1955, p. 47.

[9] H. P. Kramer, "A generalized sampling theorem," *Bull. Am. Math. Soc.* vol. 63, p. 117, 1957.

[10] E. Parzen, "A simple proof and some extensions of the sampling theorem," Department of Statistics, Stanford Univ., CA, Tech. Report 7, Dec. 1956.

[11] J. Kukaszewicz and H. Steinhaus, "On measuring by comparison," *Zastos. Mat.*, vol. 2, pp. 225–231, 1955; *Math. Reviews*, vol. 17, p. 757, 1956.

[12] M. P. Schützenberger, "Contribution aux applications statistiques de la théorie de l'information," *Pub. de l'Inst. de Statis.* de l'Université de Paris, vol. 3, Fasc. 1-2 pp. 56–69, 1954.

[13] P. E. Fleischer, "Sufficient conditions for achieving minimum distortion in quantizer," *IEEE Int. Convention Record*, part I, vol. 12, pp. 104–111, 1964.

5

Reprinted from *IEEE Trans. Inform. Theory* **IT-6**(1):7–12 (1960)

Quantizing for Minimum Distortion*

JOEL MAX†

Summary—**This paper discusses the problem of the minimization of the distortion of a signal by a quantizer when the number of output levels of the quantizer is fixed. The distortion is defined as the expected value of some function of the error between the input and the output of the quantizer. Equations are derived for the parameters of a quantizer with minimum distortion. The equations are not soluble without recourse to numerical methods, so an algorithm is developed to simplify their numerical solution. The case of an input signal with normally distributed amplitude and an expected squared error distortion measure is explicitly computed and values of the optimum quantizer parameters are tabulated. The optimization of a quantizer subject to the restriction that both input and output levels be equally spaced is also treated, and appropriate parameters are tabulated for the same case as above.**

* Manuscript received by the PGIT, September 25, 1959. This work was performed by the Lincoln Lab., Mass. Inst. Tech., Lexington, Mass., with the joint support of the U. S. Army, Navy, and Air Force.

† Lincoln Lab., Mass. Inst. Tech., Lexington, Mass.

IN MANY data-transmission systems, analog input signals are first converted to digital form at the transmitter, transmitted in digital form, and finally reconstituted at the receiver as analog signals. The resulting output normally resembles the input signal but is not precisely the same since the quantizer at the transmitter produces the same digits for all input amplitudes which lie in each of a finite number of amplitude ranges. The receiver must assign to each combination of digits a single value which will be the amplitude of the reconstituted signal for an original input anywhere within the quantized range. The difference between input and output signals, assuming errorless transmission of the digits, is the quantization error. Since the digital transmission rate of any system is finite, one has to use a quantizer which sorts the input into a finite number of ranges, N. For a given N, the system is described by specifying the end points, x_k, of the N input ranges, and an output level, y_k, corresponding to each input range. If the amplitude probability density of the signal which is the quantizer input is given, then the quantizer output is a quantity whose amplitude probability density may easily be determined as a function of the x_k's and y_k's. Often it is appropriate to define a distortion measure for the quantization process, which will be some statistic of the quantization error. Then one would like to choose the N y_k's and the associated x_k's so as to minimize the distortion. If we define the distortion, D, as the expected value of $f(\epsilon)$, where f is some function (differentiable), and ϵ is the quantization error, and call the input amplitude probability density $p(x)$, then

$$D = E[f(s_{\text{in}} - s_{\text{out}})]$$
$$= \sum_{i=1}^{N} \int_{x_i}^{x_{i+1}} f(x - y_i)p(x)\, dx$$

where $x_{N+1} = \infty$, $x_1 = -\infty$, and the convention is that an input between x_i and x_{i+1} has a corresponding output y_i.

If we wish to minimize D for fixed N, we get necessary conditions by differentiating D with respect to the x_i's and y_i's and setting derivatives equal to zero:

$$\frac{\partial D}{\partial x_j} = f(x_j - y_{j-1})p(x_j) - f(x_j - y_j)p(x_j) = 0 \qquad j = 2, \cdots, N \tag{1}$$

$$\frac{\partial D}{\partial y_j} = -\int_{x_j}^{x_{j+1}} f'(x - y_j)p(x)\, dx = 0 \qquad j = 1, \cdots, N \tag{2}$$

(1) becomes (for $p(x_j) \neq 0$)

$$f(x_j - y_{j-1}) = f(x_j - y_j) \qquad j = 2, \cdots, N \tag{3}$$

(2) becomes

$$\int_{x_j}^{x_{j+1}} f'(x - y_j)p(x)\, dx = 0 \qquad j = 1, \cdots, N. \tag{4}$$

We may ask when these are sufficient conditions. The best answer one can manage in a general case is that if all the second partial derivatives of D with respect to the x_i's and y_i's exist, then the critical point determined by conditions (3) and (4) is a minimum if the matrix whose ith row and jth column element is

$$\left.\frac{\partial^2 D}{\partial p_i\, \partial p_j}\right|_{\text{critical point}},$$

where the p's are the x's and y's, is positive definite. In a specific case, one may determine whether or not the matrix is positive definite or one may simply find all the critical points (*i.e.*, those satisfying necessary conditions) and evaluate D at each. The absolute minimum must be at one of the critical points since "end points" can be easily ruled out.

The sort of f one would want to use would be a good metric function, *i.e.*, $f(x)$ is monotonically nondecreasing

$$f(0) = 0$$
$$f(x) = f(-x).$$

If we require that $f(x)$ be *monotonically increasing* (with x) then (1) implies

$$| x_j - y_{j-1} | = | x_j - y_j | \qquad j = 2, \cdots, N$$

which implies (since y_{j-1} and y_j should not coincide) that

$$x_j = (y_j + y_{j-1})/2 \qquad j = 2, \cdots, N$$

(x_j is halfway between y_j and y_{j-1}).

We now take a specific example of $f(x)$ to further illuminate the situation.

Let $f(x) = x^2$

(3) implies

$$x_j = (y_j + y_{j-1})/2 \quad \text{or} \quad y_j = 2x_j - y_{j-1} \qquad j = 2, \cdots, N, \tag{5}$$

(4) implies

$$\int_{x_j}^{x_{j+1}} (x - y_j)p(x)\, dx = 0 \qquad j = 1, \cdots, N. \tag{6}$$

That is, y_j is the centroid of the area of $p(x)$ between x_j and x_{j+1}.

Because of the complicated functional relationships which are likely to be induced by $p(x)$ in (6), this is not a set of simultaneous equations we can hope to solve with any ease. Note, however, that if we choose y_1 correctly we can generate the succeeding x_i's and y_i's by (5) and (6), the latter being an implicit equation for x_{j+1} in terms of x_j and y_j.

A method of solving (5) and (6) is to pick y_1, calculate the succeeding x_i's and y_i's by (5) and (6) and then if y_N is the centroid of the area of $p(x)$ between x_N and ∞, y_1 was chosen correctly. (Of course, a different choice is appropriate to each value of N.) If y_N is not the appropriate centroid, then of course y_1 must be chosen again. This search may be systematized so that it can be performed on a computer in quite a short time.[1]

This procedure has been carried out numerically on the IBM 709 for the distribution $p(x) = 1/\sqrt{2\pi}\, e^{-x^2/2}$, under the restriction that $x_{N/2+1} = 0$ for N even, and $y_{(N+1)/2} = 0$ for N odd. This procedure gives symmetric results, *i.e.*,

[1] Obtaining *explicit* solutions to the quantizer problem for a nontrivial $p(x)$ is easily the most difficult part of the problem. The problem may be solved analytically where $p(x) = 1/\sqrt{2\pi}\, e^{-x^2/2}$ only for $N = 1$, $N = 2$. For $N = 1$, $x_1 = -\infty$, $y_1 = 0$, $x_2 = +\infty$. For $N = 2$, $x_1 = -\infty$, $y_1 = -\sqrt{2/\pi}$, $x_2 = 0$, $y_2 = \sqrt{2/\pi}$, $x_3 = +\infty$, ($\sqrt{2/\pi}$ is the centroid of the portion of $1/\sqrt{2\pi}\, e^{-x^2/2}$ between the origin and $+\infty$.) For $N \geq 3$, some sort of numerical estimation is required. A somewhat different approach, which yields results somewhat short of the optimum, is to be found in V. A. Gamash, "Quantization of signals with non-uniform steps," *Electrosvyaz*, vol. 10, pp. 11–13; October, 1957.

if a signal amplitude x is quantized as y_k, then $-x$ is quantized as $-y_k$. The answers appear in Table I on page 11.

An attempt has been made to determine the functional dependence of the distortion on the number of output levels. A log-log plot of the distortion vs the number of output levels is in Fig. 1. The curve is not a straight line. The tangent to the curve at $N = 4$ has the equation $D = 1.32\ N^{-1.74}$ and the tangent at $N = 36$ has the equation $D = 2.21\ x^{-1.96}$. One would expect this sort of behavior for large N. When N is large, the amplitude probability density does not vary appreciably from one end of a single input range to another, except for very large amplitudes, which are sufficiently improbable so that their influence is slight. Hence, most of the output levels are very near to being the means of the end points of the corresponding input ranges. Now, the best way of quantizing a uniformly distributed input signal is to space the output levels uniformly and to put the end points of the input ranges halfway between the output levels, as in Fig. 2, shown for $N = 1$. The best way of producing a quantizer with $2N$ output levels for this distribution is to divide each input range in half and put the new output levels at the midpoints of these ranges, as in Fig. 3. It is easy to see that the distortion in the second case is $\frac{1}{4}$ that in the first. Hence, $D = kN^{-2}$ where k is some constant. In fact, k is the variance of the distribution.

If this sort of equal division process is performed on each input range of the optimum quantizer for a normally distributed signal with N output levels where N is large then again a reduction in distortion by a factor of 4 is expected. Asymptotically then, the equation for the tangent to the curve of distortion vs the number of output levels should be $D = kN^{-2}$ where k is some constant.

Commercial high-speed analog-to-digital conversion equipment is at present limited to transforming equal input ranges to outputs midway between the ends of the input ranges. In many applications one would like to know the best interval length to use, *i.e.*, the one yielding minimum distortion for a given number of output levels, N. This is an easier problem than the first, since it is only two-dimensional (for $N \geqq 2$), *i.e.*, D is a function of the common length r of the intervals and of any particular output level, y_k. If the input has a symmetric distribution and a symmetric answer is desired, the problem becomes one dimensional. If $p(x)$ is the input amplitude probability density and $f(x)$ is the function such that the distortion D is $E[f(s_{out} - s_{in})]$, then, for an even number $2N$ of outputs,

$$D = 2\sum_{i=1}^{N-1}\int_{(i-1)r}^{ir} f\left(x - \left[\frac{2i-1}{2}\right]r\right)p(x)\,dx + 2\int_{(N-1)r}^{\infty} f\left(x - \left[\frac{2N-1}{2}\right]r\right)p(x)\,dx. \quad (7)$$

For a minimum we require

$$\frac{dD}{dr} = -\sum_{i=1}^{N-1}(2i-1)\int_{(i-1)r}^{ir} f'\left(x - \left[\frac{2i-1}{2}\right]r\right)p(x)\,dx - (2N-1)\int_{(N-1)r}^{\infty} f'\left(x - \left[\frac{2N-1}{2}\right]r\right)p(x)\,dx = 0. \quad (8)$$

A similar expression exists for the case of an odd number of output levels. In either case the problem is quite susceptible to machine computation when $f(x)$, $p(x)$ and N are specified. Results have been obtained for $f(x) = x^2$, $p(x) = 1/\sqrt{2\pi}\ e^{-x^2/2}$, $N = 2$ to 36. They are indicated in Table II on page 12.

A log-log plot of distortion vs number of output levels appears in Fig. 1. This curve is not a straight line. The tangent to the curve at $N = 36$ has the equation $D = 1.47\ N^{-1.74}$. A log-log plot of output level spacing vs number of outputs for the equal spacing which yields lowest distortion is shown in Fig. 4. This curve is also not a straight line. Lastly, a plot of the ratio of the distortion for the optimum quantizer to that for the optimum equally spaced level quantizer can be seen in Fig. 5.

Key to the Tables

The numbering system for the table of output levels, y_i, and input interval end points, x_i, for the minimum mean-squared error quantization scheme for inputs with a normal amplitude probability density with standard deviation unity and mean zero is as follows:

For the number of output levels, N, even, x_1 is the first end point of an input range to the right of the origin. An input between x_i and x_{i+1} produces an output y_i.

For the number of output levels, N, odd, y_1 is the smallest non-negative output. An input between x_{i-1} and x_i produces an output y_i.

This description, illustrated in Fig. 6, is sufficient because of the symmetry of the quantizer. The expected squared error of the quantization process and informational entropy of the output of the quantizer are also tabulated for the optimal quantizers calculated.[2] (If p_k is the probability of the kth output, then the informational entropy is defined as $-\sum_{k=1}^{N} p_k \log_2 p_k$.)

Table II also pertains to a normally distributed input with standard deviation equal to unity. The meaning of the entries is self-explanatory.

[2] The values of informational entropy given show the minimum average number of binary digits required to code the quantizer output. It can be seen from the tables that this number is always a rather large fraction of $\log_2 N$, and in most cases quite near 0.9 $\log_2 N$. In the cases where $N = 2^n$, n an integer, a simple n binary digit code for the outputs of the quantizer makes near optimum use of the digital transmission capacity of the system.

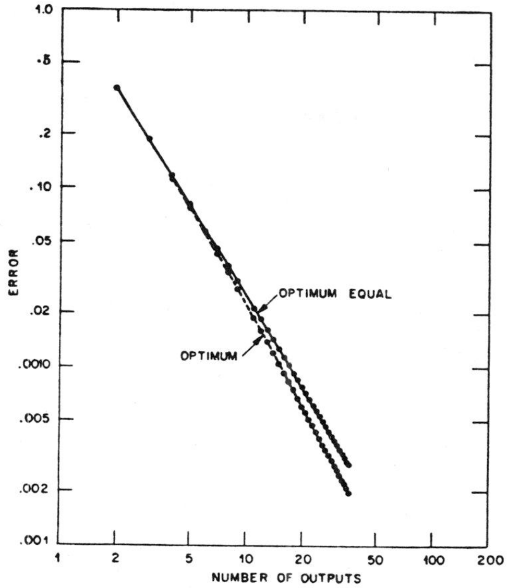

Fig. 1—Mean squared error vs number of outputs for optimum quantizer and optimum equally spaced level quantizer. (Minimum mean squared error for normally distributed input with $\sigma = 1$.)

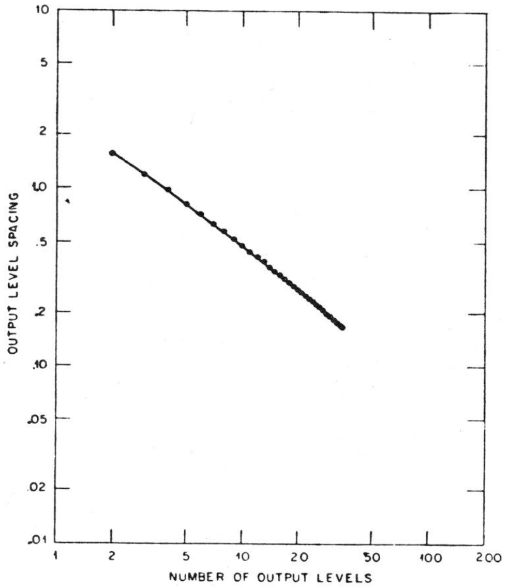

Fig. 4—Output level spacing vs number of output levels for equal optimum case. (Minimum mean squared error for normally distributed input with $\sigma = 1$.)

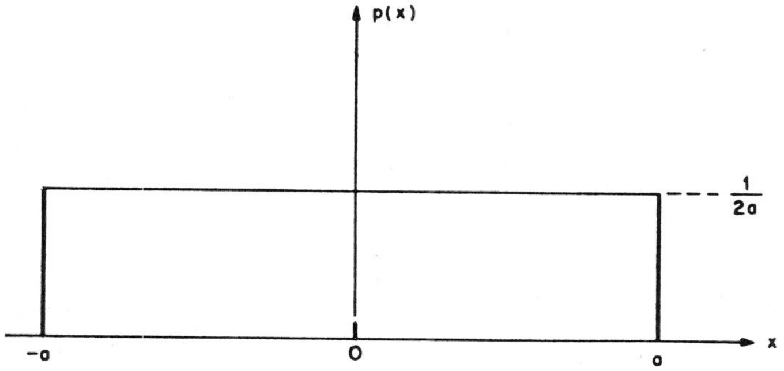

Fig. 2—Optimum quantization for the uniformly distributed case, $N = 1$. (Short strokes mark output levels and long strokes mark end points of corresponding input ranges.)

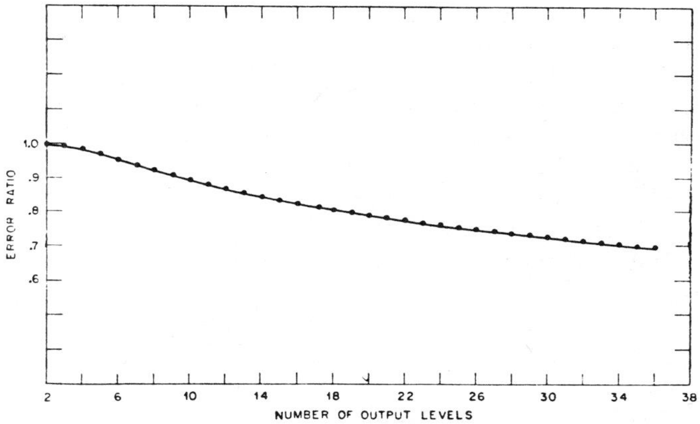

Fig. 5—Ratio of error for optimum quantizer to error for optimum equally spaced level quantizer vs number of outputs. (Minimum mean squared error for normally distribured input with $\sigma = 1$).

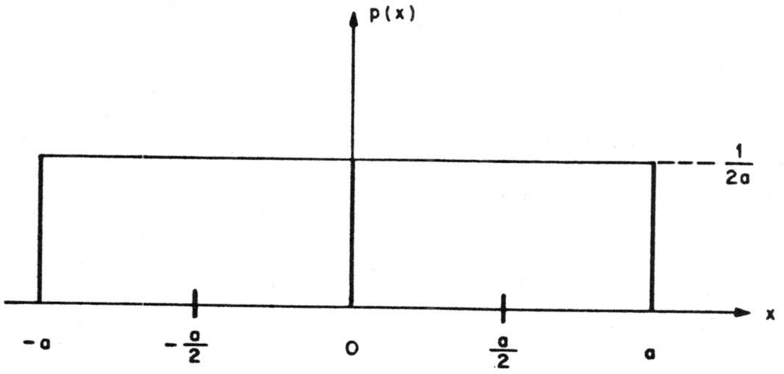

Fig. 3—Optimum quantization for the uniformly distributed case, $N = 2$. (Short strokes mark output levels and long strokes mark end points of corresponding input ranges.)

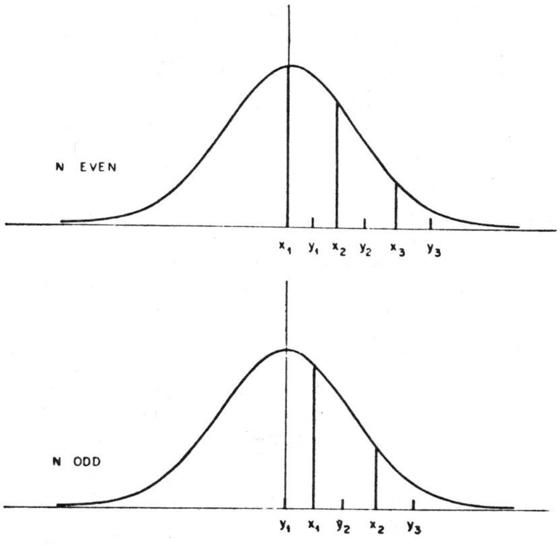

Fig. 6—Labeling of input range end points and output levels for the optimum quantizer. (Short strokes mark output levels and long strokes mark input range end points.)

TABLE I
Parameters for the Optimum Quantizer

	$N = 1$		$N = 2$		$N = 3$	
	x_j	y_j	x_j	y_j	x_j	y_j
$j = 1$	—	0.0	0.0	0.7980	0.0	1.224
2					0.6120	
Error	1.000		0.3634		0.1902	
Entropy	0.0		1.000		1.536	

	$N = 4$		$N = 5$		$N = 6$	
	x_j	y_j	x_j	y_j	x_j	y_j
$j = 1$	0.0	0.4528	0.3823	0.0	0.0	0.3177
2	0.9816	1.510	1.244	0.7646	0.6589	1.000
3				1.724	1.447	1.894
Error	0.1175		0.07994		0.05798	
Entropy	1.911		2.203		2.443	

	$N = 7$		$N = 8$		$N = 9$	
	x_j	y_j	x_j	y_j	x_j	y_j
$j = 1$	0.2803	0.0	0.0	0.2451	0.2218	0.0
2	0.8744	0.5606	0.5006	0.7560	0.6812	0.4436
3	1.611	1.188	1.050	1.344	1.198	0.9188
4		2.033	1.748	2.152	1.866	1.476
5						2.255
Error	0.04400		0.03454		0.02785	
Entropy	2.647		2.825		2.983	

	$N = 10$		$N = 11$		$N = 12$	
	x_j	y_j	x_j	y_j	x_j	y_j
$j = 1$	0.0	0.1996	0.1837	0.0	0.0	0.1684
2	0.4047	0.6099	0.5599	0.3675	0.3401	0.5119
3	0.8339	1.058	0.9656	0.7524	0.6943	0.8768
4	1.325	1.591	1.436	1.179	1.081	1.286
5	1.968	2.345	2.059	1.693	1.534	1.783
6				2.426	2.141	2.499
Error	0.02293		0.01922		0.01634	
Entropy	3.125		3.253		3.372	

	$N = 13$		$N = 14$		$N = 15$	
	x_j	y_j	x_j	y_j	x_j	y_j
$j = 1$	0.1569	0.0	0.0	0.1457	0.1369	0.0
2	0.4760	0.3138	0.2935	0.4413	0.4143	0.2739
3	0.8126	0.6383	0.5959	0.7505	0.7030	0.5548
4	1.184	0.9870	0.9181	1.086	1.013	0.8512
5	1.623	1.381	1.277	1.468	1.361	1.175
6	2.215	1.865	1.703	1.939	1.776	1.546
7		2.565	2.282	2.625	2.344	2.007
8						2.681
Error	0.01406		0.01223		0.01073	
Entropy	3.481		3.582		3.677	

	$N = 16$		$N = 17$		$N = 18$	
	x_j	y_j	x_j	y_j	x_j	y_j
$j = 1$	0.0	0.1284	0.1215	0.0	0.0	0.1148
2	0.2582	0.3881	0.3670	0.2430	0.2306	0.3464
3	0.5224	0.6568	0.6201	0.4909	0.4653	0.5843
4	0.7996	0.9424	0.8875	0.7493	0.7091	0.8339
5	1.099	1.256	1.178	1.026	0.9680	1.102
6	1.437	1.618	1.508	1.331	1.251	1.400
7	1.844	2.069	1.906	1.685	1.573	1.746
8	2.401	2.733	2.454	2.127	1.964	2.181
9				2.781	2.504	2.826
Error	0.009497		0.008463		0.007589	
Entropy	3.765		3.849		3.928	

	$N = 19$		$N = 20$		$N = 21$	
	x_j	y_j	x_j	y_j	x_j	y_j
$j = 1$	0.1092	0.0	0.0	0.1038	0.09918	0.0
2	0.3294	0.2184	0.2083	0.3128	0.2989	0.1984
3	0.5551	0.4404	0.4197	0.5265	0.5027	0.3994
4	0.7908	0.6698	0.6375	0.7486	0.7137	0.6059
5	1.042	0.9117	0.8661	0.9837	0.9361	0.8215
6	1.318	1.173	1.111	1.239	1.175	1.051
7	1.634	1.464	1.381	1.524	1.440	1.300
8	2.018	1.803	1.690	1.857	1.743	1.579
9	2.550	2.232	2.068	2.279	2.116	1.908
10		2.869	2.594	2.908	2.635	2.324
11						2.946
Error	0.006844		0.006203		0.005648	
Entropy	4.002		4.074		4.141	

	$N = 22$		$N = 23$		$N = 24$	
	x_j	y_j	x_j	y_j	x_j	y_j
$j = 1$	0.0	0.09469	0.09085	0.0	0.0	0.08708
2	0.1900	0.2852	0.2736	0.1817	0.1746	0.2621
3	0.3822	0.4793	0.4594	0.3654	0.3510	0.4399
4	0.5794	0.6795	0.6507	0.5534	0.5312	0.6224
5	0.7844	0.8893	0.8504	0.7481	0.7173	0.8122
6	1.001	1.113	1.062	0.9527	0.9122	1.012
7	1.235	1.357	1.291	1.172	1.119	1.227
8	1.495	1.632	1.546	1.411	1.344	1.462
9	1.793	1.955	1.841	1.681	1.595	1.728
10	2.160	2.366	2.203	2.000	1.885	2.042
11	2.674	2.982	2.711	2.406	2.243	2.444
12				3.016	2.746	3.048
Error	0.005165		0.004741		0.004367	
Entropy	4.206		4.268		4.327	

	$N = 25$		$N = 26$		$N = 27$	
	x_j	y_j	x_j	y_j	x_j	y_j
$j = 1$	0.08381	0.0	0.0	0.08060	0.07779	0.0
2	0.2522	0.1676	0.1616	0.2425	0.2340	0.1556
3	0.4231	0.3368	0.3245	0.4066	0.3921	0.3124
4	0.5982	0.5093	0.4905	0.5743	0.5537	0.4719
5	0.7797	0.6870	0.6610	0.7477	0.7202	0.6354
6	0.9702	0.8723	0.8383	0.9289	0.8936	0.8049
7	1.173	1.068	1.025	1.121	1.077	0.9824
8	1.394	1.279	1.224	1.328	1.273	1.171
9	1.641	1.510	1.442	1.556	1.487	1.374
10	1.927	1.772	1.685	1.814	1.727	1.599
11	2.281	2.083	1.968	2.121	2.006	1.854
12	2.779	2.480	2.318	2.514	2.352	2.158
13		3.079	2.811	3.109	2.842	2.547
14						3.137
Error	0.004036		0.003741		0.003477	
Entropy	4.384		4.439		4.491	

	$N = 28$		$N = 29$		$N = 30$	
	x_j	y_j	x_j	y_j	x_j	y_j
$j = 1$	0.0	0.07502	0.07257	0.0	0.0	0.07016
2	0.1503	0.2256	0.2182	0.1451	0.1406	0.2110
3	0.3018	0.3780	0.3655	0.2913	0.2821	0.3532
4	0.4556	0.5333	0.5154	0.4396	0.4255	0.4978
5	0.6132	0.6930	0.6693	0.5912	0.5719	0.6460
6	0.7760	0.8589	0.8287	0.7475	0.7225	0.7990
7	0.9460	1.033	0.9956	0.9100	0.8788	0.9586
8	1.126	1.218	1.172	1.081	1.043	1.127
9	1.319	1.419	1.362	1.263	1.217	1.306
10	1.529	1.640	1.570	1.461	1.404	1.501
11	1.766	1.892	1.804	1.680	1.609	1.717
12	2.042	2.193	2.077	1.929	1.840	1.964
13	2.385	2.578	2.417	2.226	2.111	2.258
14	2.871	3.164	2.899	2.609	2.448	2.638
15				3.190	2.926	3.215
Error	0.003240		0.003027		0.002834	
Entropy	4.542		4.591		4.639	

Cont'd next page

TABLE I, *Cont'd*

	$N = 31$		$N = 32$		$N = 33$	
	x_i	y_i	x_i	y_i	x_i	y_i
$j = 1$	0.06802	0.0	0.0	0.06590	0.06400	0.0
2	0.2045	0.1360	0.1320	0.1981	0.1924	0.1280
3	0.3422	0.2729	0.2648	0.3314	0.3218	0.2567
4	0.4822	0.4115	0.3991	0.4668	0.4530	0.3868
5	0.6254	0.5528	0.5359	0.6050	0.5869	0.5192
6	0.7730	0.6979	0.6761	0.7473	0.7245	0.6547
7	0.9265	0.8481	0.8210	0.8947	0.8667	0.7943
8	1.088	1.005	0.9718	1.049	1.015	0.9392
9	1.259	1.170	1.130	1.212	1.171	1.091
10	1.444	1.347	1.299	1.387	1.338	1.252
11	1.646	1.540	1.482	1.577	1.518	1.424
12	1.875	1.753	1.682	1.788	1.716	1.612
13	2.143	1.997	1.908	2.029	1.940	1.821
14	2.477	2.289	2.174	2.319	2.204	2.060
15	2.952	2.665	2.505	2.692	2.533	2.347
16		3.239	2.977	3.263	3.002	2.718
17						3.285
Error	0.002658		0.002499		0.002354	
Entropy	4.685		4.730		4.773	

	$N = 34$		$N = 35$		$N = 36$	
	x_i	y_i	x_i	y_i	x_i	y_i
$j = 1$	0.0	0.06212	0.06043	0.0	0.0	0.05876
2	0.1244	0.1867	0.1816	0.1209	0.1177	0.1765
3	0.2495	0.3122	0.3036	0.2423	0.2359	0.2952
4	0.3758	0.4394	0.4272	0.3650	0.3552	0.4152
5	0.5043	0.5691	0.5530	0.4895	0.4762	0.5372
6	0.6355	0.7020	0.6819	0.6166	0.5996	0.6620
7	0.7705	0.8391	0.8146	0.7471	0.7261	0.7903
8	0.9104	0.9818	0.9523	0.8820	0.8567	0.9231
9	1.057	1.131	1.096	1.023	0.9923	1.062
10	1.211	1.290	1.248	1.170	1.134	1.207
11	1.375	1.460	1.411	1.327	1.285	1.362
12	1.553	1.646	1.587	1.495	1.445	1.528
13	1.749	1.853	1.781	1.679	1.619	1.710
14	1.971	2.090	2.001	1.883	1.812	1.913
15	2.232	2.375	2.260	2.119	2.030	2.146
16	2.559	2.743	2.584	2.401	2.287	2.427
17	3.025	3.307	3.048	2.767	2.609	2.791
18				3.328	3.070	3.349
Error	0.002220		0.002097		0.001985	
Entropy	4.815		4.856		4.895	

TABLE II

PARAMETERS FOR THE OPTIMUM EQUALLY SPACED LEVEL QUANTIZER

Number Output Levels	Output Level Spacing	Mean Squared Error	Informational Entropy
1	—	1.000	0.0
2	1.596	0.3634	1.000
3	1.224	0.1902	1.536
4	0.9957	0.1188	1.904
5	0.8430	0.08218	2.183
6	0.7334	0.06065	2.409
7	0.6508	0.04686	2.598
8	0.5860	0.03744	2.761
9	0.5338	0.03069	2.904
10	0.4908	0.02568	3.032
11	0.4546	0.02185	3.148
12	0.4238	0.01885	3.253
13	0.3972	0.01645	3.350
14	0.3739	0.01450	3.440
15	0.3534	0.01289	3.524
16	0.3352	0.01154	3.602
17	0.3189	0.01040	3.676
18	0.3042	0.009430	3.746
19	0.2909	0.008594	3.811
20	0.2788	0.007869	3.874
21	0.2678	0.007235	3.933
22	0.2576	0.006678	3.990
23	0.2482	0.006185	4.045
24	0.2396	0.005747	4.097
25	0.2315	0.005355	4.146
26	0.2240	0.005004	4.194
27	0.2171	0.004687	4.241
28	0.2105	0.004401	4.285
29	0.2044	0.004141	4.328
30	0.1987	0.003905	4.370
31	0.1932	0.003688	4.410
32	0.1881	0.003490	4.449
33	0.1833	0.003308	4.487
34	0.1787	0.003141	4.524
35	0.1744	0.002986	4.560
36	0.1703	0.002843	4.594

6

Reprinted from pages 104–111 of *IEEE Int. Conv. Rec., 1964,* part 1, 393p.

SUFFICIENT CONDITIONS FOR ACHIEVING MINIMUM DISTORTION IN A QUANTIZER

P. E. Fleischer

Abstract

A quantizer is said to be optimum when the intervals and the representative values - one in each interval - are chosen to produce minimum error power. Necessary conditions for a quantizer to be optimum have previously been obtained. The computations required to satisfy these conditions are iterative and necessitate the use of a digital computer for all but the most trivial problems. When certain rapidly convergent methods are used, the result may correspond to a maximum or a saddle point of the error power curve instead of a minimum.

In this paper a very simple sufficient condition for checking the minimality of a quantizer structure is derived. A strengthened form of this condition lends considerable insight into the existence of unique optimum quantizers. Thus, it is shown that signals having Gaussian, Laplacian, or Rayleigh probability densities admit unique optimum quantizers.

1. Introduction

In order to transform an analog signal into a form suitable for transmission by PCM, it is necessary to sample and quantize the signal. When a PCM signal is received, it is decoded and filtered to recover the original signal. A simplified diagram of such an operation is shown in Fig. 1. Assuming perfect coding, transmission, and decoding, the remaining system errors are due to sampling and quantization. It is easy to show that if the signal and the quantizing noise are not correlated the contributions of these two sources to the mean square error add. Furthermore, it has been shown by Lloyd(1) that if a band-limited signal is sampled at the Nyquist rate, quantized, and then reconstructed by means of an ideal low-pass filter, the system error becomes equal to the quantizer error. Thus, a study of quantization noise by itself is of interest.

In this paper only the noise due to quantization will be considered. After reviewing previous work which established necessary conditions for minimum quantizer noise, a sufficient condition will be presented. This condition affords a quick test of minimality once the necessary conditions are satisfied. A strong form of the sufficiency condition can be used to establish the uniqueness of many quantizers.

2. Preliminary Analysis

For purposes of this discussion, quantization is defined as the subdivision of the input signal's amplitude range into N nonoverlapping intervals, R_k, and the assignment of N representative amplitudes, x_k. Whenever the signal sample falls into R_k, the representative value x_k is transmitted. It will be noted that quantization is a memoryless operation and, thus, only the input's amplitude probability density, $p(x)$, is of interest. The diagram of Fig. 2 will be of help in defining the symbols to be used. Note that the 2N-1 variables $x_1, x_2, \ldots, x_N$ and $e_1, e_2, \ldots, e_{N-1}$ completely define the N-level quantizer associated with the probability density $p(x)$ which has $-\infty \leq e_o \leq x \leq e_N \leq +\infty$ as its support.

The performance of a quantizer will be measured by the error power,

$$P = \sum_{k=1}^{N} \int_{e_{k-1}}^{e_k} (x-x_k)^2 p(x)dx. \qquad (1)$$

The use of a mean square error criterion makes sense only if the probability density in question has a finite second moment; i.e., the signal has finite power. This hypothesis will henceforth be made.

The following necessary conditions for minimizing P have been obtained by Lloyd(1) and Max(2);

$$\int_{e_{k-1}}^{e_k} (x-x_k)p(x)dx = 0$$

$$k = 1,2,3,\ldots,N, \qquad (2)$$

(i.e., x_k is the centroid of R_k) and

$$e_k = \frac{(x_k+x_{k+1})}{2}$$

$$k = 1,2,3,\ldots,N-1 \qquad (3)$$

These equations follow directly when appropriate partial derivatives of P are set equal to zero. An explicit solution for the x_k and e_k has not been obtained except in the trivial case where $p(x)$ = constant. There are two schemes of iteration, however, which lead to a solution. In the first scheme one starts with an assumed set of end-points. Equation (2) then yields a set of centroids whence, from (3), a new set of end-points is obtained, and so on. This method has the advantage that P is decreasing at each step; however, as shown by Goldstein,[3] it converges very slowly, even when N is moderately large, say N = 16. The second scheme starts with an assumed x_1; use of (2) then determines e_1, whence x_2 is found by means of (3) and so on down to x_N. If x_N is indeed the centroid of the last interval, one is finished. Otherwise, a suitable adjustment must be made in x_1, and the whole process repeated. This method has been programmed by Goldstein and Reading;[4] it does converge rapidly. Its disadvantage is that during the iteration one does not really know whether the error power is decreasing. It is quite possible, therefore, to converge to a local maximum or a saddle point of P. The need for a quick test to detect these situations led to the work which is described in the next section.

The problem can also be handled by a dynamic programming approach[5] where any error criterion is usable. Such a program, however, would require a very large computer memory for speed and accuracy.

3. A Sufficient Condition

Inspection of (1) reveals immediately that the noise power P is a function of the 2N-1 variables $x_1, x_2, \ldots, x_N$ and $e_1, e_2, \ldots, e_{N-1}$. The simplest test for minimality would formulate the matrix of second derivatives of the noise power as given in (1). The method to be presented here is much neater and simpler to use. The following corollary of (2) and (3) is crucial:

> If the representative values, x_k, are arbitrarily preassigned, the necessary and sufficient condition for P to be a minimum is that the end-points of the regions be chosen according to (3).*

Consider now

$$P_o(x_1, x_2, \ldots, x_N) = \sum_{k=1}^{N} \int_{\varepsilon_{k-1}}^{\varepsilon_k} (x-x_k)^2 p(x)dx \quad (4)$$

where the endpoints are constrained:

$$\varepsilon_o = e_o$$

$$\varepsilon_k = \frac{(x_k + x_{k+1})}{2} \qquad k = 1,2,\ldots,N-1 \qquad (5)$$

$$\varepsilon_N = e_N.$$

As defined above, P_o is a function only of the N representative levels. In view of the corollary, for an arbitrary set $\{x_k\}$,

$$P_o(x_1, \ldots, x_N) \leq P(x_1, \ldots, x_N, e_1, \ldots, e_{N-1}), \quad (6)$$

with equality if and only if the e_k satisfy (3), i.e., $e_k = \varepsilon_k$. Thus, in looking for a minimum of P, it is sufficient to minimize P_o. This reduces the number of variables to N.

Now, a (necessary and) sufficient condition for P_o to have a minimum at a point in the N-dimensional space $(x_1, x_2, \ldots, x_N)$ is that

$$\frac{\partial P_o}{\partial x_i} = 0, \qquad i = 1,2,\ldots,N \qquad (7)$$

and that the N×N matrix of second derivatives

$$G = [g_{ij}] = \left[\frac{\partial^2 P_o}{\partial x_i \partial x_j}\right] \qquad (8)$$

be positive (semi)definite at that point.[6]

The differentiation embodied in (7) is easy to perform. Thus, differentiating (4) subject to the constraints (5) yields the following after cancellations:

$$\frac{\partial P_o}{\partial x_i} = -2 \int_{\varepsilon_{i-1}}^{\varepsilon_i} (x-x_i)p(x)dx \qquad (9)$$

$$i = 1,2,\ldots,N.$$

*In fact, this statement is true for any symmetric error criterion which is monotonically increasing for $c > 0$. See Max.[2]

It will be noted that setting $\frac{\partial P_o}{\partial x_i} = 0$ is equivalent to satisfying the necessary condition (2). This contributes no new information. The sufficiency condition for a minimum resides in the positive definiteness of the second derivative matrix. Straightforward manipulation leads to the following matrix. Assuming that p(x) is a continuous function, one obtains

$$\begin{bmatrix} 2a_1-b_1 & -b_1 & 0 & 0 & & \\ -b_1 & 2a_2-b_1-b_2 & -b_2 & 0 & & \\ 0 & -b_2 & 2a_3-b_2-b_3 & -b_3 & & \\ 0 & 0 & \cdot & \cdot & & \\ \cdot & \cdot & \cdot & \cdot & & \\ & & & \cdot & -b_{N-1} \\ & & & -b_{N-1} & 2a_N-b_{N-1} \end{bmatrix} \quad (10)$$

where

$$a_i = \int_{\varepsilon_{i-1}}^{\varepsilon_i} p(x)dx \qquad i = 1,2,\ldots,N \quad (11)$$

and

$$b_i = \frac{x_{i+1}-x_i}{2} p(\varepsilon_i) \qquad i = 1,2,\ldots,N-1 \quad (12)$$

The matrix of second derivatives, G, is symmetric and has nonzero entries only on the main diagonal and the two contiguous diagonals. The entries in the matrix are simple to calculate and the special form of the matrix allows a very rapid numerical check of its positive definiteness. To do this, the following theorem(7) is used:

Theorem: A necessary and sufficient condition for the N×N matrix G to be positive definite is that the determinant of G, as well as its N-1 principal minors* be positive.

Consider now the matrix $G' = [g'_{ij}]$ which is obtained as follows:

$$g'_{1j} = g_{1j}$$

$$g'_{2j} = g_{2j} + \frac{b_1}{g'_{11}} g'_{1j}$$

$$\vdots$$

$$g'_{kj} = g_{kj} + \frac{b_{k-1}}{g'_{k-1,k-1}} g'_{k-1,j}$$

$$\vdots$$

$$g'_{Nj} = g_{Nj} + \frac{b_{N-1}}{g'_{N-1,N-1}} g'_{N-1,j}$$

It can be readily confirmed that the new matrix is of the superdiagonal form. Furthermore, it follows from elementary determinant theory that the kth principal minor of G is given by:

$$M_k = g'_{11}g'_{22}\cdots g'_{kk} \qquad k = 1,\ldots,N$$

Thus, a necessary and sufficient condition for G to be positive definite is that $g'_{kk} > 0$, $k = 1,\ldots,N$.

4. Consequences of the Sufficient Condition

In the previous section, it was shown that a quantizer is optimal if and only if the relations (5) and (7) are satisfied and the matrix (8) is positive definite.** Expressed in words, (5) and (7) require that the representative values be the centroids of their regions while the endpoints of the regions bisect the adjacent representative values. Unfortunately, these relationships do not provide much insight into the structure of optimum quantizers. The complicated functional relationships imposed by the requirement that the matrix (8) be positive definite further becloud the issue. While these problems cannot be entirely surmounted, it is possible to derive conditions under which the quantizer noise power has a unique stationary point which is a minimum. This also provides considerable insight into the general question. Before deriving this sufficient condition, an example*** is given where the stationary point is not unique. Consider a two-level quantizer for the probability density given in Fig. 3, where both lobes are symmetrical.

*The kth order principal of the square matrix G is the determinant obtained by deleting the last N-k rows and columns of G. The determinant itself may be considered to be its own Nth order principal minor.

**The case where (8) is positive semi-definite is not considered, since this would necessitate the use of higher order derivatives. As a matter of practical interest, this situation is uninteresting.

***This example is a modification of one given by Lloyd(1) and attributed to J. L. Kelly, Jr.

The quantizer drawn satisfies the necessary conditions (5) and (7); yet it is obviously not an optimal solution when $A \gg B$. The optimal solution is indicated at the top of the figure.

The following will be used in the derivation of the sufficient condition:

Lemma: A sufficient condition for a matrix, $G = [g_{ij}]$, of the form (10) to be positive definite is that

$$\sigma_i = \sum_j g_{ij} > 0, \qquad i = 1,2,3,\dots,N.$$

A proof and slight generalization of this version of a well-known theorem[8] is given in the Appendix.

Let us now consider a typical row-sum σ_i. Referring to (10), (11), and (12), we have

$$\sigma_i = \Sigma\, g_{ij} = 2[a_i - b_i - b_{i-1}]$$

$$= 2\int_{\varepsilon_{i-1}}^{\varepsilon_i} p(x)dx - \frac{x_{i+1}-x_i}{2}\, p(\varepsilon_i) - \frac{x_i - x_{i-1}}{2}\, p(\varepsilon_{i-1}) \tag{13}$$

The above holds for $i = 1,2,\dots,N$, provided we make the interpretation:

$$b_o = b_N = 0 \tag{14}$$

This is a natural choice, since it is equivalent to $p(\varepsilon_o) = p(\varepsilon_N) = 0$, which follows from the hypothesized continuity of $p(x)$.

The condition $\sigma_i > 0$, $i = 1,2,\dots,N$, is easily interpreted in terms of Fig. 4 which shows a typical interval. It requires that in each interval the cross-hatched area be less than the area under the curve between corresponding limits. Intuitively, it seems rather likely that $\sigma_i > 0$ for a well-behaved $p(x)$. Since x_i is the centroid of the ith region, it is likely to be closer to that end of the curve where the probability is higher. In that case it will give a lower estimate of the area than the trapezoidal rule. The following precise statement can be made:

Theorem: A sufficient condition for $\sigma_i > 0$ is that, in the region defined, the probability density obey the relation:

$$\frac{d^2}{dx^2} \ln p(x) < 0. \tag{15}$$

Proof: In order to eliminate writing subscripts, the following equivalent symbols will be used:

$\sigma = \sigma_i$ = row-sum of ith interval

$\alpha = \varepsilon_i$ = lower limit of ith interval

$\beta = \varepsilon_{i+1}$ = upper limit of ith interval

$g = x_i$ = centroid of ith interval.

In terms of these variables, the relation for σ becomes:

$$\frac{1}{2}\sigma = \int_\alpha^\beta p(x)dx - (\beta-g)p(\beta) - (g-\alpha)p(\alpha) \tag{16}$$

Integration by parts yields the following:

$$\frac{1}{2}\sigma = \beta p(\beta) - \alpha p(\alpha) - \int_\alpha^\beta xp'(x)dx - (\beta-g)p(\beta) - (g-\alpha)p(\alpha)$$

$$= g[p(\beta) - p(\alpha)] - \int_\alpha^\beta xp'(x)dx$$

$$= \frac{\int_\alpha^\beta xp(x)dx}{\int_\alpha^\beta p(x)dx}\int_\alpha^\beta p'(x)dx - \int_\alpha^\beta xp'(x)dx$$

$$= \frac{1}{\int_\alpha^\beta p(x)dx}\left[\int_\alpha^\beta xp(x)dx \int_\alpha^\beta p'(x)dx - \int_\alpha^\beta xp'(x)dx \int_\alpha^\beta p(x)dx\right]. \tag{17}$$

These integrals may now be interpreted as double integrals leading to the following equalities:

$$\sigma\left[\int_\alpha^\beta p(x)dx\right] = 2\int_\alpha^\beta\!\!\int_\alpha^\beta [xp(x)p'(y) - yp'(y)p(x)]dxdy$$

$$= 2\int_\alpha^\beta\!\!\int_\alpha^\beta (x-y)p(x)p'(y)dxdy$$

$$= 2\int_\alpha^\beta\!\!\int_\alpha^\beta (y-x)p(y)p'(x)dxdy$$

$$= \int_\alpha^\beta\!\!\int_\alpha^\beta (x-y)[p(x)p'(y) - p(y)p'(x)]dxdy. \tag{18}$$

In view of the fact that $\int_\alpha^\beta p(x)dx > 0$, it follows that a sufficient condition for $\sigma > 0$ is that

$$[p(x)p'(y) - p(y)p'(x)] > 0$$

whenever

$$(x-y) > 0. \tag{19}$$

Dividing the inequality by the positive quantity $p(x)p(y)$, the following is obtained:

$$\frac{p'(y)}{p(y)} > \frac{p'(x)}{p(x)} \quad \text{whenever} \quad x > y \tag{20}$$

This is equivalent to

$$\frac{d}{dx}[\ell n\, p(x)] = \text{decreasing} \tag{21}$$

and finally,

$$\frac{d^2}{dx^2}[\ell n\, p(x)] < 0 \qquad \text{Q.E.D.} \tag{15}$$

An equivalent sufficient condition, which follows immediately from (15), is:

$$[p'(x)]^2 > p(x)p''(x). \tag{22}$$

Suppose now that a given probability density obeys inequality (15) throughout its range. Then, the matrix (10) is positive definite for arbitrary sets of (x_k) and (e_k) which obey the necessary conditions (2) and (3).

Therefore, all stationary points of the noise power curve must be minima. It can then be shown that this implies a unique minimum. More precisely, the following is true:

Theorem: Let a given continuous probability density of finite second moment obey the equality $[\ell n\, p(x)]'' < 0$ over its total support. Then, the noise power of an N-level quantizer (N arbitrary) has a unique stationary point.

Details of the proof are given in the appendix and in a paper not yet published.

5. Examples

Simple computation shows that the Gaussian distribution has a unique quantizer:

$$\frac{d^2}{dx^2}\left[\ell n \frac{1}{\sqrt{2\pi}} e^{-x^2/2}\right] = -1 < 0. \tag{23}$$

The case of Laplacian probability density

$$p(x) = \left[\frac{1}{2\sqrt{2}} e^{-\frac{|x|}{\sqrt{2}}}\right] \quad -\infty < x < \infty \tag{24}$$

requires more care. It is easy to show that

$$\frac{d^2}{dx^2}\left[\ell n \frac{1}{2\sqrt{2}} e^{-\frac{|x|}{2}}\right] = 0, \qquad x \neq 0, \tag{25}$$

i.e., the inequality degenerates to an equality. In order to show that a unique quantizer exists, consider the following family of functions:

$$p(x,\alpha) = e^{-|x|}\left[1 - \alpha e^{-\frac{1-\alpha}{\alpha}|x|}\right]. \tag{26}$$

It is clear that in the limit as $\alpha \to 0$, $p(x,\alpha)$ approaches the Laplacian distribution of (23) uniformly (disregarding the scale factor). This family of curves is shown in Fig. 5. Consider now the following:

$$\frac{d}{dx}\ell n\, p(x,\alpha) = \frac{d}{dx}\left[-|x| + \ell n\left(1-\alpha e^{-\frac{1-\alpha}{\alpha}|x|}\right)\right]$$

$$= (\operatorname{sgn} x)\left[-1 + \frac{(1-\alpha)e^{-\frac{1-\alpha}{\alpha}|x|}}{1-\alpha e^{-\frac{1-\alpha}{\alpha}|x|}}\right]. \tag{27}$$

Certainly, for all $\alpha < 1$, the above expression is strictly monotonically decreasing. Therefore, in accordance with (21), the probability of (26) possesses a unique optimal quantizer regardless of how small α is. Since the noise power is a continuous function of α, the same conclusion will be true in the limit.

It is also easy to show that both the Rayleigh distribution:

$$p_3(x) = 0 \qquad x \le 0$$

$$p_3(x) = \frac{x}{\sigma^2} e^{-\frac{x^2}{2\sigma^2}} \qquad x > 0 \qquad (28)$$

and the χ^2 distribution:

$$p_4(x) = 0 \qquad x \le 0$$

$$p_4(x) = k_1 x^{(n/2)-1} e^{-x/2} \qquad x > 0 \qquad (29)$$

satisfy (15) and thus have unique quantizers.

A very general class of unimodal probability densities is the Pearson system.(9)* It includes the normal and the χ^2 distributions, as special cases. The class is defined by the following differential equation:

$$\frac{p'(x)}{p(x)} = [\ln p(x)]' = \frac{x+a}{b_2x^2+b_1x+b_o} = \frac{x+a}{D(x)} \qquad (30)$$

Applying the criterion of uniqueness yields the following:

$$\frac{d^2}{dx^2} \ln p(x) = \frac{d}{dx} \frac{x+a}{b_2x^2+b_1x+b_o}$$

$$= \frac{(b_2x^2+b_1x+b_o) - (x+a)(2b_2x+b_1)}{D^2(x)} < 0 \qquad (31)$$

This is equivalent to

$$b_2(x+a)^2 - D(-a) > 0 \qquad (32)$$

for all x on the support of the probability density generated. If $b_2 < 0$, the condition can be satisfied only in a neighborhood of $x = -a$. If, however, $b_2 \ge 0$, it is sufficient for uniqueness to have $D(-a) < 0$.

6. Summary

In the first part of this paper, a quick numerical method for checking the optimality of a given quantizer has been given.

By combining the necessary conditions for stationarity with a strengthened form of the sufficient conditions for a minimum, a condition under which a quantizer possesses a unique stationary point has been found. This condition is overly restrictive. Thus, a matrix may be positive definite even if not all its row-sums are positive. Furthermore, for lack of information on the locations of the $\{x_n\}$, which satisfy the necessary conditions, the row-sum condition had to be proved for every possible set of $\{x_n\}$. The sufficient condition that emerges is seen to define a class of function having the weak convexity property:

$$\frac{d^2p}{dx^2} < \frac{1}{p(x)} \left(\frac{dp}{dx}\right)^2 \qquad (30)$$

Unfortunately, there is no indication of the extent to which this class can be extended.

---

*This was pointed out to the author by M. R. Aaron.

References

1. S. P. Lloyd, "Least Squares Quantization in PCM", Unpublished paper, 1957.
2. J. Max, "Quantizing for Minimum Distortion," Transactions of P.G.I.T., March 1960, pp. 7-12.
3. A. J. Goldstein, Unpublished work.
4. A. J. Goldstein and E. S. Reading, "Computer Program for Optimal Quantizer Structure," Unpublished memorandum.
5. J. D. Bruce, "Optimum Quantization for a General Error Criterion," M.I.T.-R.L.E. Quarterly Progress Report, No. 69, 1963, pp. 135-141.
6. R. Bellman, "Introduction to Matrix Analysis," McGraw-Hill Book Company, Inc., 1960, p. 60.
7. Ibid., p. 74.
8. Ibid., pp. 294-295.
9. H. Cramer, "Mathematical Methods of Statistics," Princeton University Press, 1951, pp. 248-249.

Appendix 1

A Sufficient Condition for Positive Definiteness

The following theorem quoted in the main text will be proved here:

Theorem: A sufficient condition for the matrix G to be positive definite is $\sigma_i > 0$, $i = 1,2,\dots,N$.

Proof: The symbols used above are defined in the text (10) - (13). To prove this theorem, consider the quadratic form corresponding to G:

$$Q(\chi) = \chi^* G \chi \qquad (A1)$$

This may be written as:

$$Q(\chi) = \sum_{i=1}^{N} \sigma_i x_i^2 + \sum_{i=1}^{N-1} b_i(x_{i+1}-x_i)^2. \qquad \text{(A2)}$$

Clearly, $Q(\chi) \geq 0$. Equality occurs only if all $\sigma_i = 0$. This certainly proves the theorem. It is interesting to note that if all $\sigma_i = 0$, the matrix is positive semidefinite. The multiplicity of the zero eigenvalue is 1, since the corresponding eigenvector must satisfy the N-1 equations:

$$x_1 = x_2 = x_3 = \ldots = x_N. \qquad \text{(A3)}$$

Appendix 2

On the Uniqueness of the Minimum

The following theorem, quoted in the text, is proved here:

Theorem: Let a given continuous probability density of finite second moment obey the inequality $[\ell n\ p(x)]'' < 0$. Then the noise power of an N-level quantizer (N arbitrary) has a unique stationary point. This point is a relative and absolute minimum.

Proof: The following Lemma will be needed:

Let C_1 be a connected open region in N-dimensional Euclidian space and let C be a convex closed region in C_1. Let P(x) be a function defined in C_1 which has the following properties:

1. Grad P(x) exists and is continuous in C_1.
2. At every stationary point (a point where grad P(x) = 0), the function attains a strict local minimum.
3. At every point on the boundary of C there exists a vector pointing into C along which the directional derivative of P(x) is negative.

Then, in the region C, P(x) possesses a unique stationary point. The point is interior to C; it is a relative and absolute minimum of P(x) in C.

A proof of this Lemma will be given in a forthcoming paper.

We will consider the support of the probability density to be arbitrary but finite. Since the second moment of the probability is finite, the possible effect of this on the noise power can be made arbitrarily small. In the text it was shown that for an N-level quantizer, the minimal (with respect to end points) quantizing noise power is given by:

$$P_o = \sum_{k=1}^{N} \int_{\varepsilon_{k-1}}^{\varepsilon_k} (x-x_k)^2 p(x)dx \qquad \text{(A2-1)}$$

where

$$\varepsilon_o = \text{left end-point of support}$$

$$\varepsilon_k = \frac{x_k+x_{k+1}}{2} \qquad \text{(A2-2)}$$

$$\varepsilon_N = \text{right end-point of support.}$$

P_o is a function of the N variables $x_1, x_2, \ldots, x_N$. Although on physical grounds it is clear that $x_1 < x_2 < \ldots < x_N$, the definition (A2-1) and the subsequent operations are perfectly valid without that assumption. The necessary conditions for minimizing P_o have been shown to be

$$\frac{\partial P_o}{\partial x_k} = -2 \int_{\varepsilon_{k-1}}^{\varepsilon_k} (x-x_k)p(x)dx = 0. \qquad \text{(A2-3)}$$

Note that (A2-2) and (A2-3) imply that

$$e_o \leq x_1 \leq x_2 \leq \ldots \leq x_N \leq e_N. \qquad \text{(A2-4)}$$

To summarize, P_o and grad P_o are defined at all points of an N-dimensional Euclidian space. The gradient is continuous, since p(x) is continuous. The gradient does not vanish outside the convex region C, defined in (A2-4).* At critical points inside or on C, the matrix of second derivatives exists and is positive definite (since $[\ell n\ p(x)]'' < 0$). Thus, no critical points lie outside C and all critical points inside or on C are strict local minima.

Consider next the gradient at points on the boundary of C. Such points correspond to one or more equal signs occurring in (A2-4). Suppose, for example, that

$$x_{i-1} < x_i = x_{i+1} = \ldots = x_k < x_{k+1} \qquad \text{(A2-5)}$$

where

$$i \geq 2, \qquad k \leq N-1, \qquad k > i.$$

*This region is a simplex. Its outside consists of points which fail to satisfy all the relations in (A2-4). Thus, all the components of the gradient, (A2-3), cannot vanish.

Then, it follows from (A2-3) that

$$\frac{\partial P_o}{\partial x_i} > 0$$

$$\frac{\partial P_o}{\partial x_k} < 0 \qquad \text{(A2-6)}$$

and, if $k - i > 1$,

$$\frac{\partial P_o}{\partial x_j} = 0 \qquad i < j < k \qquad \text{(A2-7)}$$

Clearly, by the continuity of the partial derivatives, the directional derivative in the direction: $\Delta x_i < 0$ and

$$\frac{\Delta x_j}{\Delta x_k} = \frac{j-i}{k-i} \qquad i \leq j < k \qquad \text{(A2-8)}$$

is negative. Furthermore, this direction "points into the region."

A similar construction is possible for each "group" of equalities. The extension to the case when some of the representative points lie at one of the end-points (e_o or e_N) is also immediate.

This shows that all the hypotheses of the lemma are satisfied. Since the gradient does not vanish outside C, the uniqueness of the minimum has been shown. This completes the proof.

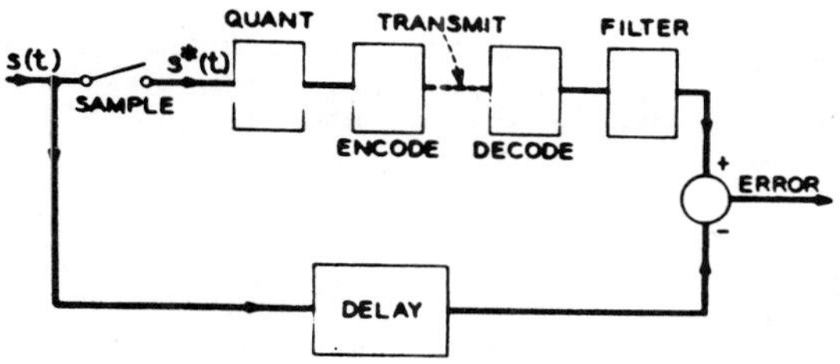

FIGURE 1.

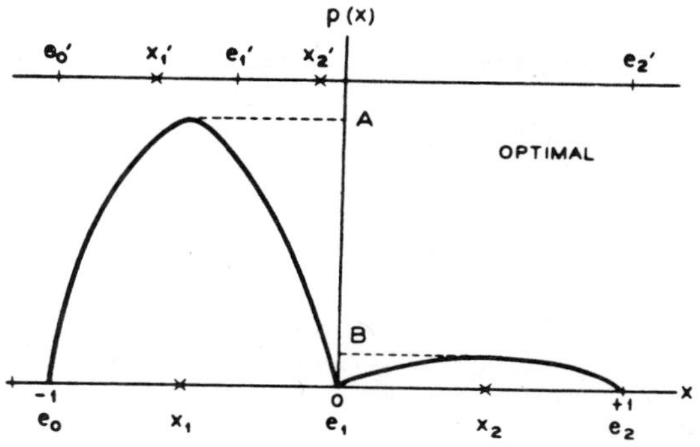

FIGURE 3.

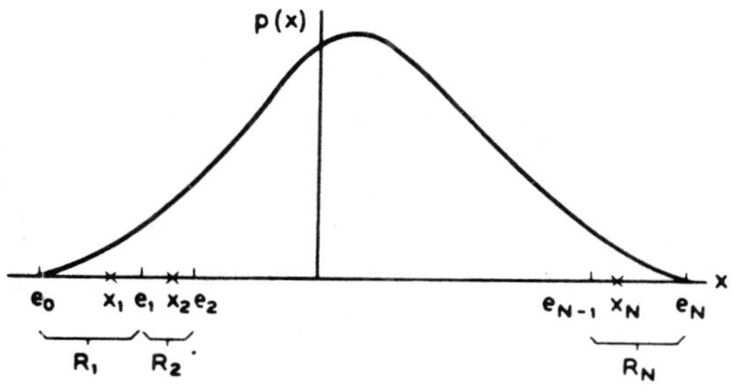

FIGURE 2.

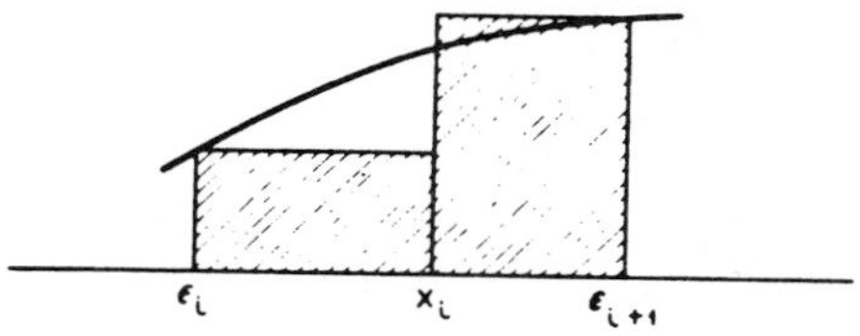

FIGURE 4.

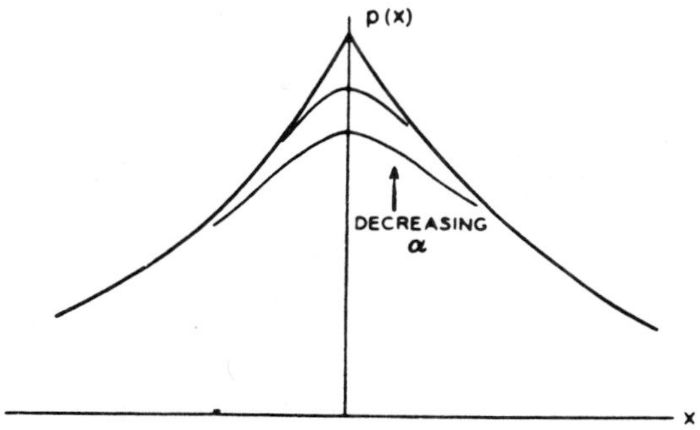

FIGURE 5.

Reprinted from pages 118-124 of *IEEE Int. Conv. Rec., 1964*, part 1, 393p.

ON THE OPTIMUM QUANTIZATION OF STATIONARY SIGNALS*

James D. Bruce

Department of Electrical Engineering and Research Laboratory of Electronics
Massachusetts Institute of Technology
Cambridge, Massachusetts

Summary -- In this paper the exact expression for the quantization error as a function of the parameters that define the quantizer, the error-weighting function, and the amplitude probability density of the quantizer-input signal is presented. An algorithm is developed that permits us to determine the specific values of the quantizer parameters that define the optimum quantizer. This algorithm, which is based on a modified form of dynamic programming, is valid for both convex and nonconvex error-weighting functions. Examples of optimum quantizers designed with this algorithm for a representative speech sample are presented. The performance of these quantizers is compared with that of the uniform quantizer.

I. INTRODUCTION TO QUANTIZATION

Quantization is the nonlinear, zero-memory operation of converting a continuous signal to a discrete signal that assumes only a finite number of levels (N). Such a conversion occurs whenever it is necessary to represent physical quantities numerically. In quantization one of the primary concerns is faithful reproduction, with respect to some fidelity criterion, of the quantizer-input signal at its output. This suggests that for a fixed number of levels (N) it should be possible to minimize the quantization error by adjusting the quantizer characteristic. Figure 1 illustrates the input-output characteristic of such a nonuniform, N-level quantizer.

Much of the interest in quantization today is due to the advent of pulse-code modulation. In particular, it is due to the possibility of transmitting telephone signals by using this mode of modulation. One of the first investigators was Bennett[1] who concluded that, with speech, it would be advantageous to taper the steps of the quantizer in such a manner that finer steps are available for weak signals. This implies that for a given number of levels coarser quantization will occur near the peaks of large signals. Tapered quantization is equivalent to inserting complementary nonlinear, zero-memory transducers in the signal path before and after a uniform quantizer. This system of quantization, which is sometimes called "companding," has been extensively studied by a number of investigators[2,3] each of whom has assumed that a large number of levels is to be used in the uniform quantizer. Other investigators such as Max[4] have attacked a similar problem, that of selecting the parameters defining the optimum quantizer when it is not desirable to assume that a large number of levels can be used. Max has been able to present a partial solution to this problem for a restricted class of error-weighting functions and input signals.

In this paper, the expression for the quantization error as a function of the quantizer parameters, the error-weighting function, and the probability density of the quantizer-input signal is presented. Our objective is to develop an algorithm that will specify the quantizer that minimizes some measure of the error, that is, the difference between the quantizer's input signal and its output. The measure of the error is taken to be the expected value of some non-negative function, called the error-weighting function, of the error. In general, we shall assume that this function is neither symmetric nor convex.

II. FORMULATION OF THE QUANTIZATION PROBLEM

Figure 1 shows the input-output characteristic of an N-level quantizer. We see that the quantizer output is y_k when the input signal x is in the range $x_{k-1} \leq x < x_k$. The x_k are called the transition values; that is, x_k is the value of the input signal at which there is a transition in the output from y_k to y_{k+1}. The y_k are called the representation values.

In communication systems it is generally desired that the quantized signal be an instantaneous replica of the input signal. Clearly, this demands more than the quantizer can accomplish. There will be an error which will be denoted e,

$$e = x - Q(x). \tag{1}$$

An appropriate mean value of e will be taken as a measure of how well the quantizer performs with respect to the demands. This measure of the error is given by

$$\mathscr{E} = \int_{-\infty}^{\infty} g[\xi - Q(\xi)]\, p_x(\xi)\, d\xi. \tag{2}$$

Here, $p_x(\xi)$ is the amplitude probability density of the quantizer input signal x; $g[\xi - Q(\xi)]$ is a function of the error, which is called the error-weighting function. We do not require the error-weighting function to be either convex or symmetric.

Now, in order to relate the parameters of the quantizer to the error (we shall call the measure of the error $\mathscr{E}$ simply the error when there is no chance of confusion), we introduce the explicit

*This work was supported in part by the U. S. Army, Navy, and Air Force under Contract DA 36-039-AMC-03200(E); and in part by the National Science Foundation (Grant G-16526), the National Institutes of Health (Grant MH-04737-03), and the National Aeronautics and Space Administration (Grant NsG-496).

expression for the characteristic of the quantizer

$$Q(\xi) = y_k \qquad x_{k-1} \leq \xi < x_k \qquad k = 1, 2, \ldots, N \tag{3}$$

into Eq. (2). Doing this, we obtain for the error

$$\mathscr{E} = \sum_{i=0}^{N-1} \int_{x_i}^{x_{i+1}} d\xi \, [g(\xi - y_{i+1}) \, p_x(\xi)]. \tag{4}$$

By definition, x_o will be equal to X_ℓ, the greatest lower bound to the input signal, and x_N will be equal to X_u, the least upper bound to the input signal; x_o and x_N are therefore constants.

From Eq. (4) it is clear that the error $\mathscr{E}$ is a function of the quantizer parameters $(x_1, x_2, \ldots, x_{N-1}; y_1, y_2, \ldots, y_N)$; that is,

$$\mathscr{E} = \mathscr{E}(x_1, x_2, \ldots, x_{N-1}; y_1, y_2, \ldots, y_N). \tag{5}$$

The problem before us then is to determine the particular x_i $(i = 1, 2, \ldots, N-1)$ and y_j $(j = 1, 2, \ldots, N)$, the quantities that we shall call X_i and Y_j, which minimize the error $\mathscr{E}$, Eq. (4). Such a minimization is subject to the constraints

$$\left.\begin{aligned} X_\ell = \quad & x_o \leq x_1 \\ & x_1 \leq x_2 \\ & \vdots \\ & x_{N-2} \leq x_{N-1} \\ & x_{N-1} \leq x_N = X_u \end{aligned}\right\} \tag{6}$$

which are explicit in Fig. 1 and Eq. (3). These constraints restrict the region of variation along the error surface $\mathscr{E}$ to a region of that surface such that every point in that region defines a quantizer characteristic $Q(x)$, which is a single-valued function of the input signal x. Such a set of constraints is necessary if the quantizer-input signal is to uniquely specify the quantizer-output signal.

The problem of determining the optimum quantizer then is equivalent to the problem of determining the coordinates of the absolute minimum of the error surface defined by Eq. (4) within the region of variation specified by Eq. (6).

III. DETERMINATION OF THE OPTIMUM QUANTIZER

We have just indicated that the problem of designing the optimum quantizer is equivalent to the problem of determining the coordinates of the absolute minimum of the error surface within the region of variation. From elementary calculus it is known that the absolute minimum of the error surface will either be within the region of variation and therefore at a relative minimum of the error surface, or on the boundary defining the region of variation. Given an arbitrary input signal and an arbitrary error-weighting function, we therefore do not know whether the absolute minimum is at a relative minimum or on the boundary. This implies that the technique used to determine the optimum quantizer should be a technique that searches for the absolute minimum within the region of variation rather than for relative extrema. The method of dynamic programming[5,6] is such a technique.

In order to apply the technique of dynamic programming to the problem of selecting the optimum quantizer (that is, to determining the absolute minimum of the error surface within the region of variation), it is necessary to define three sets of functionals: the error functionals $\{\epsilon_i(x_i)\}$; the transition-value decision functionals $\{X_i(x)\}$; and the representation-value decision functionals $\{Y_i(x)\}$. Each set of functionals has members for $i = 1, 2, \ldots, N$. These three sets of functionals are defined in the following manner:

$$\left.\begin{aligned}
\epsilon_1(x_1) &= \min_{\substack{y_1 \\ X_\ell = x_o \leq x_1 \leq X_u}} \left\{ \int_{x_o}^{x_1} d\xi \, [g(\xi - y_1) \, p_x(\xi)] \right\} \\
\epsilon_2(x_2) &= \min_{\substack{x_1, y_2 \\ X_\ell \leq x_1 \leq x_2 \leq X_u}} \left\{ \epsilon_1(x_1) + \int_{x_1}^{x_2} d\xi \, [g(\xi - y_2) \, p_x(\xi)] \right\} \\
&\;\;\vdots \\
\epsilon_i(x_i) &= \min_{\substack{x_{i-1}, y_i \\ X_\ell \leq x_{i-1} \leq x_i \leq X_u}} \left\{ \epsilon_{i-1}(x_{i-1}) + \int_{x_{i-1}}^{x_i} d\xi \, [g(\xi - y_i) \, p_x(\xi)] \right\} \\
&\;\;\vdots \\
\epsilon_N(x_N) &= \min_{\substack{x_{N-1}, y_N \\ X_\ell \leq x_{N-1} \leq x_N \leq X_u}} \left\{ \epsilon_{N-1}(x_{N-1}) + \int_{x_{N-1}}^{x_N} d\xi \, [g(\xi - y_N) \, p_x(\xi)] \right\}
\end{aligned}\right\} \tag{7}$$

$$\left.\begin{aligned}
X_1(x) &= X_\ell, \text{ a constant;} \\
X_2(x) &= \text{the value of } x_1 \text{ in the coordinate pair } (x_1, y_2) \text{ that minimizes } \left\{ \epsilon_1(x_1) + \int_{x_1}^{x_2} d\xi \, [g(\xi - y_2) \, p_x(\xi)] \right\}, \; x_2 = x; \\
&\;\;\vdots \\
X_N(x) &= \text{the value of } x_{N-1} \text{ in the coordinate pair } (x_{N-1}, y_N) \text{ that minimizes } \left\{ \epsilon_{N-1}(x_{N-1}) + \int_{x_{N-1}}^{x_N} d\xi \, [g(\xi - y_N) \, p_x(\xi)] \right\}, \; x_N = x.
\end{aligned}\right\} \tag{8}$$

$$\left.\begin{aligned}
Y_1(x) &= \text{the value of } y_1 \text{ that minimizes} \\
&\quad \left\{\int_{x_o}^{x_1} d\xi\,[g(\xi-y_1)\,p_x(\xi)]\right\},\quad x_1 = x; \\
Y_2(x) &= \text{the value of } y_2 \text{ in the coordinate pair} \\
&\quad (x_1,y_2) \text{ that minimizes } \left\{\epsilon_1(x_1) + \right. \\
&\quad \left.\int_{x_1}^{x_2} d\xi\,[g(\xi-y_2)\,p_x(\xi)]\right\},\quad x_2 = x; \\
&\quad \vdots \\
Y_N(x) &= \text{the value of } y_N \text{ in the coordinate pair} \\
&\quad (x_{N-1},y_N) \text{ that minimizes } \left\{\epsilon_{N-1}(x_{N-1}) + \right. \\
&\quad \left.\int_{x_{N-1}}^{x_N} d\xi\,[g(\xi-y_N)\,p_x(\xi)]\right\},\quad x_N = x.
\end{aligned}\right\} \tag{9}$$

Before it is possible to make use of these three sets of functionals to determine the optimum quantizer it is necessary to obtain a working knowledge of their meaning. This is best achieved through a detailed investigation of the error functionals. We begin by considering the first error functional, $\epsilon_1(x_1)$. Basically, this functional is concerned with the integral

$$\int_{x_o=X_\ell}^{x_1} d\xi\,[g(\xi-y_1)\,p_x(\xi)] \tag{10}$$

which may be interpreted with the aid of Fig. 2. This term is the value of the error that results when that portion of the signal lying in the range $X_\ell = x_o \leq \xi < x_1$ is represented by y_1. Then, from Eq. (7), $\epsilon_1(x_1)$ is seen to be an enumeration of the minimum value of this error for all x_1 in the range $X_\ell \leq x_1 \leq X_u$. $Y_1(x)$ is an enumeration of the value of y_1 which minimizes this error, Eq. (10), for $x = x_1$, $X_\ell \leq x_1 \leq X_u$. If g is a continuous function of its argument, the y_1 that minimizes (10) for a particular value of x_1 is easily determined by using the methods of elementary calculus for determining relative extrema, since y_1 is an unconstrained variable. If g is not continuous from a computational point of view, the y_1 that minimizes (10) can be found by searching among the possible values of y_1; that is, by evaluating (10) for all y_1 in the interval $-\infty \leq y_1 \leq \infty$ and selecting as the optimum value that y_1 yielding the minimum value for (10).

Having considered the first member of the error functionals, we now turn our attention to the second error functional, $\epsilon_2(x_2)$. Whereas the first error functional was concerned with the error resulting when that portion of the signal lying in the interval $X_\ell = x_o \leq \xi < x_1$ was represented by a single value y_1, the second error functional is concerned with the error resulting when that portion of the signal lying in the range $X_\ell = x_o \leq \xi < x_1$ is represented by the value y_1 and that portion of the signal lying in the range $x_1 \leq \xi < x_2$ is represented by the value y_2. The error that results from such a representation is

$$\int_{x_o=X_\ell}^{x_1} d\xi\,[g(\xi-y_1)\,p_x(\xi)] + \int_{x_1}^{x_2} d\xi\,[g(\xi-y_2)\,p_x(\xi)]. \tag{11}$$

If this representation is to have minimum error, we must determine the specific values of y_1, y_2, and x_1 that will minimize the value of (11). Now let us observe that only the first of the two terms of Eq. (11) involves the variable y_1 and that the minimum value of this integral with respect to y_1 has been enumerated as $\epsilon_1(x_1)$. Therefore, the minimization of

$$\epsilon_1(x_1) + \int_{x_1}^{x_2} d\xi\,[g(\xi-y_2)\,p_x(\xi)] \tag{12}$$

with respect to x_1 and y_2 for $X_\ell \leq x_1 \leq x_2 \leq X_u$ is equivalent to the minimization of (11) with respect to y_1, y_2, and x_1. Referring to Eq. (7), we see that $\epsilon_2(x_2)$ is an enumeration of the minimum value of the error that results when the portion of the signal lying in the range $X_\ell = x_o \leq \xi < x_1$ is represented by the value y_1 and that portion of the signal lying in the range $x_1 \leq \xi < x_2$ is represented by the value y_2, $X_\ell \leq x_1 \leq x_2 \leq X_u$. $X_2(x)$ and $Y_2(x)$ are, respectively, the value of x_1 and the value of y_2 in the coordinate pair (x_1,y_2) that minimize (12) for $x_2 = x$.

The usual method of determining the value of the coordinate pair (x_1,y_2) that minimizes (12) for a particular value of x_2 is by a direct search through all of the possible values of x_1 and y_2. That is, we determine the optimum coordinate pair (x_1,y_2) by evaluating (12) for all x_1 in the interval $X_\ell = x_o \leq x_1 \leq x_2$, and all y_2 in the interval $-\infty \leq y_2 \leq \infty$, and then selecting as the optimum or best coordinate pair that (x_1,y_2) which yields the minimum for Eq. (12). This search is a two-dimensional search. If the error-weighting function g is a continuous function of its argument, however, this two-dimensional search in x_1 and y_2 can be reduced to a one-dimensional search in x_1. This reduction in dimensionality is accomplished by using the result that, since y_2 is unconstrained, the y_2 that minimizes (12) can be determined by using the methods of calculus (for determining relative extrema) for particular values of x_1 and x_2, $x_1 \leq x_2$. Since this permits the optimum y_2 to be determined when the end points x_1 and x_2 are specified, for a specific x_2 the search is then reduced to one in only x_1. (A different presentation of this material has been given elsewhere.[7])

The exact meaning of the remaining error and decision functionals should be evident from the discussion that we have just presented in connection with $\epsilon_1(x_1)$ and $\epsilon_2(x_2)$. Therefore, we are ready to make use of these three sets of functionals to determine the optimum N-level quantizer.

First, we note that when $x_N = X_u$ the entire signal range is being considered for quantization into (N) levels. Then from our discussion and Eq. (7) $\epsilon_N(X_u)$ is the quantization error for the optimum N-level quantizer. Turning our attention to the transition-value and representation-value decision functionals, from the definition of $X_N(x)$, we find that the $(N-1)^{th}$ transition value is

$$X_{N-1} = X_N(X_u)$$

and from the definition of $Y_N(x)$ the $(N)^{th}$ representation value is

$$Y_N = Y_N(X_u).$$

At this point in the decision process the signal in the interval $X_{N-1} \leq \xi \leq X_u$ has been assigned the value Y_N. That portion of the signal lying in the interval $X_\ell \leq \xi < X_{N-1}$ remains to be assigned to quantization intervals. But from the way the error and decision functionals have been defined the next transition value, the $(N-2)^{th}$ transition value, is given by

$$X_{N-2} = X_{N-1}(X_{N-1})$$

and the next representation value is

$$Y_{N-1} = Y_{N-1}(X_{N-1}).$$

Clearly, this process of making the decisions one level at a time can be continued until we finally obtain

$$Y_1 = Y_1(X_1)$$

which is the last parameter needed to completely define the optimum quantizer.

IV. AN EXAMPLE

Consider the design of optimum quantizers for a speech sample as an example of the application of this algorithm. Two sentences, "Joe took father's shoe bench out. She was waiting at my lawn,"[8] which contain most of the important phonemes and have a frequency spectrum roughly typical of conversational speech were chosen as the quantizer-input sample. The center portion of the amplitude probability density is shown in Fig. 3. It should be noted that this short-sample (the two-sentence sample was approximately five seconds long) amplitude probability density is almost identical to published data[9] on long-sample amplitude densities.

After selecting an input signal for quantization, it is necessary to select a specific error-weighting function. For the purposes of this example we shall consider two error-weighting functions, $g(e) = e^2$ (mean-square error) and $g(e) = |e|$ (mean-absolute error). The quantization algorithm has been programmed on the IBM 7094 digital computer for these two error-weighting functions. Typical of the results that are obtained through the application of this algorithm are the minimum mean-square and minimum mean-absolute eight-level quantizers. Table 1 presents the parameters defining these optimum quantizers, together with the parameters defining the eight-level uniform quantizer. A comparison of the columns of Table 1 or Figs. 4, 5, and 6 shows that the optimum quantizers tend to place more levels in the regions of high probability. Figure 7 compares the mean-square error, minimum mean-absolute error, and uniform quantizers as a function of the number of quantization levels.

An interesting way of seeing how the optimum quantizers achieve their reduction in error is to consider the probability density of the error signal. Figures 8, 9, and 10 picture the amplitude probability density of the quantizer error signal for the eight-level uniform quantizer, the eight-level minimum mean-square-error quantizer, and the eight-level minimum mean-absolute-error quantization, respectively. These figures demonstrate that the optimum quantizer achieves its reduction in error by a manipulation of the shape of the amplitude probability density of the error signal. The exact nature of this manipulation has been discussed elsewhere.[7]

V. CONCLUSION AND REMARKS

Basically, the purpose of this paper is to present an algorithm that provides a method of obtaining the optimum quantizer, given the amplitude probability density of the signal and a suitably chosen error-weighting function. Several observations concerning the nature of this algorithm may be made.

First, because of the nature of the minimization problem that is due to the possibility of boundary solutions, it is necessary to obtain an algorithm that searches for the absolute minimum within the region of variation, rather than one that searches for relative minima. The algorithm thus obtained is applicable for both convex and nonconvex error-weighting functions and both discrete and continuous amplitude probability densities. One also observes from the formulation of the error functionals that after an initial set of computations the computation time required to calculate the parameters specifying the N-level optimum quantizer is directly proportional to (N).

Another advantage of this algorithm, not discussed in the present paper, is that it is easily extended to permit us to design optimum quantizers for the case in which the signal to be

Table 1. Eight-level quantizer parameters.

	Uniform quantizer	Minimum mean-square-error quantizer	Minimum mean-absolute-error quantizer
y_1	-11.20	-10.01	-7.60
x_1	-9.60	-8.05	-5.85
y_2	-8.00	-6.01	-4.27
x_2	-6.40	-4.75	-3.05
y_3	-4.80	-3.38	-2.00
x_3	-3.20	-2.35	-1.35
y_4	-1.60	-1.32	-0.80
x_4	0.00	-0.65	-0.35
y_5	1.60	0.04	0.00
x_5	3.20	0.85	0.45
y_6	4.80	1.64	0.70
x_6	6.40	2.65	1.65
y_7	8.00	3.68	2.40
x_7	9.60	5.25	3.65
y_8	11.20	6.79	4.60

quantized is a message signal contaminated by noise. In this extended algorithm[7] the noise is not constrained to be independent of the signal and the contamination is not constrained to be additive.

Acknowledgment

The author wishes to thank Professor Amar G. Bose of the Massachusetts Institute of Technology for his many helpful comments and suggestions concerning the work reported in this paper and to acknowledge the cooperation of the Computation Center, Massachusetts Institute of Technology.

References

1. W. R. Bennett, Spectra of quantized signals, Bell System Tech. J. 27, 446-472 (1948).
2. B. Smith, Instantaneous companding of quantized signals, Bell System Tech. J. 36, 653-709 (1957).
3. R. F. Purton, A survey of telephone speech-signal statistics and their significance in the choice of a P. C. M. companding law, Proc. I. E. E. (London) 109 Part B, 60-66 (1962).
4. J. Max, Quantizing for minimum distortion, IRE Trans. Vol. IT-6, pp. 7-12, 1960.
5. R. Bellman and J. Dreyfus, Applied Dynamic Programming (Princeton University Press, Princeton, New Jersey, 1962).
6. R. Bellman and B. Kotkin, On the Approximation of Curves by Linear Segments Using Dynamic Programming – II, Memorandum RM-2978-PR, Rand Corporation, Santa Monica, California, February 1962.
7. J. D. Bruce, An Investigation of Optimum Quantization (in preparation as an Sc.D. Thesis, Department of Electrical Engineering, Massachusetts Institute of Technology).
8. J. D. Egan, Articulation Testing Methods II, Harvard Psycho-Acoustic Laboratory Report, OSRD No. 3802, November 1944.
9. W. R. Davenport, Jr., A Study of Speech Probability Distributions, Technical Report 148, Research Laboratory of Electronics, M. I. T., August 1950.

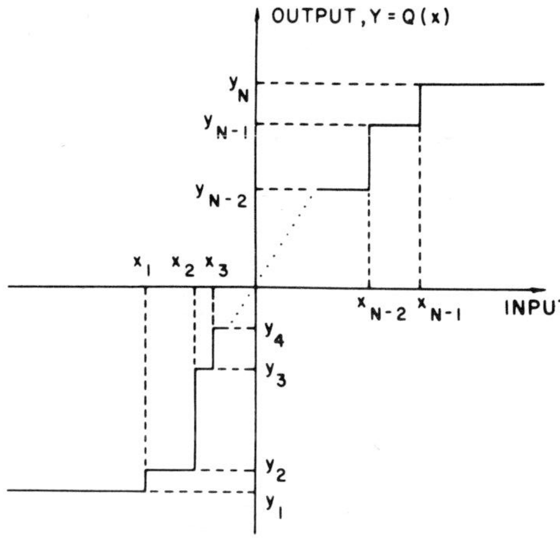

Fig. 1. Input-output characteristic of the N-level quantizer.

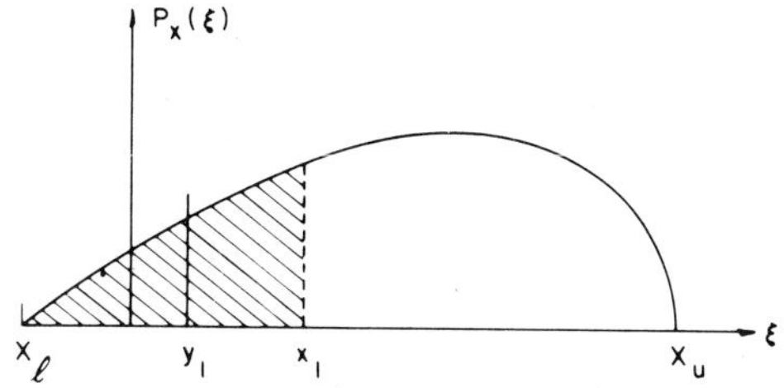

Fig. 2. Amplitude probability density of the input signal illustrating the calculation of the first error functional, $\epsilon_1(x_1)$.

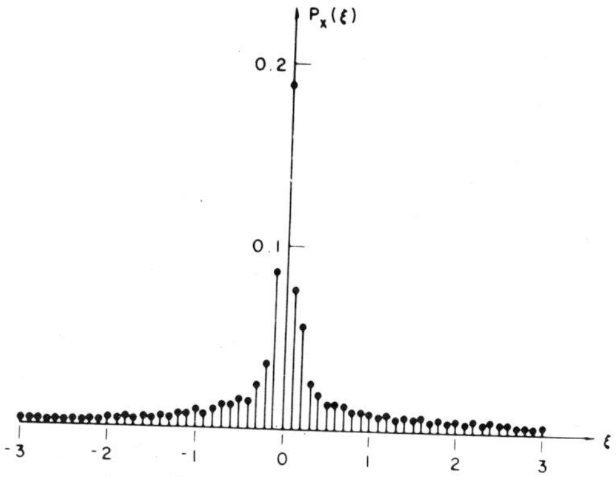

Fig. 3. Central portion $(-3 \leqslant \xi \leqslant 3)$ of the amplitude probability density of the speech sample. (The original speech sample x(t) is bounded in such a way that $-12.8 \leqslant x(t) \leqslant 12.8$ for all t.)

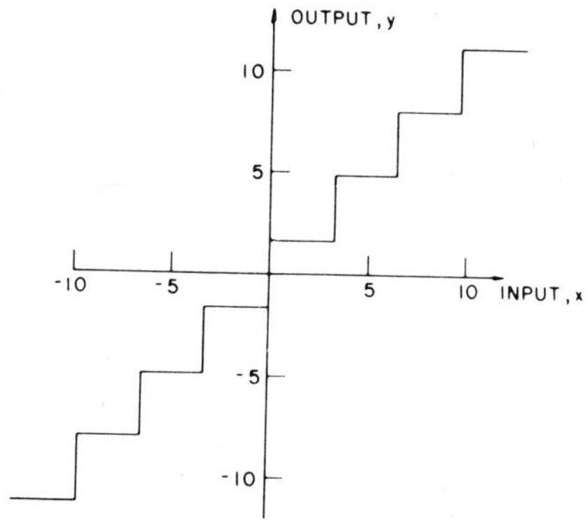

Fig. 4. Eight-level uniform quantizer.

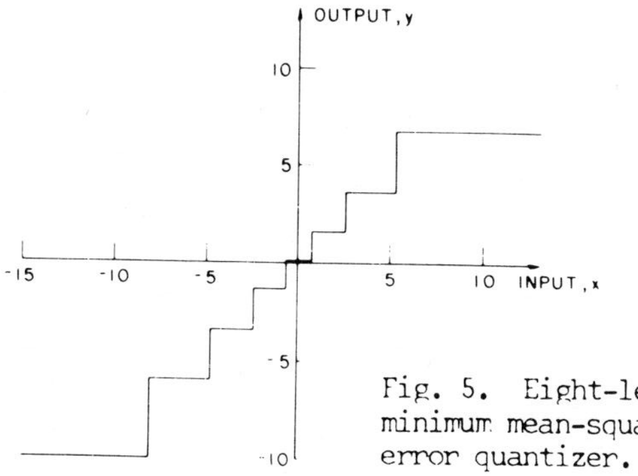

Fig. 5. Eight-level minimum mean-square-error quantizer.

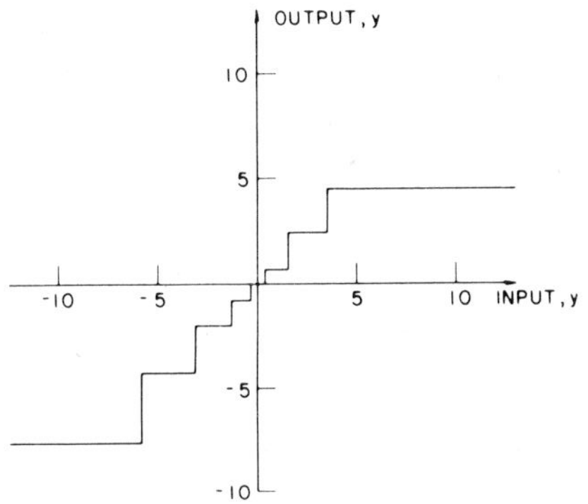

Fig. 6. Eight-level minimum mean-absolute-error quantizer.

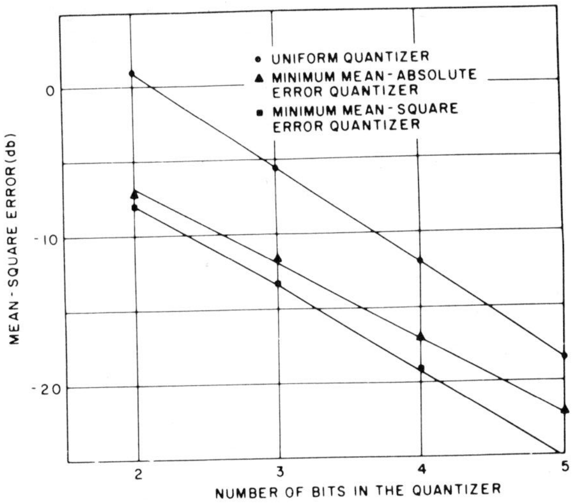

Fig. 7. Mean-square error versus the number of quantization levels in bits for uniform quantization, minimum mean-square-error quantization, and minimum mean-absolute-error quantization. (Zero db is the mean-square value of the input signal.)

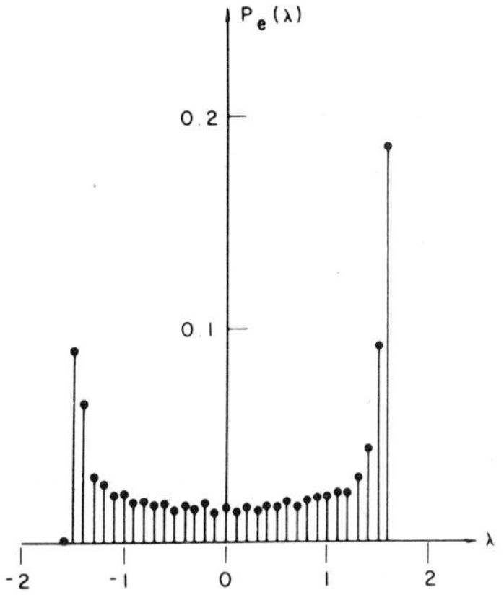

Fig. 8. Amplitude probability density of the quantizer error signal, eight-level uniform quantization.

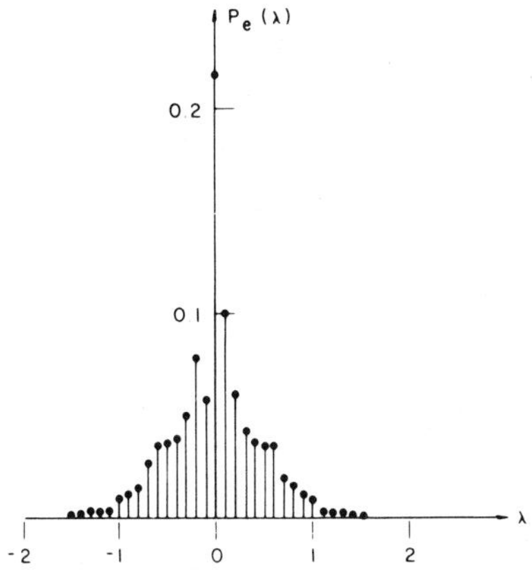

Fig. 9. Amplitude probability density of the quantizer error signal, eight-level minimum mean-square-error quantization.

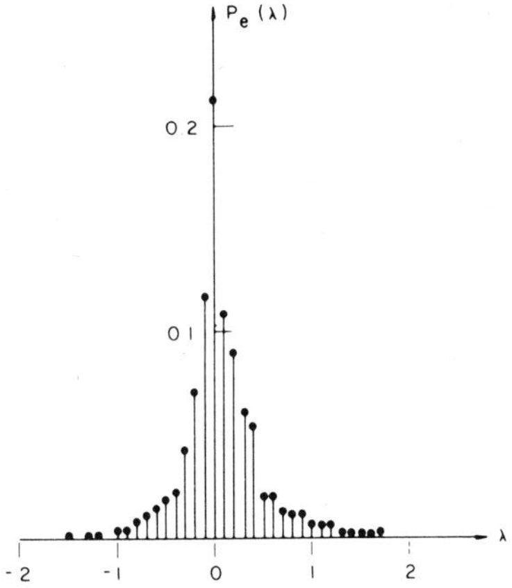

Fig. 10. Amplitude probability density of the quantizer error signal, minimum mean-absolute-error quantization.

8

Reprinted from *IEEE Trans. Commun.* **COM-30**(1):298–301 (1982)

A NOTE ON THE COMPUTATION OF OPTIMAL MINIMUM MEAN-SQUARE ERROR QUANTIZERS

J. A. Bucklew and N. C. Gallagher, Jr.

Abstract—**This paper considers the problems associated with computing optimal minimum mean-square error quantizers. Most computational methods in current use are iterative. These iterative schemes are extremely sensitive to initial conditions. Various methods of obtaining good initial conditions are presented and discussed.**

I. INTRODUCTION

In his classic paper of 1960, Max presents an iterative scheme for the computation of one-dimensional minimum mean-squared error quantization characteristics [1]. In addition, he solves for the optimum Gaussian quantizer for up to 36 output levels. In [2], Gallagher uses Max's method in the computation of optimum Rayleigh quantizer parameters, and in [3] Paez and Glisson use the same method to compute the optimum Laplacian quantizer later recomputed by Adams and Giesler [4]. Max's algorithm is very simple to program into a digital computer, and we view this simplicity as a good reason for using his method. However, one problem that arises with this algorithm is its failure to always converge to the optimum solution when the number of quantizer output levels is large. The reason for this is that the initial guess for starting the iteration must be increasingly precise as the number of quantizer levels becomes large. So, for a 64-level quantizer, Max's algorithm will not converge to the optimum solution unless the initial guess for the first output level is very close to the true value. This difficulty has prompted others to employ more sophisticated optimization methods in the solution for optimum quantizers. For example, Pearlman and Senge [5] use a vector space optimization technique that is a combination of the steepest descent and Newton–Raphson methods to solve for the optimum Rayleigh quantizer. It is not our purpose to detract from this and similar methods that do work well, but in our view, if the starting point problem can be solved, Max's method is the preferred method of solution. In Section II we discuss several methods for choosing the iteration's initial condition very accurately, and we have demonstrated convergence of Max's algorithm for at least 10 000 output levels and present numerical examples in Section III.

Paper approved by the Editor for Data Communication Systems of the IEEE Communications Society for publication without oral presentation. Manuscript received January 5, 1981; revised April 27, 1981. This work was supported by the Air Force Office of Scientific Research under Grant AFOSR 78-3605.

J. A. Bucklew is with the Department of Electrical and Computer Engineering, University of Wisconsin, Madison, WI 53706.

N. C. Gallagher, Jr. is with the School of Electrical Engineering, Purdue University, West Lafayette, IN 47907.

II. THE COMPUTATION OF OPTIMUM ONE-DIMENSIONAL QUANTIZERS

A common method for implementing one-dimensional quantizers is the companding method as discussed by Smith [6]. The companding method is straightforward: the input signal x with probability density $p(x)$ first enters the invertible nonlinearity $g(x)$, called the compressor; then it goes into a uniform quantizer over the range [0, 1], and upon reconstruction it passes through the expansion nonlinearity $g^{-1}(x)$. For minimum mean-squared error quantization, the asymptotically optimum compressor function is given by

$$g(x) = \left[\int_{-\infty}^{\infty} [p(y)]^{1/3} dy\right]^{-1} \int_{-\infty}^{x} [p(y)]^{1/3}\, dy. \qquad (1)$$

In Max's classic 1960 paper an iterative method is presented whereby the exact quantizer parameters can be computed for finite N.

Max's algorithm provides a method for the solution of the equations

$$e_i = (y_i + y_{i-1})/2, \qquad i = 2, \cdots, N \qquad (2a)$$

and

$$\int_{e_i}^{e_{i+1}} (x - y_i) p(x)\, dx = 0, \qquad i = 1, \cdots, N \qquad (2b)$$

where the output levels of the quantizer are denoted y_1, y_2, $\cdots$, y_N and the internal breakpoints as e_1, e_2, $\cdots$, e_{N+1}. Typically, endpoint values e_1 and e_{N+1} are known *a priori* and the first step of Max's procedure is to choose a value for y_1 with which to solve (2b) for the value e_2. We then use this value in (2a) to find y_2 and use this to find e_3 in (2b), and so on. The last integral over (e_N, e_{N+1}) can be used to determine the accuracy of the initial guess for y_1. If the last integral is zero within a specified error, we use the computed parameters to specify the quantizer; if not, we make a new guess for y_1 and begin the procedure again. Details on how to modify the initial guess for y_1 are not specified by Max.

We have computed quantizers using Max's method for several densities. It has been our observation that the convergence properties of Max's algorithm are greatly dependent on the initial guess for y_1. Let y_{1N} denote the first output level for an optimum N level quantizer. Intuitively, if the first guess at y_{1N} (call it $\hat{y}_{1N}$) is very close to $y_{1(N+1)}$, then Max's algorithm tries to converge to the $N+1$ level quantizer. A consideration of Max's method indicates that the first N steps of the algorithm are the same for the N or $N+1$ level quantizers. Although never reported in the literature, it is our understanding that this phenomenon has been widely observed [7].

As an aside, we remark that the conditions presented in (2) are not sufficient conditions to specify the optimum quantizer; they are only necessary. However, in 1965 Fleisher [8] showed that if

$$\frac{d^2}{dx^2}[\ln p(x)] < 0$$

then the expressions in (2) are both necessary and sufficient for the specification of the minimum mean-squared error quantizer, and their solution provides us with the unique optimum quantizer.

We now describe two similar methods for generating a good initial condition. First, note that the initial condition can be a guess at the value for y_1 or a guess for the value of any y_i, $i = 1, \cdots, N$ wherever we choose to begin the iteration. The first method is a modified version of an estimation method by Panter and Dite [9] and Roe [10]. The second method employs a companding model to produce the iteration starting point. Both methods grow more precise as the number of quantization levels N increases. Each method, however, requires computation to generate an initial value; the complexity of this computation varies depending on the distribution of the variable to be quantized.

In the first method we use the asymptotic level density $\lambda(x)$ for the minimum mean-squared error quantizer. $\lambda(x)\Delta x$ is approximately the ratio of the number of output levels in a region Δx about x to the total number of output levels N. This function is the first derivative of the compressor function $g(x)$ in (1):

$$\lambda(x) = g'(x) = [p(x)]^{1/3}\left[\int_{-\infty}^{\infty} [p(y)]^{1/3}\,dy\right]^{-1}. \tag{3}$$

Smith [6] shows that this function has the property that for adjacent output levels y_i and y_{i+1},

$$y_{i+1} - y_i \cong \frac{1}{N\lambda(y)}, \qquad \text{for } y \in [y_i, y_{i+1}] \tag{4}$$

when the number of output levels is large. As an aside, we remark that our compressors always have unity range. Smith allows more generality in his formulas. The best way to illustrate the use of (4) is through an example. Suppose that $p(x)$ is a zero-mean symmetric density (no Dirac delta functions), that N is even, and that a unique optimum quantizer exists. The initial condition for the Max iteration is a guess for first output level greater than zero. We will call this level $y_{N/2}$. We first make the observation that the output levels must be symmetric about the origin. Also, for large N, the distance between the breakpoint at zero and $y_{N/2}$ approximately equals

$$y_{\frac{N}{2}} \cong \frac{1}{2N\lambda(y_{\frac{N}{2}})}. \tag{5}$$

The solution of this equation provides the initial guess for $y_{N/2}$. This basic procedure can be used with modifications for N even or odd with most common probability densities. Some numerical examples are provided in the next section.

The second method uses the companding function to work backwards from the known uniform quantizer over [0, 1] in order to estimate the initial output level. In fact, the method provides a reasonable approximation to the entire quantizer. An N level uniform quantizer on [0, 1] has output levels

$$\hat{y}_i = \frac{2i-1}{2N}, \qquad i = 1, \cdots, N. \tag{6}$$

Therefore, the compander approximation is simply

$$y_i \cong g^{-1}(\hat{y}_i) = g^{-1}\left(\frac{2i-1}{2N}\right). \tag{7}$$

For the purpose of identification, we will refer to the first method of (5) as the λ-approximation and the second as the g-approximation. In hindsight these two methods seem obvious; however, they have apparently not been widely used.

III. NUMERICAL EXAMPLES

In this section we provide some examples using the λ- and g-approximations to estimate the initial input interval endpoint of a Max quantizer. The asymptotically optimum mean-square error companding characteristic is given by

$$\frac{\int_{-\infty}^{x} p(y)^{1/3}\,dy}{\int_{-\infty}^{\infty} p(y)^{1/3}\,dy} = g(x)$$

where $p(y)$ is our input probability density.

The first example we consider is when $p(y)$ is the Gaussian unit variance, zero mean, probability density: $g(x)$ is then given by $\frac{1}{2}(1 + \text{erf}\,(x/\sqrt{6}))$; hence, $g^{-1}(y) = \sqrt{6}\,\text{erf}^{-1}(2y-1)$. Using this equation, our expression for the initial positive input interval endpoint of an N output level quantizer is $x_{1\lambda} = \sqrt{6}\,\text{erf}^{-1}\,(2(N/2+1)-1)$.

The λ-approximation requires us to solve the equations (using a standard Newton-Raphson search)

$$x_{1\lambda} = \frac{1}{N\lambda(x_{1\lambda})} \qquad \text{for } N \text{ even}$$

$$x_{1\lambda} = \frac{1}{2N\lambda(x_{1\lambda})} \qquad \text{for } N \text{ odd}$$

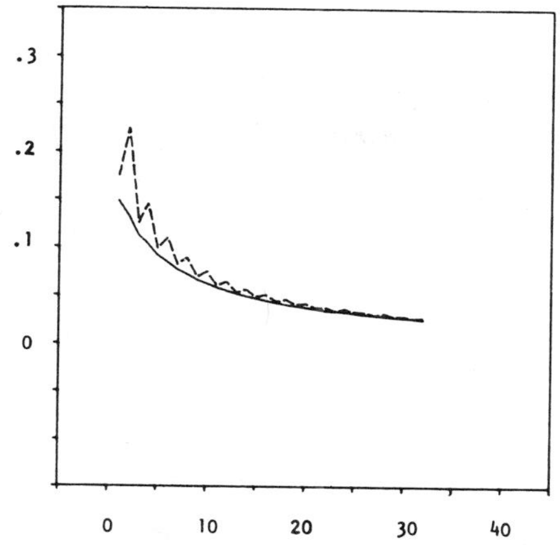

Fig. 1. P_g (solid line) and P_λ (dotted line) plotted as a function of N for the Gaussian density.

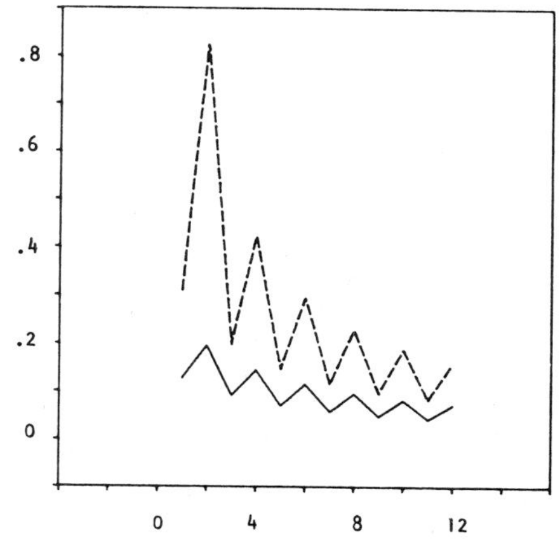

Fig. 2. P_g (solid line) and P_λ (dotted line) plotted as a function of N for the Laplacian density.

where

$$\lambda(x_{1\lambda}) = \frac{(2\pi)^{1/6}}{(6\pi)^{1/2}} p(x_{1\lambda})^{1/3}.$$

Since Max tabulated the actual values of the input interval endpoints, we may compute the quantities

$$P_g \triangleq \left| \frac{x_{1g} - x_{\text{act}}}{x_{\text{act}}} \right|$$

and

$$P_\lambda \triangleq \left| \frac{x_{1\lambda} - x_{\text{act}}}{x_{\text{act}}} \right|$$

for various values of N where x_{act} is the actual tabulated value.

In Fig. 1 we see P_g (solid line) and P_λ (dotted line) plotted as a function of N for values of N from 5 to 36. As may be seen from the figure, the g-approximation is better for all these vaues of N. Furthermore, the λ-approximation does not have a solution for $N = 4$, which is an additional drawback of using this approximation in low N regions.

We now perform the same computations for the Laplacian $(p(y) = \exp\{-|y|\}/2)$ and Rayleigh $(p(y) = y \exp\{-y^2/2\})$ probability densities. In Fig. 2 we plot P_g (solid line) and P_λ (dotted line) for values of N from 5 to 16 for the Laplacian density. Again, the g-approximation is best for all values of N and, furthermore, the λ-approximation has no solution when $N = 4$.

In Fig. 3 we see plots of P_g (solid line) and P_λ (dotted line) for values of N from 2 to 36 for the Rayleigh distribution. For every value except $N = 2$, the g-approximation is better than the λ-approximation. The plot of P_g is noisy because calculation of x_{1g} for this density required a large numerical integration which was very sensitive to the number of samples used in the summation.

We should note that Max quantizers have been computed for the Rayleigh and the Gaussian densities using both $x_{1\lambda}$ and x_{1g} as the estimate for the initial interval endpoint. With no convergence problems, quantizers of 10 000 and 200 output levels have been computed for the Gaussian and Rayleigh probability densities, respectively. In practice, we find that

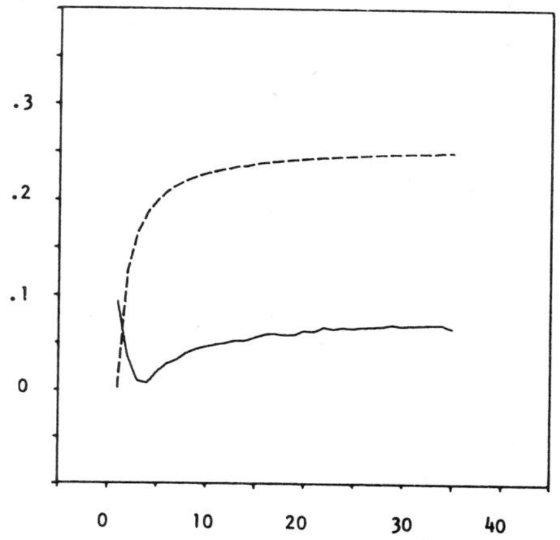

Fig. 3. P_g (solid line) and P_λ (dotted line) plotted as a function of N for the Rayleigh density.

both methods give sufficiently good estimates to allow quick convergence to the correct quantizer. A typical value is 200 iterations for a 1000 level Gaussian quantizer with the last level specified to 10^{-5} accuracy. We conclude that the x_{1g} estimate is a better approximation in most cases, but the $x_{1\lambda}$ estimate is often substantially easier to compute.

REFERENCES

[1] J. Max, "Quantizing for minimum distortion," *IRE Trans. Inform. Theory*, vol. IT-6, pp. 7–12, Mar. 1960.
[2] N. C. Gallagher, "Optimum quantization in digital holography," *Appl. Opt.*, vol. 17, pp. 109–115, Jan. 1, 1978.
[3] M. D. Paez and T. H. Glisson, "Minimum mean-squared error quantization in speech PCM and DPCM systems," *IEEE Trans. Commun.*, vol. COM-20, pp. 225–230, Apr. 1972.
[4] W. C. Adams and C. E. Giesler, "Quantizing characteristic for signals having Laplacian amplitude probability density function," *IEEE Trans. Commun.*, vol. COM-26, pp. 1295–1297, Aug. 1978.
[5] W. A. Pearlman and G. H. Senge, "Optimal quantization of the Rayleigh probability distribution," *IEEE Trans. Commun.*, vol. COM-27, pp. 101–112, Jan. 1979.
[6] B. Smith, "Instantaneous companding of quantized signals," *Bell Syst. Tech. J.*, vol. 36, pp. 653–709, May 1957.
[7] E. Delp and J. A. Bucklew, mutual correspondence, 1977.
[8] P. E. Fleisher, "Sufficient conditions for achieving minimum distortion in a quantizer," in *IEEE Int. Conv. Rec.*, 1964, pp. 104–111.
[9] P. F. Panter and W. Dite, "Quantization distortion in pulse-count modulation with nonuniform spacing of levels," *Proc. IRE*, vol. 39, pp. 44–48, Jan. 1951.
[10] G. M. Roe, "Quantizing for minimum distortion," *IRE Trans. Inform. Theory*, vol. IT-10, pp. 384–385, Oct. 1964.

9

Reprinted from *IEEE Trans. Inform. Theory* **IT-26**(5):610–613 (1980)

SOME PROPERTIES OF UNIFORM STEP SIZE QUANTIZERS

J. A. Bucklew and N. C. Gallagher, Jr.

***Abstract*—Some properties of the optimal mean-square error uniform quantizer are treated. It is shown that the mean-square error (mse) is given by the input variance minus the output variance. Furthermore $\lim_{N\to\infty}\text{mse}/(\Delta^2/12) > 1$, where N is the number of output levels and Δ (a function of M) is the step size of the uniform quantizer, with equality when the support of the random variable is contained in a finite interval. A class of probability densities is given for which the above limit is greater than one. It is shown that $\lim_{N\to\infty} N^2\cdot\text{mse}=(b-a)^2/12$, where $(b-a)$ is the measure of the smallest interval that contains the support of the input random variable.**

In many problems arising in the evaluation or design of a control or communication system it is necessary to predict the performance of a uniform quantizer. Uniform quantizers are of interest because they are usually the simplest to implement and because many noise processes in physical systems may be considered as the noise produced by a uniform quantizing operation. For example, the final position of a stepping motor or the line drawn by the pen of a computer plotting device under a continuous control may be considered to be corrupted by a uniform quantizing operation.

Because of the importance of these quantizers several authors have considered their properties. Widrow [1] shows that under certain conditions the quantization noise is uniformly distributed. Gish and Pierce [2] show that asymptotically the uniform quantizer is optimum in the sense of minimizing the output entropy subject to a fixed mean-square error. Morris and Vandelinde [3] show the uniform quantizer to be minimax. Sripad and Snyder [4] later extended Widrow's work to give a sufficient condition for the quantization error to be uniform and uncorrelated with the input.

We now prove some additional properties of these quantizers when they are designed to minimize the mean-square error (mse). We may write down the analytic expression for the quantizer characteristic $g(x)$ as

$$g(x)=\begin{cases} a, & \text{if } x<q,\\ a+(i+1)\Delta, & \text{if } q+i\Delta\leqslant x<q+(i+1)\Delta,\\ & \text{for } i=0,\cdots,N-3\\ a+(N-1)\Delta, & \text{if } x>(N-2)\Delta+q,\end{cases} \tag{1}$$

where N is the number of output levels. We see that if x is less than q or greater than $q+(N-2)\Delta$, x is truncated to a or $a+(N-1)\Delta$, respectively. An important parameter of interest is the measure of the nontruncation region, $(N-2)\Delta$.

The quantizer characteristic ($g(x)$ must be optimized with respect to three parameters, q which fixes its position along the x axis, a which fixes its position along the y axis, and Δ (a function of N) which specifies the step size of the quantizer. Because it makes little sense to speak of minimizing the mean-square error of a random variable with infinite variance, we will always assume $\int_{-\infty}^{\infty}x^2f(x)\,dx<\infty$.

Property 1: The minimum mean-square error uniform quantizer preserves the mean of the input random variable.

Proof: Suppose $g(x)$ is the optimum uniform quantizer. Then

$$\frac{\partial}{\partial\epsilon}\int(x-g(x)+\epsilon)^2f(x)\,dx|_{\epsilon=0}=0, \tag{2}$$

Manuscript received April 23, 1979; revised October 25, 1979. This work was supported by the Air Force Office of Scientific Research under Grant AFOSR 78-3605. This paper was presented at the 1979 Allerton Conference on Information Sciences and Systems, Monticello, IL, October 10–12, 1979.

J. A. Bucklew was with the School of Electrical Engineering, Purdue University, West Lafayette, IN. He is now with the Electrical and Computer Engineering Department, University of Wisconsin, Madison, WI 53706.

N. C. Gallagher, Jr. is with the School of Electrical Engineering, Purdue University, West Lafayette, IN 47907.

which implies

$$\int xf(x)\,dx=\int g(x)f(x)\,dx. \tag{3}$$

□

Property 2: For the optimum uniform quantizer

$$a=q-\Delta/2.$$

Proof: Suppose $g(x)$ is the optimum uniform quantizer. Then

$$0=\frac{\partial}{\partial\epsilon}\int(g(x-\epsilon)-x)^2f(x)\,dx|_{\epsilon=0} \tag{4}$$

$$\begin{aligned}&=\frac{\partial}{\partial\epsilon}\Bigg[\sum_{i=0}^{N-3}(a+(i+1)\Delta)^2\int_{q+\epsilon+i\Delta}^{q+\epsilon+(i+1)\Delta}f(x)\,dx+a^2\int_{-\infty}^{q+\epsilon}f(x)\,dx\\ &\quad+(a+(N-1)\Delta)^2\int_{q+\epsilon+(N-2)\Delta}^{\infty}f(x)\,dx\Bigg]\\ &\quad-2\Bigg[\sum_{i=0}^{N-3}(a+(i+1)\Delta)\int_{q+\epsilon+i\Delta}^{q+\epsilon+(i+1)\Delta}xf(x)\,dx+a\int_{-\infty}^{q+\epsilon}xf(x)\,dx\\ &\quad+(a+(N-1)\Delta)\int_{q+\epsilon+(N-2)\Delta}^{\infty}xf(x)\,dx\Bigg]\Bigg|_{\epsilon=0}\end{aligned} \tag{5}$$

$$\begin{aligned}&=\Bigg[\sum_{i=0}^{N-3}(a+(i+1)\Delta)^2[f(q+\epsilon+(i+1)\Delta)-f(q+\epsilon+i\Delta)]\\ &\quad+a^2f(q+\epsilon)-(a+(N-1)\Delta)^2f(q+\epsilon+(N-2)\Delta)\Bigg]\\ &\quad-2\Bigg[\sum_{i=0}^{N-3}(a+(i+1)\Delta)[(q+\epsilon+(i+1)\Delta)f(q+\epsilon+(i+1)\Delta)\\ &\quad-(q+\epsilon+i\Delta)f(q+\epsilon+i\Delta)]+a(q+\epsilon)f(q+\epsilon)\\ &\quad-(a+(N-1)\Delta)(q+\epsilon+(N-2)\Delta)f(q+\epsilon+(N-2)\Delta)\Bigg]\Bigg|_{\epsilon=0}.\end{aligned} \tag{6}$$

Simplifying this expression we obtain

$$(\Delta+2a-2q)\sum_{i=0}^{N-2}f(q+i\Delta)=0.$$

The solution $\sum_{i=0}^{N-2}f(q+i\Delta)=0$ corresponds to a trivial solution because without affecting the mean-square error, we may always arbitrarily set $f(q+i\Delta)=0, i=0,\cdots,N-2$. Hence $\Delta+2a-2q=0$ which is what we wish to prove. □

Property 3: The mean-square error of an optimum uniform quantizer is given by the input variance minus the output variance.

Proof:

$$\begin{aligned}\text{mse}&=E(g(x)-x)^2\\ &=E\{x^2\}-2E\{xg(x)\}+E\{g(x)^2\}.\end{aligned} \tag{7}$$

We wish to optimize this expression with respect to Δ. Using

$a=q-\Delta/2$ we first obtain

$$E\{xg(x)\}=\sum_{i=0}^{N-3}\left(q+\left(i+\frac{1}{2}\right)\Delta\right)^2\int_{q+i\Delta}^{q+(i+1)\Delta}xf(x)\,dx$$
$$+(q-\Delta/2)\int_{-\infty}^{q}xf(x)\,dx+\left(q+\left(N-\frac{3}{2}\right)\Delta\right)$$
$$\cdot\int_{q=(N-2)\Delta}^{\infty}xf(x)\,dx \tag{8}$$

and

$$E\{g(x)^2\}=\sum_{i=0}^{N-3}\left(q+\left(i+\frac{1}{2}\right)\Delta\right)^2\int_{q+i\Delta}^{q+(i+1)\Delta}f(x)\,dx$$
$$+\left(q-\frac{\Delta}{2}\right)^2\int_{-\infty}^{q}f(x)\,dx+\left(q+\left(N-\frac{3}{2}\right)\Delta\right)^2$$
$$\cdot\int_{q+(N-2)\Delta}^{\infty}f(x)\,dx. \tag{9}$$

Substitute (9) and (10) into (8); take the partial derivative with respect to Δ and set the result equal to zero. We find that

$$E\{xg(x)\}+qE\{g(x)\}=E\{g(x)^2\}+qE\{x\}, \tag{10}$$

but $E\{g(x)\}=E\{x\}$ for the optimum quantizer. Hence $E\{xg(x)\}+E\{g(x)^2\}$ and

$$\text{mse}=E\{x^2\}-E\{g(x)^2\} \tag{11}$$

which together with Property 1 completes the proof. □

Sripad and Snyder [4] show that a sufficient condition for $x-g(x)$ to be uniform and uncorrelated with x is

$$\phi_x\left(\frac{2\pi n}{\Delta}\right)=\dot{\phi}_x\left(\frac{2\pi n}{\Delta}\right)=0 \qquad \text{for } n=\pm1,\pm2,\cdots, \tag{12}$$

where $\phi_x(\omega)$ is the characteristic function of the input random variable x and $\dot{\phi}_x(\omega)=d\phi_x(\omega)/d\omega$. Frequently in the analysis of a system corrupted by a uniform quantizing operation it is assumed that the quantization noise is uncorrelated with (or sometimes independent of) the input. The next property demonstrates that this cannot be done with the optimum uniform quantizer.

Property 4: Suppose the input probability density is Riemann-integrable. Then the quantization noise is never uncorrelated with the input for the optimum uniform quantizer.

Proof: Without loss of generality assume $E\{X\}=0$. Suppose the converse holds. This implies

$$E\{(x-g(x))x\}=E\{x^2\}-E\{g(x)x\}=0, \tag{13}$$

but from Property 3

$$E\{xg(x)\}=E\{g(x)^2\},$$
$$E\{x^2\}-E\{g(x)^2\}=0. \tag{14}$$

But, again from Property 3, the left side of (14) is the mean-square error. This is a contradiction, since a Riemann-integrable probability density function necessarily implies that the mean-square error for any finite number of output levels is greater than zero (i.e., $f(x)$ has no delta functions). □

We now state an obvious property which will be used in several subsequent proofs.

Property 5: The mean-square error for the optimal uniform quantizer approaches zero as the number of output levels approaches infinity.

Proof: The mean-square error is given by $E\{(g(x)-x)^2\}$, and for this to approach zero it is sufficient that $g(x)$ approach x in mean-square. Consider a quantizer with the parameters $\Delta=1/\sqrt{N-2}$ and $q=-(N-2)\Delta/2$. The width of the nontruncation region is $(N-2)\Delta=\sqrt{N-2}$. Hence as N becomes large the width of the nontruncation region approaches infinity and delta approaches zero. It is a simple matter to show that $\lim_{N\to\infty}g(x)=x$ everywhere. Since $(g(x)-x)^2\leqslant x^2+\Delta^2$ and $\int_{-\infty}^{\infty}(x^2+\Delta^2)f(x)\,dx<\infty$, this implies

$$\lim_{N\to\infty}\int_{-\infty}^{\infty}(g(x)-x^2)f(x)\,dx=\int_{-\infty}^{\infty}\lim_{N\to\infty}(g(x)-x)^2f(x)\,dx=0$$

by the Lebesgue dominated convergence theorem. This quantizer is in general suboptimal, which implies that an optimal quantizer must have even smaller mean-square error for each N, and hence its error must also go to zero. □

As a consequence of the above property, it is easy to show $\lim_{n\to\infty}\Delta=0$ for the optimal uniform quantizer.

Let (a,b) be the smallest interval such that $\int_a^b f(x)\,dx=1$. Note that either $|a|$ or $|b|$ may be infinite.

Property 6: Suppose $f(x)$ is Riemann-integrable. Then, for the optimum uniform quantizer, $\lim_{N\to\infty}(N-2)\Delta=b-a$.

Proof: Suppose $\lim_{N\to\infty}(N-2)\Delta<b-a$. This implies that for N sufficiently large we are always truncating some finite amount of probability mass, and so the mean-square error cannot go to zero. This contradicts the previous property. Hence $\lim_{N\to\infty}(N-2)\Delta\geqslant b-a$.

Suppose $\lim_{N\to\infty}(N-2)\Delta>b-a$. This makes sense only if the random variable is of finite support. So for N large enough there is no truncation error. In the Appendix it is shown that for a family of quantizers with no truncation error $\lim_{N\to\infty}\text{mse}/(\Delta^2/12)=1$ for a Riemann-integrable density function. So, for N sufficiently large, $(N-2)\Delta>C>b-a<\infty$. Then

$$1=\lim_{N\to\infty}\frac{\text{mse}}{\Delta^2/12}\leqslant\lim_{N\to\infty}\frac{\text{mse}}{C^2/12(N-2)^2},$$

or

$$\lim_{N\to\infty}(N-2)^2\text{mse}\geqslant\frac{C^2}{12}. \tag{15}$$

Consider a suboptimal quantizer whose input intervals are obtained by dividing the interval (a,b) into $N-2$ equal subintervals. Denote the mean-square error of this quantizer by mse_{SUB} and its step size by $\Delta_S=(b-a)/(N-2)$. This quantizer has no truncation error and hence

$$1=\lim_{N\to\infty}\frac{\text{mse}_{\text{SUB}}}{\Delta_S^2/12}=\lim_{N\to\infty}\frac{\text{mse}_{\text{SUB}}}{(b-a)^2/12(N-2)^2},$$

$$\lim_{N\to\infty}(N-2)^2\text{mse}_{\text{SUB}}=\frac{(b-a)^2}{12}<\frac{C^2}{12}\leqslant\lim_{N\to\infty}(N-2)^2\text{mse}, \tag{16}$$

which is a contradiction since we have found a suboptimal quantizer with a better mean-square error than the optimal one. □

Bennett [5] shows that the mean-square error of a uniform quantizer is approximately $\Delta^2/12$, assuming that the truncation error is negligible. This is not always the case and in the discussion we will give examples for which Bennett's approximation may be very poor indeed. There are some special cases where Bennett's approximation does hold. The next property deals with one such case.

Property 7: Suppose the density function is Riemann-integrable and $b-a<\infty$. Then for the optimal uniform quantizer we have

$$\lim_{N\to\infty}\frac{\text{mse}}{\Delta^2/12}=1.$$

Proof: From Property 6 $\lim_{N\to\infty}(N-2)\Delta_0=b-a<\infty$ where Δ_0 is the optimum Δ. We may design a suboptimum quantizer by dividing the interval (a,b) into $N-2$ equal subintervals and using these subintervals as the breakpoints for our quantizer. We denote the mean-square error associated with this quantizer by mse_{SUB} and the step size by $\Delta_S\triangleq(b-a)/(N-2)$. This quantizer has no truncation error. Hence from the Appendix

$$\lim_{N\to\infty} \frac{\text{mse}_{\text{SUB}}}{\Delta_S^2/12} = 1. \tag{17}$$

Now

$$\lim_{N\to\infty} \frac{\Delta_S}{\Delta_0} = \lim_{N\to\infty} \frac{(N-2)\Delta_S}{(N-2)\Delta_0} = \frac{\lim\limits_{N\to\infty} (N-2)\Delta_S}{\lim\limits_{N\to\infty} (N-2)\Delta_0} = 1,$$

implying $\lim_{N\to\infty}\Delta_S^2/\Delta_0^2 = 1$. For any quantizer whose nontruncation region covers the support of the Riemann-integrable density function in the limit as N approaches infinity, we show in the Appendix that $\lim_{N\to\infty}\text{mse}/(\Delta^2/12) \geqslant 1$. This bound is arrived at by ignoring the truncation error and is true for density functions with finite or infinite support. Then

$$\lim_{N\to\infty} \frac{\text{mse}_{\text{SUB}}}{\Delta_0^2/12} = \lim_{N\to\infty} \left(\frac{\text{mse}_{\text{SUB}}}{\Delta_S^2/12}\right)\left(\frac{\Delta_S^2/12}{\Delta_0^2/12}\right)$$

$$= \left(\lim_{N\to\infty} \frac{\text{mse}_{\text{SUB}}}{\Delta_S^2/12}\right)\left(\lim_{N\to\infty} \frac{\Delta_S^2/12}{\Delta_0^2/12}\right) = 1, \tag{18}$$

but

$$1 = \lim_{N\to\infty} \frac{\text{mse}_{\text{SUB}}}{\Delta_0^2/12} \geqslant \lim_{N\to\infty} \frac{\text{mse}_{\text{OPTIMAL}}}{\Delta_0^2/12} \geqslant 1, \tag{19}$$

or

$$\lim_{N\to\infty} \frac{\text{mse}_{\text{OPTIMAL}}}{\Delta_0^2/12} = 1,$$

which is what we wanted to prove. □

In the above property we have shown that the truncation error is negligible for the optimum uniform quantizer, if the density function has finite support. This is not true, however, for arbitrary uniform quantizers on these densities. It is easy to design a sequence of uniform quantizers (indexed by N) such that $\lim_{N\to\infty}\text{mse} = 0$, $\lim_{N\to\infty}\Delta = 0$ but $\lim_{N\to\infty}\text{mse}/(\Delta^2/12) \neq 1$.

Zador [6] shows that if $f(x)$ is Riemann-integrable and $E\{x^{2+\delta}\} < \infty$ for some $\delta > 0$ then for the optimal nonuniform quantizer

$$\lim_{N\to\infty} N^2 \cdot \text{mse} = \|f\|_{1/3}/12$$

where $\|f\|_{1/3}$ is the $L_{1/3}$ norm. This result show that for the nonuniform quantizer the mean-square error decreases like $1/N^2$ for large N. Is there a similar property for the optimum uniform quantizer? Not always.

Property 8: Suppose $f(x)$ is Riemann-integrable. Then for the optimum uniform quantizer $\lim_{N\to\infty} N^2 \cdot \text{mse} = (b-a)^2/12$.

Proof: If $b - a < \infty$ then

$$1 \geqslant \lim_{N\to\infty} \frac{\text{mse}}{\Delta^2/12} = \lim_{N\to\infty} \frac{(N-2)^2\text{mse}}{(N-2)^2\Delta^2/12}$$

$$= \frac{\lim (N-2)^2\text{mse}}{\lim N^2\Delta^2/12} \tag{20}$$

but $(N-2)^2\Delta^2 \to \infty$ which implies $\lim_{N\to\infty}(N-2)^2\text{mse} \to \infty$.

If $b - a < \infty$ then $\lim_{N\to\infty}\text{mse}/(\Delta^2/12) = 1$ or $\lim_{N\to\infty}(N-2)^2\text{mse} = \lim_{N\to\infty}N^2\text{mse} = (12)^{-1} \cdot \lim(N-2)^2\Delta^2 = (b-a)^2/12$ which completes the proof. □

Discussion

We should note that not everyone uses our definition of the optimum uniform quantizer. For example, Pearlman and Senge [7] have published tables of the optimal uniform Rayleigh quantizer. For their computations they add the constraints $a = 0$ and $q = \Delta/2$.

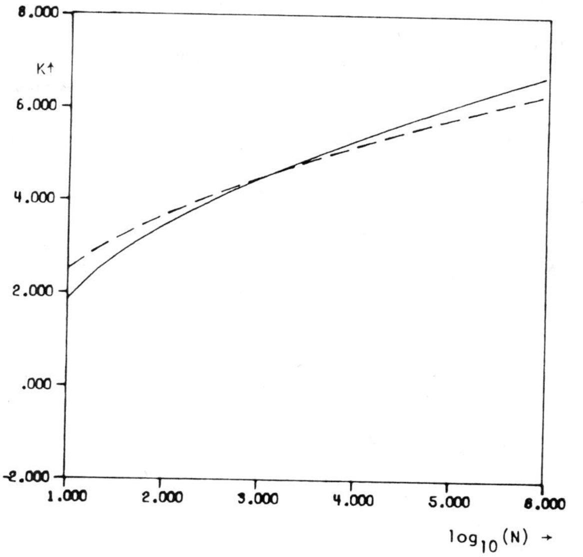

Fig. 1. K (solid line) and $D(N)$ (dashed line) plotted as a function of $\log_{10}(N)$.

It is interesting to note that Properties 1 and 3 are also shared by the optimal nonuniform quantizer as shown in [8]. As a further consequence of these two properties we find that, for the $N = 2$ case, the optimum uniform quantizer and the optimum nonuniform quantizer are identical.

Property 7 is one of the more interesting properties proved in this correspondence. A common approximation to the mean-square error of a uniform quantizer has been $\Delta^2/12$. Consider the class of density functions given by

$$f(x) = \frac{\left(1 + \frac{\delta}{2}\right)}{(1+|x|)^{3+\delta}}, \qquad -\infty < x < \infty.$$

We easily see that $\delta = \text{Sup}\{\epsilon\colon \int x^{2+\epsilon}f(x)\,dx < \infty\}$. By straightforward minimization techniques one can show for this class of densities that

$$\lim_{N\to\infty} \frac{\text{mse}}{\Delta^2/12} = 1 + \frac{2}{\delta}.$$

Property 8 is of interest because it sets forth a basic difference between uniform and nonuniform quantizers. For the nonuniform quantizer we can expect the mean-square error to be of the order of $1/N^2$. We can expect this rate of convergence to zero to hold for the uniform quantizer only if the probability density has finite support. As an example consider the Gaussian case. The Gaussian probability density is of infinite support yet has extremely light tails. We may write down an expression for the mean-square error of a Gaussian random variable and solve for the optimum Δ for a specific N. Let us set $\Delta = 2\sigma K/(N-2)$ where K is a function of N and σ is the standard deviation. We find that, for large N, K is given by the following transcendental equation:

$$\frac{2K\sqrt{\frac{\pi}{2}}}{N-2}\left[\frac{erf\left(\frac{K}{\sqrt{2}}\right)}{6} + 2\left(\frac{N-1}{2}\right)^2 erfc\left(\frac{K}{\sqrt{2}}\right)\right]$$

$$= e^{-K^2/2}\left[N - 1 + \frac{K^2}{3(N-2)}\right].$$

This equation may be solved on a computer by a standard Newton–Raphson search. In Fig. 1 plot K as a function of N for values of N from 10 to 1000000. The dotted line is put in as a reference and is given by $D(N) = 1.7\ln 36N/\pi$. It can be shown that $\lim_{N\to\infty} D(N)/K < \infty$. We conclude that the mean-square error in a uniform Gaussian quantizer is of the same or larger order than $(\ln N)/N^2$.

Appendix

Consider a sequence of quantizers $\{g_N(x)\}_{N=1}^{\infty}$, where N is the number of output levels, Δ_N is the step size, and I_N is the nontruncation region of $g_N(x)$. The measure of I_N is $(N-2)\Delta_N$. Suppose the input probability density function $f(x)$ is Riemann-integrable, and denote the support of $f(x)$ by supp f. Define $\text{mse}_N \triangleq E\{(x-g_N(x))^2\}$.

Lemma 1: Suppose $I_N \to \text{supp } f$ as $N\to\infty$ (i.e., if $x\in\text{supp } f$ then there exists an N_0 such that $x\in I_n$ for $n>N_0$) and $\lim_{N\to\infty}\Delta_n = 0$. Then $\lim_{N\to\infty}\text{mse}_N/(\Delta_N^2/12) \geqslant 1$. Furthermore if $\text{supp } f \subset I_N$ for all N and $\lim_{N\to\infty}\Delta_N = 0$ then $\lim_{N\to\infty}\text{mse}_N/(\Delta_N^2/12) = 1$.

Proof: Define

$$M_i \triangleq \sup_{x\in(q+i\Delta_N,\, q+(i+1)\Delta_N)} f(x)$$

$$m_i \triangleq \inf_{x\in(q+i\Delta_n,\, q+(i+1)\Delta_N)} f(x)$$

Then

$$\sum_{i=0}^{N-3} m_i \int_{q+i\Delta_N}^{q+(i+1)\Delta_N}\left(x-\left(q+\left(i+\frac{1}{2}\right)\Delta_N\right)\right)^2 dx \leqslant \text{mse}_N$$

and

$$\text{mse}_N \leqslant \sum_{i=0}^{N-3} M_i \int_{q+i\Delta_N}^{q+(i+1)\Delta_N}\left(x-\left(q+\left(i+\frac{1}{2}\right)\Delta_N\right)\right)^2 dx + \text{TE}_N$$

where TE_N is the truncation error. Thus

$$\frac{\Delta_N^2}{12}\sum_{i=0}^{N-3} m_i\Delta_N \leqslant \text{mse}_N \leqslant \frac{\Delta_N^2}{12}\sum_{i=0}^{N-3} M_i\Delta_N = \text{TE}_N.$$

If $I_N\to\text{supp } f$ as $N\to\infty$ and $\lim_{N\to\infty}\Delta_N = 0$ then, since $f(x)$ is Riemann-integrable, $\lim_{N\to\infty}\sum_{i=0}^{N-3} m_i\Delta_N \to 1$, which proves the first part of the lemma. If $\text{supp } f\subset I_N$ for every N then $\text{TE}_N = 0$ for every N, and since $\lim_{N\to\infty}\Delta_N = 0$ and $f(x)$ is Riemann-integrable, again $\sum_{N\to\infty}^{N-3} M_i\Delta_N \to 1$, which proves the second part of the lemma. □

References

[1] B. Widrow, "Statistical analysis of amplitude quantized sampled data systems," *Trans. AIEE Applications and Industry*, pt. II, vol. 79, pp. 555–568, Jan. 1960.

[2] H. Gish and J. N. Pierce, "Asymptotically efficient quantizing," *IEEE Trans. Inform. Theory*, vol. IT-14, pp. 676–683, Sept. 1968.

[3] J. M. Morris and V. D. Vandelinde, "Robust quantization of discrete-time signals with independent samples," *IEEE Trans. Commun.*, vol. COM-22, no. 12, pp. 1897–1901, Dec. 1974.

[4] A. B. Sripad and D. L. Snyder, "A necessary and sufficient condition for quantization errors to be uniform and white," *IEEE Trans. Acoustics, Speech, and Signal Processing*, vol. ASSP-25, pp. 442–448, Oct. 1977.

[5] W. R. Bennett, "Spectra of quantized signals," *Bell Syst. Tech. J.*, vol. 27, pp. 446–472, 1948.

[6] P. Zador, "Development and evaluation of procedures for quantizing multivariate distributions," Ph.D. dissertation, Stanford Univ., Stanford, CA, 1964.

[7] W. A. Pearlman and G. H. Senge, "Optimal quantization of the Rayleigh probability distribution," *IEEE Trans. Commun.*, vol. COM-27, pp. 101–112, Jan. 1979.

[8] J. A. Bucklew and N. C. Gallagher, Jr., "A note on optimum quantization," *IEEE Trans. Inform. Theory*, vol. IT-25, pp. 365–366, May 1979.

Part III

VECTOR SIGNAL QUANTIZERS

Editor's Comments on Papers 10 Through 20

As mentioned in the introduction, interest in vector quantization resides in the fact that additional gain in performance is achievable over scalar quantization even when the input vector consists of independent elements. To demonstrate this fact, consider the following argument. Vector quantization of a length k vector (the coordinates of a point in $\mathcal{R}^k$) induces a partitioning of $\mathcal{R}^k$ into N mutually exclusive, exhaustive regions called the quantization regions and a mapping of each of these regions into a particular value called the output point. Scalar quantization consists of the partitioning of the real line $\mathcal{R}^1$ into intervals, the only regions of interest. Considering a point in $\mathcal{R}^k$ to be a sequence of k points in $\mathcal{R}^1$ ($\mathcal{R}^k$ being the cross product of k $\mathcal{R}^1$s), scalar quantization can be envisioned as a partitioning of the Euclidean space into parallelepipeds (cross products of k intervals), with perpendicular sides due to the scalar interval partitioning. From this observation, scalar quantization is seen to be a special case of vector quantization with a region constraint. Since adding a constraint to a design problem cannot improve the performance, we see that scalar quantization can never outperform optimum vector quantization.

As a numerical example, consider the minimum mean squared error (MSE) quantization of a vector composed of two independent Gaussian random variables, each with zero mean and unit variance. Quantization to $N = 16$ levels for this case means separation of the Euclidean plane into sixteen regions. For scalar quantization, separate scalar quantizers are applied to each of the two components, x and y, having N_x and N_y levels, respectively, with $N_x N_y = 16$. A simple symmetry or variational argument demonstrates that the MSE is minimized by dividing the levels equally, four to each rectangular coordinate (see Fig. 1). A vector quantizer for the same source with better performance is depicted in Figure 2 (discussed in detail in Paper 33). Both figures display the quantization pattern on the $[-5,5]$ section of the plane.

Vector quantization as a research area began with the works of Schutzenberger

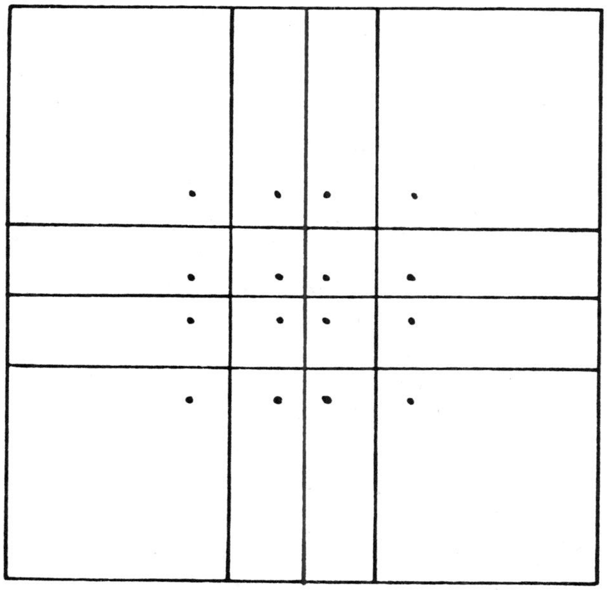

Figure 1. Rectangular quantizer. Outputs: ±1.5104, ±.4528; breakpoints: 0., ±.9816; MSE = .2350.

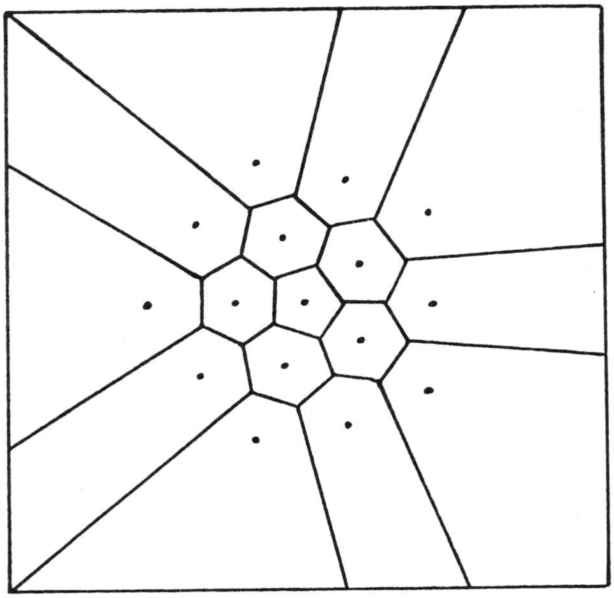

Figure 2. Vector quantizer. Output radii: .8725, 1.7123, 2.0117; angles equally distributed on [0, 2π); MSE = .2172.

(1958) and Zador (1963) over twenty years ago but has seen much renewed interest in the past eight years. The eleven papers in this section explore several topics in vector quantization, including limits on attainable performance and properties of optimum vector quantizers, design and implementation of vector quantizers, and uniform vector quantizers.

Paper 10 by Zador contains material from his dissertation and an unpublished Bell System memorandum, providing bounds on the minimum mean r th error performance of optimum vector quantizers for both the fixed number of levels case as well as the fixed entropy case (r th moment as a generalization of second moment, MSE). For a further generalization of Zador's results, see Bucklew and Wise (1982). For both cases, Zador extends the work of Schutzenberger (1958), who proved that the rate of decrease of the mean r th error as the number of levels increases is order $N^{-r/k}$. In Paper 10 Zador establishes the important result that the optimum performance of vector quantizers is separable into two terms, one due to the source density and the other due to the dimensionality of the space. This separation is helpful in imagining a companding implementation of the vector quantizers as well as for noting other properties of optimum quantizers. In Zador's work the term due to the dimensionality of the space (and the error exponent, r), $C(k,r)$, is bounded both above and below. Papers 11 and 15 provide exact values of $C(k,r)$ for $r = 2$ and $k = 1$ and 2. For performance bounds with other criteria, see also Bucklew (1981), Elias (1970), and Yamada, Tazaki, and Gray (1980).

Paper 11 by Gersho introduces several topics on the design and implementation of vector quantizers. He establishes the necessary conditions of centroids and nearest-neighbor regions for the vector case and presents a heuristic argument on performance, achieving the same optimum performance bounds as Zador does. Further, he suggests an extension of companding to the vector case and notes the problem of conformality of the compressor mapping for the bivariate case (for $k > 2$, see Paper 14). For the quantization of the uniformly distributed source in k dimensions, Gersho establishes the correspondence between the normalized inertia of space-filling polytopes and the uniform quantizer's MSE performance. This uniform quantizer consists of a tesselation of a space-filling polytope, a direct generalization of the equal-width-interval (uniform) scalar quantizer. He also presents performance values of these uniform quantizers for several different values of dimension ($k = 3$, 4, and 5), which are useful as tighter upper bounds to $C(k,2)$. In the two-dimensional case, it is well known that the minimum MSE uniform quantizer is a tesselation of hexagons on the unit square (see Fejes Toth, 1959, or Newman, 1982). In this paper Gersho conjectures that the best uniform quantizer in three dimensions is a tesselation of truncated octahedra, a result more recently proved by Barnes and Sloane (1983).

Paper 12 by Gallagher and Bucklew rederives Zador's upper bound in a simpler fashion and notes several properties of optimum vector quantizers. These include: The mean of the random vector is unchanged by the quantization process; the quantizer output's variance is smaller than the input's variance; and the MSE equals the difference between these variances (the last two noted by Smith, Paper 3, for the scalar case).

Papers 13 and 14 by Bucklew explore in depth the vector compander introduced by Gersho in Paper 11. In Paper 13 Bucklew develops the compander implementation for the vector case and presents a derivation of the resulting MSE. The separation of the quantization problem into a compressor design and the implementation of the uniform quantizer on the unit hypercube is established. Necessary conditions upon the compressor function to minimize the MSE are discussed. The difficulties of implementing optimum vector companders is considered in Paper 14. In particular, it is noted that for many multivariate densities (including the important independent Gaussian source) the compressor can not both be conformal and satisfy the minimum MSE necessary conditions of Paper 13.

Papers 15 through 17 consider the design, performance, and implementation of uniform quantizers on the unit hypercube. In Paper 15 Conway and Sloane consider the use of Voronoi regions of lattices for these uniform quantizers. The discussion includes MSE performance calculations for a variety of possible lattice quantizers with dimensions 2 through 10. The excerpt included here presents the quantization results only. For further discussion on the use of lattices and sphere packing results, see the forthcoming text by Conway and Sloane (1985). Paper 16, also by Conway and Sloane, considers the implementation details of the lattice quantizers introduced in Paper 15. As is seen in this paper, the lattice quantizers have very simple implementations (see also Conway and Sloane, 1983). Paper 17 by Rines and Gallagher describes simple implementations for the hexagonal two-dimensional quantizer (also in Rines and Gallagher,

1982) and the truncated octahedron three-dimensional uniform quantizer suggested by Gersho (Paper 11). The method, dubbed "prequantization," employs a series combination of memoryless nonlinearities, linear operators, and scalar quantizers to produce the desired space partitioning and output value mapping.

Papers 18 and 19 discuss the vector version of Lloyd's Method I quantizer design algorithm. In its vector version the algorithm converges to a local minimum of MSE by iteratively applying the following steps: (1) Holding the output point pattern fixed, choose the quantization regions as nearest-neighbor regions about the outputs; and (2) holding the regions fixed, choose each output as the centroid of its region. During this iterative procedure, the MSE is a strictly nonincreasing function. Earlier versions of this algorithm for the vector case are presented in Chen (1977) and Menez, Boeri, and Esteban (1979). The algorithm described in Paper 18 by Linde, Buzo, and Gray is constructed to work with either an exact probabilistic definition of the source or a training sequence of source outputs. Examples of quantizers for a Gaussian source with one bit per dimension ($N = 2^k$) are presented for dimensions 2 through 6. Additional examples are available in Fischer and Dicharry (1983). Paper 19 by Gray, Kieffer, and Linde is a mathematically rigorous treatment of this design algorithm; in particular, they describe conditions for convergence to a locally optimum quantizer under the training sequence situation.

The last paper in this section, Paper 20 by Gray and Karnin, demonstrates by example the lack of global optimality of the vector version of Lloyd's design technique. The examples are for the one-bit-per-dimension, independent Gaussian source quantizer.

REFERENCES

Abaya, E. F., and G. L. Wise, 1981, Convergence of Vector Quantizers with Applications to the Design of Optimal Quantizers, *Allerton Conf. Commun. Control Comput. Proc.*, pp. 79–88.

Barnes, E. S., and N. J. A. Sloane, 1983, The Optimal Lattice Quantizer in Three Dimensions, *SIAM Jour. Alg. Disc. M.* **4:**30–41.

Bucklew, J. A., 1981, Upper Bounds to the Asymptotic Performance of Block Quantizers, *IEEE Trans. Inform. Theory* **IT-27**(5):577–581.

Bucklew, J. A., and G. L. Wise, 1982, Multidimensional Asymptotic Quantization Theory with r-th Power Distortion Measures, *IEEE Trans. Inform. Theory* **IT-28**(3):239–247.

Chen, D. T. S., 1977, On Two or more Dimensional Optimum Quantizers, *IEEE Int. Conf. Acoust., Speech, and Signal Process. Proc.*, 640–643.

Conway, J. H., and N. J. A. Sloane, 1983, A Fast Encoding Method for Lattice Codes and Quantizers, *IEEE Trans. Inform. Theory* **IT-29**(6):820–824.

Conway, J. H., and N. J. A. Sloane, 1985, *The Leech Lattice, Sphere Packing and Related Topics,* Springer-Verlag, New York.

Elias, P., 1970, Bounds and Asymptotes for the Performance of Multivariate Quantizers, *Ann. Math. Stat.* **41:**1249–1259.

Fejes Toth, L., 1959, Sur la Representation d'une population infinie par un nombre fine d'elements, *Acta Mathematica,* Magyar Tudomanyos Akademia Budapest, **10:**299–304.

Fischer, T. R., and R. M. Dicharry, 1984, Vector Quantizer Design for Memoryless Gaussian, Gamma, and Laplacian Sources, *IEEE Trans. Commun.* **COM-32** (9):1065–1069.
Menez, J., F. Boeri, and D. J. Esteban, 1979, Optimum Quantizer Algorithm for Real Time Block Quantizing, *IEEE Int. Conf. Acoust., Speech, and Signal Process Proc.*, pp. 980–984.
Newman, D. J., 1982, The Hexagon Theorem, *IEEE Trans. Inform. Theory* **IT-28** (2):137–139.
Rines, K. D., and N. C. Gallagher Jr., 1982, The Design of Two-Dimensional Quantizers using Prequantization, *IEEE Trans. Inform. Theory* **IT-28**(2):232–239.
Schutzenberger, M. P., 1958, On the Quantization of Finite Dimensional Messages, *Inform. Control* **1**(5):153–158.
Yamada, Y., S. Tazaki, and R. M. Gray, 1980, Asymptotic Performance of Block Quantizers with Difference Distortion Measures, *IEEE Trans. Inform. Theory* **IT-26**(1):6–14.
Zador, P., 1963, *Development and Evaluation of Procedures for Quantizing Multivariate Distributions,* dissertation, Department of Statistics, Stanford University, Stanford, Calif., 64p.

BIBLIOGRAPHY

Abaya, E. F., and G. L. Wise, 1982, On the Existence of Optimal Quantizers, *IEEE Trans. Inform. Theory* **IT-28**(6):937–940.
Gersho, A., 1982, On the Structure of Vector Quantizers, *IEEE Trans. Inform. Theory* **IT-28**(2):157–166.
Gray, R. M., and Y. Linde, 1982, Vector Quantizers and Predictive Quantizers for Gauss-Markov Sources, *IEEE Trans. Commun.* **COM-30**(2):381–389.
Rines, K. D., N. C. Gallagher, Jr., and J. A. Bucklew, 1982, Nonuniform Multidimensional Quantizers, *Princeton Conf. Inform. Sci. Systems Proc.* pp. 43–46.
Sayood, K., and J. D. Gibson, 1982, An Algorithm for Designing Vector Quantizers, *Allerton Conf. Commun. Control, & Comput. Proc.* pp. 301–310.

10

Reprinted from *IEEE Trans. Inform. Theory* **IT-28**(2):139–149 (1982)

Asymptotic Quantization Error of Continuous Signals and the Quantization Dimension

PAUL L. ZADOR

***Abstract*—Extensions of the limiting quantization error formula of Bennet are proved. These are of the form $D_{s,k}(N, F) = N^{-\beta}B$, where N is the number of output levels, $D_{s,k}(N, F)$ is the sth moment of the metric distance between quantizer input and output, $\beta, B > 0$, $k = s/\beta$ is the signal space dimension, and F is the signal distribution. If a suitably well-behaved k-dimensional signal density $f(x)$ exists, $B = b_{s,k}[\int f^{\rho}(x)\,dx]^{1/\rho}$, $\rho = k/(s+k)$, and $b_{s,k}$ does not depend on f. For $k = 1$, $s = 2$ this reduces to Bennett's formula. If F is the Cantor distribution on [0, 1], $0 < k = s/\beta = \log 2/\log 3 < 1$ and this k equals the fractal dimension of the Cantor set [12, 13]. Random quantization, optimal quantization in the presence of an output information constraint, and quantization noise in high dimensional spaces are also investigated.**

Manuscript received October 20, 1981.
The author is with the Insurance Institute for Highway Safety, Watergate 600, Suite 300, Washington, DC 20008.
Editor's Note: This paper is a revised and expanded version of the Bell Laboratories unpublished memorandum "Topics in the asymptotic quantization of continuous random variables," Bell Laboratories, Murray Hill, NJ, February 1966.

I. Introduction

THE MEAN SQUARED ERROR due to rounding a random signal value to the nearest n place decimal fraction tends to zero as $1/12n^2$ when n increases if the signal is uniform over the unit interval. The present paper describes extensions of this result.

Consider a random signal distributed over a linear metric space E. Pick N code points y_n in E and "round" or "quantize" the point x in E to the point $Qx = y_{n_x}$ in E if y_{n_x} is the nearest neighbor of x among the y with respect to the metric distance $\| y_n - x \|$. For an arbitrary signal distribution F the mean sth power distortion due to quantization is

$$D_s(N, F) = \int_E \|x - Qx\|^s \, dF(x). \tag{1}$$

The minimum of $D_s(N, F)$ over various choices of code

points equals the mean squared rounding error for the uniform distribution, if $s = 2$.

The asymptotic behavior of the minimum mean squared error was described by Bennet in 1948 [1]. For well-behaved signal distributions over the real line, Bennett found this quantity to approach to zero as AN^{-2}. In Bennett's formula $A = [\int f(x)^{1/3}\,dx]^3$, where $f(x)$ is the signal density.

The k-dimensional version of this formula ([17, 18]; see also [6], [16], [2], [3]) shows that the minimum distortion tends to zero as $BN^{-\beta}$; $B, \beta > 0$. Here $\beta = s/k$ and $B = b_{s,k}\|f\|_\rho$, where $b_{s,k} > 0$ is a constant, $\rho = k/(k+s)$, and $\|f\|_\rho = [\int f^\rho\,dx]^{1/\rho}$.

The k-dimensional formula shows that the dimension of the space can be calculated, at least in the limit, from the minimum distortion sequence $D_s^*(N, F)$ since

$$k = s \lim_{N\to\infty} \frac{\log(1/N)}{\log D_s^*(N, F)}. \qquad (2)$$

This latter formula suggests further generalizations because in many respects, it parallels other well-known definitions for the dimension of nonstandard sets in metric spaces. Pontrjagin and Schnizelman [12] defined the dimension of a set S in terms of the smallest number $N(\delta)$ of spheres of radius less than $\delta > 0$ that cover S. This dimension is

$$k_1 = \liminf_{\delta\to 0} \frac{\log(1/N(\delta))}{\log \delta}. \qquad (3)$$

The Hansdorff–Besicovitch dimension [12] for the set S is

$$k_2 = \lim_{\delta\to 0} \inf_{\delta_i<\delta} \sum h(\delta_i), \qquad (4)$$

where $h(\delta)$ is some suitably chosen positive increasing function such that $h(\delta)$ tends to zero with δ and the infimum is taken over sets of spheres (of radii $\delta_1, \delta_2, \cdots$) covering Mandelbrot [12] refers to sets for which k_2 is less than the topological dimension as fractals. To emphasize the analogy between k and k_2 note that (1) and (4) may be written, for the case of unit density, as

$$N^{s/k} \sum_n \int_{C_n} \|x - y_n\|^s\,dx, \qquad (1')$$

$$N^{0/k} \sum_n h\left(\int_{S_n} \|x - y_n\|^0\,dx\right). \qquad (4')$$

The set C_n is the set of signals mapped into y_n by the quantizer and these sets form a partition of the signals' range. In the other formula $\{S_n\}$ is a spherical covering of the range and, since $s = 0$, the value of y_n does not matter.

Dimensions k_1 and k_2 have been calculated for many different types of nonstandard sets [12]. One of the simplest among the nonstandard sets is the Cantor set (Section III; see also [12]) for which $k_1 = k_2 = \log 2/\log 3$. It is proved in Section III that Bennett's rule applies to the Cantor set and in this case the quantization dimension has the same value as the other dimensions.[1]

Editor's Note: The application of Bennett's rule to the Cantor set is also considered by Bucklew and Wise in "Multidimensional asymptotic quantization theory with rth power distortion measures," in this issue.

Fractals abound in the theory of stochastic processes and appear to be gaining popularity as models for natural phenomena as well [12]. It is likely that the theory of fractals can be also exploited to further extend Bennett's rule. The reason for the interest in fractals in the present context is that the quantization noise of a fractal quantized as a fractal tends to zero much more rapidly than if the same set were quantized as an "ordinary" subset of the Euclidean space containing it.

II. Summary

Let $y^{(N)} = \{y_n\}$ be a set of points and $S = \{S_n\}$ a collection of subsets in a linear metric space E. Denote the indicator function of S_n by $\varphi_n(x)$ so that $\varphi_n(x) = 1$ if $x \in S_n$, and $\varphi_n(x) = 0$ if $x \notin S_n$.

The expression

$$Qx = \sum_{n=1}^{N} y_n \varphi_n(x) \qquad (5)$$

defines a mapping of E into itself. This mapping is called a *quantizer* if S is a partition of E in the sense that $\sum \varphi_n(x) = 1$ for all $x \in E$. Note that $Qx = y_n$, if and only if $x \in S_n$, and define the *quantizer distortion function* as the minimum distance between the signal point and the quantizer output levels:

$$u(x, y^{(N)}) = \min_{y_n \in y^{(N)}} \|x - y_n\|. \qquad (6)$$

If $\varphi_n(x) = 1$ implies $\|x - y_n\| = u(x, y^{(N)})S$ is called a Voronoi or *V-partition* [6]; we call the corresponding Q a *V-quantizer*. We define a *random quantizer* as the V-quantizer induced by the random code set $Y^{(N)}$. It will be assumed that $(Y_1, \cdots, Y_N)$ are jointly independent output levels with identical and continuous probability distributions (over E).

Quantizers split random signals into an information bearing component and a noise component. Set $a_n = \Pr[\varphi_n(X) = 1]$ for the probability that the random signal X occurs in S_n. If X is quantized by Q the *mean information output* [5] and the *mean sth power distortion* are defined as

$$H(Q, F) = -\sum a_n \log a_n \qquad (7)$$

and

$$D_s(Q, F) = E[u^s(X, Y^{(N)})]. \qquad (8)$$

($H(Q, F)$ is the first order entropy of the quantized process.)

We define further the *distortion boundary function*

$$D^*(H, N, F) = \inf_{H(Q,F)\le H} u^s(X, QX), \qquad (9)$$

and set

$$\beta_s(N) = N^\beta \inf_H D_s^*(H, N, F), \qquad (10)$$

$$\gamma_s(H) = e^{\beta H} \inf_N D_s^*(H, N, F). \qquad (11)$$

The functions D^*, β, and γ set limits on optimal quantizer performance in the presence of joint constraints for both H and N, for N only, or for H only. The latter is of interest if the quantizer output must be processed through a channel of limited capacity [5].

If the quantizer is random, $N^{\beta}D_s(Q, F)$ becomes a random variable. We write the sth moment of the limiting distribution of this random sequence as α_s and the mean of the Nth random variable in the sequence as

$$\alpha_s(N) = N^{\beta}E[D_s(Q, F)]. \quad (12)$$

In the following summary of results for the k-dimensional Euclidean space, $s = 1, 2, \cdots, \beta = s/k$, $\rho = k/(s + k)$, $B_k = \pi^{k/2}/\Gamma(1 + 2/k)$ is the volume of a k-sphere with unit radius, $f(x)$ is a measurable signal density function, $g(x)$ is a measurable output level density function, $\| f \|\rho = (\int f^{\rho}\, dx)^{1/\rho}$, and $R_{s,k}$, $a_{s,k}$, etc. are constants. All integrals are assumed to be finite Lebesgue integrals.

1) $\alpha_{s,k} = R_{s,k}\int_{E_k} g(x) f^{-\beta}(x)\, dx$.
2) $\alpha_{s,k} \geq R_{s,k}\| f \|\rho$ and $\alpha_{s,k} = R_{s,k}\| f \|\rho$, if and only if $g(x) = f^{\rho}(x)/(\int f^{\rho}(x)\, dx)$.
3) If the pair of density functions satisfies a certain condition (cf. subsections III-B and III-C), then $\lim_{N\to\infty}\alpha_s(N) = \alpha_{s,k}$.
4) Under the same conditions, $\lim_{N\to\infty}\beta_s(N) = b_{s,k}\| f \|_{\rho}$.
5) Under the same conditions, $\lim_{H\to\infty}\gamma_s(H) = c_{s,k}\| f \|_{\rho}$.
6) $\rho B_k^{-\beta} \leq c_{s,k} \leq b_{s,k} \leq R_{s,k} = \Gamma(1 + \beta)B_k^{-\beta}$.
7) $\mathrm{Lim}_{k\to\infty} k^{-s/2}c_{s,k} = \lim_{k\to\infty} k^{-s/2}b_{s,k}$ $= \lim_{k\to\infty} k^{-s/2}R_{s,k} = (2\pi e)^{-s/2}$.
8) If f_k is the joint density for k consecutive terms of a nondeterministic normal stationary sequence [4], then $\lim_{k\to\infty}\| f_k \|_{\rho} = (2\pi e\sigma^2)^{s/2}$, where σ^2 is the one-step prediction error for the sequence.

For signals following the Cantor distribution it is shown that both $\alpha_s(N)$ and $\beta_s(N)$ converge when $\beta = s\log 3/\log 2$.

Most of these results were given first in [17] and [18]. The results on Cantor sets, the approach to the study of $\alpha_s(N)$ for large N, and much of the presentation are new.

III. Random Quantization

Consider a measurable partition $\{S_1, \cdots, S_N\}$ and a set of code points $y^{(N)} = \{y_1, \cdots, y_N\}$ in a measurable metric space E, define a quantizer as the function

$$Qx = y_n \qquad \text{if } x \in S_n, \quad n = 1, \cdots, N \quad (13)$$

and measure the distortion due to Q using the metric in E:

$$u(x, y^{(N)}) = \|x - Qx\|. \quad (14)$$

The distortion function is a real valued function defined on the product E^{N+1} of $N + 1$ identical copies of E. Alternatively, this function may be viewed as functions on E parameterized by the code set $y^{(N)}$ or as functions on the code set parameterized by the signal x. The global behavior of a fixed quantizer corresponds to the first view, and the local behavior of a random quantizer corresponds to the second. However, the asymptotic behavior of the distortion function is the simplest to analyze when both the signal and the code points are picked at random and the set S_n represented by code point y_n is its Voronoi region [6]:

$$S_n \equiv \{x \in E: \|x - y_n\| \leq \|x - y_m\|, m \neq n\}. \quad (15)$$

A. Distortion Limit Distributions in E^k

Assume that the code points $\boldsymbol{Y}^{(N)} = \{\boldsymbol{Y}_1, \cdots, \boldsymbol{Y}_N\}$ in k-dimensional Euclidean space E^k are independent with common distribution

$$G(\boldsymbol{x}) = \Pr[\boldsymbol{Y}_{n1} \leq \boldsymbol{x}_1, \cdots, \boldsymbol{Y}_{nk} \leq \boldsymbol{x}_k], \quad (16)$$

where

$$\boldsymbol{x} = (x_1, \cdots, x_k) \in E^k.$$

For $u \geq 0$, $\boldsymbol{x} \in E^k$, and $S(u, \boldsymbol{x}) \equiv \{\boldsymbol{y} \in E^k: \|\boldsymbol{x} - \boldsymbol{y}\| \leq u\}$ set

$$G_1(u, \boldsymbol{x}) = \int_{S(u,\boldsymbol{x})} G(dy),$$

$$G_N(u, \boldsymbol{x}) = 1 - (1 - G_1(u, \boldsymbol{x}))^N. \quad (17)$$

$G_N(u, \boldsymbol{x})$ is the probability that there is at least one code point in the sphere of radius u centered at $\boldsymbol{x}$ among the N codes $\boldsymbol{Y}^{(N)}$.

When both signal and codes are random the characteristic function (CF) of the distortion distribution is

$$\varphi_N(\lambda, F) = E\left[e^{i\lambda u(\boldsymbol{X}, \boldsymbol{Y}^{(N)})}\right] = \int_{E^k} F(d\boldsymbol{x}) \int_0^{\infty} e^{i\lambda u} G_N(du, \boldsymbol{x}). \quad (18)$$

The CF of $N^{1/k}u(\boldsymbol{X}, \boldsymbol{Y}^{(N)})$ is $\psi_N(F) = \varphi_N(N^{1/k}, F)$.

It can be shown by an appeal to the k-dimensional version of Lebesgue's lemma [13] that if $G(d\boldsymbol{y})$ is an absolutely continuous probability measure, then for almost all $\boldsymbol{x}$

$$G_1(u, \boldsymbol{x}) = B_k u^k g(\boldsymbol{x}) + o(u^k), \quad (19)$$

where $B_1 = 2$ and $B_k = \pi^{k/2}/\Gamma(1 + 2/k)$, $k = 3, 4, \cdots$ is the volume of the unit sphere in E^k.

At the Lebesgue points of $G(dy)$, i.e., at points where (19) holds

$$\lim_{N\to\infty} G_n(uN^{-1/k}, \boldsymbol{x}) = 1 - e^{-B_k u^k g(\boldsymbol{x})}, \quad (20)$$

where $g(\boldsymbol{x})$ is the code point density at $\boldsymbol{x}$.

The right side of (20) is a proper distribution in u when $g(\boldsymbol{x}) \neq 0$. Its average with respect to the signal distribution is

$$1 - \int_{E^k} e^{-B_k u^k g(\boldsymbol{x})} F(d\boldsymbol{x}), \quad (21)$$

and this is also a proper distribution in u whenever the signal distribution is absolutely continuous with respect to the code distribution and $g(\boldsymbol{x}) = 0$ implies $f(\boldsymbol{x}) = 0$ almost everywhere. (Here $f(\boldsymbol{x})$ is the signal density.) The CF and the absolute moments of the distribution defined at

(21) are

$$\psi(\lambda, F) = kB_k \int_{E^k} \int_0^\infty u^{k-1} g(x) e^{iu\lambda - B_k u^k g(x)} \, du F(dx), \tag{22}$$

$$\alpha_s = B_k^{-\beta} \Gamma(1+\beta) \int_{E^k} g(x)^{-\beta} f(x) \, dx,$$

$$\beta = s/k, \; s = 0, 1, 2, \cdots. \tag{23}$$

To prove (22) one needs to show that the CF's ψ_N of the distributions on the left side of (20) converge to $\psi(\lambda, F)$. This can be done in a straightforward manner [17] by appeals to Levy's continuity theorem [11] and to the dominated convergence theorem [13]. The details will not be reproduced here.

The moments of the limiting normalized distortion distribution, shown at (16), can be readily calculated from the derivatives of the CF at $\lambda = 0$. The details are omitted.

B. The Convergence of the Moment Sequence

It is *not true* without additional assumptions that the mean sth power absolute moments

$$\alpha_s(N) = N^\beta E\left[u^s(X, Y^{(N)})\right] \tag{24}$$

converge to α_s. For, as Stein's example shows, it may happen that $\lim_N \alpha_s(N) > \alpha_s$ although $f(x) = g(x)$. This example is reproduced in the Appendix. On the other hand,

$$\liminf_N \alpha_s(N) \geq \alpha_s \tag{25}$$

is always true. The proof, omitted here, relies on Fatou's lemma [13].

1) Conditions for Moment Convergence in the Special Case $s = k = 1$: Write $\alpha(N) = \alpha_1(N)$, $\bar{G}(u, x) = 1 - G_1(u, x)$, and integrate by parts in (24) to show that

$$\alpha(N) = N \int_{-\infty}^\infty \int_0^\infty \bar{G}^N(u, x) \, dF(x). \tag{26}$$

Set $u_n = \alpha(n+1)[(n+1)\alpha(1)]^{-1}$, and for complex z in the unit disk define

$$u(z) = \sum_0^\infty u_n z^n, \qquad |z| < 1. \tag{27}$$

For $|x| < 1$, $u(z)$ is analytic, $u(z)$ has an integral representation, and $u(z)$ may be differentiated termwise:

$$u(z) = \frac{1}{\alpha(1)} \int_{-\infty}^\infty \int_0^\infty (1 - z\bar{G}(u, x))^{-1} \bar{G}(u, x) \, du \, dF(x), \tag{28}$$

$$u'(z) = \sum_0^\infty n u_n z^{n-1}$$

$$= \frac{1}{\alpha(1)} \int_{-\infty}^\infty \int_0^\infty \left[\frac{\bar{G}(u, x)}{1 - z\bar{G}(u, x)} \right]^2 du \, dF(x). \tag{29}$$

Note that $nu_n = 1/\alpha(1)(1 - 1/(n+1))\alpha(n+1)$ and so nu_n and the normalized-mean distortion $\alpha(n)$ converge or diverge together.

According to a theorem of Hardy and Littlewood [15], if $a_n \geq 0$ for all values of n, and as $x \to 1$

$$f(x) = \sum_{n=0}^\infty a_n x^n \sim \frac{1}{1-x},$$

then as $n \to \infty$

$$s_n = \sum_{\nu=0}^n a_\nu \sim n.$$

Assume now that the residue of $u'(z)$ is some finite $-b$ at $z = 1$,

$$\lim_{z \to 1} (z-1) u'(z) = -b, \tag{30}$$

and apply the previous theorem with $a_n = nu_n \geq 0$, $f(x) = u'(x)$ and $s_n = u_1 + \cdots + nu_n$.
This shows that

$$\lim_{n \to \infty} \frac{u_1 + \cdots + nu_n}{n} = b. \tag{31}$$

According to Littlewood's theorem [15], if $f(x) = \Sigma a_n x^n \to b$ as $x \to 1$, and $a_n = o(1/n)$, then Σa_n converges to the sum b. Hence, if $f(x) = (1-x)u'(x)$, then $a_n = (u+1)u_{n+1} - nu_n$ and $(n+1)u_{n+1} = \Sigma^{n-1} a_k \to b$, if $(n+1)u_{n+1} - nu_n = u_{n+1} - n(u_n - u_{n+1}) = o(1/n)$. It follows from (26) that u_n is a decreasing sequence and, hence, because of (31), $\limsup nu_n \leq 2b$. Therefore, $a_n = o(1/n)$ will follow if $(u_n - u_{n+1}) = o(1/n^2)$.

The difference $u_n - u_{n+1}$ is the increase in expected distortion when the $(n+1)$-level random quantizer $(Y_1^{(n+1)}, \cdots, Y_{n+1}^{(n+1)})$ is collapsed into one of the n-level random quantizers $(Y_1^{(n+1)}, \cdots, Y_{k-1}^{(n+1)}, Y_{k+1}^{(n+1)}, \cdots, Y_{n+1}^{(n+1)})$, $k = 1, \cdots, n+1$, with probability $1/(n+1)$. Regardless of its realized value, the distortion of a random signal increases with probability $1/(n+1)$ when a random $(n+1)$-level quantizer is collapsed. The amount of this increase cannot be more, on the average, than the expected distortion when, for random k, $Y_k^{(n+1)}$ is coded by the remaining n points. This expected distortion is u_n. Hence, $u_n - u_{n+1} \leq u(n)/(u+1)$, and so $u(n) = o(1/n)$ implies that $u_n - u_{n+1} = o(1/n^2)$. This proves the next lemma.

Lemma 1: If $u(z)$ has a simple pole at $z = 1$ with residue $-b$, then $-b$ is the limit of the normalized distortion sequence $\alpha(n)/\alpha(1) = nu(n)$,

$$b = \lim_{n \to \infty} nu(n) = \lim_{z \to 1} (1-z) u'(z). \tag{32}$$

We evaluate $u'(z)$ by substituting $a(u, x) = (1 - G(u, x))^{-1}$ in formula (29). Write $G^{-1}(u, x)$ for the inverse of $G(u, x)$ in its first variable at fixed x. Then

$$h(a, x) = \frac{du}{da} = \frac{1}{a^2 g\left(G^{-1}\left(\left(1 - \frac{1}{a}\right), x\right), x\right)}, \tag{33}$$

where $g(x) = dG(x)/dx$ is the code density,

$$g(u, x) = \frac{dG(u, x)}{du} = g(x+u) + g(x-u), \tag{34}$$

and

$$u'(z) = \frac{1}{\alpha(1)} \int_{-\infty}^{\infty} \int_{1}^{\infty} \frac{h(a, x)}{(z-a)^2} da\, dF(x). \tag{35}$$

At x with continuous and positive $g(x)$,

$$\lim_{a \to 1} h(a, x) = \frac{1}{2g(x)}. \tag{36}$$

Consider now the test function

$$h(a) = \frac{1}{\alpha(1)} \int_{-\infty}^{\infty} f(x) \left(h(a, x) - \frac{1}{2g(x)} \right) dx, \tag{37}$$

where, as usual, f and g are signal and code densities and $h(a, x)$ was defined at (33). From (35) and (37)

$$u'(z) = \int_{1}^{\infty} \frac{h(a)}{(z-a)^2} da + \frac{1}{2\alpha(1)} \cdot \int_{-\infty}^{\infty} \int_{1}^{\infty} \frac{f(x)}{g(x)} \frac{1}{(z-a)^2} da\, dx. \tag{38}$$

Since $\int_1^\infty (z-a)^{-2}\, da = (z-1)^{-1}$, $\alpha(1)$ times the negative residue of the second term in (38) equals the first moment of the limiting normalized distortion distribution $\alpha_1 = \frac{1}{2}\int f(x)/g(x)\, dx$.

We calculate the residue of the first term at $z = 1$. Set $z = s + \sigma i$ ($i = \sqrt{-1}$). Then

$$|I(z)| = \left| (z-1) \int_{-\infty}^{\infty} \frac{h(a)}{(z-a)^2} da \right| \le \left((s-1)^2 + \sigma^2 \right)^{1/2} \int_{-\infty}^{\infty} \frac{|h(a)|}{(a-s)^2 + \sigma^2} da. \tag{39}$$

By a result from the theory of Poisson kernels [8], if $|h(a)|$ is a bounded measurable function, then

$$\lim_{\sigma \to 0} \frac{\sigma}{\pi} \int_{-\infty}^{\infty} \frac{|h(a)|}{(a-s)^2 + \sigma^2} da = h(s), \tag{40}$$

at all points of continuity of $|h(s)|$.

It is now easy to prove our main theorem.

Theorem 1: If $f(x)$ and $g(x)$ are measurable density functions such that the function $h(a)$ defined at (37) is a) bounded, b) continuous at 1, and c) $\lim_{a \to 1} h(a) = 0$, then (32) holds with $\alpha(1) b = 2^{-1} \int f(x)/g(x)\, dx$ and so $\lim_n \alpha(n) = \alpha(1)$.

Proof: We choose a path $z = s + \sigma i$ that lies entirely in the unit disk and for which $[(s-1)^2 + \sigma^2]^{1/2}/\sigma \to 1$ as $z \to 1$. Along this path formula (40) holds because of assumptions a) and b) and then $h(1) = 0$ by c). The theorem then follows from (38) and (39).

Example: For $g(x) = \lambda e^{-\lambda x}$, $x \ge 0$ the conditions of the theorem are satisfied if

$$\int_0^\infty e^{\lambda x} f(x)\, dx < \infty.$$

Note that the main theorem can be slightly strengthened. It can be shown that if f and g are a pair of measurable densities such that $h(a)$ at (37) is bounded then

$$\frac{1}{2} \int \frac{f(x)}{g(x)} dx \le \liminf_n \alpha(n) \le \limsup_n \alpha(n) \le \alpha(1) \limsup_{a \to 1} h(a). \tag{41}$$

2) Convergence of sth Moments: The special case of $s = 1$, and $k = 1$ can be generalized to sth power distortion with $s \ge 1$ in k-dimensional Euclidean space. In fact, it can be shown that if the test function in the main theorem is replaced by the generalized test function

$$h_{s,k}(a) = \int_{E^k} \left[\frac{1}{g^k\left(\left[G^{-1}\left(1 - \frac{1}{a}\right) \right]^s, \boldsymbol{x} \right)} - \frac{1}{g(\boldsymbol{x})^{s/k}} \right] d\boldsymbol{x}, \tag{42}$$

then under the conditions of this theorem $\lim_{n \to \infty} \alpha_s(n) = \alpha_s$, where α_s is the sth moment for the limiting distribution (cf. (23)).

C. Optimal Choice for the Code Point Distribution

As the number of code points n is increased indefinitely the mean sth power distortion decreases to zero with $n^{-\beta}$, $\beta = s/k$,

$$\alpha_s(n) \sim n^{-\beta} R_{k,s} \int_{E_k} f(\boldsymbol{x})/g(\boldsymbol{x})^{\beta}\, dx, \tag{43}$$

where

$$R_{k,s} = \Gamma(1 + \beta) B_k^{-\beta} \text{ (cf. (23))}.$$

The right side of (43) may be viewed as a functional in the code density and for given signal density $f(x)$ the minimum of this functional is achieved when $g(\boldsymbol{x}) = f^\rho(\boldsymbol{x})/\int f^\rho(\boldsymbol{x})\, d\boldsymbol{x}$, where $\rho = k/(k+s)$, provided that the minimum $\| f \|_\rho = [\int_{E^k} f^\rho(\boldsymbol{x})\, d\boldsymbol{x}]^{1/\rho}$ is finite. This follows from Hölder's inequality [7], $\int fh > [\int f^\rho]^{1/\rho} [\int h^{\rho'}]^{1/\rho'}$, $1/\rho + 1/\rho' = 1$, $0 < \rho < 1$, with $\rho = k/(k+s)$, $\rho' = -\beta$ and $h = g^{-\beta}$.

Theorem 2: For every code density $g(\boldsymbol{x})$

$$\liminf_n n^\beta \alpha_s(n) \ge R_{k,s} \| f \|_\rho, \tag{44}$$

except for the optimal code density $g(\boldsymbol{x}) = f^\rho(\boldsymbol{x})/\int_{E^k} f^\rho(\boldsymbol{x})\, dx$ for which

$$\lim_n n^\beta \alpha_s(n) = R_{k,s} \| f \|_\rho.$$

D. Random Qauntization in Nonstandard Sets

Perhaps the simplest nonstandard set is the Cantor set on the [0, 1] interval. (For a detailed treatment of nonstandard sets see Mandelbrot's book on fractals [12].) The Cantor set is the closed, compact subset of [0, 1] obtained as follows: remove the open middle third (1/3, 2/3) leaving [0, 1/3) and (2/3, 1]; repeat the "removal of the middle third" on the two subsets left and continue the process indefinitely. The Cantor set is the intersection of the subsets that remain. Points in the Cantor set are represented as an infinite series of binary codes $a_1, a_2, \cdots$

$$x = \sum_1^\infty a_n 3^{-n}, \tag{45}$$

with $a_n = 0$ or $a_n = 2$.

The nth removal leaves 2^n connected subsets. Locally, that is within each connected subset, the Cantor set is symmetric with respect to the subsets midpoint. Hence if the distortion function is convex, the midpoints are the codes in the optimal 2^n level quantizer. For the optimum quantizer the minimum distortion is

$$\beta_s(2^n) = \beta_s(1)3^{-3n} \tag{46}$$

and this converges if $s + 1/d > 1$ where $d = \log 2/\log 3$ is the fractal dimension [12] of the Cantor set. Substitute $N = 2^n$ in (46). Then

$$\beta_s(N) = N^{-s/d}\beta_s(1).$$

Assume now that both signal X and code Y follow the Cantor distribution and X and Y are independent. If $\{A_n\}$ and $\{B_n\}$ denote the random binary codes for X and Y (cf. (45)), then the probability that A_k and B_k differ first at n is $P_n = 2^{-n}$. If the distortion function is convex, then the conditional mean distance between X and Y given that n is the first value with $A_n \neq B_n$ equals $\partial_s\beta_s(2^n)$, where $\partial_s > 0$ is a real constant. Hence, for large M, the mean distortion of an M-level random quantizer equals approximately,

$$\alpha_s(M) \approx \partial_s\beta_s(1)\sum_N \frac{M}{N}\left(1 - \frac{1}{N}\right)^{M-1} N^{-\beta}, \tag{48}$$

where $\beta = s/d = s\log 3/\log 2$. Substitute $x = M/N$ and let $N, M \to \infty$. It then follows that

$$\lim M^\beta \alpha_s(M) = \partial_s\beta_s(1)\int_0^\infty x^\beta e^{-x}\,dx$$
$$= \partial_s\beta_s(1)\Gamma(1+\beta). \tag{49}$$

Formulas (23) and (49) are remarkably similar despite the fact that one applies to absolutely continuous distributions in k-dimensional Euclidean space and the other to a distribution with no absolutely continuous component at all. This finding suggests that the problem of quantization is likely to be closely related to the problem of fractal sets in general.

E. Bounds in E^k for Large k

If f_k is the density of a k-dimensional normal distribution with zero mean and covariance matrix $C_k = \{C_{ij}\}$ then

$$\| f_k \|_\rho = (2\pi)^{s/2}(1+\beta)^{\frac{k(1+\beta)}{2}} |\det C_k|^{\beta/2} \tag{50}$$

where, as before, $\beta = s/k$ and $\rho = k/(s+k)$. The (differential) entropy functional, defined as $H(f_k) = -\int_{E^k} f_k \log f_k\,dx$, is readily calculated as

$$H(f_k) = \frac{k}{2}\log(2\pi e)|\det C_k|^{1/k} \tag{51}$$

and hence,

$$\| f_k \|_\rho = (2\pi)^{(s/2)(1-1/k)} e^{-\beta/2}(1+\beta)^{(s+k)/2} e^{\beta H(f_k)}. \tag{52}$$

Suppose now that f_k is the kth order joint density for a block of k terms from a nondeterministic stationary normal sequence [4]. If σ_k^2 denotes the mean squared distance of the $(k+1)$st term X_{k+1} from its projection on the space spanned by $X_1, \cdots, X_k$, then [4] $\sigma_k^2 = |\det C_k|$ and

$$\lim_k \sigma_k^2 = \sigma^2 = 2\pi\exp\left\{\frac{1}{2\pi}\int_{-\pi}^{\pi}\log f(\lambda)\,d\lambda\right\}, \tag{53}$$

where $f(\lambda)$ is the spectral density of the sequence and σ^2 is the "one-step prediction error." If $\sigma_k^2 \to \sigma^2$ then also $(\sigma_1^2 + \cdots + \sigma_k^2)/k \to \alpha$ and then $|\det C_k| = (|\det C_k| / |\det C_{k-1}|) \cdots (|\det C_2|/|\det C_1|)|\det C_1|$ implies that $|\det C_k|^{1/k} \to \sigma^2$ as $k \to \infty$.

Lemma 2: If f_k, $k = 1, 2, \cdots$ is the sequence of joint densities for a nondeterministic normal stationary sequence, then

a) $\lim_{k\to\infty} \bar{H}(f_k) \equiv \lim_{k\to\infty} \frac{1}{k}H(f_k) = \frac{1}{2}\log(2\pi e\sigma^2)$,

b) $\lim_{k\to\infty} \| f_k \|_\rho = (2\pi e\sigma^2)^{s/2}$,

c) $\lim_{k\to\infty} \| f_k \|_\rho = \exp\left\{s \lim_{k\to\infty}\bar{H}(f_k)\right\}$.

Proof: Substitute $\sigma^2 = \lim|\det C_k|^{1/k}$ at (50)–(52).

In terms of the Γ function the geometric quantization constant $R_{s,k}$ defined at (43) is $R_{k,s} = \pi^{-s/2}\Gamma(1+\beta)\Gamma(1+2/k)^\beta$. Using Stirling's formula,

$$\lim_{k\to\infty} k^{-s/2}R_{k,s} = (2\pi e)^{-s/2}. \tag{54}$$

The mean squared distortion ($s = 2$) per sample in a block of length k can be calculated from (54) and b) of the last lemma, and

$$\lim_{k\to\infty}\frac{1}{k}R_{k,s}\| f_k \|_\rho = \sigma^2. \tag{55}$$

If $\lim_{k\to\infty} k|\det C_k|^{1/k} = \gamma^2 < \infty$, then

$$\lim_{k\to\infty} R_{k,s}\| f_k \|_\rho = \gamma^s. \tag{56}$$

If the terms in the sequence $X_1, X_2, \cdots$, are independent and X_k has mean zero and variance λ_k^2, then $|\det C_k|^{1/k} = (\Pi_1^k \lambda_j)^{2/k}$ and the condition for (56) is that $\lim_{k\to\infty} k(\Pi_1^k\lambda_j)^{2/k} = \gamma^2 < \infty$. For example, $\log\lambda_k = -\log k/2k(k-1) + \log(1-1/k)/2(k-1)$ works with $\gamma = 1$.

For independent but not necessarily normal variables, c) of the lemma becomes

$$\lim_{k\to\infty} \| f_k \|_\rho = e^{sH(f)}, \tag{57}$$

where $f_k = (f)^k$ and $H(f) = -\int_{-\infty}^{\infty} f(x)\log f(x)\,dx$. This result is [7, theorem 187].

The conjecture [27] that c) of the lemma applies to stationary sequences in general without the assumption that the sequence is Gaussian [4] was proved in [6].

F. A General Representation for Mean Distortion

If $u_n = E[X^n]$, $n \equiv 1, 2, \cdots$, is the moment sequence of the positive random variable X, then $u_{n+1} - 2\lambda u_n + \lambda^2 u_{n-1} = E[(\lambda - X)^2 X^{n-1}] \geq 0$ and therefore,

$$u_{n+1}u_{n-1} > u_n^2. \tag{58}$$

If, in addition, the range of X is confined to the interval [0, 1], then u_n is a decreasing sequence. Kendall [9], [10] shows that if a decreasing sequence $u_n \geq u_{n+1} \geq \cdots$ satisfies (58) then for some nonnegative real λ's satisfying $\Sigma\lambda_j < \infty$ and $\Sigma j\lambda_j = \infty$,

$$u_n = \exp\left\{-\sum^{n} j\lambda_j - n\sum_{j>n}\lambda j\right\}. \tag{59}$$

The mean distortion u_n defined at (26) satisfies (58) and is decreasing, and so it can be represented as in (59). Kendall [9], [10] derived the representation at (59) for the purpose of characterizing infinitely divisible renewal sequences. It is possible that the theory of such sequences may have further applications to the problems of quantization.

IV. Optimal Quantization

If Q^* is an N-level quantizer and $u_*^s(x) = \|x - Q^*x\|^s$ has the property that

$$E\left[u_*^s(x)\right] = \inf_Q E\|X - QX\|^2, \tag{60}$$

where the infimum is taken over the set of all N-level quantizers Q satisfying (13), then Q^* is said to be optimal. For a given signal distribution and a specified nonnegative s the existence of at least one optimal quantizer is assured wherever $E\|X - y\|^s < \infty$, for some y. This observation follows from the usual compactness argument. However, Q^* need not be unique, for if the signal distribution is spherically symmetric, then the property of optimality is retained when a quantizer is rotated.

A. Self-Similar Signal Sets

Following Mandelbrot [12] we call a subset S of a linear space E self-similar if S is the set-theoretic union of a finite collection of its own disjoint translated contractions. If $x \in E$, $S \subset E$, and $0 < t \leq 1$, then $x + tS$ is a typical translated contraction of S and self-similarity calls for the existence of I points $x_1, \cdots, x_I$ and I real numbers $t_1, \cdots, t_I \in [0, 1]$ such that

$$S = \bigcup_1^I (x_i + t_i S),$$

$$x_i + t_i S \neq x_j + t_j S, \qquad \text{if } i \neq j.$$

I is called the rank of self-similarity.

Examples: 1) The unit cube C_1 without its outer boundary is self-similar in E^k. $I = m^k$ is a possible rank for every integer $m > 0$ and then $t_i = m^{-1}$ for every i. 2) The Cantor set (cf. (16)) is self-similar. $I = 2^m$ is a possible rank and $t_i = 3^{-m}$.

Lemma 3: Consider a set S, a sequence of set functions $D(N, \cdot)$ defined on subsets of S and an unbounded sequence of integers $1 = I_1 \leq I_2 \leq I_3 \leq \cdots$ and assume the following.

1) For every m, I_m is a rank of self-similarity for S,

$$S = \bigcup_{i=1}^{I_m} (x_{mi} + t_{mi}S),$$

and $t_{mi} = t_m$, $i = 1, \cdots, I_m$, $m = 1, 2, \cdots$.

2) $D(N, S)$ is a nonincreasing sequence, $D(1, S) \geq D(2, S) \geq \cdots$.

3) $D(N, S)$ is homogeneous and translation invariant in the sense that if $x + tS \subseteq S$ then $D(N, x + tS) = c(t)D(N, S)$.

4) $D(N, S)$ is subadditive in the sense that if $S = \cup S_i$ and $N \geq \Sigma N_i$, then $D(N, S) \leq \Sigma D(N, S)$.

5) For some $\beta > 0$, $\lim_m I_m I_{m+1}^\beta c(t_m) = 1$.

Under these assumptions $\lim_{M\to\infty} M^\beta D(M, S)$ exists and is finite.

Proof: Fix $0 < N \leq M$ and choose m so that $I_m N \leq M < I_{m+1}N$. For this m

$$D(M, S) \leq D(I_m N, S) = D\left(I_m N, \bigcup_1^{I_m} (x_{mi} + t_m S)\right)$$
$$\leq I_m c(t_m) D(N, S) \tag{61}$$

by conditions 2), 1), 3), and 4) of Lemma 3. It then follows from $M < I_{m+1}N$ that

$$M^\beta D(M, S) < I_m I_{m+1}^\beta c(t_m) N^\beta D(N, S). \tag{62}$$

Let $m, M \to \infty$, and use condition 5) of Lemma 3 to show that

$$\limsup_M M^\beta D(M, S) \leq N^\beta D(N, S). \tag{63}$$

Taking $\lim_N \inf$ on both sides of (63) completes the proof of the lemma.

For a signal which is uniformly distributed over the unit cube C_1 in E^k, we write the normalized minimum mean sth power distortion as

$$\beta_{s,k}(N) = N^\beta \inf_Q \int_{C_1} \|x - Qx\|^s \, dx. \tag{64}$$

Lemma 4: The limit

$$\beta_{s,k} = \lim_N \beta_{s,k}(N) \tag{65}$$

exists and is finite for $\beta = s/k$.

Proof: We set

$$D^*(N, S) = \inf_Q \int_S \|x - Qx\|^s \, dx \tag{66}$$

and note that by Example 1 $D^*(N, S)$ satisfies condition 1) of Lemma 3 for $I_m = m^k$ and $t_m = 1/m$. Condition 2) is obvious for D^*. Condition 3) is satisfied for $c(t) = t^{s+k}$. Condition 4) follows upon noting that the optimal placement of levels in subsets S_i of S may yield a suboptimal placement of the ΣN_i levels in C_1. As for condition 5),

$$\lim_{m \to \infty} m^k (m+1)^s m^{-s-k} = 1. \tag{67}$$

Hence the conclusion of Lemma 4, that is, (65) is true.

The corresponding result for the Cantor set was already proved in Section III-D.

B. Bounds for $b_{s,k}$

For the uniform distribution on the unit cube $\| p \|_\rho = 1$, and so the random quantization coefficient $R_{s,k}$ defined at (43) furnishes an upper bound for the optimal quantization coefficient, $b_{s,k} \le R_{s,k}$. We now derive a lower bound for $b_{s,k}$.

If $s > 0$, then a^s is strictly increasing in a. For an arbitrary measurable subset S of E^k and any point y in E^k, it then follows that

$$\int_S \| \boldsymbol{x} - \boldsymbol{y} \|^s \, dx > r_{s,k} m(S)^{1/\rho}, \tag{68}$$

unless S is the unit ball with volume B_k. In (68) $\rho = k/(k+s)$, $\beta = s/k$, $r_{s,k} = \rho B_k^{-\beta}$, and $m(s)$ is the volume of the set S.

Apply now Hölder's inequality

$$\sum a_j b_j > \left(\sum a_j^x \right)^{1/x} \left(\sum b_j^{x'} \right)^{1/x'}, \tag{69}$$

with $x = \rho$, $x' = -1/\beta$, $a_j = m(S_j)^{1/\rho}$, and $b_j = 1$, $j = 1, \cdots, N$, for the sets in a partition $\{S_j\}$ of the set S in formula (68); then

$$\sum_{j=1}^{N} \int_{S_j} \| x - y_j \|^s \, dx \ge r_{s,k} \sum_{j=1}^{N} m(S_j)^{1/\rho}$$

$$\ge N^{-\beta} r_{s,k} \left(\sum_{j=1}^{N} m(S_j) \right)^{1/\rho}$$

$$= N^{-\beta} r_{s,k} m(S)^{1/\rho}. \tag{70}$$

This proves the next lemma.

Lemma 5: For $s > 0$ and $k = 1, 2, \cdots$,

$$\rho B_k^{-\beta} = r_{s,k} \le b_{s,k} \le R_{s,k} = \Gamma(1+\beta) B_k^{-\beta}. \tag{71}$$

C. Optimal Quantization in E^k

Given I disjoint cubes C_i in E^k and I nonnegative numbers f_i the function

$$f(x) = f_i \qquad \text{if } x \in C_i \tag{72}$$

is a step signal density if in addition $\Sigma f_i m(S_i) = 1$.

For an arbitrary signal distribution F we write the normalized minimum-mean sth power distortion as

$$\beta_{s,k}(N, F) = N^\beta E\left[u_*^s(x) \right]. \tag{73}$$

Lemma 6: If f is a step signal density in E^k then

$$\lim_{N \to \infty} N^\beta \beta_{s,k}(N, f) = b_{s,k} \| f \|_\rho. \tag{74}$$

Proof: Let 1_j and $M(C_j) = 1_j^k$ be the side length and the volume of cube C_j and place $N_j = N g_j$, $\Sigma g_j = 1$, code points optimally in C_j. We observe that if C is a cube of side length 1, then $N^\beta D^*(N, C) = 1^{(s+k)} \beta_{s,k}(N)$ follows from (66) and (65).

Since the optimal placement of N_j code points in C_j may be suboptimal for f,

$$\beta_{s,k}(N, f) \le \sum g_j^{-\beta} 1_j^{(s+k)} f_j \beta_{s,k}(g_j N). \tag{75}$$

Set

$$g_j = f_j^\rho m(C_j) / \sum f_j^\rho m(C_j) \tag{76}$$

and evaluate the right side of (75) as $N \to \infty$. This shows that

$$\limsup_N \beta_{s,k}(N) \le b_{s,k} \| f \|_\rho. \tag{77}$$

The proof of the reverse inequality $\lim_N \inf \beta_{s,k}(N) \le b_{s,k} \| f \|_\rho$ involves a tedious but routine analysis of what happens at cube boundaries. Omitting these details we consider the proof is completed. (See [17].)

Relation (60) describes the rate at which the distortion of a step signal density approaches zero as $N \to \infty$. The next lemma extends the range of validity for (60).

Lemma 7: We suppose that a sequence of measurable functions f_n has the following properties.

1) The sequence is monotonically increasing, $0 \le f_1(\boldsymbol{x}) \le f_2(\boldsymbol{x}) \le \cdots$, for all $\boldsymbol{x} \in E^k$.
2) $\text{Lim}_n \int_{E^k} f_n^\rho \, d\boldsymbol{x} = \int_{E^k} f^\rho \, d\boldsymbol{x}$.
3) $\text{Lim}_{n \to \infty} \| f - f_n \|_\rho = 0$.
4) $\text{Lim}_{N \to \infty} \beta_{s,k}(N, f_n) = b_{s,k} \| f_n \|_\rho$.
5) $\text{Lim}_{N \to \infty} \alpha_{s,k}(N, f - f_n) = R_{s,k} \| f - f \|_\rho$, where $\alpha_{s,k}(N, f)$ is used to indicate the minimum of the normalized mean random sth power distortion (cf. (24)).

Then

$$\lim_{N \to \infty} \beta_{s,k}(N, f) = b_{s,k} \| f \|_\rho. \tag{78}$$

Remark: This lemma asserts that if asymptotic random quantizability is preserved by monotonic convergence, then asymptotic optimal quantizability is also preserved.

Proof:

$$\text{a)} \quad \liminf_N \beta_{s,k}(N, f) \ge b_{s,k} \| f \|_\rho.$$

It follows from $\beta_{s,k}(N, f_n) \le \beta_{s,k}(N, f_{n+1}) \le \cdots \beta_{s,k}(N, f)$ that

$$\liminf_N \beta_{s,k}(N, f) \ge \liminf_N \beta_{s,k}(N, f_n)$$

$$= \beta_{s,k} \| f_n \|_\rho. \tag{79}$$

Let $n \to \infty$ and use condition 2) to complete the proof of a).

b) $\limsup_N \beta_{s,k}(N, f) \le b_{s,k} \| f \|_\rho$.

If $N = N_1 + N_2$ then $f = f_n + (f - f_n)$ implies that $N^\beta D^*(N, f) \le N^\beta D^*(N_1, f_n) + N^\beta D^*(N_2, f - f_n)$. Set $N/N_1 = 1/(1-\epsilon)$ and $N/N_2 = 1/\epsilon$ for some $0 < \epsilon < 1$, let $N \to \infty$, and use conditions 4) and 5) to show

$$\limsup_N N^\beta D^*(N, f) \le \left(\frac{1}{1-\epsilon}\right)^\beta b_{s,k} \| f_n \|_\rho + \left(\frac{1}{\epsilon}\right)^\beta R_{s,k} \| f - f_n \|_\rho . \quad (80)$$

Let $n \to \infty$ and use condition 3) and then let $\varepsilon \to 0$ to show b). This completes the proof of the lemma.

Lemma 7 may be modified and applied to decreasing sequences. The only difference in the proof is that the random upper bound is replaced by the lower bound discussed in Section IV-B.

If $\lim N^\beta \alpha_{s,k}(N, f) = R_{s,k} \| f \|_\rho$ we call f R-quantizable, if $\lim N^\beta \beta_{s,k}(N, f) = b_{s,k} \| f \|_\rho$ we call f O-quantizable.

Theorem 3: Every R-quantizable function is O-quantizable.

Proof: By Lemma 6 step-signals are O-quantizable. If f is an R-quantizable limit of a monotone O-quantizable sequence then by Lemma 7 f is also O-quantizable. Hence, the class of step signals can be extended step-by-step to the class of continuous functions, then to the class of U-functions [13], and finally to the class of measurable functions without loss of O-quantizability. This proves the theorem.

It is open question whether Theorem 3 can be inverted. In other words, is it true that every O-quantizable function is R-quantizable as well?

It has been repeatedly observed [2], [3], [6], [16]–[18] that for large k the use of a specific random quantizer in place of the hard to calculate optimal quantizer may be justified because the ratio $b_{s,k}/R_{s,k}$ is close to 1. A full justification for this approach would require a study of the distribution of the mean distortion due to random quantizers (see discussion following (14). So far only the expected value of this distribution has been investigated. In particular, it is not known whether the range of the mean distortion extends to $\beta_{s,k}(N, f)$ for large N or whether the lower tail thins out so rapidly that, for all practical purposes, $\beta_{s,k}(N, f)$ cannot be reached even by the best among many random quantizers.

D. Quantization Dimension

It has been shown, at least for a large class of measurable signal densities in E^k and for the Cantor distribution (cf. (47)), that for some constant C

$$\lim_{n \to \infty} N^{s/d} D^*_{s,d}(N, f) = C > 0, \quad (81)$$

where d is some measure of the underlying space's dimension. Equation (81) may be solved for d

$$d = -s \lim_N \frac{\log N}{\log D^*_s(N, f)} . \quad (82)$$

This suggests the following definition for the dimension d of a measurable set located in a linear measurable metric space. Let f_s be the characteristic function of the set S. Recall that $D^*_s(N, F) = \inf_Q \int \| s - Qx \|^s \, dF$ and calculate $d_s(S)$ from (82) with $f = f_s$. One may call $d_s(S)$ the sth power quantization dimension of the set S. For many sets in E^k the quantization dimension is independent of s and is the same as the Euclidean dimension. For the Cantor set $d_s(S)$ equals the fractal dimension [12]. It is probable that in many other cases the quantization and fractal dimensions also coincide.

V. Optimal Quantization with Constrained Information Output

When data are put through a quantizer Q the result is information in the form of Qx and noise in the form of $x - Qx$. If the data follows a statistical distribution, say F, the information output [5] of the quantizer Q is

$$H_k(N, F) = -\sum_1^N F(S_n) \log F(S_n) \quad (83)$$

per input sample (cf. (13) for notation). The mean distortion associated with Q is

$$D_{s,k}(N, F) = \mathbf{E} \| X - QX \|^s . \quad (84)$$

The coding theorem [5] assures us, roughly speaking, that the information output of Q can be processed through a channel with no appreciable loss of information if the channel's capacity exceeds the information output of Q. Thus, it is of interest to determine the minimum of the distortion subject to an information output constraint.

For each N and F we may regard (83) and (84) as a mapping from the space of quantizers into a point in the (H, D) plane. Of special interest among those points in this plane that correspond to quantizers are the ones with minimal distortion. In terms of the distortion boundary function

$$D^*_{s,k}(H, N, F) = \inf_{H_k(N, F) \le H} D_{s,k}(N, F), \quad (85)$$

these correspond to (H, D^*), where D^* is short hand for the left side of (85) and N is fixed. D^* is the smallest distortion when both the number of quantizer levels, and the information output are constrained. The behavior of $D^*_{s,k}(N, F) = \inf_H D^*_{s,k}(H, N, F)$ was investigated in detail for large N in Section IV. We now turn to the asymptotic behavior of the minimum distortion subject to constrained information output

$$D^*_{s,k}(H, \cdots, F) = \inf_N D^*_{s,k}(H, N, F), \quad (86)$$

for $H \to \infty$.

We say that F as OI-quantizable if

$$\lim_{H\to\infty} e^{sH} D^*_{s,k}(H,\cdots,F) = c_{s,k} \| f \|_\rho, \tag{87}$$

with $c_{s,k} \geq 0$.

Theorem 4: If F is R-quantizable, then F is also OI-quantizable in the sense that (87) holds. For $s > 0$ and $k = 1, 2, \cdots$

$$r_{s,k} \leq c_{s,k} \leq b_{s,k}. \tag{88}$$

Proof: The proof of the first assertion is similar to the proof of Theorem 3 and will not be given in detail.

a) $c_{s,k} \geq r_{s,k}$.

If $H = -\Sigma a_i \log a_i$, where $a_i \geq 0$ and $\Sigma a_i = 1$, then for $\beta > 0$

$$a_i^{1+\beta} \geq e^{-\beta H}. \tag{89}$$

(Equation (89) follows from the theory of general means for the function pair $\psi = -\log x$ and $\chi = x^\beta$, [7]). Consider a partition $\{S_i\}$ of the unit cube and set $a_i = m(S_i)$ in (70). Then

$$\sum_{j=1}^{N} \int_{S_j} \| y_j - x \|^2 \, dx \geq r_{s,k} \sum_{j=1}^{N} m(S_j)^{1/\rho} \tag{90}$$

$$\geq r_{s,k} e^{-\beta}.$$

Multiply (90) by $e^{\beta H}$, let $H \to \infty$, use (87) and note that for the uniform density on C we have $\| f \|_\rho = 1$. This proves a).

b) $c_{s,k} \leq b_{s,k}$.

The set of quantizers with N levels is included in the set of quantizers with $H_k \leq \log N$. b) then follows from (87) and the definition of $b_{s,k}$.

General conditions for $c_{s,k} = b_{s,k}$ are unavailable. All we know is that

$$\lim_{k\to\infty} k^{-s/2} c_{s,k} = \lim_{k\to\infty} k^{-s/2} b_{s,k} = (2\pi e)^{-s/2} \tag{91}$$

due to (54) and Theorem 3.

It appears likely that the use of constraints of the type $C_1 \log N + C_2 H$ would help unify Theorems 3 and 4.

Given the distortion boundary function $D^*_{s,k}(H, N, F)$ for a signal distribution F it would be of some interest, at least from a mathematical point of view, to be able to characterize the class of signal distributions that have the same distortion boundary functions as F.

Acknowledgment

The author is indebted to Dr. M. Karnaugh, formerly of Bell Telephone Laboratories for originally proposing this avenue of research, to Prof. H. Solomon and Prof. C. Stein of Stanford University for many helpful suggestions, and to Ms. K. Whibley for typing.

Appendix

The following example shows that $\lim_{N\to\infty} \alpha_s(N) > \alpha_s$ can, in fact, happen. This example is due to Professor C. Stein [17].

Let $p(x) = g(x)$ vanish outside the interval $[0, 1]$. Then writing $m(x) = \max[x, (1-x)]$, we get by integration by parts

$$\alpha_2(N) = N^2 \int_0^1 p(x)\, dx \int_0^{m(x)} u^2 \, d\left[1 - (1 - G(x,u))^N\right]$$

$$= N^2 \int_0^1 p(x) m^2(x)\, dx$$

$$-2N^2 \int_0^1 p(x) \int_0^{m(x)} u\left[1 - (1 - G(u,x))^N\right]$$

$$= 2N^2 \int_0^1 p(x)\, dx \int_0^{m(x)} u(1 - G(u,x))^N \, du$$

$$\geq 2N^2 \int_0^{1/2} p(x)\, dx \int_0^{1-x} u(1 - G(u,x))^N \, du. \tag{A1}$$

Let now $p(x) = g(x) = 1$, for $x \in [1/(2^n + 1), 1/2^n]$, $n = 1, 2, \cdots$, and $p(x) = g(x) = a > 0$, for $x > \frac{1}{2}$, where a is chosen so that $\int_0^1 p(x)\, dx = 1$. We shall show that there is a fixed $\epsilon > 0$ with the property that for all $\delta > 0$, sufficiently small, there is a fixed integer $N(\delta)$ such that

$$I_N(\delta) = 2N^2 \int_0^\delta p(x)\, dx \int_0^{m(x)} u(1 - G(u,x))^N \, du$$

$$= N^2 \int_0^\delta p(x)\, dx \int_0^{m(x)} u^2 \, dF_n(u,x) \geq \epsilon, \tag{A2}$$

if $N \geq N(\delta)$. Since it follows from (25) that for all $\delta' > 0$

$$\liminf_N N^2 \int_{\delta'}^1 p(x)\, dx \int_0^{m(x)} u^2 \, dF_N(u,x) \geq \frac{1}{2} \int_{\delta'}^1 \frac{p(x)}{g(x)^2}\, dx, \tag{A3}$$

the assumption $\lim_N \alpha_2(N) = \alpha_2$ would imply

$$\limsup_N I_N(\delta) \leq \frac{1}{2} \int_0^{\delta'} \frac{p(x)}{g(x)^2}\, dx, \tag{A4}$$

which contradicts (25) if δ' is so small that $1/2 \int_0^{\delta'} p(x)/g(x)^2 \, dx < \epsilon/2$.

By the definition of $g(x)$ (writing $C_0, C_1, \cdots$, for constants independent of x, u, and N) we have

$$G(u,x) \leq 2 \sum_{1/(2^n+1) \leq (u+x)} 2^{-2n} \tag{A5}$$

$$\leq \sum_{1/2^n \leq 2(u+x)} 2^{-2n}$$

$$\leq C_0(x+u) \leq C_1(x^2 + u^2),$$

if $u + x \leq \frac{1}{2}$, $(C_0 = 32/3,\ C_1 = 64/3)$. Hence,

$$(1 - G(u,x))^N \geq e^{N \log(1 - G(u,x))} \tag{A6}$$

$$\geq e^{-NG(u,x)}$$

$$\geq e^{-NC_1(x^2+u^2)}.$$

Thus, for $\delta < 1/4$ we have, substituting $v = Nu^2$, that

$$I_N(\delta) \geq 2N^2 \int_0^\delta p(x) e^{-NC_1 x^2}\, dx \int_0^{1/4} u e^{-NC_1 u^2}\, du \tag{A7}$$

$$= N \int_0^\delta p(x) e^{-NC_1 x^2}\, dx \int_0^{N/16} e^{-C_1 v}\, dv.$$

Let $N(\delta) = \delta^{-2}$, then for $N \geq N(\delta)$, $N^{-1/2} \leq \delta$ and so

$$I_N(\delta) \geq C_2 N \int_0^{N^{-1/2}} p(x) e^{-NC_1 x^2}\, dx \tag{A8}$$

$$\geq C_2 e^{-C_1} \sum_{1/(2^n+1) \leq N^{-1/2}} \frac{1}{2^n(2^n+1)}$$

$$\geq C_3 e^{-C_1} N \sum_{\frac{1}{2^n} \geq 2N^{-1/2}} 2^{1/2n}$$

$$\geq C_3 e^{-C_1} N \left(\frac{2}{N^{-1/2}} \right)^2 = 4C_{3e}^{-C_1}.$$

This completes the proof that $\lim_N \alpha_2(N) = \alpha_2$ is impossible in this case.

References

[1] W. R. Bennett, "Spectra of quantized signals," *Bell Syst. Tech. J.*, vol. 27, 446–472, 1948.

[2] J. A. Bucklew, "Upper bounds to the asymptotic performance of block quantizers," *IEEE Trans. Inform. Theory*, vol. IT-27, pp. 577–580, Sept. 1981.

[3] J. A. Bucklew and G. L. Wise, "A note on multidimensional asymptotic quantization theory," preprint.

[4] J. L. Doob, *Stochastic Processes*. New York: Wiley, 1953.

[5] A. Feinstein, *Foundations of Information Theory*. New York: McGraw-Hill, 1958.

[6] A. Gersho, "Asymptotically optimal block quantization," *IEEE Trans. Inform. Theory*, IT-25, pp. 373–380, July 1979.

[7] G. H. Hardy, J. E. Littlewood, and G. Polya, *Inequalities*. Cambridge: University Press, 1959.

[8] E. Hille, *Analytic Function Theory, Vol. II*. Boston: Ginn, 1962.

[9] D. G. Kendall and E. F. Harding, *Stochastic Analysis*. London: Wiley, 1973.

[10] J. F. C. Kingman, *Regenerative Phenomena*. London: Wiley, 1972.

[11] E. Lukacs, *Characteristic Functions*. New York: Haffner, 1960.

[12] B. B. Mandelbrot, *Fractals, Form, Chance, and Dimension*. San Francisco: Freeman, 1977.

[13] E. J. McShane, *Integration*. Princeton, NJ: Princeton University, 1944.

[14] D. J. Newman, "The hexagon theorem," in this issue on pp. 137–139.

[15] E. C. Titchmarsh, *The Theory of Functions*. Oxford: Oxford University, 1960.

[16] Y. Yamada, S. Tazaki, and R. M. Gray, "Asymptotic performance of block quantizers with difference distortion measures," *IEEE Trans. Inform. Theory*, vol. IT-26, pp. 6–14, Jan. 1980.

[17] P. Zador, "Development and evaluation of procedures for quantizing multivariate distributions," Ph.D. dissertation, Stanford Univ., Stanford, CA, 1963.

[18] ——, "Topics in the asymptotic quantization of continuous random variables," unpublished memorandum, Bell Laboratories, Murray Hill, NJ, Feb. 1966.

Reprinted from *IEEE Trans. Inform. Theory* **IT-25**(4):373–380 (1979)

Asymptotically Optimal Block Quantization

ALLEN GERSHO, SENIOR MEMBER, IEEE

***Abstract*—In 1948 W. R. Bennett used a companding model for nonuniform quantization and proposed the formula**

$$D = \frac{1}{12N^2}\int p(x)[E'(x)]^{-2}\,dx$$

for the mean-square quantizing error where N is the number of levels, $p(x)$ is the probability density of the input, and $E'(x)$ is the slope of the compressor curve. The formula, an approximation based on the assumption that the number of levels is large and overload distortion is negligible, is a useful tool for analytical studies of quantization. This paper gives a heuristic argument generalizing Bennett's formula to block quantization where a vector of random variables is quantized. The approach is again based on the asymptotic situation where N, the number of quantized output vectors, is very large. Using the resulting heuristic formula, an optimization is performed leading to an expression for the minimum quantizing noise attainable for any block quantizer of a given block size k. The results are consistent with Zador's results and specialize to known results for the one- and two-dimensional cases and for the case of infinite block length ($k \to \infty$). The same heuristic approach also gives an alternate derivation of a bound of Elias for multidimensional quantization. Our approach leads to a rigorous method for obtaining upper bounds on the minimum distortion for block quantizers. In particular, for $k=3$ we give a tight upper bound that may in fact be exact. The idea of representing a block quantizer by a block "compressor" mapping followed with an optimal quantizer for uniformly distributed random vectors is also explored. It is not always possible to represent an optimal quantizer with this block companding model.

I. Introduction

DIGITAL CODING of analog sources is today a subject of considerable importance, yet very little is understood about optimal block quantization. On the one hand, extensive results are available for the one-dimensional (or zero-memory) quantizer. On the other hand, useful bounds are available in the limiting case where the block length approaches infinity. What is needed is a theory of quantization for finite block lengths of arbitrary size. In this note an attempt is made to apply some of the appealing features of the one-dimensional theory to the study of block quantization. A heuristically derived formula is found for the asymptotic case of high-quality quantization. This formula specializes to known results for the one- and two-dimensional cases and for the limiting case of infinite block length.

Manuscript received March 3, 1978; revised January 9, 1979. This work was supported in part by the Electronics Program of the Office of Naval Research while the author was visiting the Department of System Science at the University of California, Los Angeles. This paper was presented at the Information Theory International Symposium, Cornell University, Ithaca, NY, October 10–14, 1977.
The author is with Bell Laboratories, Murray Hill, NJ 07974.

II. Formulation

Let $\boldsymbol{X}$ be a k dimensional random vector with joint density $p(\boldsymbol{x}) = p(x_1, x_2, \cdots, x_k)$. An N point "block" quantizer is a function $\boldsymbol{Q}(\boldsymbol{x})$ which maps $\boldsymbol{x}$ in R_k into one of N output vectors or "output points" $y_1, y_2, \cdots, y_N$ each in R_k. The quantizer is specified by the values of the output points and by a partition of the space R_k into N disjoint and exhaustive regions $S_1, S_2, \cdots, S_N$, where $S_i = \boldsymbol{Q}^{-1}(y_i) \subset R_k$. Then

$$\boldsymbol{Q}(\boldsymbol{x}) = y_i, \qquad \text{if } \boldsymbol{x} \text{ is in } S_i$$

for $i = 1, 2, \cdots, N$. The term "block" quantizer is used to indicate that the quantizer operates on a "block" of k random variables, i.e., a k-dimensional random vector.

The performance of such a quantizer can be measured by the distortion:

$$D = \frac{1}{k} E\|\boldsymbol{X} - \boldsymbol{Q}(\boldsymbol{X})\|^r$$

where $\|\cdot\|$ denotes the usual l_2 norm. We assume that $E\|\boldsymbol{X}\|^r$ is finite. Note that for $r=2$, D is the familiar mean-square "per-letter" distortion measure and for $k=1$ it is the usual rth absolute moment of the quantizing error.

We wish to determine a) the minimum distortion $D_1(N)$ attainable over the set of all N point quantizers and b) the minimum distortion $D_2(H_Q)$ attainable over the set of all quantizers having a fixed output entropy H_Q where

$$H_Q = -\sum_1^N p_i \log p_i \qquad p_i = P\{\boldsymbol{X} \in S_i\}.$$

We consider only the asymptotic case of high quality quantization where N is very large in problem a) or H_Q is very large in problem b).

III. Previous Work

In 1948 W. R. Bennett [1] modeled the one-dimensional quantizer using a memoryless monotonically increasing nonlinearity $E(x)$ (called the compressor) followed by a uniform N point quantizer. This model is completely general since any finite partition of the real line into intervals can be obtained in this way using a suitable continuous compressor curve. He showed that the distortion could be approximated by the integral

$$D \cong \frac{1}{12N^2}\int_{L_1}^{L_2} \frac{p(x)}{[\lambda(x)]^2}\,dx \qquad (1)$$

where $\lambda(x) = E'(x)/(L_2 - L_1)$. The result assumes that the $N-2$ finite regions S cover the interval (L_1, L_2) and that L_1 and L_2 are appropriately chosen so that the contribution to the distortion due to the tail or "overload" regions can be neglected. The integral is based on some implicit regularity conditions on the density $p(x)$ and on the assumption that N is very large. Bennett's formula is a convenient analytical tool for optimization studies of one dimension quantization.

Several authors have pursued the problem of minimizing the distortion in one-dimensional quantization. Panter and Dite [2] found an expression for the minimum mean-square distortion ($r=2$) for large N. Lloyd [3] found optimality conditions and an algorithmic approach for finding the optimum quantizer valid for each N. Smith [4] was the first to use Bennett's formula to find the best compressor curve and the minimum distortion for optimum quantization for large N. Algazi [5] used the rth power distortion measure and showed that for large N the minimum distortion is

$$D_1(N)=\frac{1}{r+1}(2N)^{-r}\|p_1(x_1)\|_{1/(1+r)} \tag{2}$$

where $p_1(x_1)$ is the (one-dimensional) density of the random variable X_1 and

$$\|p(x)\|_\alpha=\left[\int [p(x)]^\alpha\, dx\right]^{1/\alpha}. \tag{3}$$

For the fixed entropy constraint, Gish and Pierce [6] solved the one-dimensional problem. Their result can be expressed as

$$D_2(H_Q)=\frac{1}{r+1}2^{-r}e^{-r[H_Q-H(p_1)]} \tag{4}$$

where $H(p_1)$ is the differential entropy of X_1, $H(p_1)=-\int p_1(x_1)\log p_1(x_1)\,dx_1$. Both (2) and (4) are valid only as asymptotic results for high quality quantization.

Extensions to block quantization have been studied by Zador [7], Schutzenberger [8], and Elias [9]. Schutzenberger derived an inequality bounding the distortion D for given H_Q, namely,

$$D\geqslant Ke^{-H_Q r/k}$$

where K is an unspecified constant depending on k, r, and $p(x)$. Elias defined the quantizer distortion measure $D^*=\Sigma_i^N P(S_i)[V(S_i)]^{r/k}$ where $V(S_i)$ is the k-dimensional volume of the region S_i. He showed that

$$D^*\geqslant N^{-r/k}\|p(\boldsymbol{x})\|_{k/k+r} \tag{5}$$

where $\|p(\boldsymbol{x})\|_\alpha$, the L_α norm of $p(\boldsymbol{x})$, is defined in (3) except that the integration is now k-dimensional. He also showed that for N sufficiently large there exists a quantizer with D^* arbitrarily close to this bound. Elias assumed the input vector $\boldsymbol{x}$ to be bounded so that each region S_i has finite volume. Zador, in a lengthy unpublished manuscript, found for the asymptotic case of high quality quantization that

$$D_1(N)=A(k,r)N^{-r/k}\|p(\boldsymbol{x})\|_{k/(k+r)} \tag{6}$$

and that

$$D_2(H_Q)=B(k,r)e^{-r/k[H_Q-H(p)]} \tag{7}$$

where $H(p)$ is the differential entropy of the random vector $\boldsymbol{X}$. Zador did not obtain $A(k,r)$ or $B(k,r)$ explicitly, but he showed that

$$\frac{1}{1+\beta}V_k^{-\beta}\leqslant kB(k,r)\leqslant kA(k,r)\leqslant\Gamma(1+\beta)V_k^{-\beta} \tag{8}$$

where $\beta=r/k$, $\Gamma(x)$ is the gamma function, and V_k is the volume of a unit sphere in k dimensions. A key feature of Zador's result is that $A(k,r)$ and $B(k,r)$ are independent of the density $p(\boldsymbol{x})$. Hence these functions can be studied in the simpler context of the uniform density on a unit cube in k dimensions. For $k=2$ and $r=2$, Fejes Toth [10] found the (asymptotically) minimum distortion having the form (6) with the explicit value $A(2,2)=5\sqrt{3}/108$.

Recently Gray and Gray [11] gave a simple derivation of the one-dimensional results (2) and (4) using Bennett's integral. Here we explore a similar approach but generalized to the k-dimensional case.

IV. Admissable and Optimal Polytopes

In the one-dimensional case, Bennett's integral is derived by separating the description of a quantizer into two aspects: a) a uniform quantizer that is optimal for the uniform density function and b) the compressor slope function which determines how the output points of the uniform quantizer must be redistributed to take into account the probability density function of the random variable to be quantized. Zador's expression for the minimal distortion also has the striking feature that the factor $A(k,r)$ is independent of the probability density of the random variable. Since $\|p(\boldsymbol{x})\|_\alpha=1$ if $p(\boldsymbol{x})$ is unity in a bounded region of unit volume and zero elsewhere, it follows that $A(k,r)$ is determined by the optimal quantizer for a *uniformly* distributed random variable. These observations suggest that Bennett's integral can be generalized by first considering the optimum quantizer for a uniformly distributed k-dimensional random variable and then considering the effect of a nonuniform distribution of output points on the distortion of a quantizer. We begin by exploring some relevant geometrical features of partitions in R_k.

For every finite (or countably infinite) set of points $y_1, y_2, \cdots, y_N$ in R_k a Dirichlet partition is defined with each point in S_i closer to y_i than to any other y_j for $j\neq i$. That is,

$$S_i=\{\boldsymbol{x}: \|\boldsymbol{x}-y_i\|\leqslant\|\boldsymbol{x}-y_j\| \text{ for each } j\neq i\}.$$

An optimal N-point quantizer that minimizes distortion will clearly have a Dirichlet partition. An example of a Dirichlet partition in the plane is shown in Fig. 1. In general, each bounded Dirichlet region is a polytope (bounded by segments of $k-1$ dimensional hyperplanes) and is convex. For a quantizer an effective partition would have the property that the unbounded regions or "overload" regions would make a sufficiently small contribution to the distortion. This is always possible when $E\|X\|^r<\infty$.

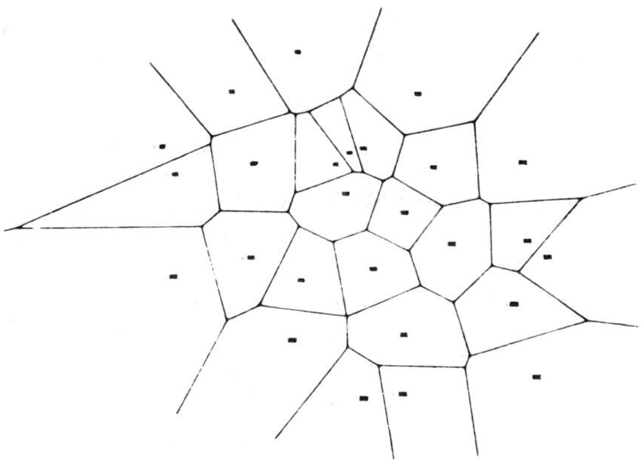

Fig. 1. Direchlet partition for Cambridge, Massachusetts, schools (from H. L. Loeb [17]).

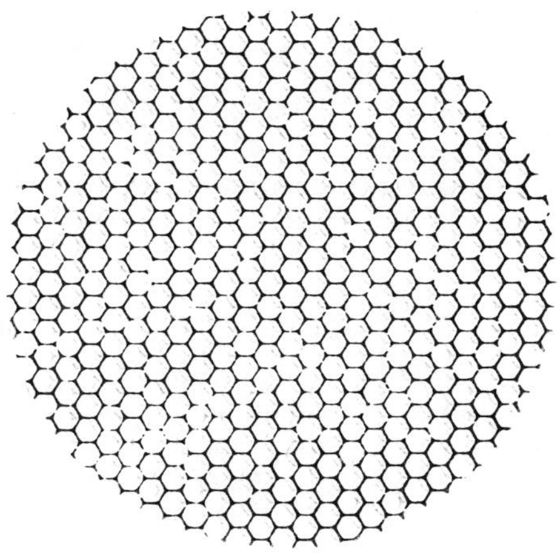

Fig. 2. Tesselation of regular hexagons.

The *centroid* $\hat{y}$ of a convex polytope H in R_k is the value of y that minimizes $\int_H \|x-y\|^r\, dx$. For $r=2$, $\hat{y}$ coincides with the usual definition for the centroid of a body with uniform mass distribution. It should be noted (see Fig. 1) that in general the points generating a Dirichlet partition are not necessarily the centroids of their respective regions. For a uniformly distributed random vector x, a quantizer will have a Dirichlet partition defined on the bounded set in R_k where $p(x)$ is positive. For the quantizer that minimizes distortion it is clearly necessary that each output point will be the centroid of the region in which it lies. The two necessary conditions for optimality, i.e., that the partition be a Dirichlet partition and that the output points be centroids, were noted for $k=1$ by Lloyd [3].

A convex polytope H is said to generate a tessellation if there exists a partition of R_k whose regions are all congruent to H. For example in the plane all triangles, quadrilaterals, and hexagons generate tessellations.

We now make the basic conjecture that for N sufficiently large the optimal (distortion-minimizing) quantizer for a random vector uniformly distributed on some convex set S will have a partition whose regions are all congruent to some polytope H, with the possible exception of regions touching the boundary of S. In other words, the optimal partition is essentially a tessellation of S. This conjecture plays a key role in the heuristic approach which follows.

We define H_k, the class of admissible polytopes in R_k as follows. A convex polytope H in R_k is in H_k if a) H generates a tessellation that is a Dirichlet partition with respect to the centroids of each region in the partition. For example, the equilateral triangle, the rectangle, and the regular hexagon are the admissible polygons in H_2. Fig. 2 illustrates a tessellation of the regular hexagon. Now we define the *normalized inertia* $I(H)$ of a polytope H as

$$I(H)=\int_H \|x-\hat{x}\|^r\, dx/[V(H)]^{1+r/k} \tag{9}$$

where $\hat{x}$ is the centroid of H and $V(H)$ is the k-dimensional volume of H. The normalization has the property that $I(\alpha H)=I(H)$ for $\alpha>0$ where the polytope $\alpha H=\{\alpha x : x\in H\}$. In other words, when the size of H is scaled, its normalized inertia remains unchanged.

We define the *coefficient of quantization*

$$C(k,r) \triangleq \frac{1}{k} \inf_{H\in H_k} I(H).$$

An *optimal polytope* H^* is an admissible polytope which attains the minimum inertia of all admissable polytopes with the same volume. Hence

$$I(H^*)=kC(k,r).$$

By calculating the normalized inertia of each admissible polygon, it can be shown that for $k=2$ and $r=2$, the optimal polytope is the regular hexagon. We conjecture that an optimal polytope exists for each k.

It is a classic isoperimetric result that every convex polytope has a greater moment of inertia with respect to its centroid than a k-dimensional sphere with the same volume. For the unit radius sphere B centered at the origin it is known that

$$\int_B \|x\|^r\, dx = \frac{k}{k+r} V_k$$

where V_k is the volume of B. Hence we have

$$I(B)=\frac{k}{k+r} V_k^{-r/k}$$

so that we have the lower bound

$$C(k,r) \geqslant \frac{1}{k+r} V_k^{-r/k}. \tag{10}$$

An upper bound on $kC(k,r)$ can be found by calculating the normalized inertia for any admissible polytope in H_k. One such choice is the k-dimensional cube (centered at the origin), which is clearly admissible. The cube has normalized inertia $k/[(r+1)2^r]$ so that

$$C(k,r) \leqslant \frac{1}{1+r} 2^{-r}. \tag{11}$$

Note that this bound is independent of the dimension k.

V. Heuristic Derivation of the Distortion Integral

Generalizing the concept of "asymptotic fractional density of quanta" introduced by Lloyd in a classic paper [3] on one-dimensional quantization, define the output point density function of a k-dimensional quantizer as

$$g_N(x)=\frac{1}{NV(S_1)}, \qquad \text{if } x\in S_i, \text{ for } i=1,2,\cdots,N.$$

where $V(S_1)$ denotes the volume of S_1. Note that $g_N(x)=0$ if x is in a region of the partition having infinite volume. In the asymptotic situation where N is very large, $g_N(x)$ can be expected to approximate closely a continuous density function $\lambda(x)$ having unit volume. Then $\lambda(x)\Delta V(x)$ may be taken as the fraction of output points located in an incremental volume element $\Delta V(x)$ containing x. Thus the volume of the quantizing region S_i associated with the output point y_i is given approximately by

$$V(S_i)\approx\frac{1}{N\lambda(y_i)} \tag{12}$$

for every bounded region S_i. Note that $N\lambda(y_i)$ is the number of points per unit volume in the neighborhood of y_i so that its reciprocal (12) is the volume per output point.

The distortion (1) can be expressed as

$$D=\frac{1}{k}\sum_{i=1}^{N}\int_{S_i}\|x-y_i\|^r p(x)\,dx. \tag{13}$$

For N large it is reasonable to assume that most of the regions S_i will be bounded sets, and the "overload" regions S_i will correspond to the tail region of the density $p(x)$. Assume the partition has been suitably chosen so that the overload distortion is negligible, treat N as the number of bounded regions, and for N large make the approximation

$$p(x)\backsim p(y_i), \qquad \text{for } x\in S_i.$$

Then we obtain

$$D=\frac{1}{k}\sum_{i=1}^{N}p(y_i)\int_{S_i}\|x-y_i\|^r\,dx. \tag{14}$$

As N becomes large the partition for any bounded region should look more and more like the partition for a uniform density, assuming $\lambda(x)$ is smoothly varying. Thus we approximate S_i by a suitably rotated, translated, and scaled optimal polytope H^*. Then

$$\int_{S_i}\|x-y_i\|^r\,dx=I(H^*)[V(S_i)]^{1+r/k} \tag{15}$$

using (9). We then have

$$D=\frac{1}{k}\sum_{r=1}^{N}p(y_i)I(H^*)[V(S_i)]^{1+r/k}, \tag{16}$$

and from (12) we obtain

$$D=N^{-\beta}C(k,r)\sum_{i=1}^{N}p(y_i)[\lambda(y_i)]^{-\beta}V(S_i). \tag{17}$$

The summation can be approximated by an integral yielding

$$D=N^{-\beta}C(k,r)\int\frac{p(y)}{[\lambda(y)]^{\beta}}\,dy. \tag{18}$$

The region of integration is actually the union of all bounded regions of the partition but may be taken to be the entire k-dimensional space since the contribution to the distortion of the overload regions will be negligible for any reasonable quantizer with sufficiently large N.

Equation (18) may be recognized as the k-dimensional version of Bennett's formula (1) for one-dimensional quantization with mean-square distortion.

VI. Minimization of the Distortion Integral

The distortion integral (18) allows the minimization of the distortion by optimizing the choice of $\lambda(x)$, the asymptotic output point density function. No reference is needed to the explicit quantizer characteristics (the output points and partition regions).

For problem a), D is to be minimized over all quantizers with N fixed. Hölder's inequality gives

$$\int p\lambda^{-\beta}\,dy\left\{\int\lambda\,dy\right\}^{\beta} \geqslant \left\{\int (p\lambda^{-\beta})^{1/(1+\beta)}\lambda^{\beta/(1+\beta)}\,dy\right\}^{1+\beta}.$$

Noting that $\int\lambda\,dy=1$, we obtain the result

$$\int p\lambda^{-\beta}dy\geqslant\|p\|_{1/(1+\beta)}$$

with equality attained only when λ is proportional to $p^{1/(1+\beta)}$. Hence the minimum value of D, referring to (18), is

$$D_1(N)=C(k,r)N^{-\beta}\|p(x)\|_{k/(k+r)}. \tag{19}$$

This is the desired result. Note that (19) coincides with Zador's result (6) when we take $A(k,r)=C(k,r)$. Furthermore using (10) we obtain a lower bound for $D_1(N)$ that coincides with Zador's lower bound.

A significant property of the optimum quantizer can now be demonstrated. Since the optimum point density λ is proportional to $p^{1/(1+\beta)}$, we observe that each term in the sum (16) reduces to a constant independent of the index i. *Therefore each region S_i of the partition makes an equal contribution to the distortion for an optimal quantizer.* This property was observed by Panter and Dite [2] for $k=1$ and by Fejes Toth [10] for $k=2$.

In problem b), D is to be minimized subject to a constraint on the quantizer output entropy H_Q. Since $p_i\backsim p(y_i)V(S_i)$ for each bounded set S_i and for large N,

$$H_Q=-\sum p(y_i)\frac{1}{N\lambda(y_i)}\log[p(y_i)/N\lambda(y_i)]$$

$$=-\sum p(y_i)\log p(y_i)\Delta V(y_i)-\sum p(y)\log\frac{1}{N\lambda(y_i)}\Delta V(y_i)$$

where $\Delta V(y_i)=1/N\lambda(y_i)$. As in the derivation of (18), the sums can be approximated by integrals, yielding

$$H_Q = H(p) - \int p(y) \log \frac{1}{N\lambda(y)} dy \tag{20}$$

where $H(p)$ is the differential entropy of the random vector X. Equation (20) reduces for $k=1$ to the corresponding one-dimensional result given by Gish and Pierce [6]. From (18) we have

$$D = C(k,r) \int e^{-\beta \log [N\lambda(y)]} p(y)\, dy. \tag{21}$$

Now applying Jensen's inequality we get

$$D \geqslant C(k,r) e^{-\beta \int p(y) \log [N\lambda(y)]\, dy}. \tag{22}$$

Applying (20) we see that

$$D \geqslant C(k,r) e^{-\beta [H_Q - H(p)]}. \tag{23}$$

The application of Jensen's inequality yields an equality when $\lambda(y)$ is a constant corresponding to a *uniform* distribution of output points. Hence the solution to problem b) is

$$D_2(H_Q) = C(k,r) e^{-\beta [H_Q - H(p)]}. \tag{24}$$

Note that (24) coincides with Zador's result (7) when we take $B(k,r) = C(k,r)$. Furthermore applying the lower bound (10) to (24) gives a bound for $D_2(H_Q)$ which coincides with Zador's lower bound (8). *It is significant to observe that for large N the optimal quantizer for a constrained entropy is very nearly, the uniform quantizer.* For $k=1$ this was noted by Gish and Pierce [6].

As an additional illustration of the use of the function $\lambda(x)$ we give a heuristic derivation of Elias' result [9]. Since

$$P(S_i) \approx p(y_i) V(S_i),$$

$$D^* \approx \sum_{i=1}^{N} p(y_i) V(S_i)^{r/k} V(S_i).$$

Using (11) gives

$$D^* \approx \sum_{i=1}^{N} p(y_i) \left[\frac{1}{N\lambda(y_i)} \right]^{r/k} \Delta V(y_i),$$

and approximating the sum for N large by an integral yields

$$D^* = N^{-\beta} \int \frac{p(y)}{[\lambda(y)]^{\beta}} dy. \tag{25}$$

The minimization of this integral as shown above leads to the result that

$$D^* = N^{-\beta} \| p(x) \|_{k/k+r} \tag{26}$$

which is Elias' lower bound.

Finally, it should be noted that the formulas (2), (4), (6), (7), (18), (19), (24), and (26), which have been written as equalities, should more correctly be taken as lower bounds on attainable distortion for any finite N. Since the minimum distortion attainable is nonincreasing as N (or H_Q) increases, the actual distortion can only be greater than these asymptotic values for any quantizer with a finite number of quantizing regions N.

VII. Special Cases and Bounds

For $k=1$, a finite interval on the real line is the only admissible polytope. The interval is therefore the optimal polytope for $k=1$. Calculating its normalized inertia gives

$$C(1,r) = \frac{1}{r+1} 2^{-r}.$$

For $r=2$, we have $C(1,2) = 1/12$ and hence our generalized Bennett integral (18) reduces to the original Bennett integral (1). For $k=1$, the minimum distortion formula (19) coincides with the known result (2) as given by Algazi [5]. Also for $k=1$, our constrained-entropy minimum distortion formula (24) reduces to the known result (4) due to Gish and Pierce [6].

For $k=2$ we have already noted that the regular hexagon is the optimal polytope. This yields the coefficient of quantization

$$C(2,2) = \frac{5}{36\sqrt{3}}.$$

A theorem by Fejes Toth [12] shows in effect that for a uniformly distributed random variable the minimum distortion for each r is obtained by a tessellation of regular hexagons. Newman [13] independently found a proof of this result for $r=2$. Their results imply that Zador's coefficient $A(2,2)$ has the value $5/(36\sqrt{3})$. Hence the complete solution for nonuniform densities $p(x_1, x_2)$ is in fact given by

$$D \sim \frac{5}{36\sqrt{3}} N^{-1} \left[\int\int \sqrt{p(x_1,x_2)}\, dx_1\, dx_2 \right]^2$$

asymptotically as $N \to \infty$, when $k=2$ and $r=2$. Unknown to Newman and Zador, Fejes Toth [10] had given a complete proof of this result. Hence our minimum distortion formula reduces to this known result for $k=2$ and $r=2$.

Fejes Toth [10] noted that the optimal partition for a given probability density $p(x_1, x_2)$ in the plane consists of "approximately" regular hexagons with the centroids distributed with a nonuniform density over the plane. An example of a hexagonal partition whose centroids are distributed nonuniformly in the plane is shown in Fig. 3. These results for $k=2$ help to clarify the role of the output point density function $\lambda(x)$ in characterizing a quantizer as used in this paper.

For $k \geqslant 3$, the minimum distortion attainable for a quantizer is not known. However we can obtain upper bounds on the quantization coefficient $C(k,r)$ as noted in Section IV by calculating the normalized inertia for any admissible polytope. Any admissible polytope generates a tessellation that can be used for the quantization of a random vector that is uniformly distributed on a unit volume region. Hence neglecting the boundary regions when N is large, the normalized inertia I' of that polytope gives the attainable distortion

$$D = \left(\frac{1}{k} I' \right) N^{-r/k}.$$

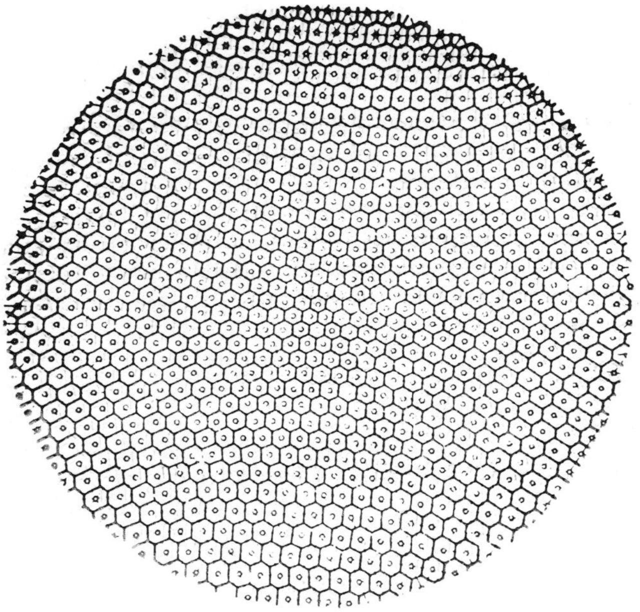

Fig. 3. Hexagonal partition for nonuniform density of points (from Fejes Toth [10]).

Hence

$$A(k,r) \leqslant \frac{1}{k} I'.$$

Therefore any upper bounds we obtain for $C(k,r)$ are in fact upper bounds for Zador's $A(k,r)$. Even though our derivation of (24) is not rigorous, these upper bounds are rigorously valid.

For $k=3$ the admissible polyhedra include the five principal parallelohedra: cube, hexagonal prism, rhombic dodecahedron, elongated dodecahedron, and the truncated octahedron. Of these five, the truncated octahedron shown in Fig. 4 and specified by the set

$$\{(x_1,x_2,x_3): |x_1|+|x_2|+|x_3|<1.5, |x_1|<1, |x_2|<1, |x_3|<1\}$$

has by direct calculation the smallest normalized inertia with

$$I=\frac{19}{64\sqrt[3]{3}}=0.23563\cdots$$

which gives the upper bound

$$C(3,2)\leqslant 0.07854 \quad \cdots$$

which is surprisingly close to the sphere lower bound (0.07697) discussed in Section IV. The truncated octahedron is clearly the best parallelohedron and is very likely the optimal polyhedron.

For $k=4$, the analog of the truncated octahedron is the admissible polytope

$$\{x: |x_1|+|x_2|+|x_3|+|x_4|\leqslant 2; |x_i|\leqslant 1, i=1,2,3,4\}$$

which by a crude Monte Carlo integration gives the bound

$$C(4,2)\leqslant 0.0766\cdots.$$

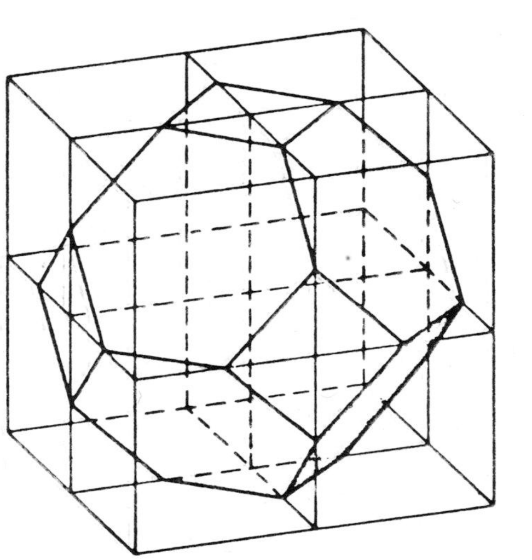

Fig. 4. Truncated octahedron imbedded in cube of side length 2, corresponding to analytical description given in text.

TABLE I
VALUES AND BOUNDS FOR QUANTIZATION COEFFICIENT $C(k,2)$

K	SPHERE LOWER BOUND	ACTUAL VALUE	POLYTOPE UPPER BOUND	CUBE UPPER BOUND	ZADOR UPPER BOUND
1	.08333	.08333(a)		.08333	.5
2	.07958	.08019(b)		.08333	.159
3	.07697		.07854(c)	.08333	.1157
4	.07503		.0766(d)	.08333	.09974
5	.07352		.0779(e)	.08333	.09133
100	.0574		.0766(f)	.08333	.05789

(a) interval
(b) regular hexagon
(c) truncated octahedron
(d) four dimensional analog of (c)
(e) cross product of truncated octahedron with an interval
(f) cross product of 25 type (d) polytopes

One technique for obtaining upper bounds is to select admissible polytopes in a higher dimensional space by forming "cross-products" or prisms using lower dimensional polytopes. For example, the cube in k-space is simply the kth cross-product of the interval, the hexagonal prism in $k=3$ is the cross product of the regular hexagon ($k=2$) with the interval ($k=1$), and the cross product of the regular hexagon with the truncated octahedron gives an admissible polytope for $k=5$. For such cross-product polytopes the normalized moment of inertia when $r=2$ is trivially obtained by summing the normalized moments of inertia of each lower dimensional polytope.

In Table I values of $C(k,2)$ are given when known together with the available lower and upper bounds.

VIII. INFINITE BLOCK LENGTH

All of the preceding results have been based on a fixed (but arbitrary) block length k with $1\leqslant k<\infty$. The results can be compared with the known performance bounds of rate distortion theory by examining the limiting case for mean-square distortion ($r=2$) letting $k \to \infty$. Since Zador's upper and lower bounds coincide asymptotically as $k \to \infty$, we use the lower bound to study the limiting behavior of $D_1(N)$. As noted by Zador, Stirling's formula gives $V_k^{-\beta} \sim (2\pi e)^{-1} k$ as $k \to \infty$. Zador also conjectured that in general

$$\|p_k(x)\|_{k/k+2} \to e^{2\bar{H}} \quad \text{as } k \to \infty \tag{27}$$

where $\bar{H}$ is the differential entropy rate of the source

$$\bar{H} = -\lim_{n \to \infty} \frac{1}{n} \int p_n(x) \log p_n(x)\, dx$$

and a subscript has been added to the joint density $p(x)$ to identify its dimensionality. In the particular case of a stationary Gaussian process, Zador showed that (27) holds. The result (27) appears to be of some fundamental theoretical interest. A proof that it holds for any stationary ergodic process is given in the Appendix. Without assuming ergodicity, the weaker result

$$\lim_{k \to \infty} \|p_k\|_{k/k+2} \geqslant e^{2\bar{H}} \tag{28}$$

will now be shown

$$\begin{aligned} \log \|p_k\|_{k/k+2} &= \frac{k+2}{k} \log \int p_k(x) e^{-[2/(k+2)] \log p_k(x)}\, dx \\ &\geqslant \frac{k+2}{k} \int p_k(x) \log e^{-[2/(k+2)] \log p_k(x)}\, dx \\ &= -\frac{2}{k} \int p_k(x) \log p_k(x)\, dx, \end{aligned}$$

using Jensen's inequality. Thus

$$\|p_k\|_{k/k+r} \geqslant e^{2H(p_k)/k} \tag{29}$$

from which (28) follows.

Now using (19) and the lower bound (10) on $C(k,r)$ gives the result for very large block length k that

$$D \geqslant \frac{1}{2\pi e} e^{-2(R-\bar{H})} \tag{30}$$

where $R = (\log N)/k$ is the *rate* or average number of nats per component of X needed to identify (or transmit) the quantizer output approximation $Q(x)$. Inverting (30) gives a more familiar result

$$R \geqslant \bar{H} - \frac{1}{2} \log (2\pi e D), \tag{31}$$

which is the generalized Shannon lower bound on the rate-distortion function (see Berger [14]). Note that (31) as derived here is only valid for the asymptotic case of small distortion (corresponding to large N).

IX. Companding Revisited

For $k=1$, Bennett introduced the "companding" model of a quantizer as a monotonically increasing nonlinear mapping $E(x)$ the compressor, followed by a uniform quantizer, and by the inverse mapping, E^{-1}. Lloyd's point density function $\lambda(x)$ is then the slope of the mapping $E(x)$. For block quantization we have introduced the concept of a point density function and derived a generalization of Bennett's integral without any reference to a mapping. To complete the connection with the one-dimensional case we can define a continuously differentiable and invertible mapping f which maps a point x in R_k into another point $f(x)$ in R_k. Then a family of k dimensional quantizers can be modeled as shown in Fig. 5 as the cascade of such a mapping with an optimal N-point uniform quantizer, let us say on the unit cube in R_k followed by the inverse mapping f^{-1}. The overall quantizer is described by

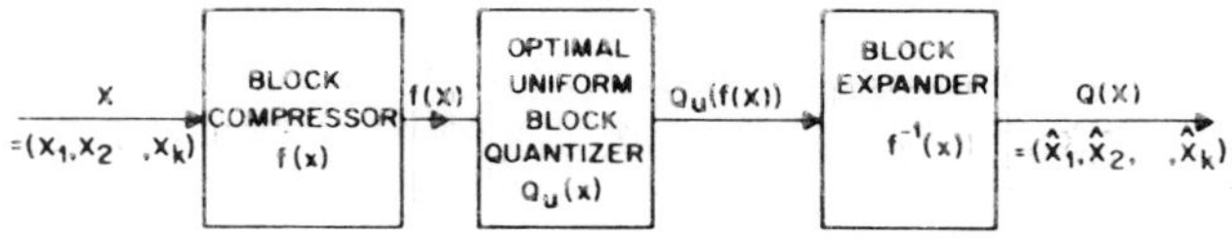

Fig. 5. Companding model of block quantization.

$$Q(x) = f^{-1} \circ Q_u \circ f(x)$$

where $Q_u(x)$ denotes the uniform quantizer. The point density function for the overall quantizer is then given by the Jacobian determinant of f. The Dirichlet partition for the uniform quantizer with regions S_i, will induce a new partition for the overall quantizer with regions $S_i^* = f^{-1}(S_i)$. In general the new partition will not be a Dirichlet partition.

The question then arises, for a given probability density $p(x)$, does there exist a mapping $f(x)$ which makes $Q(x)$ the optimum quantizer? In order to preserve the Dirichlet property it is necessary that the mapping be conformal since the line joining the centroids of two adjacent regions in a Dirichlet partition must always be perpendicular to the hyperplane separating the two regions. For $k=2$, Heppes and Szusz [15] have noted, in effect, that a necessary and sufficient condition for the existence of $f(x)$ is that the logarithm of the point density function be a harmonic function, i.e., that $\log \lambda(x)$ satisfy Laplace's equation. Since for the optimum quantizer we have shown that λ must be proportional to a power of $p(x)$, it follows that the condition is equivalent to having $\log p(x)$ satisfy Laplace's equation. This condition eliminates the joint normal density as well as any other density whose curves of constant density close. Hence there appears to be a fundamental limitation to the possibility of generalizing Bennett's companding approach to the multidimensional case.

X. Concluding Remarks

In this paper we have shown that the point density function of a quantizer, first conceptualized by Lloyd, can be generalized to the multidimensional case to provide a fruitful and intuitively satisfying way to develop the theory of optimal quantization for random vectors. With it we have heuristically generalized the classic integral of Bennett for the distortion of a quantizer which we hope will be useful for future studies of block quantization. In deriving the results of Zador, albeit heuristically, we have gone further toward a constructive theory of optimal quantization by introducing some of the salient geometrical features of the partition of space defined by a quantizer. Finally, we have pointed out the possibilities and limitations of the companding approach to modeling a quantizer for random vectors, an approach that has been of great practical importance in the one-dimensional case.

ACKNOWLEDGMENT

I would like to thank Thomas Liggett of the University of California, Los Angeles, who provided the proof given in the Appendix, Hans Witsenhausen, who introduced me to the work of Fejes Toth, and Neil Sloane, who introduced me to the polyhedra associated with sphere packings.

APPENDIX[1]

Theorem 1:

$$\lim_{k\to\infty} \|p_k(\boldsymbol{x})\|_{k/k+r} = e^{r\overline{H}}, \qquad 0<r<r_0$$

if $p_k(\boldsymbol{x})$ is the joint density of $\boldsymbol{X}^k=(X_1,X_2,\cdots,X_k)$ where $\{X_k\}$ is a stationary ergodic process and $\|p_1(x_1)\|_{1/(1+r_0)}<\infty$.

Proof: Define the random variables

$$Z_k = -\frac{1}{k}\log p_k(\boldsymbol{x}^k).$$

Then

$$\log \|p_k(\boldsymbol{x})\|_{k/k+r} = \frac{k+r}{k}\log Ee^{[kr/(k+r)]Z_k}. \tag{A.1}$$

From the Shannon-McMillan theorem for abstract alphabets [16], Z_k converges in probability to $\overline{H}$. Hence $e^{a_k Z_k}$ converges in distribution to $e^{r\overline{H}}$ where $a_k = rk/(k+r)$. If also

$$Ee^{dZ_k} < C < \infty \tag{A.2}$$

where C is independent of k, and $r<d$, then convergence in the mean would follow, namely,

$$Ee^{a_k Z_k} \to e^{r\overline{H}} \qquad \text{as } k\to\infty$$

which would complete the proof. To prove (A.2), define

$$\rho_k(\boldsymbol{X}^k) \triangleq \frac{k+r}{r}\log Ee^{a_k Z_k}.$$

Then (A.2) will follow from the fact that $\rho_k(\boldsymbol{X}^k)$ is subadditive, that is,

$$\rho_{k+l}(\boldsymbol{X}^{k+l}) \leqslant \rho_k(\boldsymbol{X}^k) + \rho_l(\boldsymbol{X}^l), \tag{A.3}$$

which is analogous to the subadditivity of the entropy. Denote $s=k+l+r$, $u=(x_1,x_2,\cdots,x_k)$, $v=(x_{k+1},x_{k+2},\ldots,x_{k+l})$, and let $f=p_{k+l}(u,v)$, $p_k=p_k(u)$, $p_l=p_l(v)$. Then

$$\begin{aligned}
\int\!\!\int f^{(k+l)/s}\,du\,dv &= \int\!\!\int \left[fp_k^{-r/(k+r)}\right]^{k/s}\left[fp_l^{-r/(l+r)}\right]^{l/s} \\
&\quad\cdot\left[p_k^{k/(k+r)}p_l^{l/(l+r)}\right]^{r/s}du\,dv \\
&\leqslant \left[\int\!\!\int fp_k^{-r/(k+r)}\,du\,dv\right]^{k/s}\left[\int\!\!\int fp_l^{-r/(l+r)}\,du\,dv\right]^{l/s} \\
&\quad\cdot\left[\int\!\!\int p_k^{k/(k+r)}p_l^{l/(l+r)}\,du\,dv\right]^{r/s} \\
&= \left[\int p_k^{k/(k+r)}\,du\right]^{(k+r)/s}\left[\int p_l^{l/(l+r)}\,dv\right]^{(l+r)/s}.
\end{aligned}$$

using the triple Hölder inequality. Hence $k^{-1}\rho_k(\boldsymbol{X}^k)$ is bounded by $\rho_1(X_1)$ so that

$$\frac{k+r}{k}\log\{Ee^{a_k Z_k}\} \leqslant K\log\|p_1(x_1)\|_{1/(1+r_0)} < \infty \tag{A.4}$$

which holds, in particular, for $r=r_0$. Given d, with $0<d<r_0$, for sufficiently large k,

$$d < \frac{kr_0}{k+r_0}.$$

Hence

$$\log Ee^{dZ_k} < \log Ee^{[kr_0/(k+r_0)]Z_k},$$

so that (A.4) implies (A.2), completing the proof of the theorem.

[1] This proof is due to Thomas Liggett.

REFERENCES

[1] W. R. Bennett, "Spectra of quantized signals," *Bell Syst. Tech. J.*, vol. 27, pp. 446–472, July, 1948.

[2] P. F. Panter and W. Dite, "Quantization in pulse-count modulation with nonuniform spacing of levels," *Proc. IRE*, vol. 39, pp. 44–48, 1951.

[3] S. P. Lloyd, "Least squares quantization in PCM," unpublished memorandum, Bell Laboratories, 1957.

[4] B. Smith, "Instantaneous companding of quantized signals," *Bell Syst. Tech. J.*, vol. 52, pp. 1037–1076, Sept. 1973.

[5] V. R. Algazi, "Useful approximations to optimum quantization," *IEEE Trans. Commun. Tech.*, vol. COM-14, pp. 297–301, 1966.

[6] H. Gish and J. N. Pierce, "Asymptotically efficient quantizing," *IEEE Trans. Inform. Theory*, vol. IT-14, pp. 676–683, Sept. 1968.

[7] P. Zador, "Asymptotic quantization of continuous random variables," unpublished memorandum, Bell Laboratories, 1966.

[8] M. P. Schutzenberger, "On the quantization of finite dimensional messages," *Inform. Contr.*, vol. 1, pp. 153–158, 1958.

[9] P. Elias, "Bounds and asymptotes for the performance of multivariate quantizers," *Ann. Math. Stat.*, vol. 41, pp. 1249–1259, 1970.

[10] L. Fejes Toth, "Sur la representation d'une population infinie par un nombre fini d'éléments," *Acta Math. Acad. Scient. Hung.*, vol. 10, pp. 299–304, 1959.

[11] R. M. Gray and A. H. Gray, Jr., "Asymptotically optimal quantizers," *IEEE Trans. Inform. Theory*, vol. IT-23, pp. 143–144, Jan. 1977.

[12] L. Fejes Toth, *Lagerungen in der Ebene auf der Kugel und im Raum. Springer-Verlag* p. 81, 1953.

[13] D. J. Newman, "The hexagon theorem," unpublished memorandum, Bell Laboratories, 1964.

[14] T. Berger, *Rate-Distortion Theory.* Englewood Cliffs, N.J.: Prentice-Hall, 1971.

[15] A. Heppes and P. Szüsz, "Bemerkung zu einer Arbeit von L. Fejes Toth, " *El. Math.*, vol. 15, pp. 134–136, 1960.

[16] A. Perez, "On the theory of information in the case of an abstract alphabet," in *Trans. First Prague Conf. on Inform. Theory, Statistical Decisions Functions, and Random Processes,* (held at Prague, Nov. 28–30, 1956). Prague: Publishing House of the Czechoslovak Academy of Sciences, 1957, pp. 209–243.

[17] H. L. Loeb, *Space Structures*, Addison-Wesley, 1975.

12

Reprinted from *IEEE Trans. Inform. Theory* **IT-28**(1):105–107 (1982)

PROPERTIES OF MINIMUM MEAN SQUARED ERROR BLOCK QUANTIZERS

N. C. Gallagher, Jr. and J. A. Buckle w

***Abstract*—Two results in minimum mean square error quantization theory are presented. The first section gives a simplified derivation of a well-known upper bound to the distortion introduced by a *k*-dimensional optimum quantizer. It is then shown that an optimum multidimensional quantizer preserves the mean vector of the input and that the mean square quantization error is given by the sum of the component variances of the input minus the sum of the variances of the output.**

I. Introduction

Block or vector quantization deals with the representation of multidimensional elements with a finite discrete set of values. The values to be quantized may naturally fall into a k-dimensional representation; typical examples are complex numbers, positional coordinates, or state vectors. In other cases, k-dimensional vectors are formed from blocks of k samples taken from one-dimensional signals. In 1964 Zador published a number of very interesting results on the properties of optimal block quantizers for the rth moment Euclidean norm distortion measure [1]. Among Zador's contributions are the derivation of both upper and lower bounds on the distortion introduced by the optimal quantizer. These bounds are derived without actually finding the optimal quantizer. Unfortunately, at some points Zador's development is not easy to follow, and alternate derivations and extensions by Gersho [2] and Yamada *et al.* [3] have recently appeared. In Section II we present an alternate derivation of Zador's random quantization upper bound not treated in either [2] or [3].

In [4] Bucklew and Gallagher show that for one-dimensional mean squared error distortion the optimum quantizer has the property that the mean value of the quantizer output equals the mean value of the input and also that the mean square quantization error equals the variance of the input minus the variance of the output. In [5] Bucklew and Gallagher prove that the same results hold for constant step-size minimum mean squared error quantizers. In Section III we extend these properties to k-dimensional optimal block quantizers.

Manuscript received December 30, 1980. This work was supported by the Air Force Office of Scientific Research under Grant AFOSR 78-3605.
N. C. Gallagher, Jr., is with the School of Electrical Engineering, Purdue University, West Lafayette, IN 47907.
J. A. Bucklew is with the Department of Electrical and Computer Engineering, University of Wisconsin, Madison, WI 53705.

II. Random Quantization Upper Bound

In [2] Gersho provides a very readable derivation of Zador's expression for quantizer distortion. To improve continuity and readability we employ Gersho's notation. The quantizer input is a k-dimensional random vector x in $\Re_k$ which is quantized to one of N levels $y_1, y_2, \cdots, y_N$ in $\Re_k$. The space $\Re_k$ is partitioned into N disjoint and exhaustive regions $S_1, S_2, \cdots, S_N$. The quantizer is defined by the function $Q(x)$ defined by $Q(x) = y_i$, if $x \in S_i$. Note that this definition does not require that $y_i \in S_i$, although in practice y_i is usually contained in S_i. The performance of the quantizer is measured by the distortion

$$D = \frac{1}{k} E\{\|X - Q(X)\|^r\}$$

where $\|\cdot\|$ denotes the usual Euclidean distance norm, the operator $E\{\cdot\}$ denotes statistical expectation, and the input X is a k-dimensional random input vector. The case where $r = 2$ is the usual mean squared distortion. The expression derived by Zador and Gersho for the minimum distortion D_0 obtained by use of the best quantizer is

$$D_0 = N^{-r/k} C(k, r) \| p(x) \|_{k/(k+r)}, \tag{1}$$

where

$$\| p(x) \|_\alpha = \left[\int [p(x)]^\alpha \, dx \right]^{1/\alpha},$$

and where the constant $C(k, r)$, called the coefficient of quantization, is independent of the density $p(x)$ and is in general unknown. This expression is an asymptotic result valid only for large N. Two special cases for which the value of $C(k, r)$ is known exactly are [2]

$$C(1, r) = \frac{1}{r+1} 2^{-r},$$

and

$$C(2, 2) = \frac{5}{36\sqrt{3}}.$$

Consider the density $p(x)$ having a constant value of one over the unit volume hypercube; then $\| p(x) \|_{k/(k+r)} = 1$. In this case (1) becomes

$$D_0 = N^{-r/k} C(k, r). \tag{2}$$

So, we see that by finding a bound on D_0 we also bound $C(k, r)$. To find this bound we choose the quantizer output levels to have a random distribution uniformly distributed over the hypercube. For a particular input value x, we find the closest output level and quantize to that value. Because this quantizer is not the optimum quantizer, the associated distortion will bound from above the distortion for the optimum quantizer.

To begin, place at random N independent uniformly distributed k-dimensional samples in the hypercube. These will be the output levels. We take the quantizer input X to have a uniform distribution over the hypercube. We also assume that N is sufficiently large so that there is a very small probability that the quantizer input is closer to an edge of the hypercube than to one of the output values. Suppose that an input value x has arrived and is sitting in the hypercube waiting to be quantized. The probability that one particular output value is within a distance ρ of this input sample is given approximately by the volume of a sphere of radius ρ about that sample point, or

$$\Pr\left(\begin{array}{l}\text{one particular output level is within } \rho \text{ of} \\ \text{the input sample}\end{array}\right) = V_k \rho^k,$$

where if V_k is volume of the unit radius sphere, then $V_k \rho^k$ is the volume of the sphere with radius ρ. We are interested in the closest output level to the input sample. To compute the probability that the closest output level is within a distance ρ of the input sample, we combine classical order statistics with the result found in [3]. By employing this approach, we compute the probability density $f(\rho)$ for the distance between the input sample and the nearest output level to be

$$f(\rho) = N[1 - V_k \rho^k]^{N-1} V_k k \rho^{k-1}.$$

Note that for large values of N this probability density goes to zero rapidly as ρ increases. By construction $\rho = \|x - y_i\|$, where x is the input value and y_i is the output value. Consequently,

$$E\{\|X - Q(X)\|^r\} = E\{\rho^r\};$$

so,

$$D = \frac{1}{k} E\{\rho^r\}$$
$$= \frac{1}{k} \int_{\text{hypercube}} \rho^{r+k-1} N[1 - V_k \rho^k]^{N-1} k V_k \, d\rho.$$

Make the change of variables $s = V_k \rho^k$ and use the fact that $s \le 1$ to write

$$D \le \frac{n}{k V_k^{r/k}} \int_0^1 s^{r/k} [1 - s]^{N-1} \, ds = \frac{N}{k V_k^{r/k}} \frac{\Gamma\left(1 + \frac{r}{k}\right) \Gamma(N)}{\Gamma\left(N + 1 + \frac{r}{k}\right)},$$

where $\Gamma(\cdot)$ is the gamma function. For large N the following approximation is valid:

$$\frac{\Gamma(N)}{\Gamma\left(N + \frac{(k+r)}{k}\right)} \simeq N^{(k+r)/k}.$$

Therefore,

$$D = \frac{N^{-r/k} \Gamma\left(1 + \frac{r}{k}\right)}{k V_k^{r/k}}.$$

Because $D \ge D_0$, we use (2) to write

$$C(k, r) \le \frac{\Gamma\left(1 + \frac{r}{k}\right)}{k V_k^{r/k}},$$

which is Zador's random quantization upper bound.

III. Moment Properties of Optimum Quantizers

In [4] and [5] it is shown that, for minimum mean squared error one-dimensional quantizers, the mean of the input equals the mean of output and the distortion equals the variance of the input minus the output variance. These properties are shown to apply with and without the equal step-size constraint. In this section we generalize these results to the k-dimensional case.

We are interested in the properties of quantizers designed to minimize the distortion defined by (2) for $r = 2$:

$$D = \frac{1}{k} E\{\|X - Q(X)\|^2\}.$$

Many constraints we might impose on the quantizer can be imposed by the functional form of $Q(x)$; for example, the k-dimensional version of the equal step-size condition might require the regions $S_1 S_2, \cdots, S_N$ to have equal volume and be congruent. We had originally employed a variational approach to obtain the results of this section; however, an alternate approach, suggested by an anonymous reviewer, provides more intuition into quantizer structure. So, we employ his method.

To begin, we define the parameters P_i and X_i as follows:

$$P_i = \int_{S_i} P(x)\, dx,$$

and

$$x_i = P_i^{-1} \int_{S_i} xP(x)\, dx. \tag{3}$$

We note that partition $\{S_i\}_{i=1}^N$ need not be the optimum partition. Consider two different quantizers defined over the partition $\{S_i\}_{i=1}^N$: one with output value X_i and one with output value Y_i. These quantizer functions are represented as $Q_0(X) = X_i$ and $Q(X) = Y_i$, respectively. It will be shown that the quantizer $Q_0(X)$ is optimum for the given partition. We have that

$$E\{\|X - Q(X)\|^2\} = \sum_{i=1}^{N} \int_{S_i} (x - x_i + x_i - y_i)^2 P(x)\, dx. \tag{4}$$

By (3), we have

$$\int_{S_i} (x - x_i)(x_i - y_i) P(x)\, dx = 0;$$

therefore, (4) becomes

$$E\{\|X - Q(X)\|^2\} = E\{\|X - Q_0(X)\|^2\} + \sum_{i=1}^{N} P_i \|x_i - y_i\|^2. \tag{5}$$

The expression in (5) illustrates that the quantizer $Q_0(X)$ produces an error no larger than any other quantizer $Q(X)$ for a given partition. Also, by (3) we see that the mean of the quantizer outputs equals the mean value of the input; this follows by

$$\sum_{i=1}^{N} P_i x_i = \sum_{i=1}^{N} \int_{S_i} xp(x)\, dx = \int xP(x)\, dx, \tag{6}$$

where the left side is the mean of the output and the right side the mean of the input. It can also be shown that the quantizer error equals the variance of the input minus the variance of the output.

Consider the input variance

$$\begin{aligned} E\{\|X - E\{X\}\|^2\} &= E\{\|X - Q_0(X) + Q_0(X) - E\{X\}\|^2\} \\ &= E\{\|X - Q_0(X)\|^2\} + E\{\|Q_0(X) - E\{X\}\|^2\}, \end{aligned} \tag{7}$$

where as before the cross terms are zero. The right side of (7) is simply the sum of the quantizer error and the output variance.

Equations (6) and (7) specify the first and second moment properties of the optimum quantizer; these properties follow regardless of the optimality of the partition. In addition, it is noteworthy that the optimum quantizer is not unique. A simple example serves to illustrate this point. Consider a two-dimensional circularly symmetric input density. Any rotation of a minimum error quantizer is also a minimum error quantizer. The same property holds for one-dimensional quantizers, where it is possible to have more than one minimum error quantizer.

VII. Summary

This correspondence contains two results dealing with the properties of k-dimensional minimum mean squared error quantizers. We have established necessary conditions for optimum quantizers. These conditions are used to show that for k-dimensional quantizers the mean value of the input is preserved in the output and that the mean squared error equals the input variance minus the output variance. Also, a simplified derivation of Zador's random quantization upper bound is developed.

References

[1] P. Zador, "Development and evaluation of procedures for quantizing multivariate distributions," Ph.D. dissertation, Stanford Univ., Stanford, CA, 1964; University Microfilms Inc. no. 64-9855.

[2] A Gersho, "Asymptotically optimal block quantization," *IEEE Trans. Inform. Theory*, vol. IT-25, pp. 373–380, July 1979.

[3] Y. Yamada, S. Tazaki, and R. M. Gray, "Asymptotic performance of block quantizers with difference distortion measures," *IEEE Trans. Inform. Theory*, vol. IT-26, pp. 6–14, Jan. 1980.

[4] J. A. Bucklew and N. C. Gallagher, "A note on optimal quantization," *IEEE Trans. Inform. Theory*, vol. IT-25, pp. 365–366, May 1979.

[5] "Some properties of uniform step size quantizers," *IEEE Trans. Inform. Theory*, vol. IT-26, pp. 610–613, Sept. 1980.

Reprinted from *IEEE Trans. Inform. Theory* **IT-27**(2):207–211 (1981)

Companding and Random Quantization in Several Dimensions

JAMES A. BUCKLEW, MEMBER, IEEE

***Abstract*—The problem of implementing multidimensional quantizers is discussed. A general equation is derived that can be used to evaluate the performance of multidimensional compandors. It is demonstrated that the optimal compandor must be conformal almost everywhere. An example is given to show that asymptotically optimal performance could be obtained through nonconformal companding schemes. Random quantizers are discussed and two techniques are evaluated for reducing memory and computation time in the implementation of such devices.**

Manuscript received February 10, 1980; revised May 5, 1980. This paper was presented at the 1980 Princeton Conference on Information Sciences and Systems, Princeton, NJ, March 1980.
The author is with the Department of Electrical and Computer Engineering, University of Wisconsin, Madison, WI 53705.

I. Introduction

THE IMPLEMENTATION of high-dimensional block quantizers has in the past received relatively little attention. In this paper we put forth two techniques, companding and random quantization, to solve this problem.

W. R. Bennett [1] was the first to model a nonuniform quantizer as a zero-memory nonlinearity followed by a uniform quantizer, in turn followed by the inverse of the first nonlinearity. This sequence of operations is generally referred to as companding. The word expresses the idea that the data is first *compressed*, then quantized, and then *expanded*. The first nonlinearity is therefore generally referred to as the "compressor" and its inverse as the "expander".

Compandors are of interest in areas other than quantization theory. In some analog systems it can be advantageous to use compandors because of signal dynamic range considerations. Analog music signals are sometimes passed through a nonlinearity that reduces the range of the large signal excursions while amplifying the small signal portions. Before playback the signal is passed through the inverse nonlinearity which tends to reduce the noise in the low signal levels. Respectable gains in the signal-to-noise ratio can be accomplished with such schemes.

The third section of this paper is an investigation of companding in several dimensions. In several dimensions the compressor characteristic is a mapping function f: $\mathbb{R}^k \to \times_{i=1}^{k}(0,1)$, where $\times$ denotes the Cartesian cross product and $\times_{i=1}^{k}(0,1)$ is a k-dimensional hypercube. In the companding approach to optimal quantization, the quantizer output levels are distributed in the hypercube. We usually choose from these output levels the nearest neighbor to $f(\boldsymbol{x})$, where $\boldsymbol{x}$ is the input data vector. The quantized output is then f^{-1} of this particular output level.

The theory will also hold for analog signal processing in several dimensions. It does not matter whether the noise is quantization noise or any other kind of additive noise as long as the noise components in each channel are uncorrelated with one another. For example, let us denote the error vectoi caused by quantization in the hypercube by $(r_1, r_2, \cdots, r_k)^T$. Then the condition that is needed is $E\{r_i r_j\} = \sigma_r^2 \delta_{ij}$, where δ_{ij} is the Kronecker delta function. In a practical sense, this assumption is not very restrictive. It may be shown, at least asymptotically (as the number of output levels in the hypercube approaches infinity), that the error vector in an optimal or random quantizer converges to an hyperspherically symmetric probability density which satisfies our above condition.

The fourth section of this paper will deal with various techniques for implementing high-dimensional random quantizers. A random quantizer is one where the output levels are samples from some k-dimensional probability distribution $\lambda(\boldsymbol{x})$. If a companding approach is desired, then $\lambda(\boldsymbol{x})$ would be defined on $\times_{i=1}^{k}(0,1)$. It is known that random quantizers (with the correct $\lambda(\boldsymbol{x})$) approach the optimum quantizer performance as k approaches infinity. The implementation schemes discussed in this section are essentially search algorithms for quantizer output levels which enable us to find a "good" output level for a particular data vector.

II. Previous Work

Many authors have considered the problem of designing an optimal quantizer subject to some difference distortion measure. Max [2] gives necessary but not sufficient conditions for the optimal one-dimensional quantizer, and Fleisher [3] provides a sufficient condition that requires certain convexity properties of the density function of the input random variable.

Panter and Dite [4] derive an expression for the expected mean square error of a minimum mean square error one-dimensional quantizer, assuming the number of output levels to be very large. Algazi [5] generalizes Panter and Dite's equation to a tth power distortion measure. Wood [6] uses some equations derived by Roe [7] to rederive Panter and Dite's result and give formulas for obtaining the asymptotic quantizer's output levels. Zador [8] generalizes the work of Panter and Dite to several dimensions and to a more general difference distortion measure. Zador's equation for the distortion error is

$$C(k,t)N^{-t/k}\|p\|_{k/(k+t)} = \frac{1}{k}E\{\|\boldsymbol{x} - Q(\boldsymbol{x})\|_2^t\}, \quad (1)$$

where

$\boldsymbol{x}$	input random vector
$Q(\boldsymbol{x})$	quantized random vector
$E\{\cdot\}$	statistical expectation operator
N	number of output levels and assumed to be large
k	dimension of $\boldsymbol{x}$
$C(k,t)$	constant depending only upon k and t
$p(\boldsymbol{x})$	probability density of $\boldsymbol{x}$
$\|p\|_\alpha$	$[\int p(\boldsymbol{x})^\alpha \, d_{x_1} \cdots d_{x_k}]^{1/\alpha}$.

Zador also shows that $\lim_{k\to\infty} C(k,2) = 1/2\pi e$. $C(1,2)$ is known to be $1/12$. Gersho [9] gives an alternate derivation of the above equation and derives new bounds for $C(k,t)$, which is known for only a few values of k and t. Yamada, *et al.* [10], extend Zador's work to more general cost functions of the error.

III. Compandor Error Derivation

Our data will be assumed to be k-dimensional samples from a probability density function $p(\boldsymbol{x})$, $\boldsymbol{x} \in \mathbb{R}^k$. Let D_p be the support of $p(\boldsymbol{x})$. Let f: $D_p \to \times_{i=1}^{k}(0,1)$ be regular and onto.

We force f to be onto; if it were not there would be code vectors in the hypercube that would never be used, and hence the quantizer would be suboptimal. We use this condition at only one point in the derivation as a constraint on the optimal compandor. All equations derived up to that point are valid without this restriction. We will sometimes represent the mapping by

$$f = (f_1(\boldsymbol{x}), f_2(\boldsymbol{x}), \cdots, f_k(\boldsymbol{x}))^T$$

Let $\boldsymbol{r} = (r_1, r_2, \cdots, r_k)^T$ be the error vector in the hypercube. As stated before, under some fairly general conditions $E\{r_i r_j\} = (E\{r^2\}\delta_{ij})/k$ where δ_{ij} is the Kronecker delta. Assuming very small distortion, a good approximation to the final error vector in the output is $(f^{-1})'(\boldsymbol{y})\boldsymbol{r}$. When $k = 1$ this is equivalent to $f(x + \Delta x) \cong f'(x)\Delta x + f(x)$. Let $\boldsymbol{y}$ be the variable in the hypercube. If $\boldsymbol{y} = \boldsymbol{f}(\boldsymbol{x})$, then

$$p_y(\boldsymbol{y}) = \frac{p_x(f^{-1}(\boldsymbol{y}))}{|f'(f^{-1}(\boldsymbol{y}))|}$$

where $|\cdot|$ indicates determinant. Therefore the mean square error (mse) of the final output may be written

$$\text{mse} = \int_{\times_{i=1}^{k}(0,1)} r^T(f^{-1})'^T(f^{-1}(y)) \cdot (f^{-1})'(f^{-1}(y)) r \frac{p_x(f^{-1}(y))\,dy}{|f'(f^{-1}(y))|}.$$

Let $x = f^{-1}(y)$. Then $dx = |(f^{-1})'(y)|\,dy$ and $|(f^{-1})'(y)| = (|f'(f^{-1}(y))|)^{-1}$ by the inverse mapping theorem [11]. Making these changes of variables, we obtain

$$\text{mse} = \int_{D_p} r^T[f'(x)]^{-1T}[f'(x)]^{-1} r p_x(x)\,dx,$$

again by the inverse mapping theorem. Write $[f'(x)]^{-1T}[f'(x)]^{-1} = \Sigma^{-1}(x)$, which is a symmetric matrix for every x. Therefore our problem is to optimize

$$\int_{D_p} r^T \Sigma^{-1}(x) r p_x(x)\,dx.$$

Using a matrix identity, we can restate the above integral as

$$\int_{D_p} \operatorname{tr}\{\Sigma^{-1}(x) r r^T\} p_x(x)\,dx.$$

Let us take the expectation over the r variable which is independent of any other quantity in the integral. (We can make a random coding argument to insure that the r variable be independent, although it is tedious to do so.)

$$E\{rr^T\} = E\left\{\begin{pmatrix} r_1^2 & r_1 r_2 & \cdots & r_1 r_n \\ r_2 r_1 & r_2^2 & & r_2 r_n \\ r_n r_1 & \cdots & \cdots & r_1^2 \end{pmatrix}\right\} = \frac{E\{r^2\}}{k} I.$$

Therefore,

$$\text{mse} = \frac{E\{r^2\}}{k} \int_{D_p} \operatorname{tr}\{\Sigma^{-1}(x)\} p_x(x)\,dx. \tag{2}$$

This expression is of interest in its own right. $E\{r^2\}/k$ is the mean square error per sample suffered by the hypercube quantization. The total error is therefore a product of two independent terms. Denote the eigenvalues of $\Sigma(x)$ by $\lambda_i^2(x)$ $(i = 1, \cdots, k)$. Then

$$\text{mse} = \frac{E\{r^2\}}{k} \sum_{i=1}^{k} \int \frac{p_x(x)}{\lambda_i^2(x)}\,dx.$$

Since our map f is onto, we have

$$\int_{D_p} |f'(x)|\,dx = \int \prod_{i=1}^{k} \lambda_i(x)\,dx = 1.$$

Let us minimize the mse subject to the above constraint. First, it is easy to show that $\lambda_i(x) = \lambda(x)$ for every i. So we must minimize

$$\int_{D_p} \frac{p(x)}{\lambda(x)^2}\,dx$$

subject to the constraint

$$\int_{\mathbf{R}^k} \lambda(x)^k\,dx = 1.$$

Writing $\beta(x) = \lambda(x)^k$, we must minimize

$$\int \frac{p(x)}{\beta(x)^{2/k}}\,dx, \qquad \text{where} \quad \int \beta(x)\,dx = 1.$$

Gersho [9] shows that the optimal $\beta(x)$ is proportional to $p(x)^{1/(1+2/k)} = p^{k/(k+2)}(x)$, which implies that $\lambda(x) = p(x)^{1/(k+2)}/(\|p\|_{k/(k+2)})^{1/(k+2)}$. Using these eigenvalues, we can set the mse $= E\{r^2\}k^{-1}\|p\|_{k/(k+2)}$. If an optimal k-dimensional uniform quantizer is implemented in the hypercube, then this equation gives the same error as Zador's optimum quantizer. The condition for the optimal compressor is that all the eigenvalues of the symmetric matrix $\Sigma(x) = [f'(x)][f'(x)]^T$ be the same. This condition implies that there exists an orthonormal matrix $\phi(x)$ such that $\phi^T(x)\Sigma(x)\phi(x) = \lambda^2(x)I$; or $\Sigma(x) = \lambda^2(x)I = [f'(x)][f'(x)]^T$, which implies that $([f'(x)])/\lambda(x)$ is an orthonormal matrix. Since $\lambda(x)$ is known in principle we could solve for $f'(x)$ for every value of x. Therefore the condition for an optimal compandor is that $[f'(x)]/cp(x)^{1/(k+2)}$ be an orthogonal matrix for almost every value of x where $c = 1/(\|p\|_{k/(k+2)})^{1/(k+2)}$.

When $k = 2$ this condition says that $f(x)$ must be conformal almost everywhere except for a set of measure zero. Gersho points out (for the two-dimensional case) that conformal maps do not exist for circularly symmetric probability densities. An illustration of this fact is the work by Heppes and Szuz [12] which shows it is not possible to tessellate a circular region with an arbitrary "surface distribution function" using regular hexagons. There must always be a "slit" where the tessellation fails. This "slit", however, is a set of measure zero. Only local conformality almost everywhere is needed, not global conformality.

We now provide an example illustrating (2). Suppose that the input probability density $p(x)$ can be written as $\prod_{i=1}^{k} p(x_i)$. Let $C = 1/\int_{-\infty}^{\infty} p(x)^\alpha\,dx$ and let the compressor function $f = (f_1(x_1), f_2(x_2), \cdots, f_k(x_k))^T$ where $f_i(x_i) = C\int_{-\infty}^{x_i} p(x)^\alpha\,dx$. With little loss of generality, we assume f is regular; f is also obviously onto. Hence,

$$[f'(x)] = \begin{bmatrix} Cp(x_1)^\alpha & 0 & \cdots & 0 \\ 0 & Cp(x_2)^\alpha & & \vdots \\ 0 & & & \\ \vdots & & \ddots & 0 \\ 0 & \cdots & 0 & Cp(x_k)^\alpha \end{bmatrix},$$

$$[f'(x)]^{-1} = \begin{bmatrix} \dfrac{1}{Cp(x_1)^\alpha} & 0 & \cdots & 0 \\ & \ddots & & \vdots \\ 0 & \dfrac{1}{Cp(x_2)^\alpha} & & \\ \vdots & & \ddots & \\ 0 & \cdots & 0 & \dfrac{1}{Cp(x_k)^\alpha} \end{bmatrix}.$$

The eigenvalues of $\Sigma^{-1}(x)$ are $1/(C^2 p(x_i)^{2\alpha})$, $i = 1, \cdots, k$, so the error may be written

$$\begin{aligned}\text{mse} &= \frac{E\{r^2\}}{k} \sum_{i=1}^{k} \int_{D_p} \frac{\prod_{j=1}^{k} p(x_j)\,dx}{C^2 P(x_i)^{2\alpha}} \\ &= \frac{E\{r^2\}}{C^2} \int_{-\infty}^{\infty} p(x)^{1-2\alpha}\,dx \\ &= E\{r^2\} \left[\int_{-\infty}^{\infty} p(x)^{\alpha}\,dx\right]^2 \left[\int_{-\infty}^{\infty} p(x)^{1-2\alpha}\,dx\right].\end{aligned}$$

Using Hölders inequality we may show that $\alpha = 1/3$ minimizes the error:

$$\text{mse} = E\{r^2\} \| p \|_{1/3}.$$

But using Zador's coefficient for the one-dimension case (see (1)) we have

$$\text{mse}_{1-\text{dim}} = \frac{\| p \|_{1/3}}{12 N^2}.$$

Therefore, this compressor characteristic gives us the same error as the optimal one-dimensional quantizer if in the hypercube we quantize with one-dimensional uniform quantizers. We can quantize in the hypercube using optimal schemes for a coefficient of

$$\text{mse} = \frac{\| p \|_{1/3}}{N^2 2\pi e}, \qquad \text{as } K \to \infty.$$

Therefore the best we may produce with this compressor characteristic is a gain of $(2\pi e)/12 \cong 1.42$ in signal to quantizing noise ratio, at the expense of implementing optimal uniform quantizers in the hypercube.

As a second example, again let $p(x) = \prod_{i=1}^{k} p(x_i)$. Suppose we choose the eigenvalues of $\Sigma(x)$ to be

$$\lambda_i^2(x) = \left[\frac{\prod_{\substack{j=1 \\ j \neq i}}^{k} p(x_i)^{1/(k+1)}}{\int_{-\infty}^{\infty} p(x)^{(k-1)/(k+1)}\,dx} \right]^2,$$

which obviously leads to a nonconformal map. Using [2], we find the mean square error for such a compressor characteristic to be

$$\text{mse} = E\{r^2\} \| p \|_{(k-1)/(k+2)} = E\{r^2\} \| p \|_{(k-1)/(k-1)+2},$$

which is the optimal coefficient for the $(k-1)$-dimensional space. This relation implies the possibility of obtaining nonconformal mapping functions that will asymptotically give optimal results.

IV. Random Quantization

A multidimensional quantizer is essentially a partition of space. If a data sample falls within a certain set in the partition, it is assigned to a particular output level. It is known that the optimal partition of the hypercube consists of polytopes (multidimensional polygons). In $\mathbb{R}^1$ there is only one possibility, the line segment. In $\mathbb{R}^2$ the optimal polytope is the hexagon. In $\mathbb{R}^3$ Gersho conjectures that it is the truncated octahedron. The optimal polytope is not known for dimensions greater than three, and hence it seems very difficult to actually design a high-dimensional optimum quantizer. Linde, *et al.* [13] present an algorithm to design such quantizers that is guaranteed to converge to a local optimum but not always a global optimum. An alternative method for obtaining an asymptotically optimal formulation would be to assign the quantizer's output levels to be random samples from some point density function $\lambda(x)$. We would then find the nearest neighbor to our data vector from these values and use that neighbor as our quantized vector. Zador's upper bound to $C(k,t)$ can be derived by considering the error for such a quantizer. Zador and Gersho both point out that as the dimension of the random quantizer becomes very large, its performance approaches that of the optimum quantizer ($\lambda(x)$ must be proportional to $p^{k/(k+2)}(x)$). There is a drawback, however, to the straightforward implementation of such a quantizer. For a fixed data rate (say b bits per sample) the number of output levels in the quantizer is an exponential function of k (2^{bk} in fact). A nearest neighbor search for the closest output level to a particular data vector could require inordinate amounts of computer time.

In this section we will discuss some optimal and suboptimal techniques to find a quantized value. In a search scheme for an optimal nearest neighbor, we can visualize an hypersphere slowly expanding about the data point until the surface of the sphere comes in contact with a quantizer output level. This output level is then the nearest neighbor. To implement such a scheme we must calculate the Euclidean distance between our data vector and every output level. A Euclidean distance calculation is cumbersome in that it requires multiplications (K per sample). Thus we need to do $NK = 2^{bk}K$ multiplications.

An alternate search scheme is to expand a hypercube about the data point until it meets an output level. To find this "largest" hypercube requires only additions. Let us calculate the mean square error of such a scheme.

The probability of an output level being in a hypercube of side length r is $\lambda(x) r^k$. Then making use of order statistics, the probability density of the smallest hypercube side out of N samples is $KN[1 - r^k \lambda(x)]^{N-1} r^{k-1} \lambda(x)$. Suppose we have searched and found the largest hypercube; there must therefore be an output level somewhere on the surface of that hypercube. If N is large, $p(x)$ varies very little in this largest hypercube, and the output level may be considered to be uniformly distributed on this largest hypercube of side r. Since there are $2k$ sides, the mean square error may then be written (given r) as

$$\text{mse}|r = \frac{2^{k-1}}{r^{k-1}} \underbrace{\int_0^{r/2} \cdots \int_0^{r/2}}_{\substack{k-1 \\ \text{integrals}}} \left[\left(\frac{r}{2}\right)^2 + n_1^2 + n_2^2 + \cdots + n_{k-1}^2 \right] dn_1 \cdots dn_k = \frac{(k+2)r^2}{12}.$$

Taking the expectation over r, we find that the mean square error is

$$\frac{k+2}{12}\int r^2 N\left[1-\lambda(x)r^k\right]^{N-1} kr^{k-1}\lambda(x)\,dr\,p(x)\,dx$$
$$=\frac{k+2}{12N^{2/k}}\Gamma\left(1+\frac{2}{k}\right)\int\frac{p(x)}{\lambda(x)^{2/k}}dx.$$

$\lambda(x)$ is optimized when it is proportional to $p(x)^{k/k+2}$, and the error becomes

$$\text{mse}=\frac{k+2}{N^{2/k}12}\Gamma\left(1+\frac{2}{k}\right)\|p(x)\|_{k/(k+2)}.$$

If k is very large then the error becomes

$$\text{mse per sample}=\frac{\|p\|_{k/(k+2)}}{N^{2/k}12}.$$

(1/12 is the one-dimensional coefficient.) By doing no multiplication, we accept a degradation of $(2\pi e)/12$ in signal to quantizing noise ratio over that of the optimal search (see the first example in the previous section).

The hypercube search still requires searching through N code vectors to find the largest hypercube. We now describe a technique that will allow us to shorten the number of codewords to be searched. Since optimal companding gives optimal performance, we will restrict our discussion to data densities contained in the hypercube. We cut our k dimensional hypercube into several smaller hypercubes. If we slice each edge of the hypercube into 2^{b_1} equal intervals using $k-1$ dimensional perpendicular hyperplanes, we will have partitioned the large hypercube into 2^{kb_1} small hypercubes or "cubelettes". In one cubelette we randomly throw $N/2^{kb_1}$ uniformly distributed k-dimensional points which will be the quantizer output levels. We then replicate this cubelette through k space 2^{kb_1} times until we have filled up our original large cube. If we assume that there is a low probability that a data vector will be closer to the edge of a cubelette than to an output level, then we can show that this quantization scheme will have the same error as if we used uniformly distributed output levels through the whole hypercube.

Using one-dimensional uniform quantizers that are extremely easy to implement, we may quickly decide which one of the cubelettes contains the data vector. We then do a sequential search on $N/2^{kb_1}=2^{k(b-b_1)}$ output levels where kb is the number of bits per k-dimensional sample.

How large may b_1 be set? Obviously if $b_1=b$ we have one output level per cubelette, so that for a given data vector there is a good chance the edge of the cubelette will be closer then the output level. One way to get an idea of the number of output levels actually needed would be to calculate the probability of hitting an edge before a data point. To make things simpler, suppose that

$$p(x)=\prod_{i=1}^{k}p(x_i),$$

and that $p(x)$ is symmetric. The probability that the data point component x_i is closer to zero or one than to a corresponding output level component y_i is $|1-2x_i|$. If $p(x)$ is symmetric, then $E|1-2x_i|=1/2$. The probability that a data vector is closer to a particular code vector than to an edge is $1/2^k$. The probability that out of N output levels the data vector is closer to the edge than to any of them is $[1-1/2^k]^N$. Of course, we want this probability to be very small. If $N=C2^k$, then for large k the probability of hitting an edge before a code vector is e^{-c}. N therefore needs to be on the order of 2^k for all of our assumptions to hold and hence b_1 may be set at $b-1$ or slightly smaller.

V. Summary

A general expression for the performance of a multidimensional compander was derived and it was proved that the optimal compressor characteristic must be conformal almost everywhere. An example was given that indicates that asymptotically optimal performance may be gained even with nonconformal maps. Random quantizers were discussed which are known to be asymptotically optimal and two techniques to reduce computer time in their implementation were evaluated.

References

[1] W. R. Bennett, "Spectra of quantized signals," *Bell. Syst. Tech. J.*, vol. 27, pp. 446–472, July 1948.

[2] J. Max, "Quantizing for minimum distortion," *IRE Trans. Inform. Theory*, vol. IT-6, pp. 7–12, Mar. 1960.

[3] P. E. Fleisher, "Sufficient conditions for achieving minimum distortion in a quantizer," *IEEE Int. Conv. Rec.*, pt. I, pp. 104–111, 1964.

[4] P. F. Panter and W. Dite, "Quantization in pulse-count modulation with nonuniform spacing of levels," *Proc. IRE*, vol. 39, pp. 44–48, 1951.

[5] V. A. Algazi, "Useful approximations to optimum quantization," *IEEE Trans. Commun. Technol.*, vol. COM-14, pp. 297–301, 1966.

[6] R. C. Wood, "On optimimum quantization," *IEEE Trans. Inform. Theory*, vol. IT-5, pp. 248–252, Mar. 1969.

[7] G. M. Roe, "Quantizing for minimum distortion," *IRE Trans. Inform. Theory*, vol. IT-10, pp. 384–385, Oct. 1964.

[8] P. Zador, "Development and Evaluation of Procedures for Quantizing Multivariate Distributions", Ph.D. dissertation, Stanford, Univ. Stanford, CA, 1964, univ. microfilm no. 64-9855.

[9] A. Gersho, "Asymptotically optimal block quantization," *IEEE Trans. Inform. Theory*, vol. IT-25, pp. 373–380, July 1979.

[10] Y. Yamada, S. Tazaki, and R. M. Gray, "Asymptotic performance of block quantizers with difference distortion measures," *IEEE Trans. Inform. Theory*, vol. IT-26, pp. 6–14, Jan. 1980.

[11] W. Fleming, *Functions of Several Variables*. New York: Springer, 1977.

[12] A. Heppes and P. Szüsz, "Bemerkung zu einer Arbeit von L. Fejes Toth," *El. Math.*, vol. 15, pp. 134–136, 1960.

[13] Y. Linde, A. Buzo, and R. M. Gray, "An algorithm for vector quantizer design," *IEEE Trans. Comm.*, vol. COM-28, pp. 84–95, Jan. 1980.

14

Reprinted from *IEEE Trans. Inform. Theory* **IT-29**(2):279 (1983)

A Note on Optimal Multidimensional Companders

JAMES A. BUCKLEW, MEMBER, IEEE

Abstract—Companding is a widely used method of implementing nonuniform quantizers. There are, however, problems in extending this technique to the multidimensional case. It is known, for example, that not all input probability densities can be quantized optimally in this fashion. In this note we characterize the class of probability density functions that can be optimally quantized with a multidimensional companding-type implementation.

Companding [1] is a method of implementing nonuniform quantizers. In essence one models a nonuniform quantizer as a zero memory nonlinearity followed by an optimal uniform quantizer in turn followed by the inverse of the first nonlinearity. In [2] a mean-squared error expression is derived for the asymptotic performance (small distortion—high-information rate) to be expected for a multidimensional compander. The function f that maps the domain of the input k-dimensional random vector, having probability density $p(x)$, into the domain of the uniform quantizer is called (for historical reasons) the compressor. The condition derived in [2] for the compressor function to be optimal is that the compressor f must be one-to-one and satisfy two conditions almost everywhere:

1) $f\colon \mathbf{R}^k \to \mathbf{R}^k$ must be conformal, i.e., $\|[f'(x)h]\| = |f'(x)||h|$,

and

2) $|f'(x)| = p(x)^{k/k+2} \Big/ \int_{\mathbf{R}^k} p(x)^{k/k+2}\,dx$,

where $h \in \mathbf{R}^k$ is arbitrary and $|f'(x)|$ denotes the Jacobian of f.

Suppose we consider the problem of finding a function f that satisfies conditions 1) and 2) everywhere. The one-dimensional case ($k = 1$) is easy. Smith in 1957 [3] derives the optimum compressor as

$$f(x) = \frac{\int_{-\infty}^{x} p(\alpha)^{1/3}\,dx}{\int_{-\infty}^{\infty} p(\alpha)^{1/3}\,dx}.$$

Heppes and Szüsz [4] show that when $k = 2$, conditions 1) and 2) can be satisfied if and only if $\log p(x)$ satisfies Laplace's equation. The purpose of this note is to consider the cases when $k > 2$.

Hartman [5], in an extension of a theorem due to Liouville, showed that if $k > 2$ and $f\colon \mathbf{R}^k \to \mathbf{R}^k$ is conformal and continuously differentiable, then f is a restriction of a Möbius transformation. By a Möbius transformation we mean a translation ($f(x) = x + a$), a magnification ($f(x) = rx$, $r > 0$), an orthogonal transformation (f is linear and $|f(x)| = |x|$ $x \in \mathbf{R}^n$), a reflection through reciprocal radii ($f(x) = a + (r^2(x-a)/|x-a|^2)$), or a combination of these elementary transformations. A Möbius transformation can always be written in one of the forms [6] $f(x) = rTX + a$ or $f(x) = I(rTX + a)$, where $r > 0$, $a \in \mathbf{R}$, T is an orthogonal mapping, and I is a reflection through reciprocal radii. It is then easy to calculate for the first case $|f'(x)| =$ constant and for the second case $|f'(x)| = \alpha|x - x_0|^{-2}$, where x_0 is a constant vector in $\mathbf{R}^k$ and α is a real constant. We therefore have shown that conditions 1) and 2) can be satisfied everywhere for dimensions greater than two only if $p(x) =$ constant or $p(x) =$ constant $|x - x_0|^{-(k+2)2/k}$. We make the following remarks:

a) In [2] f was constrained to be a mapping from $\mathbf{R}^k \to X_{i=1}^k[0,1]$, where $X_{i=1}^k[0,1]$ is the k-dimensional unit cube. We did not make this constraint in the derivation considered above. We note, however, from a generalization of Sard's theorem [7] that

$$m(f(\mathbf{R}^k)) \leqslant \int_{\mathbf{R}^k} |f'(x)|\,dx = 1.$$

In addition, since conformal functions map balls into balls, we have that the range space of f is bounded and has at most unit measure.

b) As a matter of engineering practicality, it seems that the result derived above destroys the idea of using companders to implement optimum quantizers in higher dimensions. However, there is still the possibility that forcing conditions 1) and 2) to hold only *almost everywhere* could allow more freedom.

c) In [2] an example is given showing how almost optimal performance can be achieved with nonconformal compressor functions. This currently seems to be the area of greatest promise if one desires to retain the ease and versatility of the companding idea.

References

[1] W. R. Bennet, "Spectra of quantized signals," *Bell Syst. Tech. J.*, vol. 27, pp. 446–472, July 1948.

[2] J. A. Bucklew, "Companding and random quantization in several dimensions," *IEEE Trans. Inform. Theory*, vol. IT-27, no. 2, Mar. 1981.

[3] B. Smith, "Instantaneous companding of quantized signals," *Bell Syst. Tech. J.*, vol. 52, pp. 653–709, May 1957.

[4] A. Heppes and P. Szüsz, "Bemerkung zu einer Arbeit von L. Ferjes-Toth," *El. Math.*, vol. 15, pp. 134–136, 1960.

[5] P. Hartman, "On isometries and on a theorem of Liouville," *Math. Z.*, vol. 69, pp. 202–210, 1958.

[6] J. Väisälä, *Lectures on n-Dimensional Quasiconformal Mappings*. New York: Springer–Verlag, 1971.

[7] J. T. Schwartz, *Nonlinear Functional Analysis*. New York: Courant Institute of Mathematical Sciences, New York University, 1965.

Manuscript received March 12, 1982; revised June 16, 1982.
The author is with the Electrical and Computer Engineering Department, University of Wisconsin, Madison, WI 53705.

Reprinted from pages 211-212, 216, and 225-226 of *IEEE Trans. Inform. Theory*
IT-28(2):211-226 (1982)

Voronoi Regions of Lattices, Second Moments of Polytopes, and Quantization

J. H. CONWAY AND N. J. A. SLOANE, FELLOW, IEEE

Abstract—If a point is picked at random inside a regular simplex, octahedron, 600-cell, or other polytope, what is its average squared distance from the centroid? In n-dimensional space, what is the average squared distance of a random point from the closest point of the lattice A_n (or D_n, E_n, A_n^* or D_n^*)? The answers are given here, together with a description of the Voronoi (or nearest neighbor) regions of these lattices. The results have applications to quantization and to the design of signals for the Gaussian channel. For example, a quantizer based on the eight-dimensional lattice E_8 has a mean-squared error per symbol of $0.0717\cdots$ when applied to uniformly distributed data, compared with $0.08333\cdots$ for the best one-dimensional quantizer.

I. Quantization; Codes for Gaussian Channel

A. Introduction

The motivation for this work comes from block quantization and from the design of signals for the Gaussian channel. Let us call a finite set of points $y_1,\cdots,y_M$ in n-dimensional Euclidean space $\mathbb{R}^n$ a *Euclidean code*. An n-dimensional *quantizer* with outputs $y_1,\cdots,y_M$ is the function $Q\colon \mathbb{R}^n \to \mathbb{R}^n$ which sends each point $x \in \mathbb{R}^n$ into $Q(x) =$ closest codepoint y_i (in case of a tie, pick that y_i with the smallest subscript). If x has probability density function $p(x)$, the mean-squared error per symbol of this quantizer is

$$E(n, M, p, \{y_i\}) = \frac{1}{n}\int_{\mathbb{R}^n} \|x - Q(x)\|^2 p(x)\,dx,$$

where $\|x\| = (x \cdot x)^{1/2}$. Around each codepoint y_i is its *Voronoi region* $V(y_i)$ (see [49]), consisting of all points of the underlying space which are closer to that codepoint than to any other. More precisely, we define $V(y_i)$ to be the closed set

$$V(y_i) = \{x \in \mathbb{R}^n\colon \|x - y_i\| \le \|x - y_j\| \text{ for all } j \ne i\}.$$

(Voronoi regions are also called Dirichlet regions, Brillouin zones, Wigner–Seitz cells, or nearest neighbor regions.) If x is an interior point of $V(y_i)$, the quantizer replaces x by $Q(x) = y_i$. Then we may write

$$E(n, M, p, \{y_i\}) = \frac{1}{n}\sum_{i=1}^{M}\int_{V(y_i)} \|x - y_i\|^2 p(x)\,dx. \quad (1)$$

Given n, M, and $p(x)$ one wishes to find the infimum

$$E(n, M, p) = \inf_{\{y_i\}} E(n, M, p, \{y_i\})$$

over all choices of $y_1,\cdots,y_M$. Zador ([53]; see also [6], [7], [24], [52]) showed under quite general assumptions about $p(x)$ that

$$\lim_{M\to\infty} M^{2/n}E(n, M, p) = G_n\left(\int_{\mathbb{R}^n} p(x)^{n/(n+2)}\,dx\right)^{(n+2)/n}, \quad (2)$$

where G_n depends only on n. Zador also showed that

$$\frac{1}{(n+2)\pi}\Gamma\left(\frac{n}{2}+1\right)^{2/n} \le G_n \le \frac{1}{n\pi}\Gamma\left(\frac{n}{2}+1\right)^{2/n}\Gamma\left(1+\frac{2}{n}\right). \quad (3)$$

Asymptotically the upper and lower bounds in (3) agree, giving

$$G_n \to \frac{1}{2\pi e} = 0.0585498\cdots \text{ as } n \to \infty. \quad (4)$$

Since the probability density function $p(x)$ only appears in the last term of (2), we may choose any convenient $p(x)$ when attempting to find G_n. From now on we assume that the input x is uniformly distributed over a large region in n-dimensional space, and we can usually avoid edge effects by passing to a limiting situation with infinitely many y_i. With this assumption the mean-squared error is minimized if each codepoint y_i lies at the centroid of the corresponding Voronoi region $V(y_i)$ (see [24]). It is known that, for an optimal one-dimensional quantizer with a uniform input distribution, the points y_i should be uniformly spaced along the real line; correspondingly

$$G_1 = \frac{1}{12} = 0.08333\cdots. \quad (5)$$

Similarly for the optimal two-dimensional quantizer it is known that the points y_i should form the hexagonal lattice

Manuscript received December 7, 1980; revised November 4, 1981.
J. H. Conway is with the Department of Pure Mathematics and Mathematical Statistics, University of Cambridge, 16 Mill Lane, Cambridge CB2 1SB, England.
N. J. A. Sloane is with the Mathematics and Statistics Research Center, Bell Laboratories, Murray Hill, NJ 07974.

A_2 (described in Section III-A); correspondingly

$$G_2 = \frac{5}{36\sqrt{3}} = 0.0801875 \cdots, \qquad (6)$$

(see [21], [23], [24], [37]).

In three dimensions Gersho [24] conjectures that the optimal quantizer is based on the body-centered cubic lattice A_3^*, and that

$$G_3 = \frac{19}{192\sqrt[3]{2}} = 0.0785433 \cdots. \qquad (7)$$

Similarly in four dimensions he conjectures that the optimal quantizer is based on the lattice D_4, and that

$$G_4 = 0.076602 \qquad (8)$$

(obtained by Monte Carlo integration). Furthermore, he conjectures that, in all dimensions, any optimal quantizer is such that for large M the Voronoi regions $V(y_i)$ are all congruent, to some polytope P say. For such a quantizer we obtain, from (1) and (2),

$$G_n = \frac{1}{n} \frac{\int_P \|x - \hat{x}\|^2 \, dx}{\left(\int_P dx \right)^{(n+2)/n}}, \qquad (9)$$

where $\hat{x}$ is the centroid of P. The expression on the right makes sense for any polytope and will be denoted by $G(P)$: we refer to it as the *dimensionless second moment of* P. It is also convenient to have symbols for the *volume*, *unnormalized second moment*, and *normalized second moment* of P: these are

$$\operatorname{vol}(P) = \int_P dx,$$

$$U(P) = \int_P \|x - \hat{x}\|^2 \, dx,$$

and

$$I(P) = \frac{U(P)}{\operatorname{vol}(P)},$$

respectively. Then

$$G(P) = \frac{1}{n} \frac{U(P)}{\operatorname{vol}(P)^{1+2/n}} = \frac{1}{n} \frac{I(P)}{\operatorname{vol}(P)^{2/n}}.$$

If Gersho's conjecture is correct then G_n may be determined from

$$G_n = \min_P G(P), \qquad (10)$$

taken over all n-dimensional space-filling polytopes. Whether or not the conjecture is true, any value of $G(P)$ for a space-filling polytope is an upper bound to G_n. Furthermore (1), (2) and (9) allow us to interpret G_n and $G(P)$ as *mean-squared quantization errors per symbol*, assuming a uniform input distribution to the quantizer.

In the second application, the same Euclidean code $y_1, \cdots, y_M$ is used as a code for the Gaussian channel. Now the Voronoi regions are the decoding regions: all points x in the interior of $V(y_i)$ are decoded as y_i. If the codewords are equally likely and all the Voronoi regions $V(y_i)$ are congruent to a polytope P, the probability of correct decoding is proportional to

$$\int_P e^{-x \cdot x} \, dx.$$

The description of the Voronoi regions given in Section III thus makes it possible to calculate this probability exactly for many Euclidean codes. These results will be described elsewhere.

B. Summary of Results

In Sections II and III we compute $G(P)$ for a number of important polytopes (not just space-filling ones), including all regular polytopes (see Theorem 4). The three- and four-dimensional polytopes are compared in Tables I and II. The chief tools are Dirichlet's integral (Theorem 1), an explicit formula for the second moment of an n-simplex (Theorem 2), and a recursion formula giving the second moment of a polytope in terms of its cells (Theorem 3).

In Section III we study lattices, in particular the root lattices A_n, D_n, E_6, E_7, E_8, and their duals (defined in Section III-A). We determine the Voronoi regions for these lattices, and their second moments. The second moment gives the average squared distance of a point from the lattice. The *maximum* distance of any point of the underlying space from the lattice is its *covering radius*. The covering radii of these lattices were mostly already known (see for example [2], [4], [31]), but for completeness we rederive them. The final section (Section IV) compares the quantization errors of the different lattices—see Table V and Fig. 20. E_8 is the clear winner.

It is worth mentioning that for most of these lattices there are very fast algorithms for finding the closest lattice point to an arbitrary point; these are described in a companion paper [12]. The sizes of the spherical codes obtained from these lattices have been tabulated in [47].

Although we have tried to keep this paper as self-contained as possible, some familiarity with Coxeter's book [16] will be helpful to the reader. $\mathbb{Z}$ denotes the integers, $\mathbb{Q}$ the rationals and $\mathbb{R}$ the reals.

[*Editor's Note:* Material has been omitted at this point.]

A. Definition of the Root Lattices

For $n \geq 1$, A_n is the n-dimensional[2] lattice consisting of the points $(x_0, x_1, \cdots, x_n)$ in $\mathbb{Z}^{n+1}$ with $\Sigma x_i = 0$. The dual A_n^* consists of the union of $n+1$ cosets of A_n:

$$A_n^* = \bigcup_{i=0}^{n} ([i] + A_n),$$

where

$$[i] = \left(\frac{-j}{n+1}, \frac{-j}{n+1}, \cdots, \frac{-j}{n+1}, \frac{i}{n+1}, \cdots, \frac{i}{n+1}\right)$$
$$= \left(\left(\frac{-j}{n+1}\right)^i, \left(\frac{i}{n+1}\right)^j\right) \tag{28}$$

and $i + j = n + 1$. For $n = 1$ and 2, $A_n^* \cong A_n$ (i.e., they differ only by a rotation and change of scale).

For $n \geq 2$, D_n consists of the points $(x_1, x_2, \cdots, x_n)$ in $\mathbb{Z}^n$ with Σx_i even. In other words, if we color the integer lattice points alternately red and blue in a checkerboard coloring, D_n consists of the red points. The dual D_n^* is the union of four cosets of D_n:

$$D_n^* = \bigcup_{j=0}^{3} ([i] + D_n),$$

where

$$[0] = (0^n), \quad [1] = (\tfrac{1}{2}^n),$$
$$[2] = (0^{n-1}, 1), \quad [3] = (\tfrac{1}{2}^{n-1}, -\tfrac{1}{2}).$$

Also $D_2 \cong A_1 \oplus A_1$, $D_3 \cong A_3$, and $D_4^* \cong D_4$. Equivalently, D_n may be obtained by applying Construction A of [32] or [45] to the even weight code of length n. Similarly D_n^* is obtained by applying Construction A to the dual code $\{0^n, 1^n\}$.

There are many possible definitions of the lattices E_6, E_7, and E_8 (see the references given at the beginning of this section). We shall use the following: E_8 is the union of D_8 and the coset

$$(\tfrac{1}{2}, \tfrac{1}{2}, \tfrac{1}{2}, \tfrac{1}{2}, \tfrac{1}{2}, \tfrac{1}{2}, \tfrac{1}{2}, \tfrac{1}{2}) + D_8.$$

In other words E_8 consists of the points $(x_1, \cdots, x_8)$ with $x_i \in \mathbb{Z}$ and Σx_i even, together with the points $(y_1, \cdots, y_8)$ with $y_i \in \mathbb{Z} + \frac{1}{2}$ and Σy_i even. E_7 is a subspace of dimension 7 in E_8, consisting of the points $(u_1, \cdots, u_8) \in E_8$ with $u_7 = -u_8$. E_6 is a subspace of dimension 6 in E_8, consisting of the points $(u_1, \cdots, u_8) \in E_8$ with $u_6 = u_7 = -u_8$.

[2] The subscript gives the dimension of the lattice.

[*Editor's Note:* Material has been omitted at this point.]

TABLE V
DIMENSIONLESS SECOND MOMENT $G(\Lambda)$.

n	sphere bound	best lattice known Λ	$G(\Lambda)$	Zador bound
1	.0833	A_1	.0833	.500
2	.0796	A_2	.0802	.159
3	.0770	A_3^*	.0785	.116
4	.0750	D_4	.0766	.100
5	.0735	D_5^*	.0756	.091
6	.0723	E_6	.0743	.086
7	.0713	E_7	.0732	.082
8	.0704	E_8	.0717	.080
9	.0697	D_9^*	.0747	.078
10	.0691	D_{10}^*	.0747	.076

IV. Comparison of Quantizers

Comparing the different lattices analyzed in this section we see that the best quantizers found so far in dimensions 1–10 are the following:

dimension	lattice	dimension	lattice
1	$A_1(\cong A_1^*)$	6	E_6
2	$A_2(\cong A_2^*)$	7	E_7
3	$A_3^*(\cong D_3^*)$	8	$E_8(= E_8^*)$
4	$D_4(\cong D_4^*)$	9	D_9^*
5	D_5^*	10	D_{10}^* .

The values of the dimensionless second moment $G(\Lambda)$, which is our measure of the mean-squared quantization error per symbol, are shown in Table V and Fig. 20, together with Zador's bounds (3). It is known that A_1 and A_2 are optimal, and it is tempting to make the following conjecture.

Conjecture

The best *lattice* quantizer in $\mathbb{R}^n$—that with the lowest $G(\Lambda)$—is the dual of the densest lattice packing.

Certainly E_6^*, E_7^*, and the Leech lattice should be investigated.

It is worth drawing attention to the remarkably low value of the mean-squared error for E_8 (see Fig. 20). Furthermore there is a fast algorithm [12] available for performing the quantization with this lattice (and in fact for any of the lattices described here).

Note added in proof: It has recently been shown that the body-centered cubic lattice A_3^* is the optimal three-dimensional lattice quantizer for uniformly distributed data: see E. S. Barnes and N. J. A. Sloane, "The optimal lattice

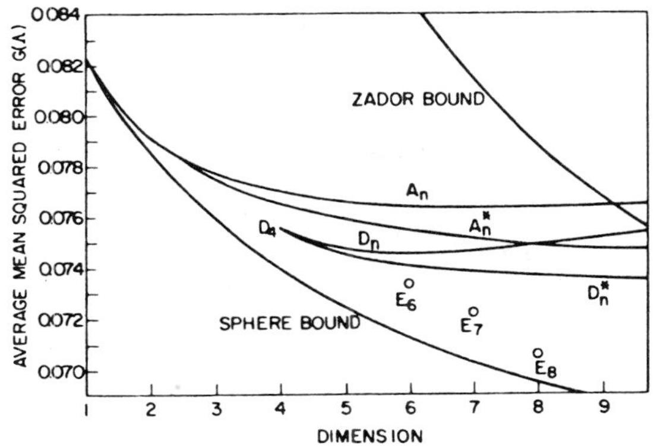

Fig. 20. Comparison of mean-squared quantization error per symbol, $G(\Lambda)$, for different lattices Λ in dimensions 1–9.

quantizer in three dimensions," *SIAM J. Discrete and Algebraic Methods*, to appear.

ACKNOWLEDGMENT

During the early stages of this work we were greatly helped by several discussions with Allen Gersho. Some of the calculations were performed on the MACSYMA system [35]. We should also like to thank E. S. Barnes and H. S. M. Coxeter for their comments on the manuscript.

REFERENCES

[1] A. D. Alexandrow, *Konvexe Polyeder.* Berlin: Akademie-Verlag, 1958.
[2] E. S. Barnes, "Construction of perfect and extreme forms II," *Acta Arithmetica*, vol. 5, pp. 205–222, 1959.
[3] C. T. Benson and L. C. Grove, *Finite Reflection Groups.* Tarrytown-on-Hudson, NY: Bogden and Quigley, 1971.
[4] M. N. Bleicher, "Lattice coverings of n-space by spheres," *Can. J. Math.*, vol. 14, pp. 632–650, 1962.
[5] N. Bourbaki, *Groupes et Algèbres de Lie, Chapitres 4, 5 et 6.* Paris: Hermann, 1968.
[6] J. A. Bucklew, "Upper bounds to the asymptotic performance of block quantizers," *IEEE Trans. Inform. Theory*, vol. IT-27, no. 5, pp. 577–581, Sept. 1981.
[7] J. A. Bucklew and G. L. Wise, "A note on multidimensional asymptotic quantization theory," *Proc. 18th Allerton Conf. on Commun., Control, and Computing*, 1980, pp. 454–463.
[8] J. W. S. Cassels, *An Introduction to the Geometry of Numbers.* New York: Springer-Verlag, 1971.
[9] J. H. Conway, R. A. Parker, and N. J. A. Sloane, "The covering radius of the Leech lattice," *Proc. Royal Soc.*, 1982, to be published.
[10] J. H. Conway and L. Queen, "On computing the characters of Lie groups," in preparation.
[11] J. H. Conway and N. J. A. Sloane, "On the enumeration of lattices of determinant one," *J. Number Theory*, to be published.
[12] ——, "Fast quantizing and decoding algorithms for lattice quantizers and codes," *IEEE Trans. Inform. Theory*, this issue, pp. 227–232.
[13] H. S. M. Coxeter, "Wythoff's construction for uniform polytopes," *Proc. London Math. Soc.*, Ser. 2, vol. 38, pp. 327–339, 1935; reprinted in [15].
[14] ——, "Extreme forms," *Can. J. Math.*, vol. 3, pp. 391–441, 1951.
[15] ——, *Twelve Geometric Essays.* Carbondale: Southern Illinois Univ., 1968.
[16] ——, *Regular Polytopes.* New York: Dover, 3rd ed., 1973.
[17] ——, "Polytopes in the Netherlands," *Nieuw Archief voor Wiskunde* (3), vol. 26, pp. 116–141, 1978.
[18] H. S. M. Coxeter and W. O. J. Moser, *Generators and Relations for Discrete Groups.* New York: Springer-Verlag, 4th ed., 1980.
[19] H. M. Cundy and A. P. Rollett, *Mathematical Models.* Oxford: The Univ., 1961.
[20] E. L. Elte, "The semiregular polytopes of the hyperspaces," Doctor of Math. and Science Thesis, State-Univ. of Gronigen, The Netherlands, 13 May 1912.
[21] L. Fejes Tóth, "Sur la représentation d'une population infinie par un nombre fini d'éléments," *Acta Math. Acad. Sci. Hungar.*, vol. 10, pp. 299–304, 1959.
[22] ——, *Regular Figures.* Oxford: Pergamon, 1964.
[23] ——, *Lagerungen in der Ebene, auf der Kugel und im Raum.* New York: Springer-Verlag, 2nd ed., 1972.
[24] A. Gersho, "Asymptotically optimal block quantization," *IEEE Trans. Inform. Theory*, vol. IT-25, no. 4, pp. 373–380, July 1979.
[25] I. J. Good and R. A. Gaskins, "The centroid method of numerical integration," *Numer. Math.*, vol. 16, pp. 343–359, 1971.
[26] B. Grünbaum, *Convex Polytopes.* New York: Wiley, 1967.
[27] M. Hazewinkel *et al.*, "The ubiquity of Coxeter–Dynkin diagrams (an introduction to the A-D-E problem)," *Nieuw Archief voor Wiskunde* (3), vol. 25, pp. 257–307, 1977.
[28] D. Hilbert and S. Cohn-Vossen, *Geometry and The Imagination.* New York: Chelsea, 1952.
[29] A. Holden, *Shapes, Space, and Symmetry.* New York: Columbia Univ., 1971.
[30] J. E. Humphreys, *Introductioı. to Lie Algebras and Representation Theory.* New York: Springer-Verlag, 1972.
[31] G. Kaur, "Extreme quadratic forms for coverings in four variables," *Proc. Nat. Inst. Sci. India*, vol. 32A, pp. 414–417, 1966.
[32] J. Leech and N. J. A Sloane, "Sphere packing and error-correcting codes," *Can. J. Math.* vol. 23, pp. 718–745, 1971.
[33] A. L. Loeb, *Space Structures: Their Harmony and Counterpoint.* Reading, MA: Addison-Wesley, 1976.
[34] L. A. Lyusternik, *Convex Figures and Polyhedra.* New York: Dover, 1963.
[35] Mathlab Group, *MACSYMA Reference Manual.* Cambridge, MA: Lab. Comp. Sci., M.I.T., version 9, 1977.
[36] P. McMullen and G. C. Shephard, *Convex Polytopes and the Upper Bound Conjecture*, London Math. Soc. Lecture Note Series 3. Cambridge: The Univ., 1971.
[37] D. J. Newman, "The hexagon theorem," *IEEE Trans. Inform. Theory*, this issue, pp. 137–139.
[38] H. V. Niemeier, "Definite quadratische Formen der Dimension 24 und Diskriminante 1," *J. Number Theory*, vol. 5, pp. 142–178, 1973.
[39] O. T. O'Meara, *Introduction to Quadratic Forms.* New York: Springer-Verlag, 1971.
[40] A. Pugh, *Polyhedra: A Visual Approach.* Berkeley: Univ. of Calif., 1976.
[41] J. Riordan, *An Introduction to Combinatorial Analysis.* New York: Wiley, 1958.
[42] J. Riordan, *Combinatorial Identities.* New York: Wiley, 1968.
[43] J. Satterly, "The moments of inertia of some polyhedra," *Math. Gazette*, vol. 42, pp. 11–13, 1958.
[44] N. J. A. Sloane, *A Handbook of Integer Sequences.* New York: Academic, 1973.
[45] N. J. A. Sloane, "Binary codes, lattices and sphere packings," in *Combinatorial Surveys* (Proc. 6th British Combinatorial Conf.), P. J. Cameron, Ed. London and New York: Academic, 1977, pp. 117–164.
[46] N. J. A. Sloane, "Self-dual codes and lattices," in *Relations Between Combinatorics and Other Parts of Mathematics*, Proc. Symp. Pure Math., vol. 34, Providence RI: Amer. Math. Soc., 1979, pp. 273–308.
[47] N. J. A. Sloane, "Tables of sphere packings and spherical codes," *IEEE Trans. Inform. Theory*, vol. IT-27, no. 3, pp. 327–338, May 1981.
[48] B. M. Stewart, *Adventures among the Toroids.* Published by the author, 4494 Wausau Rd., Okemos, MI, 1970.
[49] G. Voronoi, "Récherches sur les paralléloèdres primitifs I, II," *J. reine angew. Math.*, vol. 134, pp. 198–287, 1908, and vol. 136, pp. 67–181, 1909.
[50] M. J. Wenninger, *Polyhedron Models.* Cambridge: The Univ., 1971.
[51] E. T. Whittaker and G. N. Watson, *A Course of Modern Analysis.* Cambridge: The Univ., 4th ed., 1963.
[52] Y. Yamada, S. Tazaki, and R. M. Gray, "Asymptotic performance of block quantizers with difference distortion measures," *IEEE Trans. Inform. Theory*, vol. IT-26, no. 1, pp. 6–14, Jan. 1980.
[53] P. Zador, "Topics in the asymptotic quantization of continuous random variables," *IEEE Trans. Inform. Theory*, this issue, pp. 139–149.

Reprinted from *IEEE Trans. Inform. Theory* **IT-28**(2):227–232 (1982)

Fast Quantizing and Decoding Algorithms for Lattice Quantizers and Codes

J. H. CONWAY AND N. J. A. SLOANE, FELLOW, IEEE

Abstract—For each of the lattices $A_n (n \geq 1)$, $D_n (n \geq 2)$, E_6, E_7, E_8, and their duals a very fast algorithm is given for finding the closest lattice point to an arbitrary point. If these lattices are used for vector quantizing of uniformly distributed data, the algorithm finds the minimum distortion lattice point. If the lattices are used as codes for a Gaussian channel, the algorithm performs maximum likelihood decoding.

I. Introduction

THE SO-CALLED *root lattices* are the n-dimensional lattices A_n $(n \geq 1)$, $D_n (n \geq 2)$, and $E_n (n = 6, 7, 8)$ defined in Section II. These lattices and their duals give rise to the densest known sphere packing and coverings in dimensions $n \leq 8$, and they can be used as the basis for efficient block quantizers for uniformly distributed inputs and to construct codes for a band-limited channel with Gaussian noise (see [6], [9], [11], [16]). Around each lattice point is its *Voronoi region*, consisting of all points of the underlying space which are closer to that lattice point than to any other. (Voronoi regions are also called Dirichlet regions, Brillouin zones, Wigner–Seitz cells, or nearest neighbor regions.) If the lattice is used as a quantizer, all the points in the Voronoi region around the lattice point x are represented by x; while if the lattice is used as a code for a Gaussian channel, all the points in the Voronoi region around x are decoded as x. In the preceding paper [6] we found the Voronoi regions for most of the root lattices and their duals, as well as the mean-squared quantization error when these lattices are used to quantize uniformly distributed data.

In the present paper we give very fast and simple algorithms which, for any of the lattices A_n, D_n, E_n, and their duals (as well as many other lattices), find the closest lattice point to an arbitrary point of the space. In other words the algorithms find which Voronoi region the given point belongs to. The algorithms can therefore be used either for vector quantizing or for channel decoding. The running time of the algorithms for D_n, D_n^*, E_n, and E_n^* is proportional to n, while for A_n and A_n^* it is proportional to $n \log n$ and $n^2 \log n$, respectively.

Although considerable work has been done in the past on n-dimensional quantizers and codes (see [1], [3], [8], [9], [14], [16]–[18], and the references given there), these algorithms appear to be new.

II. Definition of Root Lattices and their Duals

If $a_1, \cdots, a_n$ are linearly independent vectors in m-dimensional real Euclidean space $\mathbb{R}^m$ with $m \geq n$, the set of all vectors

$$x = u_1 a_1 + \cdots + u_n a_n,$$

where $u_1, \cdots, u_n$ are arbitrary integers, is called an n-dimensional *lattice* Λ. The *dual* (or *reciprocal*) lattice Λ^* consists of all points y in the subspace of $\mathbb{R}^m$ spanned by $a_1, \cdots, a_n$ such that the inner product $x \cdot y = x_1 y_1 + \cdots + x_m y_m$ is an integer for all $x \in \Lambda$. Most of the lattices considered here are contained in their duals, so that there are coset representatives $r_0, \cdots, r_{d-1}$ such that

$$\Lambda^* = \bigcup_{i=0}^{d-1} (r_i + \Lambda). \tag{1}$$

The number d is called the *determinant* of Λ. The *norm* (or energy) of a vector $x = (x_1, \cdots, x_m) \in \mathbb{R}^m$ is

$$\|x\| = \sqrt{x \cdot x} = \left(\sum x_i^2\right)^{1/2}.$$

For further information about lattices see for example [2], [4]–[7], [11], [12], [15], [16].

For $n \geq 1$, A_n is the n-dimensional[1] lattice consisting of the points $(x_0, x_1, \cdots, x_n)$ having integer coordinates that sum to zero. (Thus A_n is an n-dimensional lattice described by $(n + 1)$-dimensional coordinates.) If two lattices A and B differ only by a rotation and change of scale, we say they are equivalent and write $A \cong B$. Then A_1 is equivalent to the one-dimensional lattice of integer points, $\mathbb{Z}$, and A_2 is equivalent to the familiar two-dimensional hexagonal lattice (see Fig. 1).

The dual lattice A_n^* consists of the union of $n + 1$ cosets of A_n:

$$A_n^* = \bigcup_{i=0}^{n} (r_i + A_n), \tag{2}$$

Manuscript received May 7, 1981; revised November 3, 1981.

J. H. Conway is with the Department of Pure Mathematics and Mathematical Statistics, University of Cambridge, 16 Mill Lane, Cambridge CB2 1SB, England.

N. J. A. Sloane is with the Mathematics and Statistics Research Center, Bell Laboratories, Murray Hill, NJ 07974.

[1] The subscript gives the dimension of the lattice.

where, for $i = 0, 1, \cdots, n$ and $j = n + 1 - i$,

$$r_i = \left(\frac{-j}{n+1}, \frac{-j}{n+1}, \cdots, \frac{-j}{n+1}, \frac{i}{n+1}, \cdots, \frac{i}{n+1}\right)$$
$$= \left(\left(\frac{-j}{n+1}\right)^i, \left(\frac{i}{n+1}\right)^j\right). \qquad (3)$$

Then $A_1^* \cong A_1 \cong \mathbb{Z}$ and $A_2^* \cong A_2$. Also A_3 ($\cong D_3$) is the face-centered cubic lattice and A_3^* ($\cong D_3^*$) is the body-centered cubic lattice. It is known that A_2 is the optimal two-dimensional quantizer for a uniformly distributed input, and it is conjectured that A_3^* is optimal in three dimensions.

For $N \geq 2$, D_n consists of the points $(x_1, x_2, \cdots, x_n)$ having integer coordinates with an even sum. The dual D_n^* is the union of four cosets of D_n:

$$D_n^* = \bigcup_{i=0}^{3} (r_i + D_n), \qquad (4)$$

where

$$r_0 = (0^n), \quad r_1 = \left(\tfrac{1}{2}^n\right),$$
$$r_2 = (0^{n-1}, 1), \quad r_3 = \left(\tfrac{1}{2}^{n-1}, -\tfrac{1}{2}\right).$$

Also $D_2 \cong A_1 \oplus A_1 \cong D_2^*$, $D_3 \cong A_3$, and $D_4^* \cong D_4$.

For example the vectors of small norm in D_4 consist of all permutations and sign changes of the following:

type	number	norm²
(0, 0, 0, 0)	1	0
(±1, ±1, 0, 0)	24	2
(±2, 0, 0, 0)	8	4
(±1, ±1, ±1, ±1)	16	4
(±2, ±1, ±1, 0)	96	6
(±2, ±2, 0, 0)	24	8
...	...	...

(5)

(See also [16, Table V].)

There is a second, equivalent definition of D_n^* which is sometimes easier to use: D_n^* consists of the points of the n-dimensional integer lattice $\mathbb{Z}^n$, together with the translate of $\mathbb{Z}^n$ by the vector $(\frac{1}{2}, \frac{1}{2}, \cdots, \frac{1}{2})$, i.e.,

$$D_n^* = \mathbb{Z}^n \cup \left(\left(\tfrac{1}{2}^n\right) + \mathbb{Z}^n\right). \qquad (6)$$

The most convenient definitions of E_6, E_7, and E_8 are the following.[2] E_8 is the union of D_8 and the coset

$$\left(\tfrac{1}{2}, \tfrac{1}{2}, \tfrac{1}{2}, \tfrac{1}{2}, \tfrac{1}{2}, \tfrac{1}{2}, \tfrac{1}{2}, \tfrac{1}{2}\right) + D_8.$$

In other words E_8 consists of the points $(x_1, \cdots, x_8)$ with $x_i \in \mathbb{Z}$ and Σx_i even, together with the points $(y_1, \cdots, y_8)$ with $y_i \in \mathbb{Z} + \frac{1}{2}$ and Σy_8 even. Also $E_8^* = E_8$.

[2] A slightly different (although equivalent) definition of E_7 was used in [6]. The present definition leads to a simpler algorithm in Section VII.

E_7 is a subspace of dimension 7 in E_8, consisting of the points $(u_1, \cdots, u_8)$ in E_8 with $\Sigma u_i = 0$. Equivalently

$$E_7 = A_7 \cup \left(\left(-\tfrac{1}{2}^4, \tfrac{1}{2}^4\right) + A_7\right). \qquad (7)$$

The dual E_7^* is given by

$$E_7^* = E_7 \cup \left(\left(-\tfrac{3}{4}^2, \tfrac{1}{4}^6\right) + E_7\right)$$
$$= \bigcup_{i=0}^{3} (s_i + A_7), \qquad (8)$$

where

$$s_i = \left(\left(\frac{-j}{4}\right)^{2i}, \left(\frac{i}{4}\right)^{2j}\right), \qquad i + j = 4. \qquad (9)$$

Finally E_6 is a subspace of dimension 6 in E_8, which we may take for example to consist of the points $(u_1, \cdots, u_8)$ in E_8 with $\Sigma u_i = 0$ and $u_1 + u_8 = 0$. However this lattice and its dual do not appear to be as important as the others for the applications considered here, and we shall not discuss them in as much detail.

III. Finding the Closest Point of the n-Dimensional Integer Lattice $\mathbb{Z}^n$

The algorithm for finding the closest point of the integer lattice $\mathbb{Z}^n$ to an arbitrary point $\boldsymbol{x} \in \mathbb{R}^n$ is particularly simple, and serves to introduce the notation. For a real number x, let

$$f(x) = \text{closest integer to } x.$$

In case of a tie, choose the integer with the smallest absolute value. For $\boldsymbol{x} = (x_1, \cdots, x_n) \in \mathbb{R}^n$, let

$$f(\boldsymbol{x}) = (f(x_1), \cdots, f(x_n)).$$

For later use we also define $g(\boldsymbol{x})$, which is the same as $f(\boldsymbol{x})$ except that the *worst* component of $\boldsymbol{x}$—that furthest from an integer—is rounded the *wrong way*. In case of a tie, the component with the lowest subscript is rounded the wrong way.

More formally, for $x \in \mathbb{R}$ we define $f(x)$ and the function $w(x)$ which rounds the wrong way as follows. (Here m is an integer.)

If $x = 0$,	then $f(x) = 0$, $w(x) = 1$.
If $0 < m \leq x \leq m + \frac{1}{2}$,	then $f(x) = m$, $w(x) = m + 1$.
If $0 < m + \frac{1}{2} < x < m + 1$,	then $f(x) = m + 1$, $w(x) = m$.
If $-m - \frac{1}{2} \leq x \leq -m < 0$,	then $f(x) = -m$, $w(x) = -m - 1$.
If $-m - 1 < x < -m - \frac{1}{2}$,	then $f(x) = -m - 1$, $w(x) = -m$.

(10)

(Ties are handled so as to give preference to points of

smaller norm.) We also write

$$x = f(x) + \delta(x),$$

so that $|\delta(x)| \le 1/2$ is the distance from x to the nearest integer.

Given $\boldsymbol{x} = (x_1, \cdots, x_n) \in \mathbb{R}^n$, let k $(1 \le k \le n)$ be such that

$$|\delta(x_k)| \le |\delta(x_i)| \qquad \text{for all } 1 \le i \le n$$

and

$$|\delta(x_k)| = |\delta(x_i)| \Rightarrow k \le i.$$

Then $g(\boldsymbol{x})$ is defined by

$$g(\boldsymbol{x}) = (f(x_1), f(x_2), \cdots, w(x_k), \cdots, f(x_n)).$$

Algorithm 1—To Find the Closest Point of $\mathbb{Z}^n$ to $\boldsymbol{x}$: Given $\boldsymbol{x} \in \mathbb{R}^n$, the closest point of $\mathbb{Z}^n$ is $f(\boldsymbol{x})$. (If $\boldsymbol{x}$ is equidistant from two or more points of $\mathbb{Z}^n$, this procedure finds the one with the smallest norm.)

To see that the procedure works, let $\boldsymbol{u} = (u_1, \cdots, u_n)$ be any point of $\mathbb{Z}_n$. Then

$$\|\boldsymbol{u} - \boldsymbol{x}\|^2 = \sum_{i=1}^{n} (u_i - x_i)^2,$$

which is minimized by choosing $u_i = f(x_i)$ for $i = 1, \cdots, n$. Because of (10) ties are broken correctly, favoring the point with the smallest norm.

IV. Finding the Closest Point of D_n

Algorithm 2—To Find the Closest Point of D_n to $\boldsymbol{x}$: Given $\boldsymbol{x} \in \mathbb{R}^n$, the closest point of D_n is whichever of $f(\boldsymbol{x})$ and $g(\boldsymbol{x})$ has an even sum of components (one will have an even sum, the other an odd sum). If $\boldsymbol{x}$ is equidistant from two or more points of D_n this procedure produces a nearest point having the smallest norm.

This procedure works because $f(\boldsymbol{x})$ is the closest point of $\mathbb{Z}^n$ to $\boldsymbol{x}$ and $g(\boldsymbol{x})$ is the next closest. $f(\boldsymbol{x})$ and $g(\boldsymbol{x})$ differ by one in exactly one coordinate, and so precisely one of $\Sigma f(x_i)$ and $\Sigma g(x_i)$ is even and the other is odd. Again (10) implies that ties are broken correctly.

Example: Find the closest point of D_4 to $\boldsymbol{x} = (0.6, -1.1, 1.7, 0.1)$. We compute

$$f(\boldsymbol{x}) = (1, -1, 2, 0)$$

and

$$g(\boldsymbol{x}) = (0, -1, 2, 0),$$

since the first component of $\boldsymbol{x}$ is the furthest from an integer. The sum of the components of $f(\boldsymbol{x})$ is $1 - 1 + 2 + 0 = 2$, which is even, while that of $g(\boldsymbol{x})$ is $0 - 1 + 2 + 0 = 1$, which is odd. Therefore $f(\boldsymbol{x})$ is the point of D_4 closest to $\boldsymbol{x}$.

To illustrate how ties are handled, suppose

$$\boldsymbol{x} = (\tfrac{1}{2}, \tfrac{1}{2}, \tfrac{1}{2}, \tfrac{1}{2}).$$

In fact (see (5)) $\boldsymbol{x}$ is now equidistant from eight points of D_4, namely $(0,0,0,0)$, any permutation of $(1,1,0,0)$, and $(1,1,1,1)$. The algorithm computes

$$f(\boldsymbol{x}) = (0,0,0,0), \quad \text{sum} = 0, \quad \text{even},$$
$$g(\boldsymbol{x}) = (1,0,0,0), \quad \text{sum} = 1, \quad \text{odd},$$

and selects $f(\boldsymbol{x})$. Indeed $f(\boldsymbol{x})$ does have the smallest norm of the eight neighboring points.

V. Finding the Closest Point of a Coset or of a Dual Lattice

A procedure Φ for finding the closest point of a lattice Λ to a given point $\boldsymbol{x}$ can be easily converted to a procedure for finding the closest point of a coset $\boldsymbol{r} + \Lambda$ to $\boldsymbol{x}$. For if $\Phi(\boldsymbol{x})$ is the closest point of Λ to $\boldsymbol{x}$,

$$\Phi(\boldsymbol{x} - \boldsymbol{r}) + \boldsymbol{r}$$

is the closest point of $\boldsymbol{r} + \Lambda$ to $\boldsymbol{x}$.

Suppose further that L is a lattice (or in fact any set of points) which is a union of cosets of Λ:

$$L = \bigcup_{i=0}^{d-1} (\boldsymbol{r}_i + \Lambda).$$

Then Φ can be used as the basis for the following procedure for finding the closest point of L.

Algorithm 3—To Find the Closest Point of L (A Union of Cosets of a Lattice) to a Given Point $\boldsymbol{x}$: Given $\boldsymbol{x}$, compute

$$\boldsymbol{y}_i = \Phi(\boldsymbol{x} - \boldsymbol{r}_i) + \boldsymbol{r}_i$$

for $i = 0, 1, \cdots, d-1$. Compare each of $\boldsymbol{y}_0, \cdots, \boldsymbol{y}_{d-1}$ with $\boldsymbol{x}$ and choose the closest.

In view of (1), (2), (4), (8) this algorithm reduces the problem of finding the closest point of the dual lattices A_n^*, D_n^* and E_6^*, E_7^*, E_8^* to that of finding the closest point of the original lattices. Alternatively, a faster algorithm for D_n^* is obtained from (6) and Algorithm 1. We illustrate Algorithm 3 by applying it first to D_5^* using definition (4), and then to D_3^* using definition (6).

For the first example we observe from (4) that D_5^* is the union of four cosets of D_5 with coset representatives

$$\boldsymbol{r}_0 = (0, 0, 0, 0, \quad 0),$$
$$\boldsymbol{r}_1 = (\tfrac{1}{2}, \tfrac{1}{2}, \tfrac{1}{2}, \tfrac{1}{2}, \quad \tfrac{1}{2}),$$
$$\boldsymbol{r}_2 = (0, 0, 0, 0, \quad 1),$$
$$\boldsymbol{r}_3 = (\tfrac{1}{2}, \tfrac{1}{2}, \tfrac{1}{2}, \tfrac{1}{2}, -\tfrac{1}{2}).$$

To find the closest point of D_5^* to

$$\boldsymbol{x} = (0.1, 0, -0.3, 0.4, 0.8)$$

we proceed as follows:

$$f(\boldsymbol{x}) = (0,0,0,0,1), \quad \text{sum} = 1, \quad \text{odd},$$
$$g(\boldsymbol{x}) = (0,0,0,1,1), \quad \text{sum} = 2, \quad \text{even},$$

therefore

$$\boldsymbol{y}_0 = (0,0,0,1,1).$$

Also

$$x - r_1 = (-0.4, -0.5, -0.8, -0.1, 0.3),$$

$$f(x - r_1) = (0, 0, -1, 0, 0), \quad \text{sum} = -1, \quad \text{odd},$$

$$g(x - r_1) = (0, -1, -1, 0, 0), \quad \text{sum} = -2, \quad \text{even},$$

therefore

$$y_1 = g(x - r_1) + r_1 = (0.5, -0.5, -0.5, 0.5, 0.5);$$

$$x - r_2 = (0.1, 0, -0.3, 0.4, -0.2),$$

$$f(x - r_2) = (0, 0, 0, 0, 0), \quad \text{sum} = 0, \quad \text{even},$$

$$g(x - r_2) = (0, 0, 0, 1, 0), \quad \text{sum} = 1, \quad \text{odd},$$

therefore

$$y_2 = f(x - r_2) + r_2 = (0, 0, 0, 0, 1);$$

and

$$x - r_3 = (-0.4, -0.5, -0.8, -0.1, 1.3),$$

$$f(x - r_3) = (0, 0, -1, 0, 1), \quad \text{sum} = 0, \quad \text{even},$$

$$g(x - r_3) = (0, -1, -1, 0, 1), \quad \text{sum} = -1, \quad \text{odd},$$

therefore

$$y_3 = f(x - r_3) + r_3 = (0.5, 0.5, -0.5, 0.5, 0.5).$$

The final step is to see which of $y_0, \cdots, y_3$ is closest to x. We compute

$$\|x - y_0\|^2 = 0.5,$$

$$\|x - y_1\|^2 = 0.55,$$

$$\|x - y_2\|^2 = 0.3,$$

$$\|x - y_3\|^2 = 0.55.$$

Thus

$$y_2 = (0, 0, 0, 0, 1)$$

is the point of D_5^* closest to x.

For the second example we use (6) to define the body-centered cubic lattice D_3^* as the union of two cosets of $\mathbb{Z}^3$ with coset representatives

$$r_0 = (0, 0, 0),$$

$$r_1 = (\tfrac{1}{2}, \tfrac{1}{2}, \tfrac{1}{2}).$$

To find the closest point of D_3^* to

$$x = (0.2, 0.5, 0.8),$$

we compute

$$y_0 = f(x) = (0, 0, 1),$$

$$x - r_1 = (-0.3, 0, 0.3),$$

$$f(x - r_1) = (0, 0, 0),$$

$$y_1 = r_1 + f(x - r_1) = (0.5, 0.5, 0.5).$$

The final step is to find which of y_0 and y_1 is closer to x:

$$\|x - y_0\|^2 = 0.33, \quad \|x - y_1\|^2 = 0.18.$$

Thus $y_1 = (0.5, 0.5, 0.5)$ is the closest point of D_3^* to x.

Obviously the second definition of D_n^* leads to the faster algorithm. For any n it is only necessary to compute $f(x)$, $f(x - r_1)$, and to calculate the squared norms $\|x - y_0\|^2$ and $\|x - y_1\|^2$.

VI. Finding the Closest Point of E_8

Since E_8 is the union of two cosets of D_8 the discussion in the previous section leads to the following procedure.

Algorithm 4—To Find the Closest Point of E_8 to x: Given $x = (x_1, \cdots, x_8) \in \mathbb{R}^8$.

Compute $f(x)$ and $g(x)$, and select whichever has an even sum of components; call it y_0.

Compute $f(x - \tfrac{1}{2})$ and $g(x - \tfrac{1}{2})$, where

$$\tfrac{1}{2} = (\tfrac{1}{2}, \tfrac{1}{2}, \tfrac{1}{2}, \tfrac{1}{2}, \tfrac{1}{2}, \tfrac{1}{2}, \tfrac{1}{2}, \tfrac{1}{2}),$$

and select whichever has an even sum of components; add $\tfrac{1}{2}$ and call the result y_1.

Compare y_0 and y_1 with x and choose the closest.

For example, to find the closest point of E_8 to

$$x = (0.1, 0.1, 0.8, 1.3, 2.2, -0.6, -0.7, 0.9),$$

we compute

$$f(x) = (0, 0, 1, 1, 2, -1, -1, 1), \quad \text{sum} = 3, \quad \text{odd},$$

$$g(x) = (0, 0, 1, 1, 2, 0, -1, 1), \quad \text{sum} = 4, \quad \text{even},$$

and take $y_0 = g(x)$. Also

$$x - \tfrac{1}{2} = (-0.4, -0.4, 0.3, 0.8, 1.7, -1.1, -1.2, 0.4),$$

$$f(x - \tfrac{1}{2}) = (0, 0, 0, 1, 2, -1, -1, 0), \quad \text{sum} = 1, \quad \text{odd},$$

$$g(x - \tfrac{1}{2}) = (-1, 0, 0, 1, 2, -1, -1, 0), \quad \text{sum} = 0, \quad \text{even},$$

and so

$$y_1 = g(x - \tfrac{1}{2}) + \tfrac{1}{2} = (-0.5, 0.5, 0.5, 1.5, 2.5, -0.5, -0.5, 0.5).$$

Finally,

$$\|x - y_0\|^2 = 0.65, \quad \|x - y_1\|^2 = 0.95$$

and we conclude that $y_0 = g(x)$ is the closest point to x.

VII. Finding the Closest Point of A_n, A_n^*, E_7, and E_7^*

Algorithm 5—To Find the Closest Point of A_n to x:

Step 1: Given $x \in \mathbb{R}^{n+1}$, compute $s = \Sigma x_i$ and replace x by

$$x' = x - \frac{s}{n+1}(1, 1, \cdots, 1).$$

Step 2: Calculate $f(x') = (f(x_0'), \cdots, f(x_n'))$ and the *deficiency* $\Delta = \Sigma f(x_i')$.

Step 3: Sort the x_i' in order of increasing value of $\delta(x_i')$ (defined in Section III). We obtain a rearrangement of the numbers $0, 1, \cdots, n$, say $i_0, i_1, \cdots, i_n$, such that

$$-\tfrac{1}{2} \le \delta(x'_{i_0}) \le \delta(x'_{i_1}) \cdots \le \delta(x'_{i_n}) \le \tfrac{1}{2}.$$

Step 4: If $\Delta = 0$, $f(\boldsymbol{x}')$ is the closest point of A_n to $\boldsymbol{x}$.
If $\Delta > 0$, the closest point is obtained by subtracting 1 from the components $f(x'_{i_0}), \cdots, f(x'_{i_{\Delta-1}})$.
If $\Delta < 0$, the closest point is obtained by adding 1 to the components $f(x'_{i_n}), f(x'_{i_{n-1}}), \cdots, f(x'_{i_{n-\Delta+1}})$.

Remarks: We know from Section III that $f(\boldsymbol{x})$ is the closest point of $\mathbb{Z}^{n+1}$ to $\boldsymbol{x}$. The procedure described here finds the closest point of A_n because it makes the smallest changes to the norm of $f(\boldsymbol{x}')$ needed to make $\Sigma f(x'_i)$ vanish.

Step 1 projects $\boldsymbol{x}$ onto $\boldsymbol{x}'$, the closest point of the hyperplane $\Sigma x_i = 0$. Since A_n is by definition contained in this hyperplane it may be possible to assume that $\boldsymbol{x}$ already lies there, in which case Step 1 can be omitted.

The only substantial amount of computation needed is for the sort in Step 3, which takes $O(n \log n)$ steps [10], [13]. However Step 3 can be omitted if $\boldsymbol{x}$ is expected to be close to A_n. In this case Δ will be small, and Steps 3 and 4 can be replaced by the following:

Step 3': If $\Delta = 0$, $f(\boldsymbol{x}')$ is the closest point of A_n to $\boldsymbol{x}$.

If $\Delta > 0$, find the Δ components of $\boldsymbol{x}'$, say $x'_{i_0}, \cdots, x'_{i_{\Delta-1}}$, for which $\delta(x'_i)$ is as small (i.e., as close to $-\frac{1}{2}$) as possible. The closest point of A_n is obtained by subtracting one from the components $f(x'_{i_0}), \cdots, f(x'_{i_{\Delta-1}})$ of $f(\boldsymbol{x})$.

If $\Delta < 0$, find the $|\Delta|$ components of $\boldsymbol{x}'$, say $x'_{i_n}, x'_{i_{n-1}}, \cdots, x'_{i_{n-\Delta+1}}$, for which $\delta(x'_i)$ is as large (i.e., as close to $\frac{1}{2}$) as possible. The closest point of A_n is obtained by adding 1 to the components of $f(x'_{i_n}), \cdots, f(x'_{i_{n-\Delta+1}})$ of $f(\boldsymbol{x})$.

In any case $|\Delta|$ cannot exceed $n/2$. However if Δ is expected to be large the first version of the algorithm is preferable.

The closest point of A_n^* can be found by Algorithm 3, using the fact that A_n^* is the union of $n+1$ cosets of A_n (see (2)).

For example the hexagonal lattice A_2 is shown in Fig. 1, together with ordinary two-dimensional coordinates (u_1, u_2) for the points. The three-dimensional coordinates (x_0, x_1, x_2) with $x_0 + x_1 + x_2 = 0$ that we have used are obtained by multiplying (u_1, u_2) on the right by the matrix

$$M = \begin{pmatrix} 1 & 0 & -1 \\ \dfrac{1}{\sqrt{3}} & \dfrac{-2}{\sqrt{3}} & \dfrac{1}{\sqrt{3}} \end{pmatrix}.$$

Conversely the u-coordinates may be obtained from the x-coordinates by

$$(u_1, u_2) = (x_0, x_1, x_2) \cdot \tfrac{1}{2} M^{\mathrm{tr}}.$$

For example the points $(0, 0)$, $(1, 0)$, $(1/2, \sqrt{3}/2)$, $(-1/2, \sqrt{3}\,2)$ have x-coordinates $(0, 0, 0)$, $(1, 0, -1)$, $(1, -1, 0)$, $(0, -1, 1)$, respectively. To find the closest point of A_n to the point P with coordinates

$$(u_1, u_2) = (0.4, -0.4)$$

we first find the x-coordinates of P, which are

$$\boldsymbol{x} = (x_0, x_1, x_2) = (0.169, 0.462, -0.631).$$

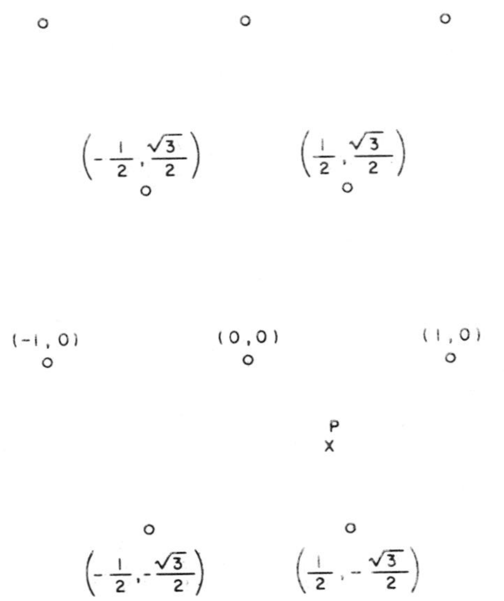

Fig. 1. The hexagonal lattice A_2.

Step 1 of the algorithm can be omitted, since $x_0 + x_1 + x_2 = 0$ holds automatically. Step 2 produces

$$f(\boldsymbol{x}) = (0, 0, -1),$$

with deficiency $\Delta = -1$. At Step 3 we find

$$\delta(x_0) = 0.169 < \delta(x_2) = 0.369 < \delta(x_1) = 0.462.$$

At Step 4 we add 1 to $f(x_1)$, obtaining

$$(0, 1, -1)$$

which is the closest point of A_2. The u-coordinates for this point are

$$(0, 1, -1) \cdot \frac{1}{2} M^{\mathrm{tr}} = \left(\frac{1}{2}, -\frac{\sqrt{3}}{2}\right)$$

(see Fig. 1).

Since $A_3 \cong D_3$, Algorithm 2 is preferable to Algorithm 5 for finding the closest point of the face-centered cubic lattice. Finally E_7 and E_7^* can be handled via the algorithm for A_7, using (7), (8), and Algorithm 3.

Acknowledgment

During the early stages of this work we had some helpful discussions with Allen Gersho.

References

[1] T. Berger, "Optimum quantizers and permutation codes," *IEEE Trans. Inform. Theory*, vol. IT-18, no. 6, 759–765, Nov. 1972.

[2] N. Bourbaki, *Groupes et Algèbres de Lie, Chapitres 4, 5 et 6.* Paris: Hermann, 1968.

[3] J. A. Bucklew, "Companding and random quantization in several dimensions," *IEEE Trans. Inform. Theory*, vol. IT-27, no. 2, 207–211, March 1981.

[4] J. W. S. Cassels, *An Introduction to the Geometry of Numbers.* New York: Springer-Verlag, 1971.

[5] J. H. Conway and N. J. A. Sloane, "On the enumeration of lattices of determinant one," *J. Number Theory*, to be published.

[6] J. H. Conway and N. J. A. Sloane, "Voronoi regions of lattices, second moments of polytopes, and quantization," *IEEE Trans. Inform. Theory*, this issue on pp. 211–226.

[7] H. S. M. Coxeter, *Regular Polytopes*, 3rd ed. New York: Dover, 1973.
[8] N. C. Gallagher, Jr., and J. A. Bucklew, "Some recent developments in quantization theory," in *Proc. 12th Annu. Sympos. System Theory*, Virginia Beach, VA, May 19–20, 1980.
[9] A. Gersho, "Asymptotically optimal block quantization," *IEEE Trans. Inform. Theory*, vol. IT-25, no. 4, 373–380, July 1979.
[10] D. E. Knuth, *The Art of Computer Programming*, vol. 3, *Sorting and Searching*. Reading, MA: Addison-Wesley, 1973.
[11] J. Leech and N. J. A. Sloane, "Sphere packing and error-correcting codes," *Canad. J. Math.*, 23, 718–745, 1971.
[12] H. V. Niemeier, "Definite quadratische Formen der Dimension 24 und Diskriminante 1," *J. Number Theory*, 5, 142–178, 1973.
[13] E. M. Reingold, J. Nievergelt, and N. Deo, *Combinatorial Algorithms: Theory and Practice*. Englewood Cliffs, NJ: Prentice-Hall, 1977.
[14] K. D. Rines and N. C. Gallagher, Jr., "The design of multidimensional quantizers using prequantization," in *Proc. 18th Allerton Conf. Commun., Control, and Computing*, Univ. of Illinois, Monticello, IL, Oct. 8–10, 1980, pp. 446–453.
[15] N. J. A. Sloane, "Binary codes, lattices and sphere packings," in *Combinatorial Surveys* (Proc. 6th British Combinatorial Conference), P. J. Cameron, Ed. London and New York: Academic, 1977, pp. 117–164.
[16] ——, "Tables of sphere packings and spherical codes," *IEEE Trans. Inform. Theory*, vol. IT-27, no. 3, 327–338, May 1981.
[17] Y. Yamada, S. Tazaki, and R. M. Gray, "Asymptotic performance of block quantizers with difference distortion measures," *IEEE Trans. Inform. Theory*, vol. IT-26, no. 1, 5–14, Jan. 1980.
[18] P. Zador, "Topics in the asymptotic quantization of continuous random variables," *IEEE Trans. Inform. Theory*, this issue on pp. 139–149.

17

Reprinted from pages 446–453 of *Allerton Conf. Commun. Control Comput., 18th, Proc.*, October 8–10, 1980, Urbana/Champaign, Ill., 1005p.

THE DESIGN OF MULTIDIMENSIONAL QUANTIZERS USING PREQUANTIZATION

Kerry D. Rines and Neal C. Gallagher, Jr.
Purdue University

ABSTRACT

A novel approach to the design of multidimensional quantizers is presented. This technique is used to design optimum uniform multidimensional quantizers that can be operated in real time. The quantizers are easily implemented using zero memory nonlinearities, linear transformations and univariate uniform step size quantizers.

I. INTRODUCTION

There is considerable interest in the use of multidimensional quantizers for the encoding of analog sources. Much of this interest has been generated from a theoretical standpoint. The multivariate quantization results of Zador [1] point to the advantages of multidimensional quantizers over univariate quantizers at high bit rates. Simply stated, the results indicate that the optimum per sample distortion decreases as the dimension of the quantizer increases. Therefore the potential exists to improve the performance of digital encoders by replacing univariate quantizers with multidimensional quantizers.

Recently the design of optimum multidimensional quantizers has been addressed. Computer algorithms for designing optimum quantizers of two or more dimensions have been presented by many authors, such as Linde et al [2]. The optimum quantizers are implemented using a search procedure to choose, from a specified output set, the output that is the smallest distance from the input. This implementation of the optimum quantizer may be difficult or impossible to operate in real time at high bit rates. In contrast the univariate uniform step size quantizer is a zero memory device that can be operated in real time. To date the easy implementation and real time operation of the univariate uniform step size quantizer has outweighed the theoretical advantages of using multidimensional quantizers in the design of digital encoders.

In this paper we present a novel approach to the design of multidimensional quantizers called prequantization. The design is illustrated in Figure 1 where a zero memory nonlinearity called a prequantizer precedes a specified multidimensional quantizer.

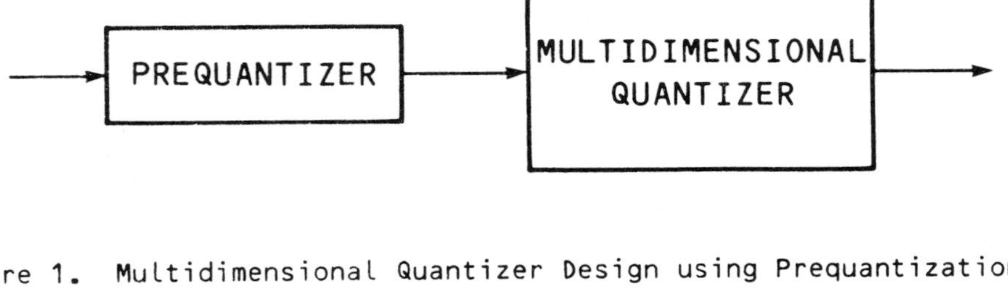

Figure 1. Multidimensional Quantizer Design using Prequantization.

This design is similar in some respects to the companding design of nonuniform univariate quantizers first proposed by Bennett [3]. In the univariate case a nonuniform quantizer may be difficult to implement directly. However, with companding we can design a nonuniform quantizer using a uniform step size quantizer, an invertible nonlinearity and the inverse nonlinearity. Similarly, prequantization can be used to design many multidimensional quantizers. Prequantization enables us to design these quantizers using a simple multidimensional quantizer, which is easy to implement and operate in real time, along with a zero memory nonlinearity. We illustrate the usefulness of prequantization with three examples.

In a recent paper Gersho [4] considers the partitioning of optimum uniform multidimensional quantizers. He states that the optimum uniform two-dimensional quantizer is the hexagonal quantizer. In three dimensions, Gersho argues that the truncated octahedral quantizer is very likely to be the optimum uniform three-dimensional quantizer. The analog of the truncated octahedron is considered for four dimensions. The resulting quantizer is not known to be optimal for four dimensions, but does have a lower per sample distortion than the three dimensional truncated octahedral quantizer. In this paper we present the designs for these three quantizers using prequantization. In each case the design is easy to implement and the quantizer can operate in real time. The real time operation of these quantizers for high bit rates is a significant result and demonstrates the important practical applications for prequantization. We begin in section II with a discussion of the prequantization design procedure.

II. PREQUANTIZATION

The design of k-dimensional quantizers using prequantization is illustrated in Figure 1. The design consists of a nonlinearity called a prequantizer preceding a specified k-dimensional quantizer. The implementation of this design approach takes place in two steps. First a k-dimensional quantizer meeting a specified criterion is chosen. In this paper we are interested in real time operation, therefore we specify that the quantizer be able to operate in real time. Examining Figure 1, we require that the real time (specified) quantizer have the same set of output values as the quantizer we wish to design. This is the only constraint placed on the choice of the real time quantizer. Free to choose from all quantizers satisfying the output constraint, we choose a real time quantizer that is easy to implement. The ability to exercise some control over the choice of the k-dimensional quantizer is one of the advantages of this design procedure.

The second step in the implementation is the design of the prequantizer. The role of the prequantizer is to complete the mapping of the input variables into the desired output values. The real time k-dimensional quantizer can be characterized by the mapping of its input space into its output values. This mapping is usually described by a partitioning of the input space, where all the input vectors contained within one partition are mapped into the same output vector. Since the real time quantizer is chosen based only on its output values, we do not expect its partitioning to be the same as the partitioning of the quantizer being designed. It is the prequantizer which is used to obtain the partitioning specified by the desired quantizer design. The prequantizing function maps a partition specified by the quantizer being designed into a partition of the real time quantizer that corresponds to the specified output. Once the prequantizing function is determined the k-dimensional quantizer design is complete. We illustrate the design procedure with a simple example.

Consider the design of a univariate quantizer with input x and output $\hat{x}$ as described in (1).

$$\tilde{x} = n\,\Delta \quad ; \quad n\,\Delta - \frac{\Delta}{4} \leq x < n\,\Delta + \frac{3\Delta}{4}. \tag{1}$$

Using the prequantization procedure, we first choose a quantizer that is easy to implement and has the same output set as given in (1). We choose the uniform step size quantizer given by

$$\hat{y} = n\,\Delta \quad ; \quad n\,\Delta - \frac{\Delta}{2} \leq y < n\,\Delta + \frac{\Delta}{2}. \tag{2}$$

We now determine the prequantizing function that must precede the quantizer in (2) to complete the design. Observe that quantizing $y = x - \frac{\Delta}{4}$ in (2) is identical to quantizing x in (1). Thus the prequantizing function is simply $f(x) = x - \frac{\Delta}{4}$ and the design of the quantizer in (1) is complete.

III. HEXAGONAL QUANTIZATION

Gersho has argued that the optimum uniform two-dimensional quantizer is the hexagonal quantizer. The design of a hexagonal quantizer using prequantizing is given here. First we attempt to find a two-dimensional quantizer that can be easily implemented and has the same set of output values as the hexagonal quantizer. One quantizer meeting these requirements is a scaled version of the diamond quantizer given below.

Let the inputs to the two-dimensional quantizer be x and y. The variables x and y are first encoded into two new variables w and z by the linear transformation,

$$\begin{aligned} w &= x + \sqrt{3}\,y \\ z &= x - \sqrt{3}\,y. \end{aligned} \tag{3}$$

The variables w and z are quantized separately by univariate quantizers with a uniform step size Δ. The outputs of the two-dimensional quantizer are then obtained using the linear transformation,

$$\begin{aligned} \hat{x} &= \frac{1}{2}(\hat{w} + \hat{z}) \\ \hat{y} &= \frac{1}{2\sqrt{3}}(\hat{w} - \hat{z}). \end{aligned} \tag{4}$$

The position of this quantizer in the hexagonal quantizer design is shown in Figure 2 and the partitioning of the scaled diamond quantizer is given in Figure 3. Having chosen the two-dimensional quantizer given in (3) and (4) we now turn to the design of the prequantizer.

The prequantizer must map the hexagonal region corresponding to each output into the scaled diamond shaped region corresponding to that same output. Consider the hexagonal partitioning shown in Figure 4.

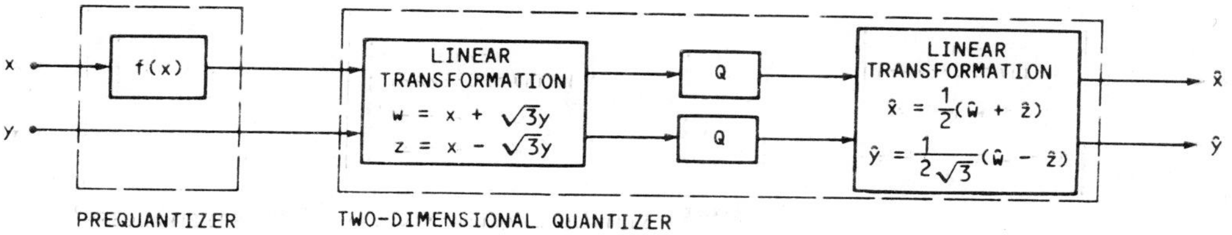

Figure 2. Prequantization design for the hexagonal quantizer. The quantizer Q has uniform step-size Δ.

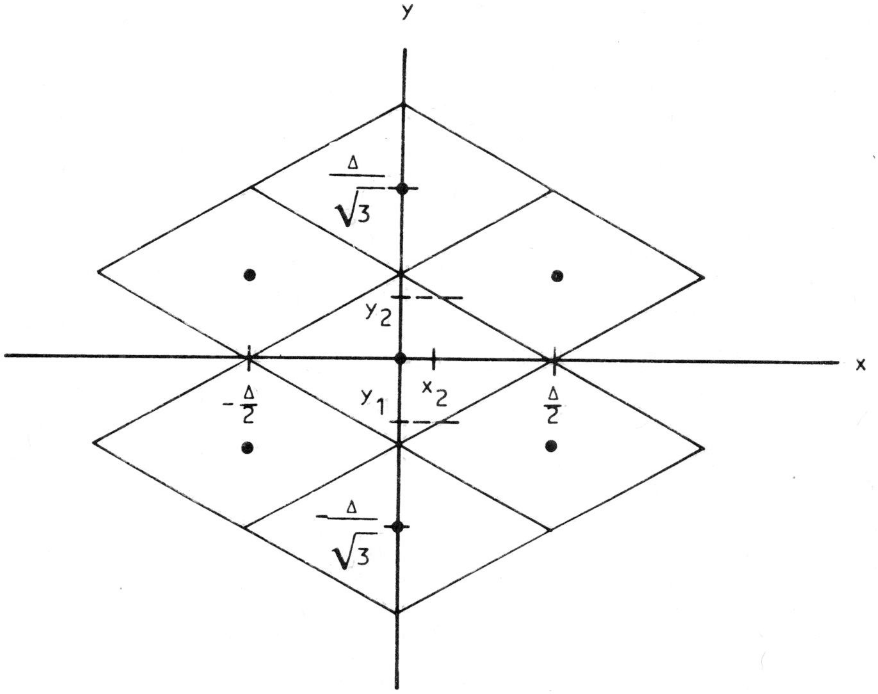

Figure 3. Partitioning of the scaled diamond quantizer.

Assume x is fixed and the pair (x,y) is contained within a given hexagonal partition. We now pose the question, does there exist a value x' such that the pair (x',y) is contained within the corresponding diamond partition for all values of y? This approach is illustrated with the following example. Let $x = x_1$ as shown in Figure 3 and let y be in the range $-\frac{\Delta}{2\sqrt{3}}$ to $\frac{\Delta}{2\sqrt{3}}$. In Figure 4 we observe that the hexagonal quantizer output will be (0,0) for all input pairs in the set $\{(x_1,y) : y_1 \le y \le y_2\}$. Similarly in Figure 3 we observe that the scaled diamond quantizer output will be (0,0) for all input pairs in the set $\{(x_2, y) : y_1 \le y \le y_2\}$. Therefore if $x_2 = f(x_1)$, the quantizer in Figure 2 will behave like the hexagonal quantizer for all input pairs in the set $\{(x_1,y) : \frac{-\Delta}{2\sqrt{3}} \le y \le \frac{\Delta}{2\sqrt{3}}\}$. In fact, we can show that the quantizer

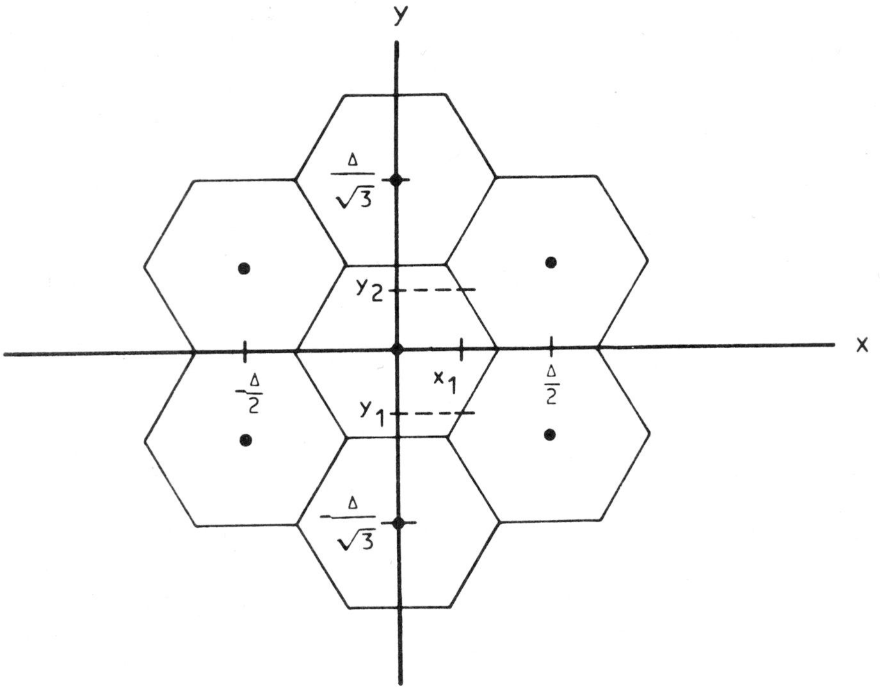

Figure 4. Partitioning of the hexagonal quantizer.

in Figure 2 behaves like the hexagonal quantizer for all inputs in the set $\{(x_1,y) : -\infty \leq y \leq \infty\}$ when $x_2 = f(x_1)$. Repeating this example for all possible values of x_1, we obtain a prequantizing function that maps the hexagonal region corresponding to each output into the scaled diamond shaped region corresponding to that same output. The prequantizing function is given in (5).

$$f(x) = n\frac{\Delta}{2} \quad ; \; n\frac{\Delta}{2} - \frac{\Delta}{3} \leq x \leq n\frac{\Delta}{2} + \frac{\Delta}{3} \tag{5}$$
$$= 3x - (2n+1)\frac{\Delta}{2} \; ; \; n\frac{\Delta}{2} + \frac{\Delta}{3} \leq x \leq (n+1)\frac{\Delta}{2} - \frac{\Delta}{3}.$$

IV. RESULTS IN HIGHER DIMENSIONS

In this section we present the design of the optimum (or near optimum) uniform quantizers for three and four dimensions. Each of these quantizers use in their designs a two-dimensional quantizer termed the diamond quantizer. The algorithm for the diamond quantizer is as follows. Let the inputs to the two-dimensional quantizer be x and y. The variables x and y are first encoded into two new variables w and z by the linear transformation,

$$w = x + y$$
$$z = x - y. \tag{6}$$

The variables w and z are quantized separately by univariate quantizers with a uniform step size Δ. The outputs of the diamond quantizer are then obtained from a linear transformation of the quantized variables $\hat{w}$ and $\hat{z}$ given by

$$\hat{x} = \frac{1}{2}(\hat{w} + \hat{z})$$

$$\hat{y} = \frac{1}{2}(\hat{w} - \hat{z}). \qquad (7)$$

The outputs $\hat{x}$ and $\hat{y}$ will be multiples of $\frac{\Delta}{2}$ for all possible inputs. A useful property of the diamond quantizer is that if either input x or y is a multiple of $\frac{\Delta}{2}$, its quantized value $\hat{x}$ or $\hat{y}$ will be that same multiple of $\frac{\Delta}{2}$. Therefore if the output of one diamond quantizer $\hat{x}$ is used as the input to a second diamond quantizer, the output of the second diamond quantizer will also be $\hat{x}$. Using this property we are able to design quantizers of higher dimensions by cascading diamond quantizers. The results of these designs are now given.

Gersho states that the truncated octahedral quantizer is very likely the optimum three dimensional quantizer. This quantizer is defined by a tessellation of a truncated octahedron specified by the set $\{(x_1,x_2,x_3) : |x_1|+|x_2|+|x_3| < \frac{3\Delta}{2}\ ;\ |x_i| < \frac{\Delta}{2},\ i=1,2,3\}$. The design of this quantizer is given in Figure 5.

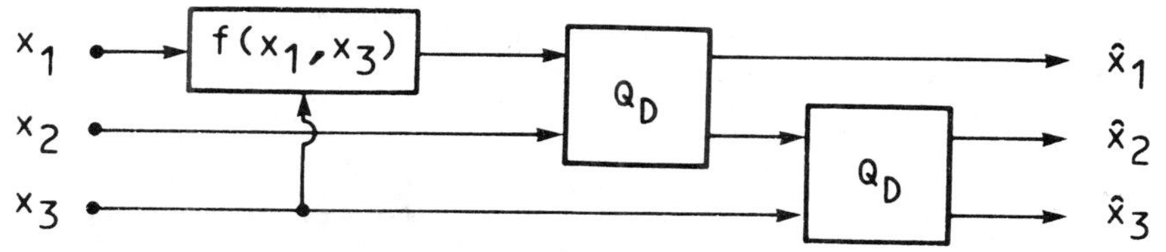

Figure 5. The truncated octahedral quantizer design using prequantization. Q_D is the diamond quantizer.

The prequantizing function is given in (8) where $e = |x_3| \bmod (0,\frac{\Delta}{2})$. For $e \le \frac{\Delta}{4}$,

$$\begin{aligned} f(x_1,x_3) &= n\frac{\Delta}{2} &&;\ n\frac{\Delta}{2} - \frac{\Delta}{4} + e \le x_1 \le n\frac{\Delta}{2} + \frac{\Delta}{4} - e \\ &= x_1 - \frac{\Delta}{4} + e &&;\ n\frac{\Delta}{2} + \frac{\Delta}{4} - e \le x_1 \le (n+1)\frac{\Delta}{2} \\ &= x_1 + \frac{\Delta}{4} - e &&;\ (n-1)\frac{\Delta}{2} \le x_1 \le n\frac{\Delta}{2} - \frac{\Delta}{4} + e. \end{aligned} \qquad (8)$$

A similar result is obtained for $\frac{\Delta}{4} \le e \le \frac{\Delta}{2}$.

The four dimensional analog of the truncated octahedral quantizer is defined by the tessellation of the polytope specified by the set $\{(x_1,x_2,x_3,x_4) : |x_1|+|x_2|+|x_3|+|x_4| < 2\Delta\ ;\ |x_i| \le \frac{\Delta}{2},\ i=1,2,3,4\}$. For convenience we will call this quantizer the 4-d uniform quantizer. The design of the 4-d uniform quantizer is shown in Figure 6.

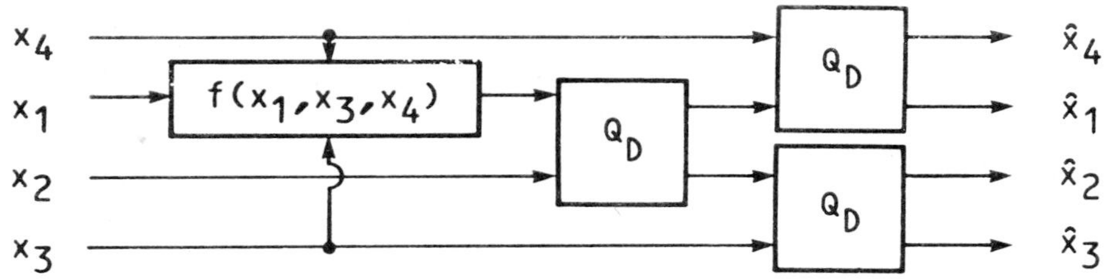

Figure 6. The 4-d uniform quantizer using prequantization. Q_D is the diamond quantizer.

The prequantizing function is given in (9) where $z = |x_3| \bmod(0,\frac{\Delta}{2})$, $w = |x_4| \bmod(0,\frac{\Delta}{2})$ and $e = z+w$. For $e \leq \frac{\Delta}{2}$,

$$\begin{aligned} f(x_1,x_3,x_4) &= n\,\frac{\Delta}{2} &&; (n-1)\,\frac{\Delta}{2} + e \leq x_1 \leq (n+1)\frac{\Delta}{2} - e \\ &= x_1 - \frac{\Delta}{2} + e &&; (n+1)\,\frac{\Delta}{2} - e \leq x_1 \leq (n+1)\,\frac{\Delta}{2} \\ &= x_1 + \frac{\Delta}{2} - e &&; (n-1)\,\frac{\Delta}{2} \leq x_1 \leq (n-1)\,\frac{\Delta}{2} + e. \end{aligned} \tag{9}$$

A similar result is obtained for $\frac{\Delta}{2} \leq e \leq \Delta$.

A comparison of the normalized mean-squared error performance of the uniform univariate and multidimensional quantizers is given in Table 1. The results were obtained by computer simulation using 30,000 samples uniformly distributed $(-\frac{1}{2},\frac{1}{2})$. The output alphabet of each quantizer was assigned one hundred quantization levels per input sample.

Dimension	Quantizer	nmse ($\times 10^{-5}$)
1	uniform step-size	9.99
2	hexagonal	9.66
3	truncated octahedral	9.48
4	4-d uniform	9.17

V. DISCUSSION

In this paper we have presented a new approach to the design of multidimensional quantizers. The usefulness of the prequantization approach has been demonstrated by the design of three optimum (or near optimum) uniform multidimensional quantizers. In each example the quantizer can be implemented using a zero memory nonlinearity, linear transformations, and univariate uniform step-size quantizers. As a result the computation time of each quantizer is independent of the output alphabet size. Therefore, these quantizers are both easy to implement and are able to operate in real time even at very high bit rates.

The prequantization design approach is also compatible with the design of nonuniform multidimensional quantizers. In [4] Gersho generalizes the companding technique for the design of nonuniform univariate quantizers to the design of nonuniform multidimensional quantizers. Bucklew [5] shows that an optimum k-dimensional quantizer can be designed using an optimum uniform k-dimensional quantizer, which is preceded by a multivariate invertible nonlinearity and followed by the inverse nonlinearity. Therefore the nonlinear prequantizing function used in optimum uniform k-dimensional quantizers is compatible and may even be of an advantage when the companding approach is applied to multidimensional quantizers.

ACKNOWLEDGMENT

The authors gratefully acknowledge the support of the Air Force Office of Scientific Research under grant AFOSR 78-3605.

REFERENCES

[1] P. Zador, Development and Evaluation of Procedures for Quantizing Multivariate Distributions, Ph.D. Dissertation, Stanford University, 1964, University Microfilm No. 64-9855.

[2] Y. Linde, A. Buzo, R.M. Gray, "An Algorithm for Vector Quantizer Design," IEEE Trans. Comm., Vol. COM-28, pp.84-95, January 1980.

[3] W.R. Bennett, "Spectra of Quantized Signals," B.S.T.J., Vol. 27, pp. 446-472, July, 1948.

[4] A. Gersho, "Asymptotically Optimal Block Quantization," IEEE Trans. on Inform. Theory, Vol. IT-25, pp. 373-380, July, 1979.

[5] J. A. Bucklew, "Companding and Random Quantization in Several Dimensions," to be published.

Reprinted from *IEEE Trans. Commun.* **COM-28**(1):84–95 (1980)

An Algorithm for Vector Quantizer Design

YOSEPH LINDE, MEMBER, IEEE, ANDRÉS BUZO, MEMBER, IEEE, AND ROBERT M. GRAY, SENIOR MEMBER, IEEE

***Abstract*—An efficient and intuitive algorithm is presented for the design of vector quantizers based either on a known probabilistic model or on a long training sequence of data. The basic properties of the algorithm are discussed and demonstrated by examples. Quite general distortion measures and long blocklengths are allowed, as exemplified by the design of parameter vector quantizers of ten-dimensional vectors arising in Linear Predictive Coded (LPC) speech compression with a complicated distortion measure arising in LPC analysis that does not depend only on the error vector.**

INTRODUCTION

An efficient and intuitive algorithm for the design of good block or vector quantizers with quite general distortion measures is developed for use on either known probabilistic source descriptions or on a long training sequence of data. The algorithm is based on an approach of Lloyd [1], is not a variational technique, and involves no differentiation; hence it works well even when the distribution has discrete components, as is the case when a sample distribution obtained from a training sequence is used. As with the common variational techniques, the algorithm produces a quantizer meeting necessary but not sufficient conditions for optimality. Usually, however, at least local optimality is assured in both approaches.

We here motivate and describe the algorithm and relate it to a number of similar algorithms for special cases that have appeared in both the quantization and cluster analysis literature. The basic operation of the algorithm is simple and intuitive in the general case considered here and it is clear that variational techniques are not required to develop nor to apply the algorithm.

Several of the algorithm's basic properties are developed using heuristic arguments and demonstrated by example. In a companion theoretical paper [2], these properties are given precise mathematical statements and are proved using arguments from optimization theory and ergodic theory. Those results will occasionally be quoted here to characterize the generality of certain properties.

Paper approved by the Editor for Data Communication Systems of the IEEE Communications Society for publication after presentation in part at the 1979 International Telemetering Conference, Los Angeles, CA, November 1978 and the 1979 International Symposium on Information Theory, Gringano, Italy, June 1979. Manuscript received May 22, 1978; revised August 21, 1979. This work was supported by Air Force Contract F44620-73-0065, F49620-78-C-0087, and F49620-79-C-0058 and by the Joint Services Electronics Program at Stanford University, Stanford, CA.

Y. Linde is with the Codex Corporation, Mansfield, MA.

A. Buzo was with Stanford University, Stanford, CA and Signal Technology Inc., Santa Barbara, CA. He is now with the Instituto de Ingenieria, National University of Mexico, Mexico City, Mexico.

R. M. Gray is with the Information Systems Laboratory, Stanford University, Stanford, CA 94305.

In particular, the algorithm's convergence properties are demonstrated herein by several examples. We consider the usual test case for such algorithms—quantizer design for memoryless Gaussian sources with a mean-squared error distortion measure, but we design and evaluate block quantizers with a rate of one bit per symbol and with blocklengths of 1 through 6. Comparison with recently developed lower bounds to the optimal distortion of such block quantizers (which provide strict improvement over the traditional bounds of rate-distortion theory) indicate that the resulting quantizers are indeed nearly optimal and not simply locally optimal. We also consider a scalar case where local optima arise and show how a variation of the algorithm yields a global optimum.

The algorithm is also used to design a quantizer for 10-dimensional vectors arising in speech compression systems. A complicated distortion measure is used that does not simply depend on the error vector. No probabilistic model is assumed, and hence the quantizer must be designed based on a training sequence of real speech. Here the convergence properties for both length of the training sequence and the number of iterations of the algorithm are demonstrated experimentally. No theoretical optimum is known for this case, but our system was used to compress the output of a traditional 6000 bit/s Linear Predictive Coded (LPC) speech system down to a rate of 1400 bits/s with only a slight loss in quality as judged by untrained listeners in informal subjective tests. To the authors' knowledge, direct application of variational techniques have not succeeded in designing block quantizers for such large block lengths and such complicated distortion measures.

BLOCK QUANTIZERS

An N-level k-dimensional quantizer is a mapping, q, that assigns to each input vector, $\boldsymbol{x} = (x_0, \cdots, x_{k-1})$, a reproduction vector, $\hat{\boldsymbol{x}} = q(\boldsymbol{x})$, drawn from a finite reproduction alphabet, $\hat{A} = \{y_i;\ i = 1, \cdots, N\}$. The quantizer q is completely described by the reproduction alphabet (or codebook) $\hat{A}$ together with the partition, $S = \{S_i;\ i = 1, \cdots, N\}$, of the input vector space into the sets $S_i = \{\boldsymbol{x}\colon q(\boldsymbol{x}) = y_i\}$ of input vectors mapping into the ith reproduction vector (or codeword). Such quantizers are also called block quantizers, vector quantizers, and block source codes.

DISTORTION MEASURES

We assume the distortion caused by reproducing an input vector $\boldsymbol{x}$ by a reproduction vector $\hat{\boldsymbol{x}}$ is given by a nonnegative distortion measure $d(\boldsymbol{x}, \hat{\boldsymbol{x}})$. Many such distortion measures have been proposed in the literature. The most common for reasons of mathematical convenience is the squared-error distortion.

$$d(x,\hat{x}) = \sum_{i=0}^{k-1} |x_i - \hat{x}_i|^2 \tag{1}$$

Other common distortion measures are the l_ν, or Holder norm,

$$d(x,\hat{x}) = \left\{ \sum_{i=0}^{k-1} |x_i - \hat{x}_i|^\nu \right\}^{1/\nu} \triangleq \|x - \hat{x}\|_\nu, \tag{2}$$

and its νth power, the νth-law distortion:

$$d(x,\hat{x}) = \sum_{i=0}^{k-1} |x_i - \hat{x}_i|^\nu = \|x - \hat{x}\|_\nu{}^\nu. \tag{3}$$

While both distortion measures (2) and (3) depend on the νth power of the errors in the separate coordinates, the measure of (2) is often more useful since it is a distance or metric and hence satisfies the triangle inequality, $d(x, \hat{x}) \leqslant d(x, y) + d(y, \hat{x})$, for all y. The triangle inequality allows one to bound the overall distortion easily in a multi-step system by the sum of the individual distortions incurred in each step. The usual ν^{th}-law distortion of (3) does not have this property. Other distortion measures are the l_∞, or Minkowski norm,

$$d(x,\hat{x}) = \max_{0 \leqslant i \leqslant k-1} |x_i - \hat{x}_i|, \tag{4}$$

the weighted-squares distortion,

$$d(x,\hat{x}) = \sum_{i=0}^{k-1} w_i |x_i - \hat{x}_i|^2, \tag{5}$$

where $w_i \geqslant 0$, $i = 0, \cdots, k-1$, and the more general quadratic distortion

$$d(x,\hat{x}) = (x - \hat{x})B(x - \hat{x})^t$$
$$= \sum_{i=0}^{k-1} \sum_{j=0}^{k-1} B_{i,j}(x_i - \hat{x}_i)(x_j - \hat{x}_j), \tag{6}$$

where $B = \{B_{i,j}\}$ is a $k \times k$ positive definite symmetric matrix.

All of the previously described distortion measures have the property that they depend on the vectors x and $\hat{x}$ only through the error vector $x - \hat{x}$. Such distortion measures having the form $d(x, \hat{x}) = L(x - \hat{x})$ are called difference distortion measures. Distortion measures not having this form but depending on x and $\hat{x}$ in a more complicated fashion have also been proposed for data compression systems. Of interest here is a distortion measure of Itakura and Saito [3, 4] and Chaffee [5, 32] which arises in speech compression systems and has the form

$$d(x,\hat{x}) = (x - \hat{x})R(x)(x - \hat{x})^t, \tag{7}$$

where for each x, $R(x)$ is a positive definite $k \times k$ symmetric matrix. This distortion resembles the quadratic distortion of (6), but here the weighting matrix depends on the input vector x.

We are here concerned with the particular form and application of this distortion measure rather than its origins, which are treated in depth in [3-9] and in a paper in preparation. For motivation, however, we briefly describe the context in which this distortion measure is used in speech systems. In the LPC approach to speech compression [10], each frame of sampled speech is modeled as the output of a finite-order all-pole filter driven by either white noise (unvoiced sounds) or a periodic pulse train (voiced sounds). LPC analysis has, as input, a frame of speech and produces parameters describing the model. These parameters are then quantized and transmitted. One collection of such parameters consists of a voiced/unvoiced decision together with a pitch estimate for voiced sounds, a gain term σ (related to volume), and the sample response of the normalized inverse filter $(1, a_1, a_2, \cdots, a_K)$, that is, the normalized all-pole model has transfer function or z-transform $\{\sum_{k=0}^{K} a_k z^{-k}\}^{-1}$. Other parameter descriptions such as the reflection coefficients are also possible [10].

In traditional LPC systems, the various parameters are quantized separately, but such systems have effectively reached their theoretical performance limits [11]. Hence it is natural to consider block quantization of these parameters and compare the performance with the traditional scalar quantization techniques. Here we consider the case where the pitch and gain are (as usual) quantized separately, but the parameters describing the normalized model are to be quantized together as a vector. Since the lead term is 1, we wish to quantize a vector $(a_1, a_2, \cdots, a_K) \triangleq x = (x_0, \cdots, x_{K-1})$. A distortion measure, $d(x, \hat{x})$, between x and a reproduction x, can then be viewed as a distortion measure between two normalized (unit gain) inverse filters or models. A distortion measure for such a case has been proposed by Itakura and Saito [3, 4] and by Chaffee [5, 32] and it has the form of (7) with $R(x)$ the autocorrelation matrix $\{r_x(k - j); k = 0, 1, \cdots, K-1; j = 0, 1, \cdots, K-1\}$ defined by

$$r_x(k) = \int_{-\pi}^{\pi} \left| \sum_{m=0}^{K} a_i e^{-im\theta} \right|^{-2} e^{ik\theta} d\theta/2\pi, \tag{8}$$

described by x when the input has a flat unit amplitude spectrum.

Many properties and alternative forms for this particular distortion measure are developed in [3-9], where it is also shown that standard LPC systems implicitly minimize this distortion, which suggests that it is also an appropriate distortion measure for subsequent quantization. Here, however, the important fact is that it is not a difference distortion measure; it is one for which the dependence on x and $\hat{x}$ is quite complicated.

We also observe that various functions of the previously defined distortion measures have been proposed in the literature, for example, distortion measures of the forms $\|x - \hat{x}\|^r$ and $\rho(\|x - \hat{x}\|)$, where ρ is a convex function and the norm is any of the previously defined norms. The techniques to be developed here are applicable to all of these distortion measures.

PERFORMANCE

Let $X = (X_0, \cdots, X_{k-1})$ be a real random vector described by a cumulative distribution function $F(x) = \Pr\{X_i \leq x_i; i = 0, 1, \cdots, k-1\}$. A measure of the performance of a quantizer q applied to the random vector X is given by the expected distortion

$$D(q) = Ed(X, q(X)), \tag{9}$$

where E denotes the expectation with respect to the underlying distribution F. This performance measure is physically meaningful if the quantizer q is to be used to quantize a sequence of vectors $X_n = (X_{nK}, \cdots, X_{nK+K-1})$ that are stationary and ergodic, since then the time-averaged distortion,

$$n^{-1} \sum_{i=0}^{n-1} d(X_i, q(X_i)),$$

converges with probability one to $D(q)$ as $n \to \infty$ (from the ergodic theorem), that is, $D(q)$ describes the long-run time-averaged distortion.

An alternative performance measure is the maximum of $d(x, q(x))$ over all x in A, but we use only the expected distortion (9) since, in most problems of interest (to us), it is the average distortion and not the peak distortion that determines subjective quality. In addition, the expected distortion is more easily dealt with mathematically.

OPTIMAL QUANTIZATION

An N-level quantizer will be said to be optimal (or globally optimal) if it minimizes the expected distortion, that is, q^* is optimal if for all other quantizers q having N reproduction vectors $D(q^*) \leq D(q)$. A quantizer is said to be locally optimum if $D(q)$ is only a local minimum, that is, slight changes in q cause an increase in distortion. The goal of block quantizer design is to obtain an optimal quantizer if possible and, if not, to obtain a locally optimal and hopefully "good" quantizer. Several such algorithms have been proposed in the literature for the computer-aided design of locally optimal quantizers. In a few special cases, it has been possible to demonstrate global optimality either analytically or by exhausting all local optima. In 1957, in a classic but unfortunately unpublished Bell Laboratories' paper, S. Lloyd [1] proposed two methods for quantizer design for the scalar case ($k = 1$) with a squared-error distortion criterion. His "Method II" was a straightforward variational approach wherein he took derivatives with respect to the reproduction symbols, y_i, and with respect to the boundary points defining the S_i and set these derivatives to zero. This in general yields only a "stationary-point" quantizer (a multidimensional zero derivative) that satisfies necessary but not sufficient conditions for optimality. By second derivative arguments, however, it is easy to establish that such stationary-point quantizers are at least locally optimum for ν^{th}-power law distortion measures. In addition, Lloyd also demonstrated global optimality for certain distributions by a technique of exhaustively searching all local optima. Essentially the same technique was also proposed and used in the parallel problem of cluster analysis by Dalenius [12] in 1950, Fisher [13] in 1953, and Cox [14] in 1957. The technique was also independently developed by Max [15] in 1960 and the resulting quantizer is commonly known as the Lloyd-Max quantizer. This approach has proved quite useful for designing scalar quantizers, with power-law distortion criteria and with known distributions that were sufficiently well behaved to ensure the existence of the derivatives in question. In addition, for this case, Fleischer [16] was able to demonstrate analytically that the resulting quantizers were globally optimum for several interesting probability densities.

In some situations, however, the direct variational approach has not proved successful. First, if k is not equal to 1 or 2, the computational requirements become too complex. Simple combinations of one-dimensional differentiation will not work because of the possibly complicated surface shapes of the boundaries of the cells of the partition. In fact, the only successful applications of a direct variational approach to multidimensional quantization are for quantizers where the partition cells are required to have a particular simple form such as multidimensional "cubes" or, in two dimensions, "pie slices," each described only by a radius and two angles. These shapes are amenable to differentiation techniques, but only yield a local optimum within the constrained class. Secondly, if, in addition, more complex distortion measures such as those of (4)-(7) are desired, the required computation associated with the variational equations can become exorbitant. Thirdly, if the underlying probability distribution has discrete components, then the required derivatives may not exist, causing further computational problems. Lastly, if one lacks a precise probabilistic description of the random vector X and must base the design instead on an observed long training sequence of data, then there is no obvious way to apply the variational approach. If the underlying unknown process is stationary and ergodic, then hopefully a system designed by using a sufficiently long training sequence should also work well on future data. To directly apply the variational technique in this case, one would first have to estimate the underlying continuous distribution based on the observations and then take the appropriate derivatives. Unfortunately, however, most statistical techniques for density estimation require an underlying assumption on the class of allowed densities, e.g., exponential families. Thus these techniques are inappropriate when no such knowledge is available. Furthermore, a good fit of a continuous model to a finite-sample histogram may have ill-behaved differential behavior and hence may not produce a good quantizer. To our knowledge, no one has successfully used such an approach nor has anyone demonstrated that this approach will yield the correct quantizer in the limit of a long training sequence.

Lloyd [1] also proposed an alternative nonvariational approach as his "Method I." Not surprisingly, both approaches yield the same quantizer for the special cases he considered, but we shall argue that a natural and intuitive extension of his Method I provides an efficient algorithm for the design of good vector quantizers that overcomes the problems of the variational approach. In fact, variations of Lloyd's Method I have been "discovered" several times in the literature for squared-error and magnitude-error distortion criteria for both scalar and multidimensional cases (e.g., [22], [23], [24], [31]). Lloyd's basic development, however, remains the simplest, yet it extends easily to the general case considered here.

To describe Lloyd's Method I in the general case, we first assume that the distribution is known, but we allow it to be either continuous or discrete and make no assumptions requiring the existence of derivatives. Given a quantizer q described by a reproduction alphabet $\hat{A} = \{y_i; i = 1, \cdots, N\}$ and partition $S = \{S_i; i = 1, \cdots, N\}$, then the expected distortion $D(\{\hat{A}, S\}) \triangleq D(q)$ can be written as

$$D(\{\hat{A}, S\}) = Ed(X, q(X))$$

$$= \sum_{i=1}^{N} E(d(X, y_i) \mid X \in S_i) \Pr(X \in S_i), \tag{10}$$

where $E(d(X, y_i) \mid X \in S_i)$ is the conditional expected distortion, given $X \in S_i$, or, equivalently, given $q(X) = y_i$.

Suppose that we are given a particular reproduction alphabet $\hat{A}$, but a partition is not specified. A partition that is optimum for $\hat{A}$ is easily constructed by mapping each x into the $y_i \in \hat{A}$ minimizing the distortion $d(x, y_i)$, that is, by choosing the minimum distortion or nearest-neighbor codeword for each input. A tie-breaking rule such as choosing the reproduction with the lowest index is required if more than one codeword minimizes the distortion. The partition, say $P(\hat{A}) = \{P_i; i = 1, \cdots, N\}$ constructed in this manner is such that $x \in P_i$ (or $q(x) = y_i$) only if $d(x, y_i) \leqslant d(x, y_j)$, all j, and hence

$$D(\{\hat{A}, P(\hat{A})\}) = E\left(\min_{y \in \hat{A}} d(X, y)\right) \tag{11}$$

which, in turn, implies for any partition S that

$$D(\{\hat{A}, S\}) \geqslant D(\{\hat{A}, P(\hat{A})\}). \tag{12}$$

Thus for a fixed reproduction alphabet $\hat{A}$, the best possible partition is $P(\hat{A})$.

Conversely, assume we are given a partition $S = \{S_i; i = 1, \cdots, N\}$ describing a quantizer. For the moment, assume also that the distortion measure and distribution are such that, for each set S with nonzero probability in k-dimensional Euclidean space, there exists a minimum distortion vector $\hat{x}(S)$ for which

$$E(d(X, \hat{x}(S)) \mid X \in S) = \min_{u} E(d(X, u) \mid X \in S). \tag{13}$$

Analogous to the case of a squared-error distortion measure and a uniform probability distribution, we call the vector $\hat{x}(S)$ the centroid or center of gravity of the set S. If such points exist, then clearly for a fixed partition $S = \{S_i; i = 1, \cdots, N\}$, no reproduction alphabet $\hat{A} = \{y_i; i = 1, \cdots, N\}$ can yield a smaller average distortion then the reproduction alphabet $\hat{x}(S) \triangleq \{\hat{x}(S_i); i = 1, \cdots, N\}$ containing the centroids of the sets in S since

$$D(\{\hat{A}, S\}) = \sum_{i=1}^{N} E(d(X, y_i) \mid X \in S_i) \Pr(X \in S_i)$$

$$\geqslant \sum_{i=1}^{N} \min_{u} E(d(X, u) \mid X \in S_i) \Pr(X \in S_i)$$

$$= D(\{\hat{x}(S), S\}). \tag{14}$$

It is shown in [2] that the centroids of (13) exist for all sets S with nonzero probability for quite general distortion measures including all of those considered here. In particular, if $d(x, y)$ is convex in y, then centroids can be computed using standard convex programming techniques as described, e.g., in Luenberger [17, 18] or Rockafellar [19]. In certain cases, they can be found easily using variational techniques. If the probability of a set S is zero, then the centroid can be defined in an arbitrary manner since then the conditional expectation given that S in (13) has no unique definition.

Equations (12) and (14) suggest a natural algorithm for designing a good quantizer by taking any given quantizer and iteratively improving it:

Algorithm (Known Distribution)

(0) Initialization: Given N = number of levels, a distortion threshold $\epsilon \geqslant 0$, and an initial N-level reproduction alphabet $\hat{A}_0$ and a distribution F. Set $m = 0$ and $D_{-1} = \infty$.

(1) Given $\hat{A}_m = \{y_i; i = 1, \cdots, N\}$, find its minimum distortion partition $P(\hat{A}_m) = \{S_i; i = 1, \cdots, N\}$: $x \in S_i$ if $d(x, y_i) \leqslant d(x, y_j)$ for all j. Compute the resulting average distortion, $D_m = D(\{\hat{A}_m, P(\hat{A}_m)\}) = E \min_{y \in \hat{A}_m} d(X, y)$.

(2) If $(D_{m-1} - D_m)/D_m \leqslant \epsilon$, halt with $\hat{A}_m$ and $P(\hat{A}_m)$ describing final quantizer. Otherwise continue.

(3) Find the optimal reproduction alphabet $\hat{x}(P(\hat{A}_m)) = \{\hat{x}(S_i); i = 1, \cdots, N\}$ for $P(\hat{A}_m)$. Set $\hat{A}_{m+1} \triangleq \hat{x}(P(\hat{A}_m))$. Replace m by $m + 1$ and go to (1).

If, at some point, the optimal partition $P(\hat{A}_m)$ has a cell S_i such that $\Pr(X \in S_i) = 0$, then the algorithm, as stated, assigns an arbitrary vector as centroid and continues. Clearly, alternative rules are possible and may perform better in practice. For example, one can simply remove the cell S_i and the corresponding reproduction symbol from the quantizer without affecting performance, and then continue with an $(N - 1)$ level quantizer. Alternatively, one could assign to S_i its Euclidean center of gravity or the ith centroid from the previous iteration. One could also simply reassign the reproduction vector corresponding to S_i to another cell S_j and continue the algorithm. The stated technique is given simply for convenience, since zero probability cells were not a problem in the examples considered here. They can, however, occur and in such situations alternative techniques such as those described may well work better. In practice, a simple alternative is that, if the final quantizer produced by the algorithm has a zero probability (hence useless) cell, simply rerun the algorithm with a different initial guess.

From (12) and (14). $D_m \leqslant D_{m-1}$ and hence each iteration of the algorithm must either reduce the distortion or leave it unchanged. We shall later mention some minor additional details of the algorithm and discuss techniques for choosing an initial guess, but the previously given description contains the essential ideas.

Since D_m is nonincreasing and nonnegative, it must have a limit, say D_∞, as $m \to \infty$. It is shown in [2] that if a limiting quantizer $\hat{A}_\infty$ exists in the sense $\hat{A}_m \to \hat{A}_\infty$ as $m \to \infty$ in the usual Euclidean sense, then $D(\{\hat{A}_\infty, P(\hat{A}_\infty)\}) = D_\infty$ and $\hat{A}_\infty$ has the property that $\hat{A}_\infty = \hat{x}(P(\hat{A}_\infty))$, that is, $\hat{A}_\infty$ is exactly the centroid of its own optimal partition. In the language of optimization theory, $\{\hat{A}_\infty, P(\hat{A}_\infty)\}$ is a *fixed point* under further iterations of the algorithm [17, 18]. Hence the limit quantizer (if it exists) is called a fixed-point quantizer (in contrast to a stationary-point quantizer obtained by a variational approach). In this light, the algorithm is simply a standard technique for finding a fixed point via the method of successive approximation (see, e.g., Luenberger [17, p. 272]). If $\epsilon = 0$ and the algorithm halts for finite m, then such a fixed point has been attained [2].

It is shown in [2] that a necessary condition for a quantizer to be optimal is that it be a fixed-point quantizer. It is also shown in [2] that, as in Lloyd's case, if a fixed-point quantizer is such that there is no probability on the boundary of the partition cells, that is if $\Pr(d(X, y_i) = d(X, y_j)$, for some $i \neq j) = 0$, then the quantizer is locally optimum. This is always the case with continuous distributions, but can in principle be violated for discrete distributions. It was never found to occur in our experiments, however. As Lloyd suggests, the algorithm can easily be modified to test a fixed point for this condition and if there is nonzero probability of a vector on a boundary, the strategy would be to reassign the vector to another cell of the partition and continue the iteration.

For the $N = 1$ case with a squared-error distortion criterion, the algorithm is simply Lloyd's Method I, and his arguments apply immediately in the more general case considered herein. A similar technique was earlier proposed in 1953 by Fisher [13] in a cluster analysis problem using Bayes decisions with a squared-error cost. For larger dimensions and distortion measures of the form $d(x, \hat{x}) = \|x - \hat{x}\|_2^r$, $r \geqslant 1$, the relations (12) and (14) were observed by Zador [20] and Gersho [21] in their work on the asymptotic performance of optimal quantizers, and hence the algorithm is certainly implicit in their work. They did not, however, actually propose or apply the technique to design a quantizer for fixed N. In 1965, Forgy [31] proposed the algorithm for cluster analysis for the multidimensional squared-error distortion case and a sample distribution (see the discussion in MacQueen [25]). In 1977, Chen [22] proposed essentially the same algorithm for the multidimensional case with the squared-error distortion measure and used it to design two-dimensional quantizers for vectors uniformly distributed in a circle.

Since the algorithm has no differentiability requirements, it is valid for purely discrete distributions. This has an important application to the case where one does not possess *a priori* a probabilistic description of the source to be compressed, and hence must base his design on an observed long training sequence of the data to be compressed. One approach would be to use standard density estimation techniques of statistics to obtain a "smooth" distribution of F and to then apply variational techniques. As previously discussed, we do not adopt this approach as it requires additional assumptions on the allowed densities. Instead we consider the following approach: Use the training sequence, say $\{x_k; k = 0, \cdots, n-1\}$ to form the time-average distortion

$$\frac{1}{n} \sum_{i=0}^{n-1} d(x_i, q(x_i))$$

and observe that this is exactly the expected distortion $E_{G_n} d(X, q(X))$ with respect to the sample distribution G_n determined by the training sequence, i.e., the distribution that assigns probability m/n to a vector x that occurs in the training sequence m times. Thus we can design a quantizer that minimizes the time-average distortion for the training sequence by running the algorithm on the sample distribution G_n(1,2). This yields the following variation of the algorithm:

Algorithm (Unknown Distribution)

(0) Initialization: Given N = number of levels, distortion threshold $\epsilon \geqslant 0$, an initial N-level reproduction alphabet $\hat{A}_0$, and a training sequence $\{x_j; j = 0, \cdots, n-1\}$. Set $m = 0$ and $D_{-1} = \infty$.

(1) Given $\hat{A}_m = \{y_i; i = 1, \cdots, N\}$, find the minimum distortion partition $P(\hat{A}_m) = \{S_i; i = 1, \cdots, N\}$ of the training sequence: $x_j \in S_i$ if $d(x_j, y_i) \leqslant d(x_j, y_l)$, for all l. Compute the average distortion

$$D_m = D(\{\hat{A}_m, P(\hat{A}_m)\}) = n^{-1} \sum_{j=0}^{n-1} \min_{y \in \hat{A}_m} d(x_j, y).$$

(2) If $(D_{m-1} - D_m)/D_m \leqslant \epsilon$, halt with $\hat{A}_m$ final reproduction alphabet. Otherwise continue.

(3) Find the optimal reproduction alphabet $\hat{x}(P(\hat{A}_m)) = \{\hat{x}(S_i); i = 1, \cdots, N\}$ for $P(\hat{A}_m)$. Set $\hat{A}_{m+1} \triangleq \hat{x}(P(\hat{A}_m))$. Replace m by $m + 1$ and go to (1).

Observe above that while designing the quantizer, only partitions of the training sequence (the input alphabet) are considered. Once the final codebook $\hat{A}_m$ is obtained, however, it is used on new data outside the training sequence with the optimum nearest-neighbor rule, that is, an optimum partition of k-dimensional Euclidean space.

(1) It was observed by a reviewer that application of the algorithm to the sample distribution provides a "Monte Carlo" design of the quantizer for a vector with a known distribution, that is, the design is based on samples of the random vectors rather than on an explicit distribution.

(2) During the period this paper was being reviewed for publication, two similar techniques were reported for special cases. In 1978, Capria, Westin, and Esposito [23] presented a similar technique for the scalar case using dynamic programming arguments. Their approach was for average squared-error distortion and for maximum distortion over the training sequence. In 1979, Menez, Boeri, and Esteban [24] proposed a similar technique for scalar quantization using squared-error and magnitude-error distortion measures.

If the sequence of random vectors is stationary and ergodic, then it follows from the ergodic theorem that, with probability one, G_n goes to the true underlying distribution F as $n \to \infty$. Thus if the training sequence is sufficiently long, hopefully a good quantizer for the sample distribution G_n should also be good for the true distribution F, and hence should yield good performance for future data produced by the source. All of these ideas are made precise in [2] where it is shown that, subject to suitable mathematical assumptions, the quantizer produced by applying the algorithm to G_n converges, as $n \to \infty$, to the quantizer produced by applying the algorithm to the true underlying distribution F. We observe that no analogous results are known to the authors for the density estimation/variational approach, and that independence of successive blocks is not required for these results, only block stationarity and ergodicity.

We also point out that, for finite alphabet distributions such as sample distributions, the algorithm always converges to a fixed-point quantizer in a finite number of steps [2].

A similar technique used in cluster analysis with squared-error cost functions was developed by MacQueen in 1967 [25] and has been called the k-means approach. A more involved technique using the k-means approach is the "ISODATA" approach of Ball and Hall [26]. The basic idea of finding minimum distortion partitions and centroids is the same, but the training sequence data is used in a different manner and the resulting quantizers will, in general, be different. Their sequential technique incorporates the training vectors one at a time and ends when the last vector is incorporated. This is in contrast to the previous algorithm which considers all of the training vectors at each iteration. The k-means method can be described as follows: The goal is to produce a partition $S_0 = \{S_0, \cdots, S_{N-1}\}$ of the training alphabet, $A = \{x_i; i = 0, \cdots, n-1\}$ consisting of all vectors in the training sequence. The corresponding reproduction alphabet $\hat{A}$ will then be the collection of the Euclidean centroids of the sets S_i, that is, the final reproduction alphabet will be optimal for the final partition (but the final partition may not be optimal for the final reproduction alphabet, except as $n \to \infty$). To obtain S, we first think of the each S_i as a bin in which to place training sequence vectors until all are placed. Initially, we start by placing the first N vectors in separate bins, i.e., $x_i \in S_i$, $i = 0, \cdots, N-1$. We then proceed as follows: at each iteration, a new training vector x_m is observed. We find the set S_i for which the distortion between x_m and the centroid $\hat{x}(S_i)$ is minimized and then add x_m to this bin. Thus, at each iteration, the new vector is added to the bin with the closest centroid, and hence the next time, this bin will have a new centroid. This operation is continued until all sample vectors are incorporated.

Although similar in philosophy, the k-means algorithm has some crucial differences. In particular, it is suited for the case where *only* the training sequence is to be classified, that is, where a long sequence of vectors is to be grouped in a low distortion manner. The sequential procedure is computationally efficient for grouping, but a "quantizer" is not produced until the procedure is stopped. In other words, in cluster analysis, one wishes to group things and the groups can change with time, but in quantization, one wishes to fix the groups (to get a time-invariant quantizer), and then use these groups (or the quantizer) on future data outside of the training sequence.

An additional problem is that the only theorems which guarantee convergence, in the limit of a long training sequence, require the assumption that successive vectors be independent [25], unlike the more general case for the proposed algorithm [2].

Recently Levenson *et al.* used a variation of the k-means and ISODATA algorithms with a distortion measure proposed by Itakura [4] to determine reference templates for speaker-independent word recognition [27]. They used, as a distortion measure, the logarithm of the distortion of (7) (which is a gain-optimized Itakura-Saito distortion [7]—our use of the distortion measure with unit-gain-normalized models results in no such logarithmic function). In their technique, however, a minimax rule was used to select the reproduction vectors (or cluster points) rather than finding the "optimum" centroid vector. If instead, the distortion measure of (7) is used, then the centroids are easily found, as will be seen.

CHOICE OF $\hat{A}_0$

There are several ways to choose the initial reproduction alphabet $\hat{A}_0$ required by the algorithm. One method for use on sample distributions is that of the k-means method, namely choosing the first N vectors in the training sequence. We did not try this approach as, intuitively, one would like these vectors to be well-separated, and N consecutive samples may not be. Two other methods were found to be useful in our examples. The first is to use a uniform quantizer over all or most of the source alphabet (if it is bounded). For example, if used on a sample distribution, one uses a k-dimensional uniform quantizer on a k-dimensional Euclidean cube including all or most of the points in the training sequence. This technique was used in the Gaussian examples described later.

The second technique is useful when one wishes to design quantizers of successively higher rates until achieving an acceptable level of distortion. Here we consider M-level quantizers with $M = 2^R$, $R = 0, 1, \cdots$, and continue until we achieve an initial guess for an N-level quantizer as follows:

INITIAL GUESS BY "SPLITTING"

(0) Initialization: Set $M = 1$ and define $\hat{A}_0(1) = \hat{x}(A)$, the centroid of the entire alphabet (the centroid of the training sequence, if a sample distribution is used).

(1) Given the reproduction alphabet $\hat{A}_0(M)$ containing M vectors $\{y_i; i = 1, \cdots, M\}$, "split" each vector y_i into two close vectors $y_i + \epsilon$ and $y_i - \epsilon$, where ϵ is a fixed perturbation vector. The collection $\tilde{A}$ of $\{y_i + \epsilon, y_i - \epsilon, i = 1, \cdots, M\}$ has $2M$ vectors. Replace M by $2M$.

(2) Is $M = N$? If so, set $\hat{A}_0 = \tilde{A}(M)$ and halt. $\tilde{A}_0$ is then the initial reproduction alphabet for the N-level quantization algorithm. If not, run the algorithm for an M-level quantizer on $\tilde{A}(M)$ to produce a good reproduction alphabet $\hat{A}_0(M)$, and then return to step (1).

Using the splitting algorithm on a training sequence, one starts with a one-level quantizer consisting of the centroid of the training sequence. This vector is then split into two vectors and the two-level quantizer algorithm is run on this pair to obtain a good (fixed-point) two-level quantizer. Each of these two vectors is then split and the algorithm is run to produce a good four-level quantizer. At the conclusion, one has fixed-point quantizers for $1, 2, 4, 8, \cdots, N$ levels.

EXAMPLES

Gaussian Sources

The algorithm was used initially to design quantizers for the classical example of memoryless Gaussian random variables with the squared-error distortion criterion of (1), based on a training sequence of data. The training and sample data were produced by a zero-mean, unit-variance memoryless sequence of Gaussian random variables. The initial guess was a unit quantizer on the k-dimensional cube, $\{\mathbf{x}: |x_i| \leqslant 4; i = 0, \cdots, k-1\}$. A distortion threshold of 0.1% was used. The overall algorithm can be described as follows:

(0) Initialization: Fix N = number of levels, k = block length, n = length of training sequence, $\epsilon = .001$. Given a training sequence $\{\mathbf{x}_j; j = 0, \cdots, n-1\}$. Let $\hat{A}_0$ be an N-level uniform quantizer reproduction alphabet for the k-dimensional cube, $\{\mathbf{u}: |u_i| \leqslant 4, i = 0, \cdots, k-1\}$. Set $m = 0$ and $D_{-1} = \infty$.

(1) Given $\hat{A}_m = \{\mathbf{y}_i; i = 1, \cdots, N\}$, find the minimum-distortion partition $P(\hat{A}_m) = \{S_i; i = 1, \cdots, N\}$. For example, for each $j = 0, \cdots, n-1$, compute $d(\mathbf{x}_j, \mathbf{y}_i)$ for $i = 1, \cdots, N$. If $d(\mathbf{x}_j, \mathbf{y}_i) \leqslant d(\mathbf{x}_j, \mathbf{y}_l)$ for all l, then $\mathbf{x}_j \in S_i$. Compute:

$$D_m = D(\{\hat{A}_m, P(\hat{A}_m)\}) = n^{-1} \sum_{j=0}^{n-1} \min_{\mathbf{y} \in \hat{A}_m} d(\mathbf{x}_j, \mathbf{y}).$$

(2) If $(D_{m-1} - D_m)/D_m \leqslant \epsilon = .001$, halt with final quantizer described by $\hat{A}_m$. Otherwise continue.

(3) Find the optimal reproduction alphabet $\hat{\mathbf{x}}(P(\hat{A}_m)) = \{\hat{\mathbf{x}}(S_i); i = 1, \cdots, N\}$ for $P(\hat{A}_m)$. For the squared-error criterion, $\hat{\mathbf{x}}(S_i)$ is the Euclidean center of gravity or centroid given by

$$\mathbf{x}(S_i) = \frac{1}{\| S_i \|} \sum_{j: \mathbf{x}_j \in S_i} \mathbf{x}_j,$$

where $\|S_i\|$ denotes the number of training vectors in the cell S_i. If $\|S_i\| = 0$, set $\hat{\mathbf{x}}(S_i) = \mathbf{y}_i$, the old codeword. Define $\hat{A}_{m+1} = \hat{\mathbf{x}}(P(A_m))$, replace m by $m+1$, and go to (1).

Table 1 presents a simple but nontrivial example intended to demonstrate the basic operation of the algorithm. A two-dimensional quantizer with four levels is designed, based on a short training sequence of twelve training vectors. Because of the short training sequence in this case, the final distortion is lower than one would expect and the final quantizer may not work well on new data outside of the training sequence. The tradeoffs between the length of the training sequence and the performance inside and outside the training sequence are developed more carefully in the speech example.

TABLE 1

A SIMPLE EXAMPLE

(0) <u>Initialization</u>: $N = 4$, $k = 2$, $\epsilon = .001$, $n = 12$.

Training Sequence:

$\underset{\sim}{x}_1 = (-.37449, .98719)$	$\underset{\sim}{x}_7 = (-.59161, .17968)$
$\underset{\sim}{x}_2 = (.63919, -.11875)$	$\underset{\sim}{x}_8 = (.14093, 1.76413)$
$\underset{\sim}{x}_3 = (-.83293, .60645)$	$\underset{\sim}{x}_9 = (.70898, -.35017)$
$\underset{\sim}{x}_4 = (-.70534, -1.21856)$	$\underset{\sim}{x}_{10} = (.30038, .79836)$
$\underset{\sim}{x}_5 = (-.28952, -.94821)$	$\underset{\sim}{x}_{11} = (.30165, 1.06552)$
$\underset{\sim}{x}_6 = (1.09924, .516)$	$\underset{\sim}{x}_{12} = (.37801, -.32708)$

$\hat{A}_0 = \{(2,2), (2,-2), (-2,2), (-2,-2)\}$

$= \{\underset{\sim}{y}_1, \underset{\sim}{y}_2, \underset{\sim}{y}_3, \underset{\sim}{y}_4\}$

$D_{-1} = 9.99E + 62$ (∞ on a microcomputer)

Set $m = 0$.

<u>m=0</u> (1) Find $P(\hat{A}_0) = \{S_1, S_2, S_3, S_4\}$:

$\underset{\sim}{x}_j \in S_i$ if $d(\underset{\sim}{x}_j, \underset{\sim}{y}_i) \leq d(\underset{\sim}{x}_j, \underset{\sim}{y}_m)$, all m.

$S_1 = \{\underset{\sim}{x}_6, \underset{\sim}{x}_8, \underset{\sim}{x}_{10}, \underset{\sim}{x}_{11}\}$

$S_2 = \{\underset{\sim}{x}_2, \underset{\sim}{x}_9\}$

$S_3 = \{\underset{\sim}{x}_1, \underset{\sim}{x}_3, \underset{\sim}{x}_7\}$

$S_4 = \{\underset{\sim}{x}_4, \underset{\sim}{x}_5, \underset{\sim}{x}_{12}\}$

Compute D_0:

$$D_0 = \frac{1}{12} \sum_{j=1}^{12} \min_{\underset{\sim}{y} \in \hat{A}_0} d(\underset{\sim}{x}_j, \underset{\sim}{y}) = 2.0172 .$$

(2) $(D_{-1} - D_0)/D_0 > .001$, continue.

(3) Find the optimal reproduction alphabet $\hat{A}_1 \triangleq \hat{\underset{\sim}{x}}(P(\hat{A}_0)) = \{\hat{\underset{\sim}{x}}(S_i), i = 1, \ldots, 4\}$:

$\hat{\underset{\sim}{x}}(S_1) = (\underset{\sim}{x}_6 + \underset{\sim}{x}_8 + \underset{\sim}{x}_{10} + \underset{\sim}{x}_{11})/4 = (.46055, 1.036)$

$\hat{\underset{\sim}{x}}(S_2) = (\underset{\sim}{x}_2 + \underset{\sim}{x}_9)/2 = (.674085, -.23446)$

$\hat{\underset{\sim}{x}}(S_3) = (\underset{\sim}{x}_1 + \underset{\sim}{x}_3 + \underset{\sim}{x}_7)/3 = (-.599676, .591106)$

$\hat{\underset{\sim}{x}}(S_4) = (\underset{\sim}{x}_4 + \underset{\sim}{x}_5 + \underset{\sim}{x}_{12})/3 = (-.457623, -.831283)$

Set $m = 1$. Go to (1).

<u>m=1</u> (1) Find $P(\hat{A}_1)$:

Evaluating distortions shows $P(\hat{A}_1) = P(\hat{A}_0)$ (no change in partition)

Compute D_1:

$$D_1 = \frac{1}{12} \sum_{j=1}^{12} \min_{\underset{\sim}{y} \in \hat{A}_1} d(\underset{\sim}{x}_j, \underset{\sim}{y}) = .0997308 .$$

(2) $(D_0 - D_1)/D_1 \cong 19 > .001$

(3) $\hat{A}_2 \triangleq \hat{\underset{\sim}{x}}(P(\hat{A}_1)) = \hat{A}_1$, since $P(\hat{A}_1) = P(\hat{A}_0)$ and hence $\hat{\underset{\sim}{x}}(P(\hat{A}_1)) = \hat{\underset{\sim}{x}}(P(\hat{A}_0)) = \hat{A}_1$. Thus $\hat{A}_1$ is a fixed point. Set $m=2$. Go to (1).

<u>m=2</u> (1) $P(\hat{A}_1) = P(\hat{A}_0)$ and hence $D_2 = D_1$ and hence $(D_1 - D_2)/D_2 = 0 < .001$.

Halt with final quantizer described by $\{\hat{A}_1, P(\hat{A}_1)\}$.

Note: Characters with tildes underneath appear boldface in text.

Observe that, in the example of Table 1, the algorithm could actually halt in step (1) of the $m = 1$ iteration since, if $P(\hat{A}_m) = P(\hat{A}_{m-1})$, it follows that $\hat{A}_{m+1} = \hat{\mathbf{x}}(P(\hat{A}_m)) = \hat{\mathbf{x}}(P(\hat{A}_{m-1})) = \hat{A}_m$, and hence $\hat{A}_m$ is the desired fixed point. In other words, if the quantizer stays the same for two iterations, then the two distortions are equal and an "$\epsilon = 0$" threshold is satisfied.

As a more realistic example, the algorithm was run for the scalar ($k = 1$) case with $N = 2, 3, 4, 6$ and 8, using a training sequence of 10,000 samples per quantizer output from a zero-mean, unit-variance memoryless Gaussian source. The resulting quantizer outputs and distortion were within 1% of the optimal values reported by Max [15]. No more than 20

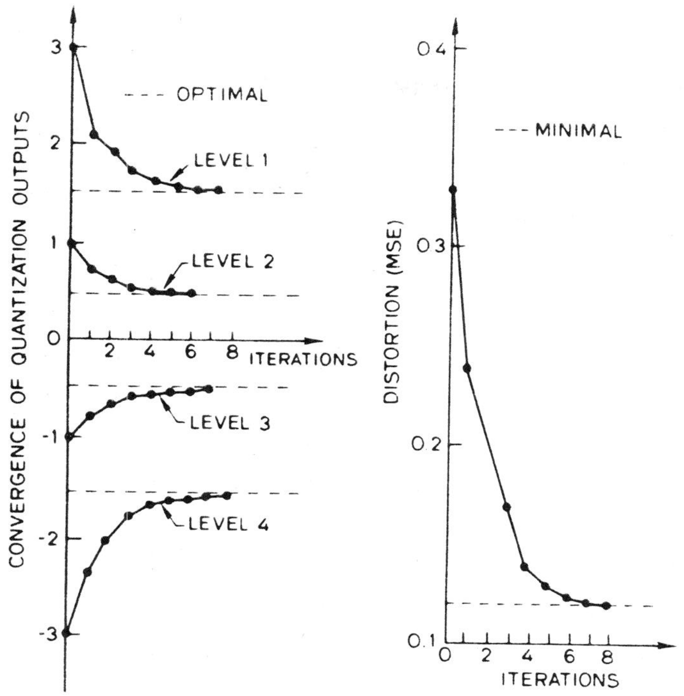

Fig. 1. The Basic Algorithm: Gaussian Source $N = 4$.

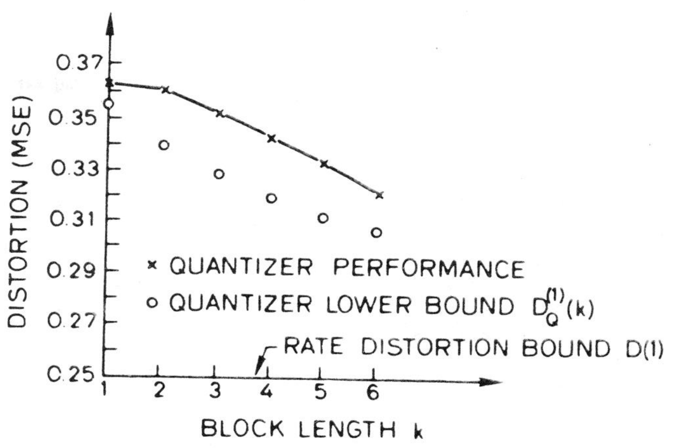

Fig. 2. Block Quantization Rate 1 bit/symbol Gaussian Source.

iterations were required for $N = 8$ and, for smaller N, the number of iterations was considerably smaller. Figure 1 describes the convergence rate of one of the tests for the case $N = 4$.

The algorithm was then tried for block quantizers for memoryless Gaussian variables with block lengths k equal to 1, 2, 3, 4, 5 and 6 and a rate of one bit per sample, so that $N = 2^k$. The distortion criterion was again the squared-error distortion measure of (1). The algorithm used a training sequence of 100,000 samples. In each case, the algorithm converged in fewer than 50 iterations and the resulting distortion is plotted in Fig. 2, together with the one bit-per-symbol scalar case as a function of block length. For comparison, the rate-distortion bound [28, p. 99] $D(R) = 2^{-2R}$ for $R = 1$ bit-per-symbol is also plotted. As expected and as shown in Fig. 2, the block quantizers outperform the scalar quantizer, but for these block lengths, the performance is still far from the rate distortion bound (which is achievable, in principle, only in the limit as $k \to \infty$). A more favorable comparison is obtained using a recent result of Yamada, Tazaki, and Gray [29] which provides a lower bound to the performance of an optimal N-level k-dimensional quantizer with a difference distortion measure when N is large. This bound provides strict improvement over the rate-distortion bound for fixed k and tends to the rate-distortion bound as $k \to \infty$. In the current case, the bound has the form

$$D_Q^{(k)}(R) = D(R) \cdot \left\{ \left(\frac{e}{1 + k/2} \right) \Gamma(1 + k/2)^{2/k} \right\},$$

where Γ is the gamma function. This bound is theoretically inappropriate for small k, yet it is surprisingly close for the $k = 1$ result, which is known to be almost optimal. For $k = 6$, $N = 2^6 = 64$ is moderately large and the closeness of the actual performance to the lower bound, compared to optimal performance provided by $D_Q{}^{(k)}(1)$, suggests that the algorithm is indeed providing a quantizer with block length six and rate one bit-per-symbol that is nearly optimal (within 6% of the optimal).

LLOYD'S EXAMPLE

Lloyd [1] provides an example where both variational and fixed-point approaches can yield locally optimal quantizers instead of a globally optimum quantizer. We next propose a slight modification of the fixed-point algorithm that indeed finds a globally optimum quantizer in Lloyd's example. We conjecture that this technique will work more generally, but we have been unable to prove this theoretically. A similar technique can be used with the stationary-point algorithm.

Instead of using samples from the source that we wish to quantize, we use samples corrupted by additive independent noise, where the marginal distribution of the noise is such that only one locally optimum quantizer exists for it. As an example, for scalar quantization with the squared-error distortion measure, we use Gaussian noise. In this case, any locally optimal quantizer is also globally optimum. Other distributions, such as the uniform or a discrete amplitude noise with an alphabet size equal to the number of quantizer output levels, can also be used.

When the noise power is much greater than the source power, the distribution of their sum is essentially the distribution of the noise. We assume that, initially, the noise power is so large that only one locally optimum quantizer exists for the sum; hence, regardless of the initial guess, the algorithm will converge to this optimum. On the next step, the noise power is reduced slightly and the quantizer resulting from the previous run is used as the initial guess. Intuitively, since the noise has been reduced by a small amount, the global optimum for the new sum should be close to that of the previous sum (we use the same source and noise samples with reduced noise power). Thus we expect that the algorithm will converge to the global optimum even though new local optimum points might have been introduced. We continue in the same manner reducing the noise gradually to zero.

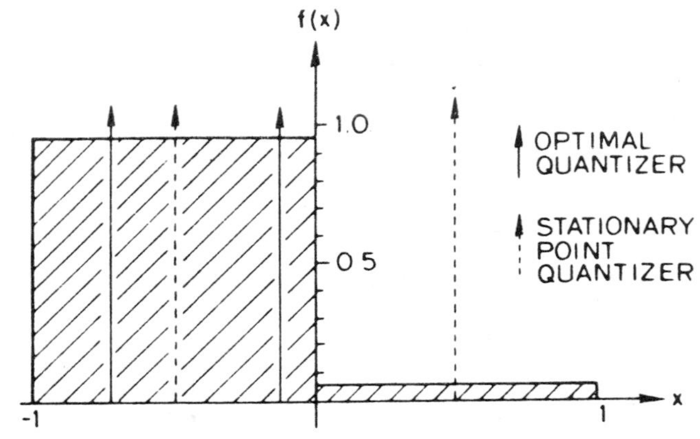

Fig. 3. The Probability Density Function.

To illustrate how the algorithm works, we use a source with a probability density function as shown in Fig. 3. In this case, there are two locally optimum two-level quantizers. One has the output levels +0.5 and −0.5 and yields a mean-squared error of 0.083, the second (which is the global optimum) has output levels −0.71 and 0.13 and yields a mean-squared error 0.048. (This example is essentially the same as one of Lloyd's [1].)

The modified algorithm was tested on a sequence of 2,000 samples chosen according to the probability density shown in Fig. 3. Gaussian noise was added starting at unity variance and reducing the variance by approximately 50% on each successive run. The initial guess was + 0.5 and − 0.5 which is the non-global optimum. Each run was stopped when the distortion was changed by less than 0.1% from its previous value.

The results are given in Fig. 4 and it is seen that, in spite of the bad initial guess, the modified algorithm converges to the globally optimum quantizer.

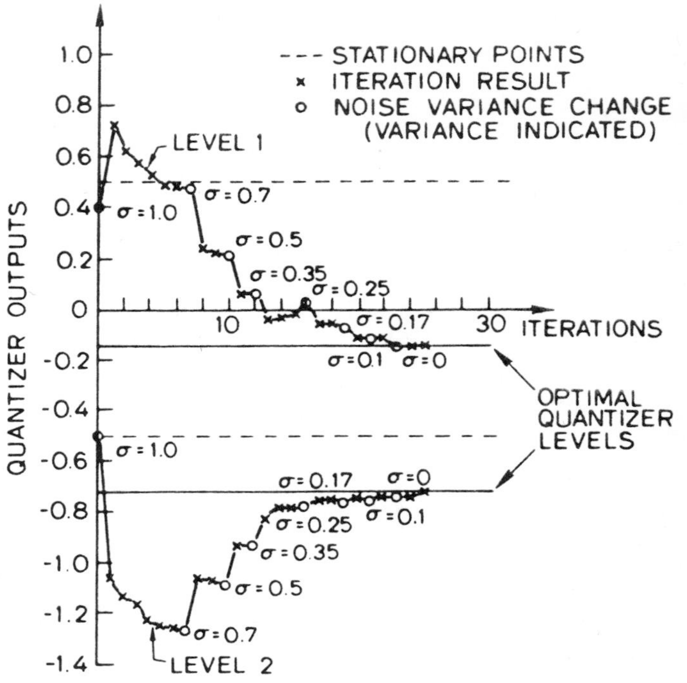

Fig. 4. The Modified Algorithm.

SPEECH EXAMPLE

In the next example, we consider the case of a speech compression system consisting of an LPC analysis of 20 ms-long speech frames producing a voiced/unvoiced decision and a pitch, a gain, and a normalized inverse filter as previously described, followed by quantization where the pitch and gain are separately quantized as usual, but the normalized filter coefficients $(a_1, \cdots, a_K) = (x_0, x_1, \cdots, x_{K-1})$ are quantized as a vector with $K = 10$, using the distortion measure of (7)-(8). The training sequence consisted of a sequence of normalized inverse filter parameter vectors(3). The LPC analysis was digital, and hence the training sequence used was already "finely quantized" to 10 bits per sample or 100 bits for each vector. The original gain required 12 bits per speech frame and the pitch used 8 bits per speech frame. The total rate of the LPC output (which we wish to further compress by block quantization) is 6000 bits/s. No further compression of gain or pitch was attempted in these experiments as our goal was to study only potential improvement when block quantizing the normalized filter parameters. The more complete problem including gain and pitch involves many other issues and is the subject of a paper still in preparation.

For the distortion measure of (7)-(8), the centroid of a subset S of a training sequence $\{x_j, j = 0, 1, \cdots, n-1\}$ is the vector u minimizing

$$\sum_{j: x_j \in S} (x_j - u) R(x_j) (x_j - u)^t .$$

We observe that the autocorrelation matrix $R(x)$ is a natural byproduct of the LPC analysis and need not be recomputed. This minimization, however, is a minimum-energy-residual minimization problem in LPC analysis and it can be solved by standard LPC algorithms such as Levinson's algorithm [10]. Alternatively, it is a much studied minimization problem in Toeplitz matrix theory [30] and the centroid can be shown via variational techniques to be

$$\hat{x}(S) = \left\{ \sum_{j: x_j \in S} R(x_j) \right\}^{-1} \sum_{j: x_j \in S} R(x_j) x_j^t . \tag{15}$$

The splitting technique for the initial guess and a distortion threshold of 0.5% were used. The complete algorithm for this example can thus be described as follows:

(0) Initialization: Fix $N = 2^R$, R an integer, where N is the largest number of levels desired. Fix $K = 10$, n = length of training sequence, $\epsilon = .005$. Set $M = 1$.

Given a training sequence $\{x_j; j = 0, \cdots, n-1\}$, set $A = \{x_j; j = 0, \cdots, n-1\}$, the training sequence alphabet. Define $\hat{A}(1) = \hat{x}(A)$, the centroid of the entire training sequence using (15) or Levinson's algorithm.

(3) The training sequence and additional test data of LPC reflection coefficients were provided by Signal Technology Inc. of Santa Barbara and were produced using standard LPC techniques on a single male speaker.

(1) (Splitting): Given $\hat{A}(M) = \{y_i, i = 1, \cdots, M\}$, split each reproduction vector y_i into $y_i + \epsilon$ and $y_i - \epsilon$, where ϵ is a fixed perturbation vector. Set $\hat{A}_0(2M) = \{y_i + \epsilon, y_i - \epsilon, i = 1, \cdots, M\}$ and then replace M by $2M$.

(2) Set $m = 0$ and $D_{-1} = \infty$.

(3) Given $\hat{A}_m(M) = \{y_1, \cdots, y_M\}$, find its optimum partition $P(\hat{A}_m(M)) = \{S_i; i = 1, \cdots, M\}$, that is, $x_j \in S_i$ if $d(x_j, y_i) \leqslant d(x_j, y_l)$, all l. Compute the resulting distortion

$$D_m = D(\{\hat{A}_m(M), P(\hat{A}_m(M))\})$$

$$= n^{-1} \sum_{j=0}^{n-1} \min_{y \in \hat{A}_m} d(x_j, y).$$

(4) If $(D_{m-1} - D_m)/D_m \leqslant \epsilon = .005$, then go to step (6). Otherwise continue.

(5) Find the optimal reproduction alphabet $\hat{A}_{m+1}(M) = \hat{x}(P(\hat{A}_m(M))) = \{\hat{x}(S_i); i = 1, \cdots, 1\}$ for $P(\hat{A}_m(M))$. Replace m by $m + 1$ and go to (3).

(6) Set $\hat{A}(M) = \hat{A}_m(M)$. The final M-level quantizer is described by $\hat{A}(M)$. If $M = N$, halt with final quantizer described by $\hat{A}(N)$. Otherwise go to step (1).

Table 2 describes the results of the algorithm for $N = 64$, and hence for one- to eight-bit quantizers trained on $n =$ 19,000 frames of LPC speech produced by a single speaker. The distortion at the end of each iteration is given and, in all cases, the algorithm converged in fewer than 14 iterations. When the resulting quantizers were applied to data from the same speaker outside of the training sequence, the resulting distortion was within 1% of that within the training sequence. A total of three and one-half hours of computer time on a PDP 11/35 was required to obtain all of these codebooks.

Figure 5 depicts the rate of convergence of the algorithm with a training sequence length for a 16-level quantizer. Note the marked difference between the distortion for 2400 frames inside the training sequence and outside the training sequence for short training sequences. For a long training sequence of over 12,000 frames, however, the distortion is nearly the same.

Tapes of the synthesized speech at 8 bits per frame for the normalized model sounded similar to those of the original LPC speech with 100 bits per frame for the normalized model (the gain and the pitch were both left at the original LPC rate of 12 and 8 bits per frame, respectively). While extensive subjective tests were not attempted, all informal listening tests judged the synthesized speech perfectly intelligible (when heard *before* the original LPC!) and the quality only slightly inferior when the two were compared. The overall compression was from 6000 bits/s to 1400 bits/s. This is not startling as existing scalar quantizers that optimally allocate bits among the parameters and optimally quantize each parameter using a spectral deviation distortion measures [11] also perform well in this range. It is, however, promising as these were preliminary results with no attempt to further compress pitch and gain (which, taken together in our system, had more than twice the bit rate of the normalized model vector quantizer). Further results on applications of the algorithm to the overall speech compression system will be the subject of a forthcoming paper [33].

TABLE 2
ITAKURA-SAITO DISTORTION VS. NUMBER OF ITERATIONS.
TRAINING SEQUENCE LENGTH = 19,000 FRAMES.

NUMBER OF LEVELS	DISTORTION	ITERATION NUMBER
2	10.33476	1
	1.98925	2
	1.78301	3
	1.67244	4
	1.55983	5
	1.49814	6
	1.48493	7
	1.48249	8
4	1.38765	1
	1.07906	2
	1.04223	3
	1.03252	4
	1.02709	5
8	0.96210	1
	0.85183	2
	0.81353	3
	0.79191	4
	0.77472	5
	0.76188	6
	0.75130	7
	0.74383	8
	0.73341	9
	0.71999	10
	0.71346	11
	0.70908	12
	0.70578	13
	0.70347	14
16	0.64653	1
	0.55665	2
	0.51810	3
	0.50146	4
	0.49235	5
	0.48761	6
	0.48507	7
32	0.44277	1
	0.40452	2
	0.39388	3
	0.38667	4
	0.38128	5
	0.37778	6
	0.37574	7
	0.37448	8
64	0.34579	1
	0.31766	2
	0.30850	3
	0.30366	4
	0.30086	5
	0.29891	6
	0.29746	7
128	0.27587	1
	0.25628	2
	0.24928	3
	0.24550	4
	0.24309	5
	0.24142	6
	0.24021	7
	0.23933	8
256	0.22458	1
	0.20830	2
	0.20228	3
	0.19849	4
	0.19623	5
	0.19479	6
	0.19386	7
	0.19319	8

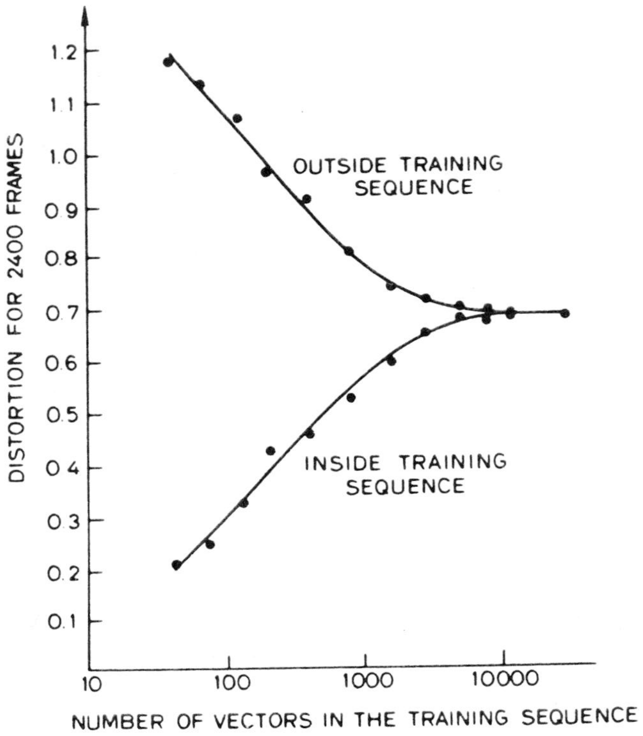

Fig. 5. Convergence with Training Sequence.

EPILOGUE

The Gaussian example of Figure 2, Lloyd's example, and the speech example were run on a PDP 11/34 minicomputer at the Stanford University Information Systems Laboratory. The simple example of Table 1 was run in BASIC on a Cromemco System 3 microcomputer. As a check, the microcomputer program was also used to design quantizers for the Gaussian case of Figure 2 using the splitting method, $k = 1, 2,$ and 3, and a training sequence of 10,000 vectors. The results agreed with the PDP 11/34 run to within one percent.

ACKNOWLEDGMENT

The authors would like to acknowledge the help of J. Markel of Signal Technology, Inc. of Santa Barbara and A. H. Gray, Jr., of the University of California at Santa Barbara in both the analysis and synthesis of the speech example.

REFERENCES

[1] Lloyd, S. P., "Least Squares Quantization in PCM's," Bell Telephone Laboratories Paper, Murray Hill, NJ, 1957.

[2] Gray, R. M., J. C. Kieffer and Y. Linde, "Locally Optimal Block Quantization for Sources without a Statistical Model," Stanford University Information Systems Lab Technical Report No. L-904-1, Stanford, CA, May 1979 (submitted for publication).

[3] Itakura, F. and S. Saito, "Analysis Synthesis Telephony Based Upon Maximum Likelihood Method," *Repts. of the 6th Internat'l. Cong. Acoust.*, Y. Kohasi, ed., Tokyo, C-5-5, C17-20, 1968.

[4] Itakura, F., "Maximum Prediction Residual Principle Applied to Speech Recognition," *IEEE Trans. ASSP*, 23, pp. 67-72, Feb. 1975.

[5] Chaffee, D. L., "Applications of Rate Distortion Theory to the Bandwidth Compression of Speech Signals," Ph.D. Dissertation, Univ. of Calif. at Los Angeles, 1975.

[6] Gray, R. M., A. Buzo, A. H. Gray, Jr., and J. D. Markel, "Source Coding and Speech Compression," *Proc. of the 1978 Internat'l. Telemetering Conf.*, pp. 371-878, 1978.

[7] Matsuyama, Y., A. Buzo and R. M. Gray, "Spectral Distortion Measures for Speech Compression," Stanford Univ. Inform. Systems Lab. Tech. Rept. 6504-3, Stanford, CA, April, 1978.

[8] Buzo, A., "Optimal Vector Quantization for Linear Predicted Coded Speech," Stanford Univ., Ph.D. Dissertation, Dept. of Elec. Engrg., August, 1978.

[9] Matsuyama, Y. A., "Process Distortion Measures and Signal Processing," Ph.D. Dissertation, Dept. of Elec. Engrg., Stanford Univ., 1978.

[10] Markel, J. D. and A. H. Gray, Jr., *Linear Prediction of Speech*, Springer-Verlag, NY 1976.

[11] Gray, A. H., Jr., R. M. Gray and J. D. Markel, "Comparison of Optimal Quantizations of Speech Reflection Coefficients," *IEEE Trans. ASSP*, Vol. 24, pp. 4-23, Feb. 1977.

[12] Dalenius, T., "The Problem of Optimum Stratification," *Skandinavisk Aktuuvielidskrift*, Vol. 33, pp. 203-213, 1950.

[13] Fisher, W. D., "On a Pooling Problem from the Statistical Decision Viewpoint," *Econometrica*, Vol. 21, pp. 567-585, 1953.

[14] Cox, D. R., "Note on Grouping," *J. of the Amer. Statis. Assoc.*, Vol. 52, pp. 543-547, 1957.

[15] Max, J., "Quantizing for Minimum Distortion," *IRE Trans. on Inform. Theory*, IT-6, pp. 7-12, March 1960.

[16] Fleischer, P., "Sufficient Conditions for Achieving Minimum Distortion in a Quantizer," *IEEE Int. Conv. Rec.*, pp. 104-111, 1964.

[17] Luenberger, D. G., *Optimization by Vector Space Methods*, John Wiley & Sons, NY, 1969.

[18] Luenberger, D. G., *Introduction to Linear and Nonlinear Programming*, Addison-Wesley, Reading, MA, 1973.

[19] Rockafellar, R. T., *Convex Analysis*, Princeton Univ. Press, Princeton, NJ, 1970.

[20] Zador, P., "Topics in the Asymptotic Quantization of Continuous Random Variables," Bell Telephone Laboratories Technical Memorandum, Feb. 1966.

[21] Gersho, A., "Asymptotically Optimal Block Quantization," *IEEE Trans. on Inform. Theory*, Vol. IT-25, pp. 373-380, 1979.

[22] Chen, D. T. S., "On Two or More Dimensional Optimum Quantizers," *Proc. 1977 IEEE Internat'l. Conf. on Acoustics, Speech, & Signal Processing*, pp. 640-643, 1977.

[23] Caprio, J. R., N. Westin and J. Esposito, "Optimum Quantization for Minimum Distortion," *Proc. of the Internat'l. Telemetering Conf.*, pp. 315-323, Nov. 1978.

[24] Menez, J., F. Boeri, and D. J. Esteban, "Optimum Quantizer Algorithm for Real-Time Block Quantizing," *Proc. of the 1979 IEEE Internat'l. Conf. on Acoustics, Speech, & Signal Processing*, pp. 980-984, 1979.

[25] MacQueen, J., "Some Methods for Classification and Analysis of Multivariate Observations," *Proc. of the Fifth Berkeley Symposium on Math., Stat. and Prob.*, Vol. 1, pp. 281-296, 1967.

[26] Ball, G. H. and D. J. Hall, "Isodata—An Iterative Method of Multivariate Analysis and Pattern Classification," in *Proc. IFIPS Congr.*, 1965.

[27] Levinson, S. E., L. R. Rabiner, A. E. Rosenberg and J. G. Wilson, "Interactive Clustering Techniques for Selecting Speaker-Independent Techniques for Selecting Speaker-Independent Reference Templates for Isolated Word Recognition," *IEEE Trans. ASSP*, Vol. 27, pp. 134-141, 1979.

[28] Berger, T., *Rate Distortion Theory*, Prentice-Hall, Englewood Cliffs, NJ, 1971.

[29] Yamada, Y., S. Tazaki and R. M. Gray, "Asymptotic Performance of Block Quantizers with Difference Distortion Measures," to appear, *IEEE Trans. on Inform. Theory*.

[30] Grenander, U. and G. Szego, *Toeplitz Forms and Their Applications*, Univ. of Calif. Press, Berkeley, 1958.

[31] Forgy, E., "Cluster Analysis of Multivarate Data: Efficiency vs. Interpretability of Classifications," Abstract, *Biometrics*, Vol. 21, p. 768, 1965.

[32] Chaffee, D. L. and J. K. Omura, "A Very Low Rate Voice Compression System," Abstract in *Abstracts of Papers, 1974, IEEE Intern. Symp. on Inform. Theory*, Notre Dame, Oct. 28-31, IEEE, 1974.

[33] Buzo, A., A. H. Gray, Jr., R. M. Gray, and J. D. Markel, "Speech Coding Based on Vector Quantization," submitted for publication.

19

Reprinted from *Inform. Control* **45**(2):178–198 (1980)

Locally Optimal Block Quantizer Design*

R. M. Gray

Information Systems Laboratory, Stanford University, Stanford, Califnornia 94305

J. C. Kieffer

Department of Mathematics, University of Missouri-Rolla, Rolla, Missouri 65401

AND

Y. Linde

Codex Corporation, Mansfield, Massachusetts 01108

Several properties are developed for a recently proposed algorithm for the design of block quantizers based either on a probabilistic source model or on a long training sequence of data. Conditions on the source and general distortion measures under which the algorithm is well defined and converges to a local minimum are provided. A variation of the ergodic theorem is used to show that if the source is block stationary and ergodic, then in the limit as $n \to \infty$, the algorithm run on a sample distribution of a training sequence of length n will produce the same result as if the algorithm were run on the "true" underlying distribution.

I. Introduction

In a recent paper (Linde *et al.* (1980)), an iterative technique for the design of block quantizers was proposed and several of its properties demonstrated by experimental examples. It is the purpose of this paper to state an prove appropriate mathematical statements of these properties. The algorithm is a simple and natural extension of an algorithm proposed by Lloyd (1957) as his "Method I" for finding optimal quantizers for a squared-error distortion measure and scalar random variables possessing a known distribution. We show that many of Lloyd's observations remain true for block quantizers with a quite general class of distortion measures that, in particular, need not depend on the

* This research was partially supported by the Joint Services Electronics Program at Stanford University.

original vector $\mathbf{x}$ and its quantized value $q(\mathbf{x})$ only through the error vector $\mathbf{x} - q(\mathbf{x})$. As in Lloyd's case, the quantizers will in general only satisfy necessary conditions for optimality and will be globally optimum only in certain special cases.

Of particular interest is the case where one lacks an a priori probabilistic description of the vectors to be quantized and must instead base the quantizer design on a long training sequence of the random process to be quantized. We here consider several aspects of the algorithm when it is run on the sample distribution implied by the training sequence. This is equivalent to using Lloyd's method to minimize a time average distortion. We show that in this case the algorithm always converges to a locally optimum quantizer. More importantly, we show that if successive blocks of a random process are stationary and ergodic, then in the limit as $n \to \infty$ the quantizer produced by the algorithm run on a training sequence of length n will converge with probability one to the quantizer produced by the algorithm when run on the "true" underlying distribution. Thus a quantizer disigned to work well on a sufficiently long training sequence will also work well for future data produced by the same source.

As discussed in some detail in Linde *et al.* (1980), the algorithm considered here is a variation on a generalization of the k-means algorithm of cluster analysis with Euclidean distortion measures [MacQueen (1967)]. MacQueen (1967) developed similar properties of convergence with sample length for block independent processes. While this paper was in review, we discovered the independent work of Sverdrup–Thygeson (1980) and Pollard (1980) that strengthen and generalize MacQueen's k-means results to distortion measures that are increasing functions of Euclidean distance. Their results are for block independent processes, but their techniques apparently extend to the block ergodic and block stationary case considered here. In particular, Pollard's (1980) results hold for more general unbounded alphabets, but for less general distortion measures.

II. Preliminaries

Let $\mathbf{X} = (X_1, \ldots, X_K)$ be a K-dimensional random vector described by a probability distribution function F on $\mathscr{R}^K$, K-dimensional Euclidean space. It will occasionally be useful to define a set $A \subseteq \mathscr{R}^K$ as the alphabet if $\Pr(\mathbf{X} \in A) = \int_A dF(\mathbf{x}) = 1$.

An N-level quantizer $Q = \{\hat{A}, \mathscr{S}\}$ for $\mathscr{R}^K$ (block quantizer, vector quantizer, K-dimensional quantizer) consists of (i) a reproduction alphabet or codebook $\hat{A} = \{\mathbf{y}_i; i = 1, 2, \ldots, N\}$, $\mathbf{y}_i \in \mathscr{R}^K$; (ii) a partition $\mathscr{S} = \{S_i; i = 1, 2, \ldots, N\}$ of the alphabet A; and (iii) a mapping q: $A \to \hat{A}$ defined by $q(\mathbf{x}) = \mathbf{y}_i$ if $\mathbf{x} \in S_i$. If an N level quantizer Q is applied to a vector $\mathbf{X}$ for which $\Pr(\mathbf{X} \in S_i) = 0$ for some i, then we can remove $\mathbf{y}_i$ from $\hat{A}$ and S_i from $\mathscr{S}$ to form i, then we can remove $\mathbf{y}_i$ from $\hat{A}$ and S_i from $\mathscr{S}$ to form an $N - 1$ level quantizer without affecting performance.

The distortion or cost $d(\mathbf{x}, q(\mathbf{x}))$ of reproducing $\mathbf{x}$ as $q(\mathbf{x})$ is assumed bo be a nonnegative real valued function that satisfies the following requirements:

(a) For any fixed $\mathbf{x} \in \mathscr{R}^K$, $d(\mathbf{x}, \mathbf{y})$ is a convex function of $\mathbf{y}$, that is, for $\mathbf{y}_1$, $\mathbf{y}_2 \in \mathscr{R}^K$, $\lambda \in (0, 1)$, $d(\mathbf{x}, \lambda\mathbf{y}_1 + (1 - \lambda)\mathbf{y}_2) \leqslant \lambda d(\mathbf{x}, \mathbf{y}_1) + (1 - \lambda)\, d(\mathbf{x}, \mathbf{y}_2)$. If the inequality is strict then $d(\mathbf{x}, \mathbf{y})$ is strictly convex in $\mathbf{y}$.

(b) For any fixed $\mathbf{x}$, if $\mathbf{y}(n) = (y_1(n),\ldots, y_K(n)) \to \infty$ as $n \to \infty$ (that is, $y_i(n)$ diverges for some i), then also $d(\mathbf{x}, \mathbf{y}(n)) \to \infty$.

(c) d is locally bounded, that is, for any bounded sets B_1, $B_2 \subseteqq \mathscr{R}^K$, $\sup_{\mathbf{x} \in B_1, \mathbf{y} \in B_2} d(\mathbf{x}, \mathbf{y}) < \infty$.

Property (a) is the key assumption, the others being technical conditions to avoid pathologies. Assumption (b) effectively prevents $d(\mathbf{x}, \mathbf{y})$ from being constant along some line, and (c) prevents $d(\mathbf{x}, \mathbf{y})$ from "blowing up" on "nice" sets. We also make the obvious assumption that $d(\mathbf{x}, \mathbf{y})$ is measurable so that integrals make sense. We next consider several examples of distortion measures meeting these requirements.

Norm Distortion Measures

Let $\| \mathbf{u} \|$ denote a norm on $\mathscr{R}^K$. Let ρ be any nonconstant convex function on $[0, \infty)$ with $\rho(0) = 0$ (this ensures that ρ is nondecreasing). Then any distortion measure of the form

$$d(\mathbf{x}, \mathbf{y}) = \rho(\| \mathbf{x} - \mathbf{y} \|) \tag{1}$$

satisfies (a)–(c). Common examples are the l_p or Holder norms

$$\| \mathbf{u} \|_p = \left\{ \sum_{i=1}^{K} | u_i |^p \right\}^{1/p},$$

the l_∞ or sup or Minkowski norm

$$\| \mathbf{u} \|_\infty = \max_{i=1,\ldots,K} | u_i |,$$

and $\rho(\alpha) = | \alpha |^r$, $r \geqslant 1$. If ρ is strictly convex, then $d(\mathbf{x}, \mathbf{y})$ is strictly convex in $\mathbf{y}$. This class includes the K-dimensional rth power distortion

$$d(\mathbf{x}, \mathbf{y}) = \| \mathbf{x} - \mathbf{y} \|_r^r = \sum_{i=1}^{K} | x_i - y_i |^r, \tag{2}$$

$r > 1$.

Inner Product Distortion Measures

Let $(\mathbf{x}, \mathbf{y})$ denote an inner product on $\mathscr{R}^K$. Let ρ be as above. The distortion measure $d(\mathbf{x}, \mathbf{y}) = \rho((\mathbf{x} - \mathbf{y}, \mathbf{x} - \mathbf{y}))$ satisfies (a)–(c) (since $(\mathbf{x}, \mathbf{x})$ is a norm on $\mathscr{R}^K$). The most important example is the inner product

$$(\mathbf{x}, \mathbf{y}) = \mathbf{x}\mathbf{B}\mathbf{y}^t = \sum_{i=1}^{K} \sum_{j=1}^{K} x_i y_i B_{ij},$$

where t denotes transpose and $\mathbf{B}$ is a $K \times K$ symmetric positive definite matrix and hence

$$d(\mathbf{x}, \mathbf{y}) = (\mathbf{x} - \mathbf{y})\, \mathbf{B}(\mathbf{x} - \mathbf{y})^t$$

is a weighted-squares distortion.

Itakura–Saito Distortion

A distortion measure arising in speech communications due to Itakura and Saito (1968), Chaffee (1975), and Magill (1973) can be written in the following form:

$$d(\mathbf{x}, \mathbf{y}) = (\mathbf{x} - \mathbf{y})\, R(\mathbf{x})(\mathbf{x} - \mathbf{y})^t, \tag{3}$$

where $R(\mathbf{x})$ is the symmetric positive definite matrix depending on $\mathbf{x}$. We are not here concerned with the origin of this distortion measure (see Itakura and Saito (1968), Chaffee (1975), Matsuyama *et al.* (1978), Gray *et al.* (1978), Gray and Markel (1976), Buzo *et al.* (1979, 1980), and Gray *et al.* (1980)), but rather with the fact that it has been proposed for use in speech compression systems. In such systems $(1, x_1, \ldots, x_K)$ are the normalized (unit gain) filter parameters produced by Linear Predictive Coding (LPC) (see, e.g., Markel and Gray (1976)). The point here is that the distortion measure is sufficiently nasty to demonstrate the power of the quantization algorithm: it is not symmetric and it is not simply a weighting of errors since $\mathbf{R}(\mathbf{x})$ depends on $\mathbf{x}$. The measure is convex is positive definite and hence for $\lambda \in (0, 1)$

$$\begin{aligned} &d(\mathbf{x}, \lambda\mathbf{y}_1 + (1 - \lambda)\mathbf{y}_2) - \lambda d(\mathbf{x}, \mathbf{y}_1) - (1 - \lambda)\, d(\mathbf{x}, \mathbf{y}_2) \\ &\quad = \lambda(1 - \lambda)\{(\mathbf{x} - \mathbf{y}_1)\, R(\mathbf{x})(\mathbf{x} - \mathbf{y}_1)^t + (\mathbf{x} - \mathbf{y}_2)\, R(\mathbf{x})(\mathbf{x} - \mathbf{y}_2)^t\} > 0. \end{aligned} \tag{4}$$

The average distortion of a quantizer Q applied to a random vector $\mathbf{X}$ with distribution F is defined by

$$\begin{aligned} D(Q, F) &= Ed(\mathbf{X}, q(\mathbf{X})) \\ &= \sum_{i=1}^{N} \int_{S_i} d(\mathbf{x}, \mathbf{y}_i)\, dF(\mathbf{x}). \end{aligned}$$

We make the standard assumption that for F there exists a reference letter $a^* \in \mathscr{R}^K$ such that

$$Ed(\mathbf{X}, a^*) < \infty. \tag{5}$$

Equation (5) ensures the existence of a quantizer Q having $D(Q, F) < \infty$.

A quantizer Q^* with N levels is said to be *optimal* for F if $D(Q^*, F) \leqslant D(Q, F)$ for all quantizers Q having N or fewer levels. A quantizer Q^* is *locally optimal* if small perturbations in its codebook or partition cannot decrease the average distortion.

III. Properties of Optimal Quantizers

Given a quantizer $Q = \{\hat{A}, \mathscr{S}\}$, $\hat{A} = \{\mathbf{y}_i; i = 1, \ldots, N\}$, observe that

$$D(\{\hat{A}, \mathscr{S}\}, F) \geqslant \int \min_i d(\mathbf{x}, \mathbf{y}_i)\, dF(\mathbf{x}),$$

that is, no partition can yield lower average distortion than the partition obtained by mapping each $\mathbf{x}$ into the $\mathbf{y}_i \in \hat{A}$ that minimizes the distortion $d(\mathbf{x}, \mathbf{y}_i)$, i.e., by using a minimum distortion or nearest neighbor mapping. Define the sets $R_i = \{\mathbf{x}: d(\mathbf{x}, \mathbf{y}_i) \leqslant d(\mathbf{x}, \mathbf{y}_j), j \neq i\}$ and define an *optimal* or *Dirichlet partition* $\mathscr{P}(\hat{A}) = \{S_i\}$ of $\hat{A}$ as the collection $\{R_i\}$ with some tie-breaking rule, for example,

$$\begin{aligned} S_1 &= R_1, \\ &\vdots \\ S_k &= R_k - \bigcup_{i<k} R_k. \\ &\vdots \end{aligned}$$

For this rule each $\mathbf{x}$ is mapped into the $\mathbf{y}_i$ such that $d(\mathbf{x}, \mathbf{y}_i)$ is minimized and, if there is a tie, then $\mathbf{x}$ is placed in the atom with the lowest index. We therefore have that for any $\hat{A} = \{\mathbf{y}_i\}$,

$$\begin{aligned} D(\{\hat{A}, \mathscr{S}\}, F) &\geqslant D(\{\hat{A}, \mathscr{P}(\hat{A})\}, F) \\ &= \int \min_i d(\mathbf{x}, \mathbf{y}_i)\, dF(\mathbf{x}). \end{aligned} \tag{6}$$

Conversely, given an $\{\hat{A}, \mathscr{S}\}$ for which (6) holds with equality then $\mathscr{S} = \{P_i\}$ must be a Dirichlet partition with probability one in the sense that $\Pr(P_i - R_i) = \Pr(P_i \cap R_i^c) = 0$, $i = 1, \ldots, N$ since otherwise there is for some i a set B of $\mathbf{x}$ of positive probability, where $d(\mathbf{x}, \mathbf{y}_i) > d(\mathbf{x}, \mathbf{y}_j)$ for some $j \neq i$, yet $q(\mathbf{x}) = \mathbf{y}_i$, and hence $d(\mathbf{x}, q(\mathbf{x})) > \min_i d(\mathbf{x}, \mathbf{y}_i)$ for $\mathbf{x} \in B$. Thus

$$\begin{aligned} D(\{\hat{A}, \mathscr{S}\}, F) &= \int d(\mathbf{x}, q(\mathbf{x}))\, dF(\mathbf{x}) \\ &= \int_{B^c} d(\mathbf{x}, q(\mathbf{x}))\, dF(\mathbf{x}) + \int_B d(\mathbf{x}, q(\mathbf{x}))\, dF(\mathbf{x}) \\ &> \int_{B^c} d(\mathbf{x}, q(\mathbf{x}))\, dF(\mathbf{x}) + \int_B \min d(\mathbf{x}, \mathbf{y}_i)\, dF(\mathbf{x}) \\ &\geqslant \int \min_i d(\mathbf{x}, \mathbf{y}_i)\, dF(\mathbf{x}) = D(\{\hat{A}, \mathscr{P}(\hat{A})\}, F), \end{aligned}$$

contradicting (6). Thus if (6) holds with equality, $\mathscr{S}$ can be made into a Dirichlet partition by redefining things on a set of total probability zero.

Next consider the case of a quantizer $Q = \{\hat{A}, \mathscr{S}\}$ where we vary $\hat{A}$ and leave $\mathscr{S} = \{S_i\}$ fixed. First observe that

$$\begin{aligned} D(Q, F) &= \sum_{i=1}^{N} \int_{S_i} d(\mathbf{x}, \mathbf{y}_i)\, dF(\mathbf{x}) \\ &\geqslant \sum_{i=1}^{N} \inf_{\mathbf{u} \in \mathscr{R}^k} \int_{S_i} d(\mathbf{x}, \mathbf{u})\, dF(\mathbf{x}). \end{aligned} \tag{7}$$

In Appendix A it is shown that provided $\Pr(\mathbf{X} \in S_i) \neq 0$, then there exists a minimum distortion letter or generalized centroid or center of gravity $\hat{x}(S_i)$ for S_i such that

$$\inf_{\mathbf{u}} \int_{S_i} d(\mathbf{x}, \mathbf{u})\, dF(\mathbf{x}) = \int_{S_i} d(\mathbf{x}, \hat{x}(S_i))\, dF(\mathbf{x}) < \infty, \tag{8}$$

and that the set of all solutions $\hat{x}(S_i)$ to (8) is convex, closed, and bounded. If $d(\mathbf{x}, \mathbf{y})$ is strictly convex in $\mathbf{y}$, the solution is unique. If the solution is not unique, then an arbitrary rule can be used to define $\hat{x}(S_i)$ from the set of possible solutions. The minimization can be carried out by standard convex programming techniques (see, e.g., Luenberger (1969, 1973) or Rockafeller (1970)). It is shown in Appendix B that if $d(\mathbf{x}, \mathbf{y})$ is differentiable in $\mathbf{y}$ then centroids are stationary points and hence the minimization defining centroids can be accomplished using variational techniques. Here, however, convexity guarantees that the only stationary point is a global minimum (unlike $D(Q, F)$). Note that $\hat{x}(S_i)$ can be viewed as a Bayes estimate of $\mathbf{X}$ given $\mathbf{X} \in S_i$ subject to a cost function d.

For any partition $\mathscr{S} = \{S_i\}$ define $\hat{x}(\mathscr{S}) = \{\hat{x}(S_i); i = 1,..., N\}$. For $\hat{A} = \{\mathbf{y}_i; i = 1,..., N\}$ we have from (7)–(8) that

$$\int_{S_i} d(\mathbf{x}, \mathbf{y}_i)\, dF(\mathbf{x}) \geqslant \int_{S_i} d(\mathbf{x}, \hat{x}(S_i))\, dF(\mathbf{x}), \tag{9}$$

$$D(\{\hat{A}, \mathscr{S}\}, F) \geqslant D(\{\hat{x}(\mathscr{S}), \mathscr{S}\}, F), \tag{10}$$

that is, $\hat{x}(\mathscr{S})$ is optimal for $\mathscr{S}$ in the sense that no other reproduction alphabet can yield lower distortion. Observe that if equality holds in (10), then equality must hold in (9) for $i = 1,..., N$ and hence the $\mathbf{y}_i$ must be minimum distortion points for the S_i.

THEOREM 1. *Given a distortion measure satisfying* (a)–(b), *then necessary conditions for a quantizer $Q = \{\hat{A}, \mathscr{S}\}$ to be optimal for F is that the partition be optimal for the reproduction alphabet ($\mathscr{S} = \mathscr{P}(\hat{A})$) and that the reproduction alphabet be optimal for the partition ($\hat{A} = \hat{x}(\mathscr{S})$).*

Proof. If $Q = \{\hat{A}, \mathscr{S}\}$ is optimal, then (6) and (10) must each hold with equality. As previously argued this means that $\mathscr{S}$ agrees with a Dirichlet partition of $\hat{A}$ with probability one and that $\hat{A} = \{\mathbf{y}_i\}$, where the $\mathbf{y}_i$ are minimum distortion points for $\mathscr{S}$.

The theorem implies that given a quantizer $Q = \{\hat{A}, \mathscr{S}\}$ we can assume without loss of generality that either $\hat{A} = \hat{x}(\mathscr{S})$ or $\mathscr{S} = \mathscr{P}(\hat{A})$, since this can only yield improved or equal performance. We opt for the latter choice as the optimal partition for a fixed reproduction does not depend on the underlying distribution. Thus we will often write $Q = \hat{A}$ and

$$D(Q, F) = D(\hat{A}, F) = \int dF(\mathbf{x}) \min_{\mathbf{y} \in \hat{A}} d(\mathbf{x}, \mathbf{y}). \tag{11}$$

Thus a quantizer Q^* is optimal for F if

$$D(Q^*, F) = \inf_{\hat{A}} D(\hat{A}, F), \tag{12}$$

where the infimum is over all reproduction alphabets having N or fewer levels. Unfortunately, however, (12) is in general an intractable computation. $D(\hat{A}, F)$ is not convex in $\hat{A}$, there may exist numerous local minima and maxima, and there may not exist an optimal quantizer, that is, a Q^* satisfying (12).

One approach to finding an optimal quantizer is to attempt to find a stationary point of $D(Q, F)$ and hence find a quantizer that at least meets necessary conditions for optimality. Theorem 1 provides an alternate approach: Define a mapping T_F mapping one reproduction alphabet $\hat{A} = \{\mathbf{y}_i\}$ into another $T_F\hat{A} \triangleq \hat{x}(\mathscr{P}(\hat{A}))$. If T_F should change points without reducing distortion, that is, if $\mathbf{y}_i$

is already a minimum distortion point for $S_i \in \mathscr{P}(\hat{A})$, then we assign $\hat{x}(S_i) = \mathbf{y}_i$ (a tie-breaking rule). Simply restating the theorem yields the following:

COROLLARY 1. *Given the assumptions of Theorem 1, a necessary condition for a quantizer $\hat{A}$ to be optimal is that it be a fixed point of T_F.*

If $\mathscr{P}(T_F\hat{A})$ should have an atom of zero probability, then this atom and the corresponding reproduction vector are removed yielding an $N-1$ level quantizer.

We have the following easy property of T_F:

LEMMA 1. *$D(T_F\hat{A}, F) \leqslant D(\hat{A}, F)$, with equality if and only if $\hat{A}$ is a fixed point.*

Proof. The inequality follows from the definition of T_F, (6), and (10). If equality holds, then for $\hat{A} = \{\mathbf{y}_i\}$, $\mathscr{P}(\hat{A}) = \{S_i\}$, we have

$$\int_{S_i} d(\mathbf{x}, \mathbf{y}_i)\, dF(\mathbf{x}) \geqslant \int_{S_i} d(\mathbf{x}, \hat{x}(S_i))\, dF(\mathbf{x}),$$

$$D(\hat{A}, F) = \sum_{i=1}^{N} \int_{S_i} d(\mathbf{x}, \mathbf{y}_i)\, dF(\mathbf{x}) = D(T_F\hat{A}, F) = \sum_{i=1}^{N} \int_{S_i} d(\mathbf{x}, \hat{x}(S_i))\, dF(\mathbf{x}), \tag{13}$$

which means that $\mathbf{y}_i$ is a minimum distortion point for S_i and hence $T_F\hat{A} = \hat{A}$ from the tie-breaking rule.

It is shown in Appendix B that if $d(\mathbf{x}, \mathbf{y})$ is differentiable in $\mathbf{y}$ and if there is no probability on the boundaries of the decision regions, that is, if

$$\Pr(d(\mathbf{X}, \mathbf{y}_i)) = d(\mathbf{X}, \mathbf{y}_j)) = 0, \qquad \text{all} \quad i \neq j, \tag{14}$$

then $\hat{A}$ is also a stationary point of $D(\hat{A}, F)$. This means that subject to suitable conditions, both fixed point and variational algorithms will yield the same set of solutions. Thus if $D(\hat{A}, F)$ should have a unique stationary point that is a global minimum as is the case for one-dimensional Gaussian and Laplace distributions with squared error distortion (Lloyd (1957), Max (1960). Fleisher (1964)), then a fixed point is also a global minimum. We note that (14) is satisfied for the examples listed if the distribution F is absolutely continuous. It also is often satisfied even for discrete distributions F, that is, even if $D(\hat{A}, F)$ is not everywhere differentiable, it is often differentiable at fixed points.

IV. THE ALGORITHM

The following is a natural generalization of Lloyd's Method I to K-dimensions and the distortion measures described.

The Fixed Point Algorithm

(0) Initialization: Given an alphabet size N, a threshold $\epsilon \geqslant 0$, an initial guess $\hat{A}_0$, $|\hat{A}_0| = N$, and a distribution F such that $\Pr(\mathbf{X} \in S_i) \neq 0$ for $S_i \in \mathscr{P}(\hat{A}_0)$ and $D(\hat{A}_0, F) < \infty$. Set $m = 1$.

(1) Given $\hat{A}_{m-1}$, form $\hat{A}_m = T_F\hat{A}_{m-1} = T_F{}^m\hat{A}_0$.

(2) Compute $D(\hat{A}_m, F)$.

(3) Compute $\Pr(\mathbf{X} \in S_i)$, $S_i \in \mathscr{P}(\hat{A}_m)$. If $\Pr(S_i) = 0$ for, say, M values of i, set $N - M \rightarrow N$ and remove those levels $\mathbf{y}_i$ from $\hat{A}_m$.

(4) If $D(\hat{A}_{m-1}, F) - D(\hat{A}_m, F) \leqslant \epsilon$, halt. Otherwise set $m + 1 \rightarrow m$ and go to (1).

Step (4) can be replaced by a percentage condition of the form

(4′) If $(D(\hat{A}_{m-1}, F) - D(\hat{A}_m, F))/D(\hat{A}_m, F) \leqslant \epsilon$, halt. Otherwise set $m + 1 \rightarrow m$ and go to (1).

Step (3) removes any zero probability atoms and the corresponding reproduction levels from the quantizer. Thus the algorithm may yield a quantizer having fewer than N levels. Perhaps surprisingly, this oddity does occur in practice. Practically, if this happens the algorithm has simply reached a point where all N levels are not needed to obtain the available reduction of distortion. For example the algorithm could be converging to a local optimum for N-levels that is achievable with only $N - 1$ levels. In practice, if this happens one would probably either insert a new level (to replace the old) in a high probability atom or simply restart the algorithm with a new initial guess.

A threshold of $\epsilon = 0$ is the most interesting case since if $D(\hat{A}_{m-1}, F) = D(T_F\hat{A}_{m-1}, F)$, then from Lemma 1 $\hat{A}_{m-1}$ is a fixed point of T_F and future iterations will leave the reproduction alphabet unchanged. Thus if this condition is ever satisfied the algorithm has actually converged to a fixed point.

Several techniques of choosing the initial guess $\hat{A}_0$ are described in Linde *et al.* (1980). The algorithm as described above is the one that we shall analyze, but certain modifications might be useful for particular special cases. For example, analogous to Lloyd's one-dimensional observation it can be shown that for differentiable and strictly convex $d(\mathbf{x}, \cdot)$, a necessary condition for $\hat{A} = \{\mathbf{y}_i\}$ to be globally optimum is that there be no probability on the boundaries of the atoms of $\mathscr{P}(\hat{A})$, that is (14) be satisfied. If F has a discrete component, this possibility is not excluded by $\hat{A}$ being a fixed point. One might therefore add a step that tests for probability on the boundaries. If there is a nonzero probability of hitting a boundary point, then strict improvement can be obtained by changing the tie-breaking rule to reassign boundary points and then continuing. This behavior never occurred in any of the simulations in Linde *et al.* (1980), however, and hence for simplicity we do not consider it further. As another example, having reached a fixed point one might perturb $\hat{A}$ to see if any further improvement is possible, e.g., "shake" the algorithm loose from a fixed point or local minimum to see if further improvement is possible.

Several related algorithms arising in both the quantizer and cluster analysis literature for the special cases of squared-error and magnitude error are discussed in Linde *et al.* (1980) and compared and contrasted with the fixed-point algorithm.

From Lemma 1, $D(T_F{}^m\hat{A}_0, F)$ is nondecreasing in m and hence (since it is nonnegative) must have a limit, say

$$D_\infty(A_0, F) \triangleq \lim_{m\to\infty} D(T_F{}^m\hat{A}_0, F). \tag{15}$$

One might hope that also $T_F{}^m\hat{A}_0 = \{\mathbf{y}_i(m)\}$ might itself converge to a fixed point $\hat{A}_\infty = \{\mathbf{y}_i\}$ in the sense that $\mathbf{y}_i(m) \to \mathbf{y}_i$ (for those i remaining in the limit, that is, not eliminated in Step 3). When such an $\hat{A}_\infty$ exists we way that the algorithm *converges* to a fixed point. Unfortunately, however, as Lloyd (1957) observed, the algorithm may not converge to a fixed point even in the one-dimensional, squared-error distortion case.

It can be shown via standard integration theorems from Luenberger (1973), p. 125, that if $\mathbf{X}$ has a bounded alphabet A and if $\hat{A}_n \to \hat{A}_\infty$, then $\hat{A}_\infty$ is indeed a fixed point. Equation (15), however, does not guarantee the existence of such a convergent sequence.

We next consider a special case where the algorithm can be proved to converge to a fixed point in a finite number of steps. The special case is that of a random vector $\mathbf{X}$ possessing a finite alphabet A. This case seems mathematically artificial, but it is prectically important for two reasons. First, if the algorithm is performed on a digital computer then even a "truly continuous" alphabet is represented as discrete, that is, as a "finely quantized" alphabet. Hopefully if "finite" is big enough (vastly larger than N), the results should be good. Second, and most important, say that the random variable $\mathbf{X}$ is described by a "true" but unknown distribution F, which for convenience we assume to be absolutely continuous and hence $\mathbf{X}$ has a continuous alphabet. Since F is not known, however, we are allowed to observe a training sequence of vectors $\mathbf{x}(k)$, $k = 1,\ldots, n$ produced by a stationary and ergodic vector source. A natural estimate of the underlying "true" distribution $F(\mathbf{x})$, $\mathbf{x} \in \mathcal{R}^K$, is the sample distribution $F_n(\mathbf{x})$ defined as follows: Given a training sequence $\mathbf{x}(k)$; $k = 1,\ldots, n$, define the finite alphabet $A_n = \{\mathbf{x}(k); k = 1,\ldots, n\} \subseteq \mathcal{R}^K$, define a probability measure μ_n on $\mathcal{R}^K$ by

$$\mu_n(F) = \sum_{k:\mathbf{x}(k)\in F} n^{-1} \tag{16}$$

(that is, μ_n assigns measure $1/n$ to each vector in the training sequence), and then let F_n be the distribution corresponding to the Lebesque–Stieltjes measure μ_n. An application of the Ergodic Theorem (e.g., Parthasarathy (1967, p. 52)) states that with probability 1, one gets a training sequence such that $F_n \to F$ as $n \to \infty$. Hence one would hope that for a long enough training sequence running the algorithm on F_n sould likely yield nearly the same quantizer as if the algorithm were run on the "true" F and that the resulting performance on future data produced by the source should be nearly that computed by the algorithm. This argument is made precise in the next section. We now return to the convergence issue for the finite alphabet case.

The following Theorem, which is proved in Appendix C, demonstrates that when the alphabet is finite, the algorithm converges to a fixed point in a finite number of iterations.

THEOREM 2. *Given a distortion measure satisfying assumptions* (a)–(c) *and a random vector* $\mathbf{X}$ *with a finite alphabet* $A \subseteqq \mathscr{R}^K$, *then for* $\epsilon = 0$ (*and hence also* $\epsilon \geqslant 0$) *the fixed point algorithm converges to a fixed point in a finite number of iterations; that is, there is a fixed point* $\hat{A}^*$ *and an* $M < \infty$ *such that* $T_F{}^M \hat{A}_0 = \hat{A}^*$. *If* $d(\mathbf{x}, \cdot)$ *is differentiable and* (14) *is satisfied, then* $\hat{A}^*$ *is a stationary point and a local minimum of* $D(\hat{A}, F)$.

From the theorem, given a distribution F with a finite alphabet A and an initial guess $\hat{A}_0$, there is a limiting reproduction alphabet or fixed point $\hat{A}_\infty = \hat{A}_\infty(\hat{A}_0, F)$ such that $T_F{}^m \hat{A}_0 \to \hat{A}_\infty$ and the limit is achieved for a finite m. In the particular case of a sample distribution F_n determined from a training sequence $\{\mathbf{x}(i);\, i = 1,\ldots, n\}$ there is an $M(n) < \infty$ such that

$$\lim_{m\to\infty} T_{F_n}^M \hat{A}_0 = T_{F_n}^{M(n)} \hat{A}_0 \triangleq \hat{A}(n), \tag{17}$$

is a fixed point for T_{F_n}.

V. Asymptotic Properties for Long Training Sequences

We next characterize the asymptotic behavior of the algorithm applied to a sample distribution F_n based on a training sequence of length n as $n \to \infty$. In particular, we make rigorous the intuition that if n is large, then likely a quantizer $\hat{A}(n)$ of (17) designed using F_n should yield a distortion $D(\hat{A}(n), F)$ when applied to the "true" source that is nearly $D_\infty(\hat{A}_0, F)$, the limiting distortion achievable if the algorithm were run forever on the "true" distribution with the same initial guess. All the proofs of this section are in Appendix D.

We here require that the distortion measure satisfy (a)–(c) and that $d(\mathbf{x}, \mathbf{y})$ be a strictly convex function of $\mathbf{y}$ so that the minimum distortion points are unique. We assume that the "true" distribution F describing the random vector $\mathbf{X}$ is absolutely continous (has a density function) and has a bounded alphabet A.

Thus the results will be valid for a Gaussian density if truncated at some point, but not for the idealized nontruncated Gaussian density.

We also require an additional technical assumption on the distortion measure:

(d) For any $\mathbf{y}, \mathbf{y}' \in \mathscr{R}^K$, $\mathbf{y} \neq \mathbf{y}'$,

$$\int_{\mathbf{x}.d(\mathbf{x},\mathbf{y})=d(\mathbf{x},\mathbf{y}')} d\mathbf{x} = 0,$$

that is, the boundaries of any Dirichlet partition have zero volume. Condition (d) is met by all of the examples of Section II.

Say that nature chooses an infinite length sequence of vectors $\{\mathbf{x}(i);\ i = 1, 2,...\}$ produced by a stationary and ergodic source (the vectors are stationary and ergodic). For each n let F_n denote the sample distribution on $\mathscr{R}^K$ included by $\{\mathbf{x}(i);\ i = 1,..., n\}$. For a fixed initial guess $\hat{A}_0$ and each n the fixed-point algorithm can be run on $\hat{A}_0$ using F_n to obtain the quantizers $T^m_{F_n}\hat{A}_0$, $m = 1, 2,...$ which converge in a finite number, say $M(n)$, of steps to a fixed point or limiting quantizer $\hat{A}(n)$ of (17).

On the other hand, if we actually knew the true distribution F describing $\mathbf{X}$ we could (in theory) run the algorithm on $\hat{A}_0$ using F to form quantizers $T_F{}^m\hat{A}_0$, $m = 1, 2,...$ with distortion converging to

$$\lim_{m\to\infty} D(T_F{}^m\hat{A}_0, F) = D_\infty(\hat{A}_0, F). \tag{18}$$

The given assumptions and Appendix B imply that if $T_F{}^m\hat{A}_0$ should actually converge to a fixed point $\hat{A}_\infty$, then $\hat{A}_\infty$ is also a stationary point.

The key "continuity" property of the algorithm with respect to the sample distributions is given in the following lemma:

LEMMA 2. *Let $d(\mathbf{x}, \mathbf{y})$ be a distortion measure satisfying* (a)–(d), *where strict convexity in* $\mathbf{y}$ *is assumed. Assume that the alphabet A is bounded. With probability one the training sequence is such that for $m = 1, 2,...$*

$$\lim_{n\to\infty} T^m_{F_n}\hat{A}_0 = T_F{}^m\hat{A}_0, \tag{19}$$

$$\lim_{n\to\infty} D(T^m_{F_n}\hat{A}_0, F) = D(T_F{}^m\hat{A}_0, F). \tag{20}$$

In the above lemma convergence is guaranteed in (19) only for those $\mathbf{y}_i \in T_F{}^m\hat{A}_0$ whose atoms in $\mathscr{P}(T_F{}^m\hat{A}_0)$ have nonzero probability, that is, (19) holds for the reproduction alphabet remaining when any such "unused" symbols are removed in Step 3.

The lemma states that with probability one the quantizer produced by m iterations on $\hat{A}_0$ using F_n converges as $n \to \infty$ to the quantizer produced by m

iterations on $\hat{A}_0$ using F. In addition, the average distortion resulting when the quantizer $T^m_{F_n}\hat{A}_0$ designed using F_n is used on the source described by the "true" distribution F (not F_n!) converges to $D(T_F{}^m\hat{A}_0, F)$, the distortion resulting when using the quantizer designed using F on F.

The previous result is used to prove the following more important result that states that with probability one the training sequence is such that as $n \to \infty$, the distortion resulting from applying the limit quantizer $\hat{A}(n)$ resulting from the algorithm using F_n on the true distribution F must be no greater than $D_\infty(\hat{A}_0, F)$, the distortion achievable in the limit if the true distribution were known.

THEOREM 3. *Given the assumptions of the previous lemma we have with probability one that*

$$\limsup_{n\to\infty} D(\hat{A}(n), F) \leqslant D_\infty(\hat{A}_0, F).$$

APPENDIX A: MINIMUM DISTORTION POINTS

LEMMA A. *Given a distortion measure d satisfying properties* (a)–(b) *of Section II and a set S with* $\Pr(\mathbf{X} \in S) \neq 0$, *then there exists a generalized centroid* $\hat{x}(S)$ *satisfying*

$$\inf_{\mathbf{u}} \int_S d(\mathbf{x}, \mathbf{u})\, dF(\mathbf{x}) = \int_S d(\mathbf{x}, \hat{x}(S))\, dF(\mathbf{x}). \tag{A.1}$$

Furthermore, the set of all vectors $\hat{x}(S)$ satisfying (A.1) *is convex, closed, and bounded. If $d(\mathbf{x}, \cdot)$ is strictly convex, the centroid is unique.*

Proof. Define for any set S such that $\Pr(\mathbf{X} \in S) \neq 0$ the function

$$\Delta_S(\mathbf{y}) = \int_S d(\mathbf{X}, \mathbf{y})\, dF(\mathbf{x}) = E\{d(\mathbf{X}, \mathbf{y}) \mid \mathbf{x} \in S\} \Pr(\mathbf{X} \in S), \qquad \mathbf{y} \in \mathscr{R}^K.$$

Δ_S is a *proper convex function* (Rockafellar (1970, p. 24) from the properties of d. From Fatou's Lemma (Ash (1972, p. 48)) and the continuity of $d(\mathbf{x}, \mathbf{y})$ in $\mathbf{y}$ (Luenberger (1969, p. 194)) we have that Δ_S is lower semicontinuous (*LSC*). A proper convex *LSC* function is said to be *closed* (Rockafellar (1970, p. 52)).

A function $\Delta_S(\mathbf{y})$ is said to have no direction of recession, (Rockafellar (1970, p. 265)) if there exists no nonzero $\mathbf{y} \in \mathscr{R}^K$ such that for all $\mathbf{u} \in \mathscr{R}^K$, $\Delta_S(\mathbf{u} + \lambda\mathbf{y})$ is nonincreasing in λ. From property (b) $d(\mathbf{x}, \mathbf{u} + \lambda\mathbf{y}) \to \infty$ as $\lambda \to \infty$ and hence $\Delta_S(\mathbf{u} + \lambda\mathbf{y})$ cannot be nonincreasing for any $\mathbf{u}$. Thus $\Delta_S(\mathbf{y})$ is a closed proper convex function with no direction of recession and hence the lemma follows from Theorems 27.2 and 27.1 of Rockafellar (1970).

Appendix B: Fixed Points and Stationary Points

Assume that the function $d_k(\mathbf{y}) = d(\mathbf{x}, \mathbf{y})$ is differentiable with respect to $\mathbf{y}$ and hence the two-sided directional derivative (or Gateaux differential) [Luenberger (1969, 1973)] exists and is given by

$$d'_{\mathbf{x}}(\mathbf{y}; \mathbf{u}) = \lim_{\alpha \to 0} \left\{ \frac{d_{\mathbf{x}}(\mathbf{y} + \alpha \mathbf{u}) - d_{\mathbf{x}}(\mathbf{y})}{\alpha} \right\}$$

$$= \inf_{\alpha > 0} \left\{ \frac{d_{\mathbf{x}}(\mathbf{y} + \alpha \mathbf{u}) - d_{\mathbf{x}}(\mathbf{y})}{\alpha} \right\} = -d'_{\mathbf{x}}(\mathbf{y}; -\mathbf{u}). \tag{B.1}$$

We also assume that $d'_{\mathbf{x}}(\mathbf{y}; \mathbf{u})$ is an integrable function of $\mathbf{x}$. This is true, for example if $\mathbf{X}$ has a bounded alphabet or if $d(\mathbf{x}, \mathbf{y})$ depends only on $\mathbf{x} - \mathbf{y}$. From (B.1) and the Monotone Convergence Theorem (Ash (1972, p. 44))

$$\Delta'_S(\mathbf{y}; \mathbf{u}) = \lim_{\alpha \downarrow 0} \frac{\Delta_S(\mathbf{y} + \alpha \mathbf{u}) - \Delta_S(\mathbf{y})}{\alpha}$$

$$= \lim_{\alpha \downarrow 0} \int_S dF(\mathbf{x}) \left\{ \frac{d_{\mathbf{x}}(\mathbf{y} + \alpha \mathbf{u}) - d_{\mathbf{x}}(\mathbf{y})}{\alpha} \right\}$$

$$= \int_S dF(\mathbf{x})\, d'_{\mathbf{x}}(\mathbf{y}; \mathbf{u}) \tag{B.2}$$

and similarly

$$\Delta'_S(\mathbf{y}; -\mathbf{u}) = \int_S dF(\mathbf{x})\, d'_{\mathbf{x}}(\mathbf{y}; -\mathbf{u})$$

$$= -\int_S dF(\mathbf{x})\, d'_{\mathbf{x}}(\mathbf{y}; \mathbf{u}) = -\Delta'_S(\mathbf{y}; \mathbf{u})$$

and therefore $\Delta_S(\mathbf{y})$ has a two-sided directional derivative and therefore (Luenberger (1969, p. 178))

$$\Delta'_S(\hat{x}(S); \mathbf{u}) = 0, \qquad \text{all } \mathbf{u} \in \mathscr{R}^K. \tag{B.3}$$

Next consider the directional derivative of $D(\hat{A})$: Let $\hat{A} = \{\mathbf{y}_i\}$, $\tilde{A} = \{\mathbf{u}_i\}$, $\mathscr{P}(\hat{A} + \alpha\tilde{A}) = \{S_i(\alpha)\}$, $\mathscr{P}(\hat{A}) = \{S_i\}$, and write

$$f(\alpha) = \frac{D(\hat{A} + \alpha\hat{A}, F) - D(\hat{A}, F)}{\alpha}$$

$$= \sum_{i=1}^{N} \frac{1}{\alpha} \left\{ \int_{S_i(\alpha)} dF(\mathbf{x})\, d_{\mathbf{x}}(\mathbf{y}_i + \alpha \mathbf{u}_i) - \int_{S_i} dF(\mathbf{x})\, d_{\mathbf{x}}(\mathbf{y}_i) \right\} \tag{B.4}$$

We have from the properties of optimal partitions, (B.1)–(B.2), and the Monotone Convergence Theorem that

$$f(\alpha) \leqslant \sum_{i=1}^{N} \int_{S_i} dF(\mathbf{x}) \left\{ \frac{d_{\mathbf{x}}(\mathbf{y}_i + \alpha\mathbf{u}_i) - d_{\mathbf{x}}(\mathbf{y}_i)}{\alpha} \right\}$$

$$\underset{\alpha\to 0}{\longrightarrow} \sum_{i=1}^{N} \Delta'_{S_i}(\mathbf{y}_i\,;\mathbf{u}_i) \tag{B.5}$$

Similarly we have from (B.1) that for $\alpha > 0$

$$f(\alpha) \geqslant \sum_{i=1}^{N} \int_{S_i(\alpha)} dF(\mathbf{x})\, d'_{\mathbf{x}}(\mathbf{y}_i\,;\mathbf{u}_i). \tag{B.6}$$

Let $1_S(\mathbf{x})$ denote the indicator function of a sets. If (14) holds then

$$\lim_{\alpha\to 0} 1_{S_i(\alpha)}(\mathbf{x}) = 1_{S_i}(\mathbf{x}), \qquad \text{a.e.,}$$

(almost everywhere) and hence since $d'_{\mathbf{x}}(\mathbf{y}_i;\mathbf{u}_i)$ is assumed integrable and $1_{S_i}(\mathbf{x}) \leqslant 1$ we have from the Dominated Convergence Theorem (Ash (1972, p. 49)) that the right-hand side of (B.6) goes to $\sum_{i=1}^{N} \Delta'_{S_i}(\mathbf{y}_i;\mathbf{u}_i)$. Thus if (14) holds, then

$$\lim_{\alpha\to 0} \frac{D(\hat{A} + \alpha\tilde{A}, F) - D(\hat{A}, F)}{\alpha} = \sum_{i=1}^{N} \Delta'_{S_i}(\mathbf{y}_i\,;\mathbf{u}_i). \tag{B.8}$$

In particular, if $\hat{A}$ is a fixed point, then (B.3) and (B.8) show that $\hat{A}$ is a stationary point of $D(\hat{A})$.

Appendix C: Proof of Theorem 2

Given the finite alphabet A of size n, let $\hat{A}_m = \{\mathbf{y}_i(m)\}$, $m = 1, 2,\ldots$ be the reproduction alphabet produced at the mth iteration of the algorithm and let $\mathscr{P}(\hat{A}_m)$ denote the corresponding Dirichlet partitions. Since A has n elements, there are 2^{nN} possible partitions of $\hat{A}$ and hence at most 2^{nN} possible optimal reproduction alphabets $\hat{x}(\mathscr{S})$. This means that the sequence $\hat{A}_m$ is uniformly bounded and hence must possess a convergent subsequence $\hat{A}_{m_k} \to \hat{A}_\infty$ as $k \to \infty$ (Rudin (1964, p. 35)) and hence $\hat{A}_\infty$ is a fixed point [Luenberger (1973, p. 125)]. Furthermore, since the set of possible reproduction alphabets is finite, this can only happen if $\hat{A}_M = A_\infty$ for some M which in turn implies that $\hat{A}_m = \hat{A}_M$, $m \geqslant M$, from the tie-breaking rule used to define T. If (14) is satisfied, then $\hat{A}_\infty$ is also a stationary point from Appendix B. In addition, if (14) is satisfied, then we can find an $\epsilon > 0$ such that if $\| \hat{A} - \tilde{A} \|_2 \leqslant \epsilon$, then $\mathscr{P}(\hat{A}) = \mathscr{P}(\tilde{A}) = \{S_i\}$ (since there are only a finite number of such partitions). This means that if $\hat{A}$ is a fixed point and $\| \hat{A} - \tilde{A} \|_2 \leqslant \epsilon$, then if $\tilde{A} = \{\tilde{\mathbf{y}}_i\}$

$$D(\tilde{A}, F) = \sum_{i=1}^{N} \int_{S_i} d(\mathbf{x}, \mathbf{y}_i)\, dF(\mathbf{x})$$

$$\geqslant \sum_{i=1}^{N} \int_{S_i} d(\mathbf{x}, \hat{x}(S_i))\, dF(\mathbf{x}) = D(\hat{A}, F),$$

which defines $\hat{A}$ as a local minimum.

Appendix D: Long Training Sequences

The following is a slight and straightforward modification of Breiman's "modified Birkhoff Theorem" [21]. It is presented without proof as the proof is almost identical to Breiman's.

Lemma D.1 (An Ergodic Theorem). *Given a stationary ergodic discrete time random process $\{\mathbf{X}(i)\}_{i=1}^{\infty}$ with alphabet $\mathscr{R}^K$ described by a probability measure μ, let f_n, $n = 1, 2,\ldots$ be a sequence of functions $f_n: \mathscr{R}^K \to [0, \infty)$ possessing a subsequence f_{n_k} that converges to a function f almost everywhere. Assume also that $|f_{n_k}(x)| \leqslant K < \infty$, all k. Then*

$$\lim_{k\to\infty} \frac{1}{n_k} \sum_{i=1}^{n_k} f_{n_k}(\mathbf{X}(i)) = Ef, \qquad \text{a.e.},$$

that is, the limit exists and equals Ef with probability one.

Corollary D.1. *Let $\{\mathbf{X}(i)\}$ be as in the previous lemma. Let f_n, $n = 1, 2,\ldots$ be a sequence of nonnegative functions possessing a subsequence f_{n_k} such that for $\mathbf{x} \in S \subset A$, $\Pr(\mathbf{X} \in S) \neq 0$, we have that $f_{n_k}(\mathbf{x}) \to_{k\to\infty} \infty$. That is, f_{n_k} diverges on a set of positive probability. We then have that*

$$\lim_{k\to\infty} \frac{1}{n_k} \sum_{i=1}^{n_k} f_{n_k}(\mathbf{X}(i)) = \infty, \qquad \text{a.e.}$$

Proof. Define for $\alpha > 0$

$$f_{n_k}^{(\alpha)}(\mathbf{x}) = f_{n_k}(\mathbf{x}), \qquad \text{if } f_{n_k}(\mathbf{x}) < \alpha,$$
$$= \alpha, \qquad \text{if } f_{n_k}(\mathbf{x}) \geqslant \alpha$$

Observe that by assumption $f_{n_k}^{(\alpha)}(\mathbf{x})\, 1_S(\mathbf{x}) \to_{k+\infty} \alpha 1_S(\mathbf{x})$ and hence from the lemma

$$\lim_{k\to\infty} \frac{1}{n_k} \sum_{i=1}^{n_k} f_{n_k}^{(\alpha)}(\mathbf{X}(i))\, 1_S(\mathbf{X}(i)) = E(\alpha 1_S(\mathbf{X})) = \alpha \Pr(\mathbf{X} \in S), \qquad \text{a.e.}$$

The left-hand side is a lower bound to

$$\liminf_{k\to\infty} \frac{1}{n_k} \sum_{i=1}^{n_k} f_{n_k}(\mathbf{X}(i))\, 1_S(\mathbf{X}(i))$$

for *all* $\alpha > 0$. Since $\Pr(\mathbf{X} \in S) > 0$, the above limit infimum is therefore bounded below by arbitrarily large positive numbers with probability one and hence diverges, proving the corollary.

We next apply these results to the sample distortion based on a training sequence.

Let F_n denote the sample distribution induced by a training sequence $\{\mathbf{X}(i); i = 1,\dots, n\}$ produced by a stationary and ergodic vector source $\{\mathbf{X}(i)\}_{i=1}^{\infty}$. By construction we have that

$$\frac{1}{n} \sum_{i=1}^{n} d(\mathbf{X}(i), \mathbf{y})\, 1_S(\mathbf{X}(i)) = \int_S d(\mathbf{x}, \mathbf{y})\, dF_n(\mathbf{x}).$$

COROLLARY D.2. *Let* $\mathbf{y}(n)$, $n = 1, 2,\dots$, $\mathbf{y}(n) \in \mathcal{R}^K$, *be a sequence of vectors po sessing a subsequence* $\mathbf{y}(n_k) \to_{k\to\infty} \mathbf{y}$ *and assume that the* $\mathbf{y}(n_k)$ *are uniformly bounded. Let* $S(n)$ *be a sequence of subsets of* $\mathcal{R}^K$ *such that* $S(n) \subset A$, *a compact set in* $\mathcal{R}^K$, *and such that there is a set* S *for which* $1_{S(n)}(\mathbf{x}) \to 1_S(\mathbf{x})$ *a.e., where* $1_F(\mathbf{x})$ *is the indicator function of the set* S. *Then*

$$\lim_{k\to\infty} \frac{1}{n_k} \sum_{i=1}^{n_k} d(\mathbf{X}(i), \mathbf{y}(n_k))\, 1_{S(n_k)}(\mathbf{X}(i)) = \int_S d(\mathbf{x}, \mathbf{y})\, dF(\mathbf{x}), \qquad \text{a.e.} \tag{D.2}$$

Proof. Define $f_{n_k}(\mathbf{x}) = d(\mathbf{x}, \mathbf{y}(n_k)) 1_{S(n_k)}(\mathbf{x})$. From the given assumptions and the continuity of $d(\mathbf{x}, \cdot)$ we have that

$$f_{n_k}(\mathbf{x}) \to d(\mathbf{x}, \mathbf{y})\, 1_S(\mathbf{x}) \triangleq f(\mathbf{x}), \qquad \text{a.e.}$$

Furthermore, since the $\mathbf{y}(n_k)$ are uniformly bounded and hence are contained in a compact sphere, we have using assumption (c) that $f_{n_k}(\mathbf{x}) \leqslant d_{\max} < \infty$, all k, and hence the corollary follows from Lemma D.1.

Proof of Lemma 2. First consider the first iteration of the algorithm on the F_n

and on F. Let $\hat{A}_0 = \{\mathbf{y}_j;\ j = 1,\ldots, N\}$, $\mathscr{P}(\hat{A}_0) = \{S_j\}$, and $T_{F_n}\hat{A}_0 = \{\mathbf{y}_j(n);\ j = 1,\ldots, N\}$, where $\mathbf{y}_j(n)$ is the unique solution to

$$\int_{S_j} d(\mathbf{x}, \mathbf{y}_j(n))\, dF_n(\mathbf{x}) = \inf_{\mathbf{u}\in\mathscr{R}^k} \int_{S_j} d(\mathbf{x}, \mathbf{u})\, dF_n(\mathbf{x}).$$

We have from Corollary D.2 and D.2) that

$$\begin{aligned}\limsup_{n\to\infty} \int_{S_j} d(\mathbf{x}, \mathbf{y}_j(n))\, dF_n(\mathbf{x}) &= \limsup_{n\to\infty} \inf_{\mathbf{u}} \int_{S_j} d(\mathbf{x}, \mathbf{u})\, dF_n(\mathbf{x}) \\ &\leqslant \inf_{\mathbf{u}} \limsup_{n\to\infty} \int_{S_j} d(\mathbf{x}, \mathbf{u})\, dF_n(\mathbf{x}) = \inf_{\mathbf{u}} \int_{S_j} d(\mathbf{x}, \mathbf{u})\, dF(\mathbf{x}) \\ &= \int_{S_j} d(\mathbf{x}, \mathbf{y}_j)\, dF(\mathbf{x}) < \infty, \end{aligned} \tag{D.4}$$

where $T_F\hat{A}_0 = \{\mathbf{y}_j;\ j = 1,\ldots, N\}$.

Say the sequence $\mathbf{y}_j(n)$ is unbounded. This means there exists a subsequence $\mathbf{y}_j(n_k)$ such that $\|\mathbf{y}_j(n_k)\|_2 \to \infty$ as $k \to \infty$ and hence from property (b) $d(\mathbf{x}, \mathbf{y}_j(n_k)) \to \infty$ for all $\mathbf{x}$. From Corollary D.1 with $f_{n_k}(\mathbf{x}) = d(\mathbf{x}, \mathbf{y}(n_k))$ and (D.2) this means that the leftmost term of (D.4) must be ∞, contradicting (D.4). Thus $\{\mathbf{y}_j(n)\}_{n=1}^{\infty}$ is a bounded sequence and hence must contain a convergent subsequence $\mathbf{y}_j(n_k) \to \mathbf{y}_j^*$ for some $\mathbf{y}_j^*$. From Corollary D.2, (D.4) and (D.2) this means that

$$\int_{S_j} d(\mathbf{x}, \mathbf{y}_j^*)\, dF(\mathbf{x}) = \int_{S_j} d(\mathbf{x}, \mathbf{y}\ )\, dF(\mathbf{x})$$

and hence by uniqueness $\mathbf{y}_j^* = \mathbf{y}_j$. By the same argument every subsequence of $\mathbf{y}_j(n)$ must contain a further subsequence that converges to $\mathbf{y}_j$ and hence (Royden (1968, p. 135)) $\mathbf{y}_j(n) \to_{n\to\infty} \mathbf{y}_j$. Since this is true for each j we have shown that

$$T_{F_n}\hat{A}_0 \triangleq \hat{A}_1^{(n)} \underset{n\to\infty}{\to} T_F\hat{A}_0 \triangleq \hat{A}_1 . \tag{D.5}$$

Now proceed by induction. Say we are given that

$$\hat{A}_m^{(n)} = T_{F_n}^m\hat{A}_0 \to \hat{A}_m = T_F{}^m\hat{A}_0 , \tag{D.6}$$

where $\hat{A}_m^{(n)} = \{\mathbf{y}_j(n)\}$ and $\hat{A}_m = \{\mathbf{y}_j\}$ consist of uniformly bounded vectors. We assume that all atoms of the Dirichlet partition $\mathscr{P}(\hat{A}_m)$ have nonzero probability (or, equivalently, that $\mathbf{y}_j \in \hat{A}_m$ corresponding to zero probability atoms have been removed). Convergence of $\hat{A}_m^{(n)}$ to $\hat{A}_m$ implies convergence of the atoms of $\mathscr{P}(\hat{A}_m^{(n)}) = \{S_j(n)\}$ to those of $\mathscr{P}(\hat{A}_m) = \{S_j\}$ except possibly on the boundaries (or "tie" regions) of atoms of $\hat{A}_m$, that is, sets of the form $\{\mathbf{x}\colon d(\mathbf{x}, \mathbf{y}_i) = d(\mathbf{x}, \mathbf{y}_j)\}$

for $i \neq j$. Since F is absolutely continuous and these boundaries have zero volume from assumption (d), the union of all such points has probability zero and hence

$$\lim_{n\to\infty} 1_{S_j(n)}(\mathbf{x}) = 1_{S_j}(\mathbf{x}), \qquad \text{a.e.} \tag{D.7}$$

Analogous to the first iteration we have using Corollary D.2 that

$$\begin{aligned}\limsup_{n\to\infty} \int_{S_j(n)} d(\mathbf{x}, \mathbf{y}_j(n))\, dF_n(\mathbf{x}) &= \limsup_{n\to\infty} \inf_{\mathbf{u}} \int_{S_j(n)} d(\mathbf{x}, \mathbf{u})\, dF_n(\mathbf{x})\\ &\leq \inf_{\mathbf{u}} \limsup_{n\to\infty} \int_{S_j(n)} d(\mathbf{x}, \mathbf{u})\, dF_n(\mathbf{x}) = \inf_{\mathbf{u}} \int_{S_j} d(\mathbf{x}, \mathbf{u})\, dF(\mathbf{x})\\ &= \int_{S_j} d(\mathbf{x}, \mathbf{y}_j)\, dF(\mathbf{x}).\end{aligned} \tag{D.8}$$

As before the $\mathbf{y}_j(n)$ must be bounded lest there exist a subsequence for which $\| \mathbf{y}_j(n_k)\|_2 \to \infty$ and hence

$$f_{n_k}(\mathbf{x}) = d(\mathbf{x}, \mathbf{y}_j(n_k))\, 1_{S_j(n_k)}(\mathbf{x}) \to \infty, \qquad \mathbf{x} \in S_j,$$

which from Corollary D.1 and (D.2) would contradict (D.8). Thus $\mathbf{y}_j(n_k) \to \mathbf{y}_j^*$, which again using Corollary D.2 must be $\mathbf{y}_j$ by uniqueness. As before this means $\mathbf{y}_j(n) \to_{n\to\infty} \mathbf{y}_j$, which with (D.7) proves that

$$T_{F_n} \hat{A}_m^{(n)} = \hat{A}_{m+1}^{(n)} \underset{n\to\infty}{\to} T_F \hat{A}_m = \hat{A}_{m+1}$$

and that the vectors in $\hat{A}_{m+1}^{(n)}$ are uniformly bounded. This completes the induction and proves (19). Equation (20) then follows from (19) and the dominated convergence theorem.

Proof of Theorem 3. By definition of limit supremum there is a subsequence $D(\hat{A}(m_k), F)$ of $D(\hat{A}(m), F)$ such that

$$\lim_{k\to\infty} D(\hat{A}(m_k), F) = \limsup_{n\to\infty} D(\hat{A}(m), F). \tag{D.9}$$

The subsequence $\hat{A}(m_k)$ has a further subsequence $\hat{A}(n_k)$ that converges in the extended real space $\bar{\mathscr{R}}^K$ where $\bar{\mathscr{R}}$ is the "compactified" real line (Royden (1968, p. 168)), that is, there is a subsequence $\hat{A}(n_k)$ and an index set $I \subseteq \{1, \ldots, N\}$ such that $\mathbf{y}_j(n_k) \to \mathbf{y}_j \in \mathscr{R}^K$ for $j \in I$ and $\| \mathbf{y}_j(n_k)\|_2 \to \infty$ for $j \notin I$. First observe that since $\| \mathbf{y}_j(n_k)\|_2 \to \infty$ implies $d(\mathbf{x}, \mathbf{y}_j(n_k)) \to \infty$, we have for each $\mathbf{x} \in A$ that

$$\begin{aligned}\lim_{k\to\infty} \min_{j\in\{1,\ldots,N\}} d(\mathbf{x}, \mathbf{y}_j(n_k)) &= \min_{j\in I} d(\mathbf{x}, \mathbf{y}_j) \qquad I \text{ not emtpy}\\ &= \infty \qquad I \text{ empty}.\end{aligned} \tag{D.10}$$

Since also for any $m = 1, 2, \ldots$

$$\limsup_{n\to\infty} D(\hat{A}(n), F_n) \leqslant \limsup_{n\to\infty} D(T_{F_n}^m \hat{A}_0, F_n) = \lim_{n\to\infty} D(T_{F_n}^m \hat{A}_0, F_n)$$
$$= D(T_F{}^m \hat{A}_0, F) < \infty. \tag{D.11}$$

I cannot be empty lest Corollary D.1 with $f_{n_k}(\mathbf{x}) = \min_j d(\mathbf{x}, \mathbf{y}_j(n_k))$ and (D.10) imply

$$D(\hat{A}(n_k), F_{n_k}) = \frac{1}{n_k} \sum_{i=1}^{n_k} \min_j d(\mathbf{x}(i), \mathbf{y}_j(n_k)) \to \infty,$$

contradicting (D.11). Define the size of the index set I as $M \geqslant 1$ and form the M-level quantizers $\tilde{A}(n) = \{\mathbf{y}_j(n); j \in I\}$. From the definition of I and $\tilde{A}(n)$ we have defining $\tilde{A} = \{\mathbf{y}_j; j \in I\}$ that $\tilde{A}(n_k) \to \tilde{A}$, that is, the M-level quantizer $\tilde{A}(n)$ has a subsequence converging to $\tilde{A}$. We now have from (D.10) and Lemma D.1 that

$$\lim_{k\to\infty} D(\hat{A}(n_k), F_{n_k}) = \lim_{k\to\infty} D(\tilde{A}(n_k), F_{n_k})$$
$$= D(\tilde{A}, F) \tag{D.12}$$

and hence from (D.11) we have for $m = 1, 2, \ldots$ that $D(\tilde{A}, F) \leqslant D(T_F{}^m \hat{A}_0, F)$. From (D.10) and the dominated convergence theorem we also have that

$$\lim_{k\to\infty} D(\hat{A}(n_k), F) = \lim_{k\to\infty} D(\tilde{A}(n_k), F) = D(\tilde{A}, F)$$

From the above formula, (D.9), (D.11), and (D.12) we have that

$$\limsup_{n\to\infty} D(\hat{A}(n), F) \leqslant D(T_F{}^m \hat{A}_0, F), \qquad m = 1, 2, \ldots,$$

which with (18) proves Theorem 3.

References

Ash, R. B. (1972), "Real Analysis and Probability," Academic Press, New York.

Breiman, L. (1957), The individual ergodic theorem of information theory, *Ann. Math. Statis.* **28**, 809–811.

Buzo, A., Gray, A. H., Jr., Gray, R. M., and Markel, J. D. (1979), A two-step speech compression system with vector quantizing, *in* "Proceedings 1979 IEEE Internat'l. Acoustics, Speech, and Signal Processing Conf.," Inst. Elec. Electronics Eng., New York.

Buzo, A., Gray, A. H., Jr., Gray, R. M., and Markel, J. D. (1980), Speech coding based upon vector Quantization, *IEEE Trans. ASSP*, in press.

Chaffee, D. L. (1975), "Applications of Rate Distortion Theory to the Bandwith Compression of Speech Signals," Ph.D. Dissertation, Univ. of Calif. at Los Angeles.

FLEISCHER, P. (1964), Sufficient conditions for achieving minimum distortion in a Quantizer, *IEEE Int. Conv. Rec.*, 104–111.

GRAY, A. H., JR. AND MARKEL, J. D. (1976), Distance measures for speech processing, *IEEE Trans. ASSP*, **24**, 380–391.

GRAY, R. M., BUZO, A., GRAY, A. H., JR., AND MARKEL, J. D. (1978), Source coding and speech compression, *in* "Proceedings, of the 1978 Internat'l. Telemetering Conf., pp. 371–878.

GRAY, R. M., BUZO, A., GRAY, A. H., JR., AND MATSUYAMA, Y. (1980), Distortion measures for speech processing, *IEEE Trans. ASSP*, in press.

ITAKURA, F. AND SAITO, S. (1968), Analysis synthesis telephony based upon maximum likelihood method, *in* "Repts 6th Internat'l. Cong. Acoust." (Y. Kohasi, Ed.), Tokyo, *C*-5-5, C17-20.

LINDE, Y., BUZO, A., AND GRAY, R. M. (1980), An algorithm for vector quantizer design, *IEEE Trans. Commun. COM*-**28**, 84–95.

LLOYD, S. P. (1957), "Least Squares Quantization in PCM's," Bell Telephone Labs Memorandum, Murray Hill, N.J.

LUENBERGER, D. G. (1969), "Optimization by Vector Space Methods," Wiley, New York.

LUENBERGER, D. G. (1973), "Introduction to Linear and Nonlinear Programming," Addison–Wesley, Reading, Mass.

MACQUEEN, J. (1967), Some methods for classification and analysis of Multivariate observations, *in* "Proceedings 5th Berkeley Symposium on Math., Statist. and Prob." Vol. 1, pp. 281–296.

MAGILL, D. T. (1973), Adaptive speech compression for Packet communication systems, *in* "Conf. Record 1973 IEEE Telecommunications Conf.," Inst. Elec. Electronics Eng., New York.

MARKEL, J. D. AND GRAY, A. H., JR. (1976), "Linear Prediction of Speech," Springer–Verlag, New York.

MATSUYAMA, Y., BUZO, A., AND GRAY, R. M. (1978), "Spectral Distortion Measures for Speech Compression," Stanford Univ. Inform. Sys. Lab. Tech. Rept. 6504-3, Stanford, Calif.

MAX, J. (1960), Quantizing for minimum distortion, IRE Trans. Inform. Theory, *IT*-**6**, 7–12.

PARTHASARATHY, K. R. (1967), "Probability Measures on Metric Spaces," Academic Press, New York.

POLLARD, D. (1980), Strong consistency of *k*-means clustering, preprint.

ROCKAFELLAR, R. T. (1970), "Convex Analysis," Princeton Univ. Press, Princeton, N.J.

ROYDEN, H. L. (1968), "Real Analysis," MacMillan, Toronto.

RUDIN, W. (1964), "Principles of Mathematical Analysis," McGraw-Hill, N.J.

SVERDRUP-THYGESON, H. (1980), Strong law of large numbers for measures of central tendency and dispersion of random variables in compact metric Spaces, *Ann. Statis.* in press.

20

Reprinted from *IEEE Trans. Inform. Theory* **IT-28**(2):256–261 (1982)

MULTIPLE LOCAL OPTIMA IN VECTOR QUANTIZERS

R. M. Gray and E. D. Karnin

Abstract—Two results are presented on vector quantizers meeting necessary conditions for optimality. First a simple generalization of well-known centroid and moment properties of the squared-error distortion measure to a weighted quadratic distortion measure with an input dependent weighting is presented. The second result is an application of the squared-error special case of the first result to a simulation study of the design of 1 bit per sample two- and three-dimensional quantizers for a memoryless Gaussian source using the generalized Lloyd technique. The existence of multiple distinct local optima is demonstrated, thereby showing that sufficient conditions for unique local optima do not exist for this simple common case. It is also shown that at least three dimensions are required for a vector quantizer to outperform a scalar quantizer for this source.

I. Introduction

Let $\{X_i\}$ be a stationary discrete time sequence of real numbers $(X_i \in \mathcal{R})$. For a fixed dimension k let $\{\boldsymbol{X}_t\}$, $\boldsymbol{X}_t = (X_{kt}, X_{kt+1}, \cdots, X_{k(t+1)-1})^t \in \mathcal{R}^k$, denote the corresponding vector sequence (t denotes transpose). An N-level vector quantizer consists of a codebook or reproduction alphabet $\hat{A} = \{y_1, \cdots, y_N\}$ and a mapping q: $\mathbf{R}^k \to \hat{A}$ or, equivalently, a partition $\mathbb{S} = (S_1, \cdots, S_N)$ of $\mathcal{R}^k$ such that $q(\boldsymbol{X}) = y_i$, if $\boldsymbol{X} \in S_i$. In fact a quantizer is usually an encoder mapping of $\mathcal{R}^k$ into binary vectors and a decoder mapping from binary vectors to $\hat{A}$, but for performance analysis only the overall mapping q is important.

The rate of a quantizer is $R = k^{-1}\log_2 N$ bits per source symbol, that is, the number of binary symbols that must be transmitted or stored for each source symbol in order for the receiver to produce $q(\boldsymbol{X})$.

Given a distortion measure d: $\mathcal{R}^k \times \hat{A} \to [0, \infty)$ assigning a distortion $d(\boldsymbol{X}, \boldsymbol{y})$ to the reproduction of $\boldsymbol{y}$ for $\boldsymbol{X}$, the performance of the quantizer q can be measured by the expected distortion

$$\begin{aligned} D(q) &= E\, d(\boldsymbol{X}, q(\boldsymbol{x})) \\ &= \sum_{i=1}^{N} E\{d(\boldsymbol{X}, y_i) \mid \boldsymbol{X} \in S_i\} \Pr(\boldsymbol{X} \in S_i), \end{aligned} \tag{1}$$

where E denotes expectation with respect to the distribution of the random process. An N-level quantizer is said to be optimal for a source, if $D(q)$ is minimized over all N-level quantizers.

As in Lloyd's Method I [1] for $k = 1$ and $d(x, y) = (x - y)^2$, two almost obvious necessary conditions for optimality are a) that $\mathbb{S}$ be optimal for $\hat{A}$, which is accomplished by using a minimum distortion or nearest neighbor selection rule

$$q(\boldsymbol{x}) = y_i, \quad \text{if} \quad d(\boldsymbol{x}, y_i) \le d(\boldsymbol{x}, y_j), \quad \text{all} \quad j \ne i, \tag{2}$$

and which results in the cells S_i being the Voroni regions or Dirchlet regions [2] of the alphabet, and b) that $\hat{A}$ should be optimal for $\mathbb{S}$, which is accomplished by choosing $\boldsymbol{y}_i$ so that

$$E\{d(\boldsymbol{X}, y_i) \mid \boldsymbol{X} \in S_i\} = \min_{\boldsymbol{u}} E\{d(\boldsymbol{X}, \boldsymbol{u}) \mid \boldsymbol{X} \in S_i\}, \tag{3}$$

$i = 1, \cdots, N$, if such vectors exist. (We assume for simplicity that $\Pr(\boldsymbol{X} \in S_i) > 0$, all i.)

Manuscript received October 3, 1981; revised November 3, 1981. This work was supported by the National Science Foundation and by the Joint Services Electronics Program at Stanford University.

The authors are with the Information Systems Laboratory, Electrical Engineering Department, Stanford University, Stanford, CA 94305.

If for the given distribution, distortion measure d, and set S, a vector $y = y(s)$ satisfying

$$E\{d(\boldsymbol{X}, \boldsymbol{y}) \mid \boldsymbol{X} \in S\} = \inf_{\boldsymbol{u}} E\{d(\boldsymbol{X}, \boldsymbol{u}) \mid \boldsymbol{X} \in S\} \tag{4}$$

exists, then it is called the (generalized) centroid of the set S (with respect to d), and we write $y = \text{cent}(S)$.

These two properties form the basis of Lloyd's iterative Method I [1] and its generalizations and analogs (see [3] for a literature survey of these algorithms): begin with an initial $\hat{A}$, find the optimal $\mathbb{S}$ for $\hat{A}$, find the optimal $\hat{A}$ for $\mathbb{S}$, and continue until some convergence criterion is met. The algorithm can be run either using the "true" expectation corresponding to a known distribution or using sample averages based on a long training sequence [4].

If the algorithm is allowed to converge completely, then the resulting quantizer will meet both necessary conditions and will, subject to some differentiability assumptions, yield a stationary point in the minimization of $D(q)$ [4].

Observe that the algorithm is well defined for a particular distribution and distortion measure if one has a constructive solution to (4). Observe also that in general the final quantizer is at best only a local optimum in the sense of yielding a local minimum of $D(q)$. Trushkin [5] provides sufficient conditions on the distribution for uniqueness of the local optimum and hence for global optimality for the special case of $k = 1$ and a distortion measure of the form

$$d(x, y) = f(x, x - y),$$

where f is a convex $\cup$ function of the error $|x - y|$. No such conditions have been found for $k > 1$.

This note focuses on aspects of these two observations. In the next section we present a simple development for the centroid computation and some useful properties of a weighted quadratic distortion measure of the form

$$d(\boldsymbol{x}, \boldsymbol{y}) = (\boldsymbol{x} - \boldsymbol{y})^t \boldsymbol{W}_x (\boldsymbol{x} - \boldsymbol{y}), \tag{5}$$

where W_x is a (strictly) positive definite weighting matrix. A distortion measure of this form—the gain-normalized Itakura–Saito distortion—was considered in [3] for speech compression applications. The centroid using a sample average and a specific W_x was given in [3] without proof. It was there stated that the result followed from variational techniques, but the details are tedious and have not been published. Section II contains a simple and general proof of this result together with some additional properties generalizing some well-known properties of the usual squared-error distortion.

The distortion measure of (5) includes the usual squared-error distortion ($\boldsymbol{W}_x = \boldsymbol{I}$, the $k \times k$ identity matrix) and simple weighted quadratic distortion measure such as the Mahalonobis distortion [6], [7] where the weighting matrix does not depend on $\boldsymbol{x}$. An additional example of possible interest is the case $\boldsymbol{W}_x = (1/\|\boldsymbol{x}\|^2)\boldsymbol{I}$, where

$$\|\boldsymbol{x}\|^2 = \sum_{i=0}^{k-1} x_i^2$$

is the energy of the vector, and

$$d(\boldsymbol{x}, \boldsymbol{y}) = \|\boldsymbol{x} - \boldsymbol{y}\|^2 / \|\boldsymbol{x}\|^2$$

can be viewed as the short term sample quantization-noise-to-signal ratio. It has been suggested in the speech waveform coding literature that average short term signal-to-noise ratios or segmented signal-to-noise ratios are a better indication of subjective quality than long run signal-to-noise ratios, that is, the average of the ratio is more important than the ratio of the averages. The above distortion measure is such a segmented ratio and will be seen to yield tractable centroids.

Note that the distortion measure of (5) is a special case of the vector analog of Trushkin's distortion measure [5], that is, it has the form $f(\boldsymbol{x}, \boldsymbol{x}-\boldsymbol{y})$ and it is convex ∪ in the second argument.

Section III confines itself to the special case of a squared-error distortion measure and a memoryless source. Lloyd's algorithm is used to design one bit per sample quantizers for dimensions $k = 2$ and 3. The original motivation for this study was a question often encountered by the first author in tutorial talks and discussions on data compression and source coding. The question can be paraphrased as follows: "I understand that Shannon's source coding theorem proves that there exist vector quantizers of a given rate that have smaller average distortion than scalar quantizers; I find this result hard to believe, however, for a memoryless source with no redundancy to remove. Can you describe a particular vector quantizer (and not an abstract existence proof) for a common special case such as a memoryless Gaussian source that yields better performance than a scalar quantizer and, I hope, has intuitive support for its superiority?"

Implicit in the question is the requirement that one can visualize the resulting code book and, hence, that it be of only two or three dimensions. An interesting and derivative question is how large a dimension k is required before a vector quantizer can outperform a scalar quantizer in this special case? In [3] $k = 2$ and $k = 3$ appear to provide slightly better performance than $k = 1$ for this example, but the training sequence was too small to draw firm conclusions. In Section III we develop several one bit per sample locally optimal quantizers for $k = 2$ and 3 and provide evidence that a dimension of at least $k = 3$ is required to outperform a scalar quantizer. The resulting code books are depicted in the figures. One firm conclusion is that many local optima exist for this source, and hence generalizations of the sufficient conditions for global optimality of [5] do not exist for $k > 1$.

II. Properties of Input-Weighted Quadratic Distortion Measures

Assume that $\boldsymbol{W_X}$ and the distribution are such that $E\{\boldsymbol{W_X}\}$ and $E\{\boldsymbol{W_X X}\}$ are finite. (The expectation of a matrix (vector) is the matrix (vector) of the component expectations.) Since $\boldsymbol{W_X}$ is positive definite for all $\boldsymbol{X}$, the matrix $E\{\boldsymbol{W_X}\}$ is positive definite and hence invertible. Hence the following vector is well defined

$$y = (E\{\boldsymbol{W_X}\})^{-1}E\{\boldsymbol{W_X X}\}. \quad (6)$$

We have immediately the following variation of the orthogonality principle:

$$E\{\boldsymbol{W_X}(\boldsymbol{X}-y)\} = E\{\boldsymbol{W_X X}\} - E\{\boldsymbol{W_X}\}y = \boldsymbol{0}. \quad (7)$$

Theorem: a) Given a distortion measure $d(\boldsymbol{x}, y) = \boldsymbol{x} - y)'\boldsymbol{W_x}(\boldsymbol{x} - y)$ with $\boldsymbol{W_x}$ positive definite for all $\boldsymbol{x}$ and a set S such that $E(\boldsymbol{W_X} \mid \boldsymbol{X} \in S\}$ and $E\{\boldsymbol{W_X X} \mid \boldsymbol{X} \in S\}$ are finite, then

$$\text{cent}(S) = E\{\boldsymbol{W_X} \mid \boldsymbol{X} \in S\}^{-1}E\{\boldsymbol{W_X X} \mid \boldsymbol{X} \in S\}. \quad (8)$$

b) Given a partition $\mathcal{S} = \{S_1, \cdots, S_N\}$ and a reproduction alphabet $\hat{A} = \{\text{cent}(S_i);\ i = 1, \cdots, N\}$, let q denote the corresponding quantizer. Then

$$E\{\boldsymbol{W_X X}\} = E\{\boldsymbol{W_X} q(\boldsymbol{X})\} \quad (9)$$

and

$$D(q) = E\{\boldsymbol{X}'\boldsymbol{W_X X}\} - E\{q(\boldsymbol{X})'\boldsymbol{W_X} q(\boldsymbol{X})\}. \quad (10)$$

Proof: a) Abbreviate $E(\cdot \mid \boldsymbol{X} \in S)$ by $E_S(\cdot)$ and define y as in (6) with E_S in place of E. Then for arbitrary $\boldsymbol{u}$

$$\begin{aligned} E_S\{(\boldsymbol{X}-\boldsymbol{u})'\boldsymbol{W_X}(\boldsymbol{X}-\boldsymbol{u})\} &= E_S\{((\boldsymbol{X}-y)+(y-\boldsymbol{u}))'\boldsymbol{W_X}((\boldsymbol{X}-y)+(y-\boldsymbol{u}))\} \\ &= E_S\{(\boldsymbol{X}-y)'\boldsymbol{W_X}(\boldsymbol{X}-y)\} + (y-\boldsymbol{u})'E_S(\boldsymbol{W_X})(y-\boldsymbol{u}) \\ &\quad + 2(y-\boldsymbol{u})'E_S\{\boldsymbol{W_X}(\boldsymbol{X}-y)\}. \end{aligned}$$

The right-most term is zero from (7) and, hence,

$$E_S\{(\boldsymbol{X}-\boldsymbol{u})'\boldsymbol{W_X}(\boldsymbol{X}-\boldsymbol{u})\} \geq E_S\{(\boldsymbol{X}-y)^T\boldsymbol{W_X}(\boldsymbol{X}-y)\}$$

with equality, if $\boldsymbol{u} = y$, characterizing y as cent(S).

b) Let f: $\mathcal{R}^k \to \mathcal{R}^k$ be an arbitrary measurable mapping and observe that (7) implies

$$\begin{aligned} &E\{f(q(\boldsymbol{X}))'\boldsymbol{W_X}(\boldsymbol{X}-q(\boldsymbol{X}))\} \\ &= \sum_{i=1}^{N} \Pr\{\boldsymbol{X} \in S_i\} f(\text{cent}(S_i))'E_{S_i}\{\boldsymbol{W_X}(\boldsymbol{X}-\text{cent}(S_i))\} = 0. \end{aligned}$$

This implies (9) and also that

$$E\{q(\boldsymbol{X})'\boldsymbol{W_X}(\boldsymbol{X}-q(\boldsymbol{X}))\} = 0,$$

which with

$$D(q) = E\{\boldsymbol{X}'\boldsymbol{W_X X}\} + E\{q(\boldsymbol{X})'\boldsymbol{W_X} q(\boldsymbol{X})\} - 2E\{q(\boldsymbol{X})'\boldsymbol{W_X X}\}$$

proves (10).

For the case of a sample distribution defined by a training sequence $T = (\boldsymbol{x}_i, i = 1, \cdots, L\}$, the centroid of a set $S \subseteq T$ is given by

$$\text{cent}(S) = \left\{\sum_{\boldsymbol{x} \in S} \boldsymbol{W_x}\right\}^{-1}\left\{\sum_{\boldsymbol{x} \in S} \boldsymbol{W_x x}\right\}, \quad (11)$$

as was stated for a special case of $\boldsymbol{W_X}$ without proof in [3]. The numerator and denominator terms of centroid are amenable to recursive computation during the minimum distortion encoding step of the algorithm. For reference the basic algorithm for this case is summarized in the flowchart of Fig. 1. The initial code book $\hat{A}_0$ can be selected in a variety of ways, e.g., from lower rate codes using the splitting technique of [3], from lower dimension codes using the product code technique of [7], or by selecting N vectors from the training sequence as in the k-means technique [8].

Computationally the most difficult part of (11) is the matrix inversion of the average weighting matrix. In some cases such as the Itakura–Saito distortion this inversion can be avoided (see, e.g., [9]).

For the case $\boldsymbol{W_x} = \boldsymbol{I}$, (8)–(10) become the familiar

$$\text{cent}(S) = E\{\boldsymbol{X} \mid \boldsymbol{X} \in S\}, \quad (12)$$

$$E\{\boldsymbol{X}\} = E\{q(\boldsymbol{X})\}, \quad (13)$$

$$e\{\|\boldsymbol{X}-q(\boldsymbol{X})\|^2\} = E\{\|\boldsymbol{X}\|^2\} - E\{\|q(\boldsymbol{X})\|^2\}, \quad (14)$$

for the squared-error case (see, e.g., [10], [14]). If $\boldsymbol{W_x}$ does not depend on $\boldsymbol{x}$, then (12) remains true, as shown in [7]. Unlike the developments of [10] and [7], however, the theorem does not assume the existence of probability density functions.

Observe that in the general case (9), the quantizer does not preserve the first moment of the input as in the special case (13).

III. Quantizing Memoryless Gaussian Sources

Let $\{\boldsymbol{X}_i\}$ be a sequence of zero-mean, unit variance memoryless Gaussian random variables and $d(\boldsymbol{x}, y) = \|\boldsymbol{x} - y\|^2$. We consider the design of quantizers of rate one bit per sample and

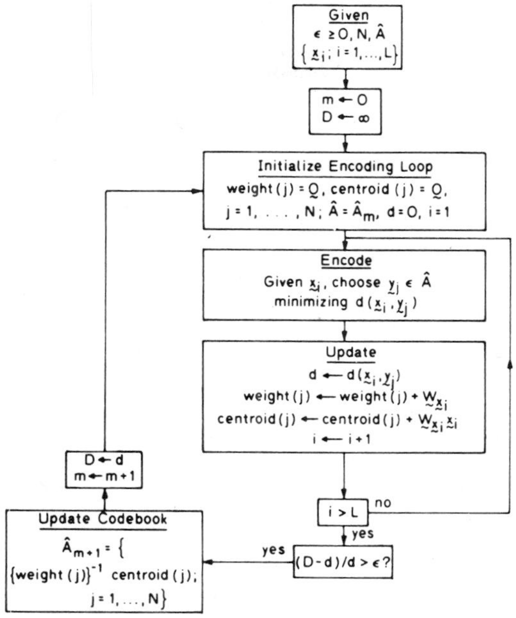

Fig. 1. Design algorithm.

dimensions 2 and 3. In place of the training sequence of 60 000 samples used in [3] we use a training sequence of 1 000 000 samples and a variety of initial codes and distortion convergence thresholds. The designs are based on a sequence of numbers produced by the polar method driven by a linear congruence random number generator [11, ch. 3]. Most of the experiments were run on a DEC VAX at the Stanford University Information Systems Laboratory, operating under the UNIX operating system.

Before describing the experiments and supporting computations, we make some simple observations regarding quantizers for the given source. Given any code book $\hat{A}$, the optimum (minimum distortion) codeword selection procedure, and any source that is rotationally symmetric (as is the current one), then clearly any rotation of the codeword constellation about the origin yields a code book with the same performance. Thus, any rotation of a locally optimum code book is also locally optimum, but we should not consider two such code books as being genuinely different.

Less obvious is the following observation. Let $\pi\colon \mathcal{R}^k \to \mathcal{R}^k$ be an invertible mapping such that $\pi(y) = y'$ is formed by permuting and possibly reflecting the components of y, that is, $y_i' = \pm y_j$ for some j. Given a code book $\hat{A} = \{y_1, \cdots, y_N\}$ one can form a new code book $\pi\hat{A} = \{\pi(y_1), \cdots, \pi(y_N)\}$. It is easy to see that π preserves distance ($\|\pi(x) - \pi(y)\| = \|x - y\|$) (and hence is an isometry). If $\mathcal{S} = \{S_1, \cdots, S_N\}$ is the minimum distortion partition for $\hat{A}$ and $\mathcal{S}' = \{S_1', \cdots, S_N'\}$ the corresponding one for $\pi\hat{A}$, then $x \in S_i$ if $\|x - y\| = \|\pi(x) - \pi(y_i)\| \le \|x - y_j\| = \|\pi(x) - \pi(y_j)\|$, for all $i \neq j$, and hence if $\pi(x) \in S_i'$ or $x \in \pi^{-1}(S_i')$, the inverse image of S_i'. (We have ignored the effects of tie-braking rules for simplicity and since this is a zero probability event for the source considered.) Thus, $S_i = \pi^{-1}(S_i')$. This fact, the fact that the Jacobian of π is one, and a change of variables of integration imply the following result.

Lemma: Given a continuous alphabet source with a distribution rotationally symmetric about the origin, let $\hat{A}$ be a code book and

$$D(\hat{A}) = E\left\{ \min_{y \in \hat{A}} d(X, y) \right\}.$$

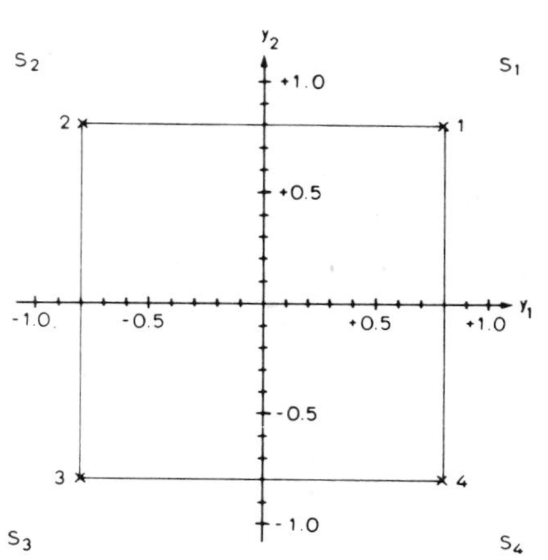

Fig. 2. Two-dimensional product code.

For any invertible permutation/reflection mapping π,

$$D(\pi\hat{A}) = D(\hat{A}).$$

From the previous discussion we should not consider two local optima as distinct if one is a rotation or permutation/reflection (or combination) of the other.

As a first test and check the algorithm was run on the well-understood $k = 1$ (scalar) case since it is well-known that there exists a unique local and, hence, global minimum for this case [1], [5]. With a convergence threshold of $\epsilon = 0.005$ the algorithm converged in three iterations to a final distortion of 0.3635. Reducing the threshold to 0.0 resulted in an additional two iterations and an unchanged final distortion to four figures. The final codewords were 0.7955 and −0.8010.

From (14) we have that

$$D(q) = E\{\|X\|^2\} - E\{\|q(X)\|^2\}$$

$$= k - \sum_{i=1}^{N} \|y_i\|^2 \Pr(X \in S_i). \tag{15}$$

From symmetry the only local code must have $y_1 = -y_2$ and hence from (15), $D(q) = 1 - |y_1|^2$ and from (12)

$$\operatorname{cent}(S_1) = E\{X \mid X \ge 0\} = \sqrt{2/\pi} \cong 0.7979, \tag{16}$$

whence

$$k^{-1}D(q) = 1 - 2/\pi \cong 0.3634 \tag{17}$$

is the theoretical optimum. This test simply pointed out the good agreement of the algorithm with the known theoretical result for the simplest possible case. If also points out the usefulness of (15) for evaluation of $D(q)$ for a given quantizer.

It is perhaps surprising that the optimal two-dimensional quantizer for this simple case does not appear to have been published (to our knowledge). The algorithm was run with a variety of initial code books. Three distinct final quantizers were found and are depicted in Figs. 2, 3, and 4. Fig. 2 is simply the product code of the two optimal one-dimensional codes and, hence, is equivalent to the scalar quantizer with distortion 0.3634. The code satisfies the necessary conditions of (1)–(3) and initial guesses near this code always converged to this code (or a rotation of it). Fig. 3 corresponds to putting both bits into one dimension and, hence, yields a distortion of 0.1175 in that dimension [1] and one in the other for an average of $k^{-1}D(q) = (0.1175 + 1.0)/2 = 0.5588$, a clearly inferior code. Fig. 4 consists of a point at zero surrounded by the points of an equilateral triangle. The only parameter required is b. For the necessary conditions (2)–(3) to be met, $(b, 0)$ must be the centroid of its

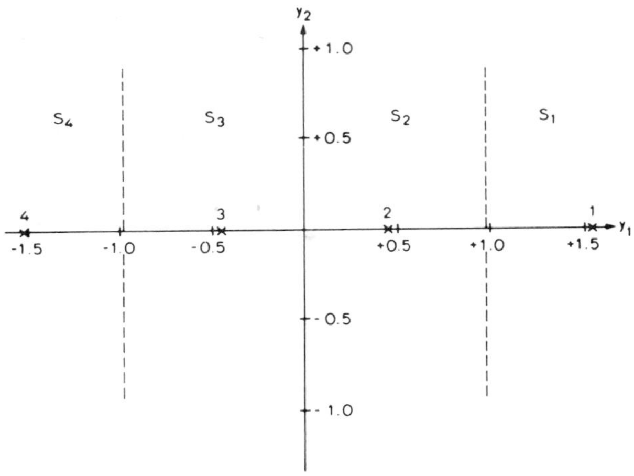

Fig. 3. One-dimensional code in two dimensions.

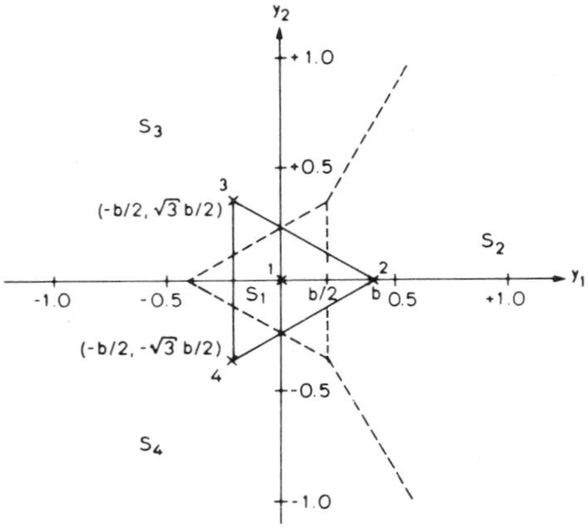

Fig. 4. Triangle-dot two-dimensional code.

Voroni region and hence

$$b = E\{X_1 \mid \boldsymbol{X} \in S_1\} = Q/\Pr(\boldsymbol{X} \in S_2),$$

where

$$\Pr(\boldsymbol{X} \in S_2) = \frac{1}{3} - 2\int_0^{\pi/3} d\theta \int_0^{b/2\cos\theta} dr \cdot re^{-r^2/2}/2\pi$$

$$= \pi^{-1}\int_0^{\pi/3} \exp(-b^2/8\cos^2\theta)\, d\theta, \qquad (18)$$

and

$$Q = 2\int_{b/2}^{\infty} dx \int_0^{\sqrt{3}x} dy \cdot x\{\exp(-(x^2+y^2)/2)\}/2\pi$$

$$= \sqrt{2/\pi}\left\{e^{-b^2/8}\Phi(\sqrt{3}b/2) + (\sqrt{3}/2)\Phi(b)\right\}, \qquad (19)$$

where

$$\Phi(\alpha) = (2\pi)^{-1/2}\int_0^{\theta} e^{-t^2/2}\, dt.$$

Thus b can be obtained by solving

$$b = \frac{\sqrt{2/\pi}\left\{e^{-b^2/8}\Phi(\sqrt{3}b/2) + \sqrt{3}\,\Phi(b)/2\right\}}{\frac{1}{\pi}\int_0^{\pi/2} \exp\{-b^2/8\cos^2\theta\}\, d\theta}. \qquad (20)$$

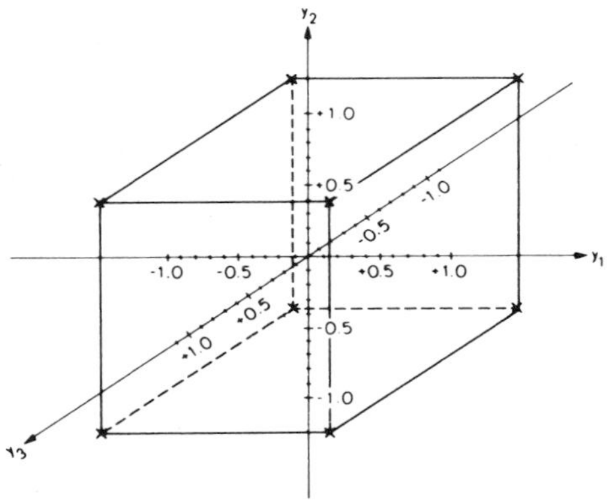

Fig. 5. Three-dimensional product code.

A numerical solution of (18) yields

$$b \cong 1.2791,$$
$$\Pr(\boldsymbol{X} \in S_1) \cong 0.2403,$$

which yields from (15)

$$k^{-1}D(q) = \frac{1}{2}\left(1 - 3b^2\Pr(\boldsymbol{X} \in S_1)\right)$$
$$\cong 0.4102.$$

The algorithm produced final distortions quite close to this numerical evaluation of the theoretical value.

The code of Fig. 4 has an intuitive potential advantage over the product scalar code in that one codeword is devoted to the highly probable vectors with small magnitudes. It was a surprise to us that the distortion resulting from this code was so much larger than that of the scalar code.

We also tried to find another local optimum by increasing the magnitudes of codewords 1 and 3 of Fig. 2 along their common axis and decreasing the magnitude of codewords 2 and 4 to produce a diamond shape. The distortion was studied both by the algorithm and by using (15) and numerical integration and varying the magnitudes of the vectors. Both experiments always yielded distortions that decreased as the code book became closer to the product code and no distinct local optimum was found.

While we have been unable to prove that Figs. 2–4 include all quantizers satisfying (2)–(3) for this example, we believe this to be the case as no other (distinct) locally optimum quantizers were found by the algorithm or by construction. We have demonstrated the existence of distinct quantizers meeting the necessary conditions of (2)–(3) and hence yielding stationary points [4] and we have therefore shown that sufficient conditions for a uniaue such point do not extend from the scalar case [5] to the two-dimensional case. We conjecture that these three quantizers are the only distinct stationary quantizers and hence that no two-dimensional vector quantizer is better than the scalar quantizer for this example.

For the three-dimensional case analytical evaluation such as (16)–(17) and (18)–(20) of the code books yielding stationary points, and the resulting distortions were too difficult and, hence, only the results of the algorithm are reported for such cases.

The product code using the optimal scalar quantizer is depicted in Fig. 5 and obviously satisfies (2)–(3). Not pictured are three-dimensional analogs of Fig. 3, e.g., codes putting 3 bits into only one or two dimensions. These codes did not occur in any of our design runs and perform notably worse than the scalar quantizer.

Fig. 6 shows the first new code book produced by the algorithm that satisfies (2)–(3).

The code book of Fig. 6 has a word at the origin (#1) surrounded by words of equal magnitude—as does the code book

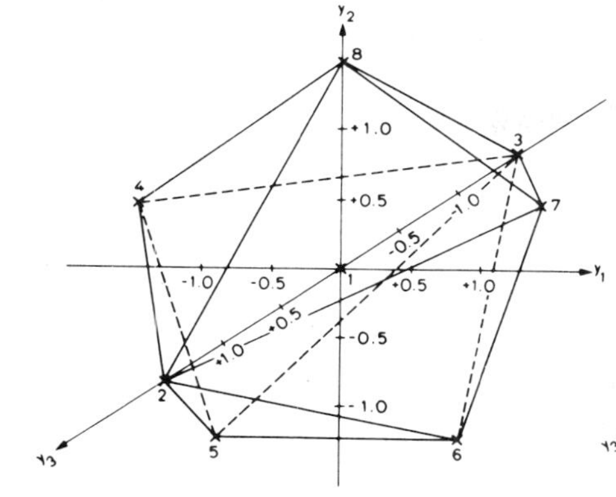

Fig. 6. Pentagon origin-poles code.

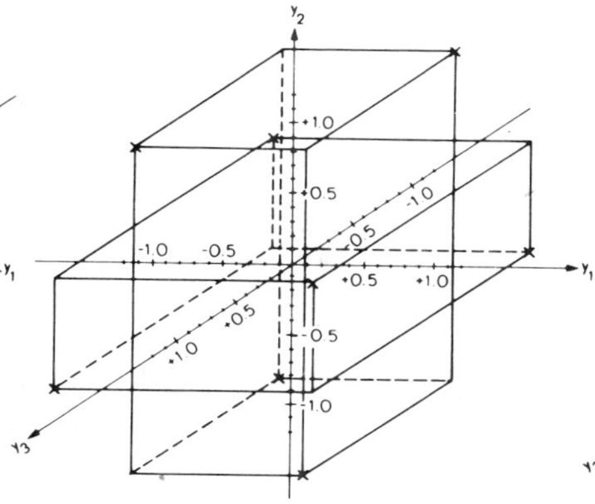

Fig. 7. Skewed product code.

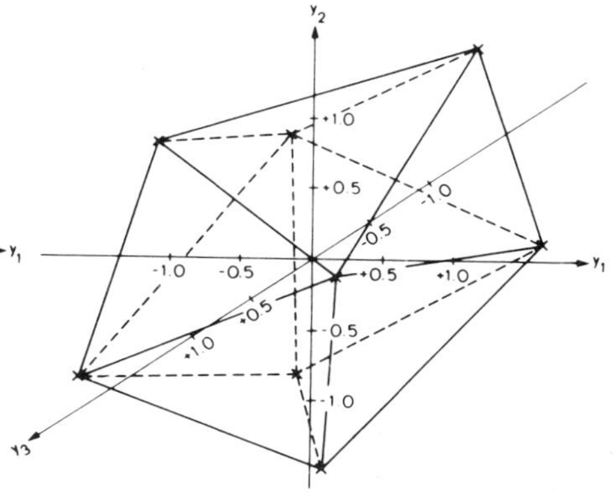

Fig. 8. Convex hull of skewed product code.

of Fig. 4. The nonzero words form a pentagon in one plane (#4, #8, #7, #6, #5) with the remaining two words being poles on an axis perpendicular to the pentagon plane, (#2 and #3). The design distortion for this code for the 1 000 000 sample training sequence was 0.3590 in contrast to the design distortion of 0.3635 for the scalar quantizer and 0.3634 for the theoretical optimal scalar quantizer, an improvement of about 1.2 percent in distortion or 0.05 dB. In fact, the algorithm first produced a rotated version of the figure with slightly better distortion (0.3588), but the code book was rotated for ease of display and the algorithm rerun to produce the code book of Fig. 6.

Because of the closeness of the distortion values, the two code books were also tested on a selection of other test sequences: a 100 000 sample sequence produced by the same random number generator (and a different seed), a 100 000 sample sequence produced with a different uniform random generator on the same computer, and a 60 000 sample sequence with a different uniform random number generator on a Z-80 microcomputer operating under CP/M. In all cases the code book of Fig. 6 outperformed the scalar code (both the theoretical optimum scalar code and the designed scalar code) by about 1 percent. Hence, the evidence all supports the code book of Fig. 6 as being slightly superior to the scalar code, but no numerical evaluation such as (18)–(20) has been done to prove this.

The code book of Fig. 6 thus provides a code that is (albeit only slightly) superior to a scalar code, and it has an intuitive advantage—a word at the origin better reproduces low energy words than can a scalar code, and the remaining seven words do an adequate job of reproducing the larger energy words. We note that in the final design iteration the origin was the most commonly used word (15 percent), the poles were the next most common (14 percent), and the pentagon vertices each occurred 11–12 percent of the time.

While the code book of Fig. 6 provides a solution to the problem of finding an intuitively reasonable code that is superior to the scalar code, it may not be the best three-dimensional code. We were able to find two codes that were slightly better. The way that the first code was found points out some odd behavior of the algorithm. In the design of the code of Fig. 6 the algorithm was first run using the splitting technique [3] and a convergence threshold of $\epsilon = 0.001$. The final code book was then used as an initial guess in a design with $\epsilon = 0.0$. The final code book and distortion were, however, unchanged to the four place accuracy quoted. When the design algorithm was run with the product scalar code as initial guess and $\epsilon = 0.001$, it immediately converged in two iterations with negligible change. When ϵ was set to 0.0, however, the code book did begin to change—possibly due to the "noise" of digital computation and a "shallow" local minimum—and eventually evolved into the "skewed product" code book of Fig. 7 and a distortion value of 0.3556, an improvement of about 2.1 percent over the scalar code and about 0.9 percent over the code of Fig. 6. Because of the lack of intuitive advantages of Fig. 7 over Fig. 5, an additional design run was made with a different uniform random generator, and a variety of tests similar to those used to compare Figs. 5 and 6 were run. The design run produced a permutation of Fig. 7 (and hence effectively the same code) with essentially the same distortion (0.3560) and in all test runs the code of Fig. 7 outperformed the scalar code by about 2 percent and the code of Fig. 6 by about 1 percent. These results suggest that the code book of Fig. 7 is indeed better than that of Fig. 6. The algorithm was also run several times on initial guesses produced by sampling vectors of the training sequence (e.g., take every mth vector until a code book is filled, this is a slight variation of the k-means initialization [8]). All resulting code books were approximately equal to a rotation or permutation of one of the previously described code books and hence Fig. 7 yields the best code book yet found. We note that the identity of the final code books is easily checked by the relative frequencies of the codewords, and that in eight such randomly initiated designs with one million samples, the code book of Fig. 7 occurred seven times (in various rotations, permutations, and reflections) and that of Fig. 6 appeared one time.

Fig. 7 depicts the codewords as vertices of two prisms [2] with square tops and bottoms, the one with the larger squares being "squashed." The codewords on the squashed prism are smaller in magnitude than the remaining ones (about 1.3 in comparison with 1.49). The vertices of the squashed prism occurred more often (14 percent) than the others (11 percent). Fig. 8 depicts the convex hull of the codeword constellation.

As an additional check, an initial code book was formed by forcing all of the faces to be squares yielding nested cubes, that is, the codewords were the vertices of nested regular tetrahedrons. The algorithm converged to the irregular code book of Fig. 7.

The authors confess to being unable to find a satisfactory intuitive explanation for the nonsymmetric and nonregular code book of Fig. 7 to be the best known three-dimensional code book.

The final code book configuration tested was suggested by N.J.A. Sloane and consisted of the product code of Fig. 5 with the code words on the top square rotated by 45° to form the twisted cube of Fig. 9. Running the algorithm with a threshold of 0.0001 led to only slight changes in the code book and a final distortion of 0.3573, between that of the code books of Figs. 6 and 7. Reducing the threshold to 0.0 and continuing the algorithm eventually resulted in the code book changing into a permutation of the code book of Fig. 7 and a final distortion of 0.3564. This suggests that the twisted cube, like the cube or product, is a shallow local minimum.

IV. Comments

We have presented evidence that the optimal scalar quantizer is superior to any two-dimensional vector quantizer for a memoryless Gaussian source and the squared-error distortion. All quan-

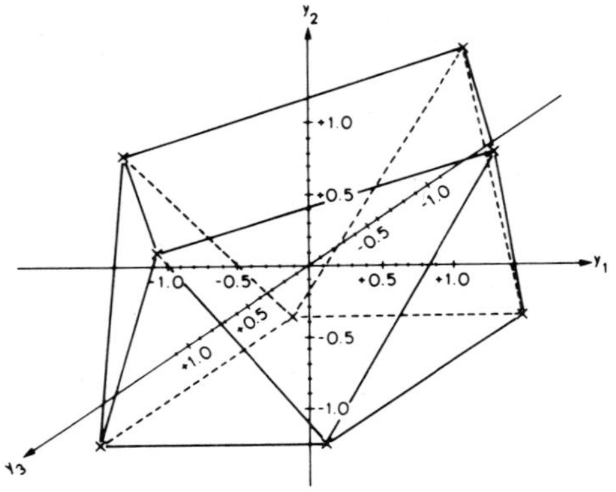

Fig. 9. Twisted cube.

tizers that we found to satisfy the necessary conditions (2)–(3) for optimality yielded larger distortion both in the design algorithm and in numerical evaluation. In three dimensions two quantizers were found to be superior to the optimal scalar quantizer by a small amount. One of these quantizers was found by running the design algorithm on a known local optimum with a zero convergence threshold. This suggests that some local optima (e.g., Figs. 6 and 7) are less sensitive to the computation "noise" of digital arithmetic approximating continuous variables than are others (e.g., Figs. 5 and 9). While the slight performance gains do not justify considering the better quantizers for practical implementation, they do make some interesting points and raise some new questions. First, the new quantizers do not yield uniformly distributed reproductions and hence are amenable to further compression using variable rate noiseless coding. Second, the geometrical structure of the new codes may be amenable to efficient coding techniques analagous to the lattice codes of [12], the prequantization ideas of [13], or the techniques suggested in [14]. Third, no analog of Lloyd's exhaustive search of stationary points for scalar quantizers exists in higher dimensions. It would be useful to develop some means of methodically finding all such quantizers in order to be able to prove which is best and gain insight from its structure.

References

[1] S. P. Lloyd, "Least squares quantization in PCM," in this issue on pp. 129–137.
[2] H. S. M. Coxeter, *Introduction to Geometry*. New York: Wiley, 1962.
[3] Y. Linde, A. Buzo, and R. M. Gray, "An algorithm for vector quantization," *IEEE Trans. on Commun.*, vol. COM-28, pp. 84–95, Jan. 1980.
[4] R. M. Gray, J. C. Kieffer, and Y. Linde, "Locally optimal block quantizer design," *Inform. and Contr.* vol. 45, pp. 178–198, May 1980.
[5] A. V. Trushkin, "Sufficient conditions for uniqueness of a locally optimal quantizer for a class of convex error weighting functions," in this issue on pp. 187–198.
[6] G. J. A. Hartigan, *Clustering Algorithms*. New York: Wiley, 1975.
[7] H. Abut, R. M. Gray, and G. Rebolledo, "Vector quantization of speech and speech-like waveforms," submitted to *IEEE Trans. Acous., Speech, Signal Processing*, to be published.
[8] J. MacQueen, "Some methods for classification and analysis of multivariate observations," *Proc. 5th Berkeley Symp. on Math., Statist., and Prob.*, vol. 1, pp. 281–296, 1973.
[9] A. Buzo, A. H. Gray, Jr., R. M. Gray, and J. D. Markel, "Speech coding based upon vector quantization," *IEEE Trans. Acous., Speech, Signal Processing*, vol. ASSP-28, pp. 562–574, Oct. 1980.
[10] N. C. Gallagher, Jr., and J. A. Bucklew, "Properties of minimum mean squared error block quantizers," *IEEE Trans. Inform. Theory*, vol. IT-28, pp. 105–107, Jan. 1982.
[11] D. E. Knuth, *The Art of Computer Programming Volume 2/Seminumerical Algorithms*. Menlo Park: Addison-Wesley, 1969.
[12] J. H. Conway and N. J. A. Sloane, "Fast quantizing and decoding algorithms for lattice quantizers and codes," in this issue on pp. 227–232.
[13] K. D. Rines and N. C. Gallagher, Jr., "The design of two-dimensional quantizers using prequantization," in this issue on pp. 232–239.
[14] A. Gersho, "On the structure of vector quantizers," in this issue on pp. 157–166.

Part IV

ROBUST SIGNAL QUANTIZERS

Editor's Comments on Papers 21 Through 26

The majority of papers on signal quantization assume a large amount of information about the statistical description of the source; in particular, the source density, $p(\mathbf{x})$ is assumed known. Unfortunately, this assumption is often unrealistic either because the source is nonstationary (the source has time-varying statistics or the quantizer is to be used for multiple sources) or because the source description is not well defined before the quantizer is designed.

Previous papers in this volume have demonstrated that scalar uniform quantizers are robust to changes in the source density. In fact, Bennett's result (ignoring overload) of $\Delta^2/12$ as the mean squared error of a uniform quantizer with stepsize Δ is independent of the source marginal density. Unfortunately, the performance is also quite poor when compared to optimum scalar quantizers. For vector quantization, as explored in the articles of Part III of this volume, the optimum design depends upon the multivariate density, which is typically not known unless the vector has a multivariate Gaussian density or has independent elements with known marginal densities. Since there exists a multitude of multivariate densities with the same marginals, optimum *scalar quantizers* are robust in the sense that their performance is only dependent upon the marginal densities and therefore robust to the actual form of the multivariate density function.

Early work in quantization, as addressed in the papers of Bennett and Smith in Part II, considered the robustness of scalar quantizers with A- and μ-law compressors. As explained by Gersho in Paper 1, the logarithmic companders attempted to rid the mean squared performance measure of its dependence on the source density's functional form and the source power yet employ some knowledge of the source statistics to help attain better performance than could uniform quantizers. The additional source information utilized was that the source's density function was unimodal with a peak at the origin; hence, placing additional levels near the origin (compressing) would improve performance. To demonstrate this fact, consider Figure 3, which displays the MSE performance of a sixteen-level μ-law compander quantizer designed for a zero-mean, unit-variance source. The compressor for this example is:

$$g(x) = \text{sgn}(x) \frac{ln(1 + \mu |x| / V)}{ln(1 + \mu)}; \mu = 255, V = 4.$$

The two sources under consideration are zero-mean Gaussian and Laplacian sources with variance σ^2. The MSE performance is expressed in dB by signal to quantizing noise ratio (SNR$=10 \log_{10}\sigma^2/$MSE). The plot shows that the μ-law quantizer is robust to variations in both the source's density function and source power, having a performance of approximately 13.6 dB for both example sources with $0.01 < \sigma^2 < 1.0$. For comparison, the SNR performance of optimum $N = 16$ Gaussian and Laplacian quantizers are 20.2 dB and 18.1 dB, respectively. As in the uniform quantizer case, A- and μ-law compressors result in performance robust to changes in the source but still have poor performance in comparison to the optimum quantizer for a specific source.

The first two papers in this section address the issue of degradation of performance when the source and quantizer are mismatched. In Paper 21 Mauersberger evaluates the MSE for a variety of situations involving mismatch of scalar quantizers and sources. In particular, he considers mismatch of the density function's variance and mismatch of the density functional form itself. The source model is the generalized Gaussian form (which includes the Gaussian, the Laplacian, and an approximation to the Gamma) with parameters of exponent value and variance. For the above-listed mismatches, he presents specific MSE results along with commentary on how the quantizer might be designed under unknown situations. For an evaluation of the variance mismatch of vector quantizers, see Yamada et al. (1984). Mauersberger also considers the quantization of a uniformly spaced discrete source (i.e., a more finely quantized process) by a quantizer designed using a continuous density function model for the source. He presents examples for which greater than four discrete source values ɔer quantizer level yield performance very near that of the continuous source. ·per 22 by Gray and Davisson addresses the scalar source/quantizer mismatch

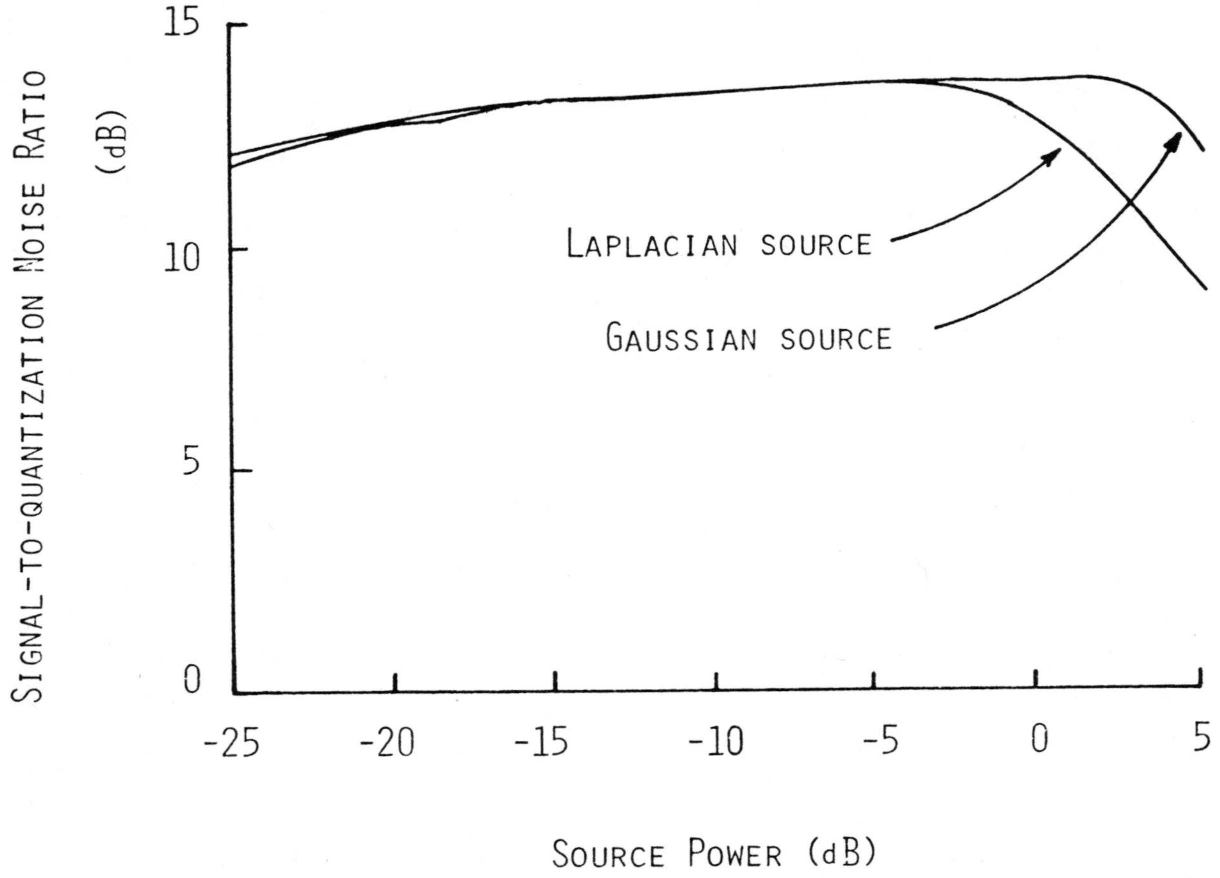

Figure 3. Performance of the μ-law compressor for Gaussian and Laplacian sources.

problem from a more theoretical viewpoint, employing distance measures between densities to evaluate bounds on mismatch performance.

The final four papers in this section are concerned with the design of robust scalar quantizers when partial information about the source is available. For these discussions, the quantizer design problem is stated in a minimax formulation. Let P be a suitably defined class of source densities with elements p, and Q be the class of N-level scalar quantizers with elements q. If $D(q,p)$ is the distortion resulting from applying quantizer q to source p, then the minimax quantizer q^* is defined to be that quantizer in Q that has the least worst-case performance over all densities in P:

$$q^* = \arg \left\{ \min_{q \in Q} \max_{p \in P} D(q,p) \right\}.$$

The reason for specifying the solution in this manner is that no matter which actual statistical description the source has (within the allowable set P), a reasonable level of performance is maintained by the minimax quantizer, q^*.

Although minimax results are often conservative designs for realistic situations, the minimax analytical procedure often leads to tractable and intuitively pleasing schemes. Papers 23 through 25 consider the minimax quantizers under several different constraints upon the class of densities *P*.

In Paper 23 Morris and VandeLinde assume that the only information available about the source is that it has finite range, the closed interval [-c,c]. For this case of a bounded input, the minimax quantizer is shown to be a uniform quantizer on [-c,c]. Papers 24 and 25 by Bath and VandeLinde explore minimax quantizers for the class of sources whose densities are unimodal about the origin and satisfy a generalized moment constraint. The constraint is of the form

$$\int_{-\infty}^{\infty} g(x)\, p(x)\, dx \leq \text{constant } C,$$

where $g(x)$ is some moment-type function (e.g., $g(x) = x^2$ specifies a class of density functions with bounded power). Paper 24 examines the small-N specification of the minimax quantizer while Paper 25 considers the compander model implementation of this minimax quantizer. A game-theoretic approach leading to similar results for this case can be found in Kazakos (1983). The final paper in this section, Paper 26 by Swaszek and Thomas, considers the compander implementation of a minimaxlike quantizer for sources that follow a prescribed histogram. An adaptive implementation is included.

REFERENCES

Kazakos, D., 1983, New Results on Robust Quantization, *IEEE Trans. Commun.* **COM-31**(8):965–974.

Yamada, Y., S. Tazaki, M. Kasahova, and T. Namekawa, 1984, Variance Mismatch of Vector Quantizers, *IEEE Trans. Inform. Theory,* **IT-30**(1):104–107.

BIBLIOGRAPHY

Morris, J. M., and V. D. VandeLinde, 1973, *Robust Quantization of Stationary Signals,* John Hopkins University Tech. Rep. 73-18, Baltimore.

Papantoni-Kazakos, P., 1980, A Robust and Efficient Quantization Scheme, *Princeton Conf. Inform. Sci. Syst. Proc.,* pp. 177–187.

Reprinted from *IEEE Trans. Inform. Theory* **IT-25**(4):381–386 (1979)

Experimental Results on the Performance of Mismatched Quantizers

WOLFGANG MAUERSBERGER

***Abstract*—Quantizers for digital coding systems are usually optimized with respect to a model of the probability density function of the random variable to be quantized. Thus a mismatch of the quantizer relative to the actual statistics of the random variable may be unavoidable. This paper presents the results of an experimental investigation of three types of mismatch. For the modeling of the source statistics, the gamma-, Laplacian-, and Gaussian-distribution are used. The optimization of the quantizers is carried out with respect to the minimum mean-square error criterion.**

I. Introduction

Digital signal coding systems often make use of memoryless quantization. A simple example is the quantizer for a pulse-code modulation (PCM) transmission system; the samples of the source signal are quantized and coded directly. On the other hand in a more sophisticated coding system there is an indirect quantization. A differential pulse-code modulation (DPCM)-coder quantizes the differences between the samples of the signal and the corresponding estimated values. In a transform coding system the components of the spectral vector are quantized and coded separately using a set of quantizers. These systems generally use memoryless quantizers; that is, the quantization characteristics are independent of the past of the signal. (Adaptive systems, which are not regarded here, modify the quantization characteristic depending on the past of the signal.)

For the optimization of a quantizer, a fidelity criterion must be defined. For example, an optimization for minimum mean square error (mse) or for maximum mutual information is possible.

Given the number of output levels and the fidelity criterion, the optimum quantization characteristic depends on the probability density function (pdf) $p_X(\xi)$ of the random variable (r.v.) X to be quantized. In designing a coder, usually the statistics of the samples are not known exactly *a priori*. Thus for the optimization of the quantizers, one has to model the shape of the pdf and estimate a value for the variance of the r.v. to be quantized. Familiar models in speech and image coding are the gamma-, Laplacian-, and Gaussian-distribution [1]–[3].

The quantizer optimization is carried out with respect to a pdf model $p_Q(\xi)$. If $p_X(\xi)$ does not match the pdf model, there is a quantizer mismatch resulting in an increase of the mse. We regard three types of quantizer mismatch.

a) Mismatch relative to the shape: Although the variances are equal the pdf $p_X(\xi)$ of the r.v. X differs from the pdf model $p_Q(\xi)$.
b) Mismatch relative to the variance: The pdf model $p_Q(\xi)$ and the pdf of X are of the same shape, but the variances differ.
c) Uniform pre-quantization: The pdf model $p_Q(\xi)$ usually corresponds to a continuous r.v. whereas the pdf of X may be of the lattice type. This is true if X is processed in a digital system. The r.v. X can be regarded as resulting from a uniform pre-quantization of a continuously distributed r.v.

The following studies were carried out experimentally by digital computation. They rely on the mse criterion.

II. Quantizer Optimization for Minimum Mean-Square Error

For optimizing the N-level quantizer Q_N, $2N-1$ parameters $\{g_1, g_2, \cdots, g_{N-1}, s_0, s_1, \cdots, s_{N-1}\}$ must be calculated; the decision levels g_i define the quantization intervals J_i, each of which contains one output level s_i (Fig. 1). There are also the two fixed decision levels $g_0 = -\infty$, $g_N = \infty$. The quantizer performs a mapping of the input r.v. X, which may assume any real value ξ, onto the discrete output r.v. $\hat{X}$

$$Q_N: \xi \rightarrow s_i \Leftrightarrow \xi \in J_i, \qquad i = 0, 1, \cdots, N-1. \tag{1}$$

The domain of $\hat{X}$ is $\{s_0, s_1, \cdots, s_{N-1}\}$. The mse is defined by

$$\overline{\epsilon_N^2} = \sum_{i=0}^{N-1} \int_{g_i}^{g_{i+1}} (\xi - s_i)^2 p_X(\xi)\, d\xi. \tag{2}$$

For each pdf there exists a quantizer Q_N minimizing the mse; in [4] Max presented an iterative algorithm for the calculation of the optimum parameters.

Manuscript received June 12, 1978; revised November 28, 1978. This work was supported in part by Deutsche Forschungsgemeinschaft, Bonn-Bad Godesberg, Germany.
The author is with Technische Hochschule Aachen, Institut für elektrische Nachrichtentechnik, 5100 Aachen, West Germany.

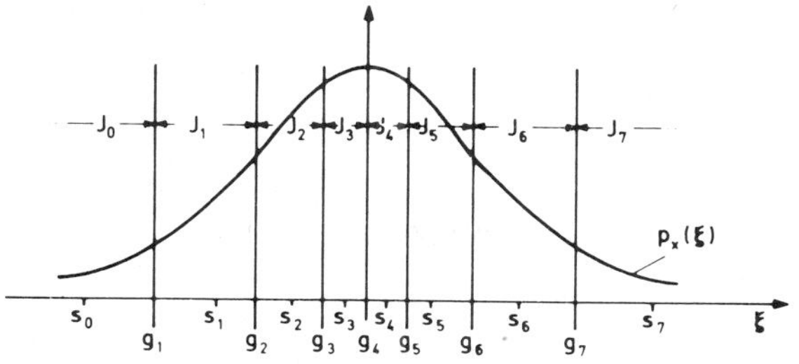

Fig. 1. Parameters defining quantization rule ($N=8$).

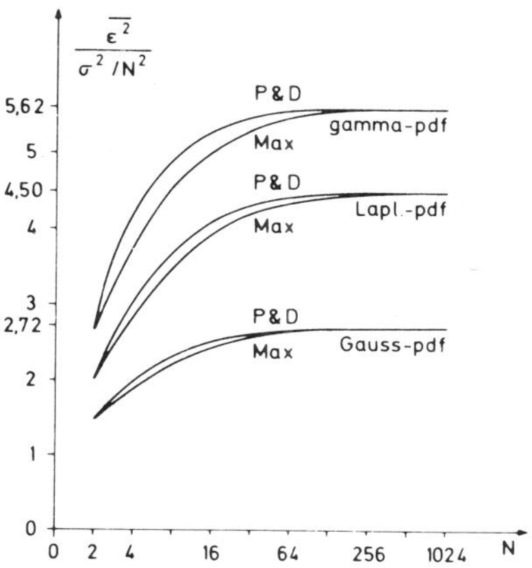

Fig. 2. Normalized mse for gamma-, Laplacian-, and Gaussian-quantizers.

An approximation of the Max quantizer for large N results in the following condition for the decision levels [5]:

$$\int_{-\infty}^{g_i} p_Q^{1/3}(\xi)\, d\xi = \frac{i}{N} \int_{-\infty}^{\infty} p_Q^{1/3}(\xi)\, d\xi; \qquad i=1,2,\cdots,N-1. \tag{3}$$

The optimum output levels are

$$s_i = \int_{g_i}^{g_{i+1}} \xi p_Q(\xi)\, d\xi \Big/ \int_{g_i}^{g_{i+1}} p_Q(\xi)\, d\xi; \qquad i=0,1,\cdots,N-1 \tag{4}$$

iff $p_Q(\xi)=p_X(\xi)$.

These rules slightly modify the quantizer proposed by Panter and Dite [6]. Compared with the Max quantizer the computation time is significantly reduced. Fig. 2 shows the mse normalized to σ_X^2/N^2 for both types of quantizers optimized for a gamma-, Laplacian-, and Gaussian-distributed r.v. The difference of the error is negligible even for small values of N. Thus in the following the quantizer optimized by (3) and (4) is considered.

Equation (2) shows that the mse depends on the parameters of the quantizer, each of which is a functional of the pdf model $p_Q(\xi)$, as well as on the pdf $p_X(\xi)$ of the r.v. which is to be quantized. This fact is symbolized by the arguments of $\overline{\epsilon_N^2}(a_Q=a_q,\ a_X=a_\xi)$: a_Q is an arbitrary parameter, characterizing the pdf model $p_Q(\xi)$, that differs from the corresponding parameter a_X of $p_X(\xi)$. For example, $\overline{\epsilon_N^2}(\sigma_Q^2=1,\ \sigma_X^2=2)$ means the mse resulting from quantizing an r.v. X with a variance of two using an N-level quantizer optimized for a pdf $p_Q(\xi)$ of the same shape with variance one. For simplicity the arguments do not include equal parameters (in this example the shape parameter). This will be evident by the context.

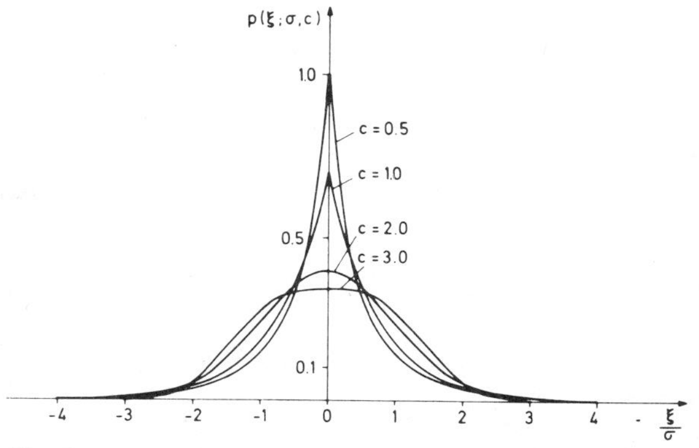

Fig. 3. Generalized Gaussian probability density function.

III. Mismatch Relative to the Shape

This type of quantizer mismatch occurs if the shape of the pdf model $p_Q(\xi)$ used for the optimization is different from the shape of the pdf $p_X(\xi)$ of the r.v. to be quantized, although the variances are equal. A great variety of shape differences are possible. Thus for a systematic investigation of this problem we use the generalized Gaussian distribution [7]:

$$p(\xi;\sigma,c) = \frac{c\cdot\eta(\sigma,c)}{2\cdot\Gamma(1/c)} \exp\left\{-\left[\eta(\sigma,c)\cdot|\xi|\right]^c\right\} \tag{5a}$$

with

$$\eta(\sigma,c) = \frac{1}{\sigma}\left\{\frac{\Gamma(3/c)}{\Gamma(1/c)}\right\}^{1/2}. \tag{5b}$$

The shape of this pdf. can be varied by means of the shape parameter c without affecting the variance (Fig. 3). The value $c=1$ yields the Laplacian-distribution, and $c=2$ yields the Gaussian-distribution. For $c\to\infty$ we get a uniform-distribution, whereas for small values of $c(c>0)$ there is a sharp peak at $\xi=0$.

The gamma-distribution

$$p_\Gamma(\xi) = \frac{1}{2}\left\{\frac{\sqrt{3}}{2\pi\sigma|\xi|}\right\}^{1/2} \exp\left\{-\sqrt{3}\,|\xi|/2\sigma\right\} \tag{6}$$

is not representable by the generalized Gaussian-distribution, but it can be approximated by choosing $c\approx0.75$.

For the studies of the mse performance of quantizers mismatched relative to the shape, we first computed the quantizer characteristics[1] $Q_{N,c=\text{gamma}}$ for the gamma-distribution (called a gamma-quantizer), $Q_{N,c=1}$ for the Laplacian-distribution (called a Laplacian-quantizer), and $Q_{N,c=2}$ for the Gaussian-distribution (called a Gaussian-

[1] $c=$ gamma means that the gamma-pdf is used instead of the generalized Gaussian distribution.

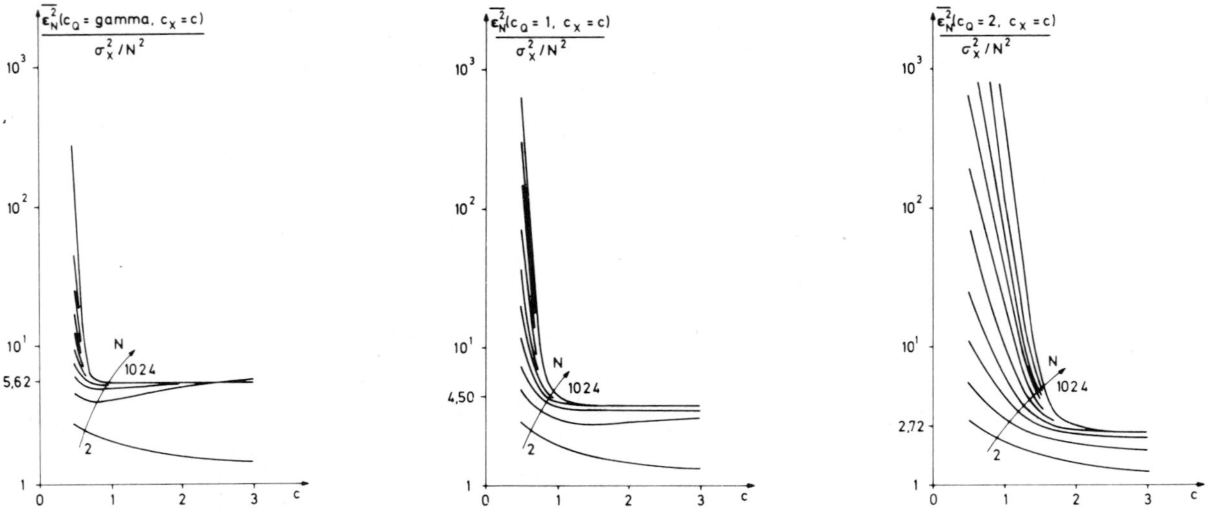

(a) (b) (c)

Fig. 4. Normalized mse for (a) gamma-, (b) Laplacian-, and (c) Gaussian-quantizers (shape mismatch).

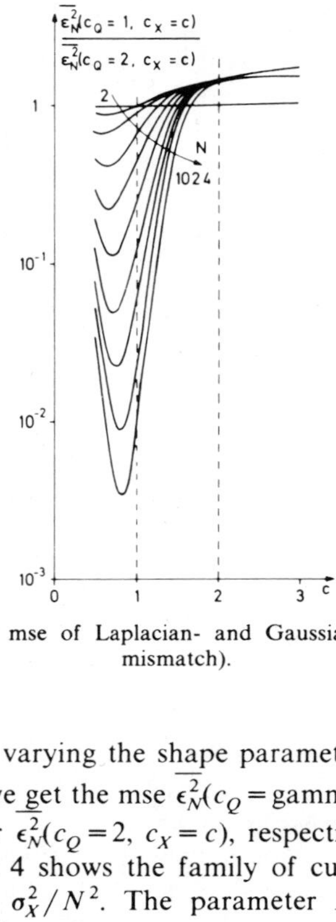

Fig. 5. Ratio of mse of Laplacian- and Gaussian-quantizers (shape mismatch).

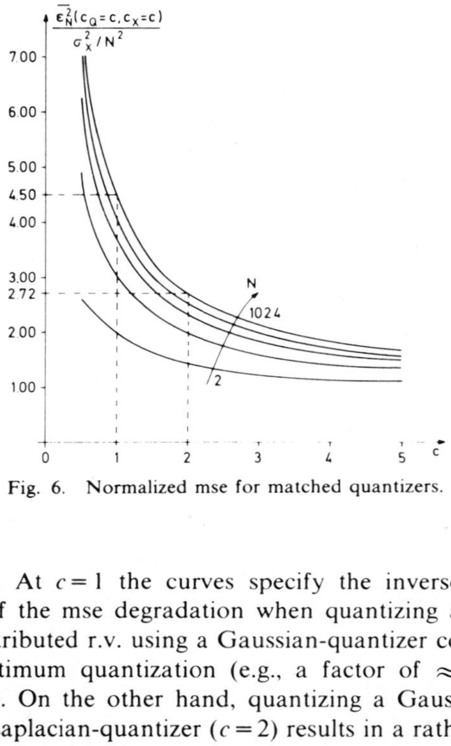

Fig. 6. Normalized mse for matched quantizers.

quantizer). By varying the shape parameter of $p_X(\xi; \sigma_X, c)$ with $\sigma_X = \sigma_Q$, we get the mse $\overline{\epsilon_N^2}(c_Q = \text{gamma}, c_X = c)$, $\overline{\epsilon_N^2}(c_Q = 1, c_X = c)$, or $\overline{\epsilon_N^2}(c_Q = 2, c_X = c)$, respectively, as a function of c. Fig. 4 shows the family of curves of the mse normalized to σ_X^2/N^2. The parameter N indicates the number of levels.

There is a common behavior for these quantizers: for $c_X < c_Q$ there is a severe increase of the mse, whereas it is approximately constant for $c_X > c_Q$. Note that in the latter region the error decreases with an increasing c_Q.

For the comparison of the Laplacian- and Gaussian-quantizers these results can be interpreted more easily by considering the ratio $\overline{\epsilon_N^2}(c_Q = 1, c_X = c)/\overline{\epsilon_N^2}(c_Q = 2, c_X = c)$ (Fig. 5). At $c = 1$ the curves specify the inverse of the factor of the mse degradation when quantizing a Laplacian-distributed r.v. using a Gaussian-quantizer compared with optimum quantization (e.g., a factor of ≈ 17 for $N = 256$). On the other hand, quantizing a Gaussian r.v. with a Laplacian-quantizer ($c = 2$) results in a rather small mse degradation which is nearly independent of the level number (approximately 1.4). For small values of N the mse is nearly independent of the mismatch.

These results also become evident from another experiment. For several shape parameters we calculated the optimum quantizing characteristics $Q_{N,c}$, and then computed the mse for an r.v. with the same variance and shape parameter. Fig. 6 shows the optimum normalized mse $\overline{\epsilon_N^2}(c_Q = c, c_X = c)/(\sigma_X^2/N^2)$. The ratio of the mse of a

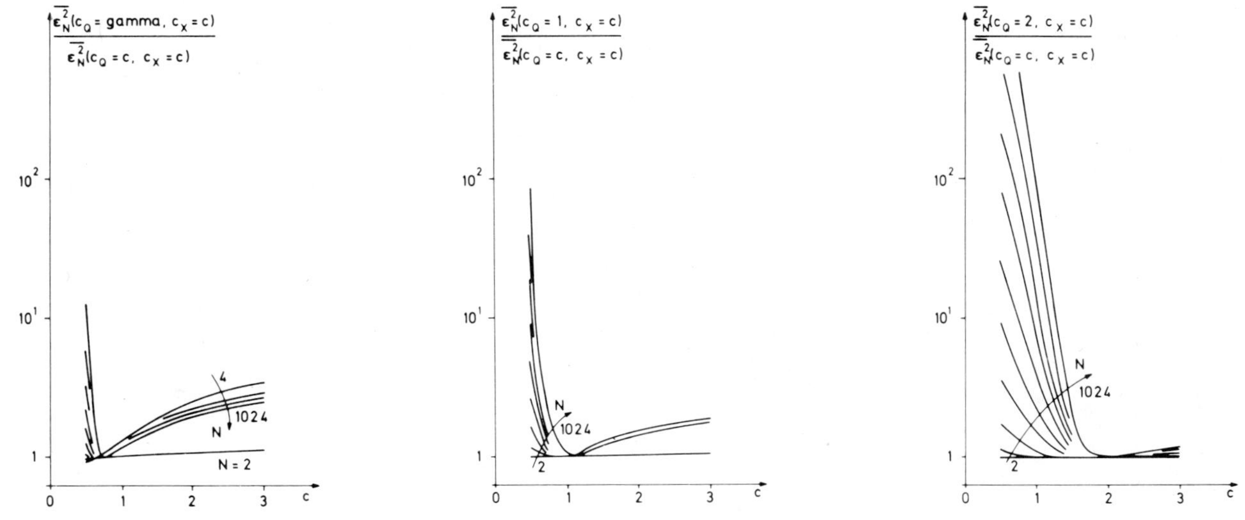

(a) (b) (c)

Fig. 7. Ratio of mse of (a) gamma-, (b) Laplacian-, and (c) Gaussian-quantizer and matched quantizer (shape mismatch).

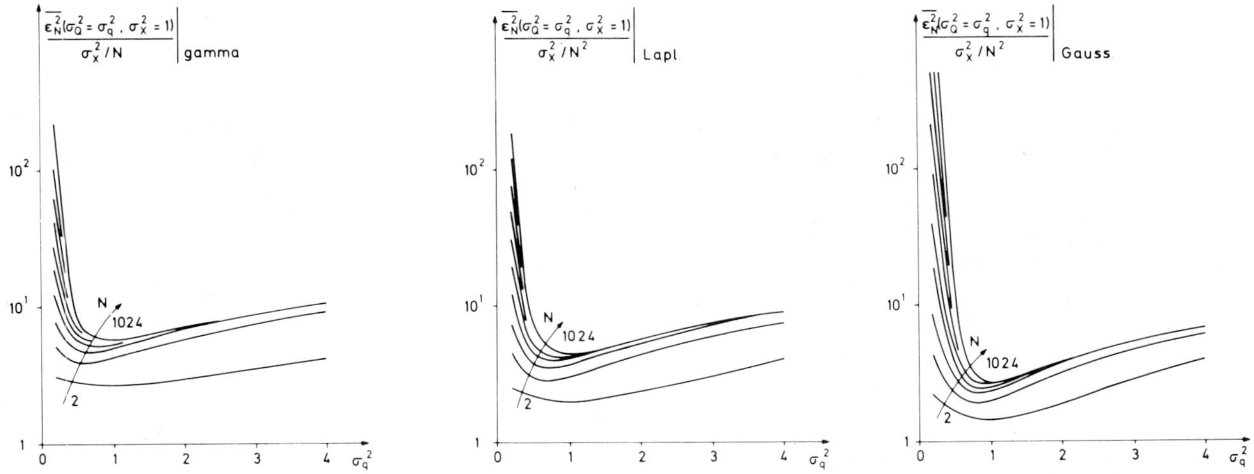

(a) (b) (c)

Fig. 8. Normalized mse of (a) gamma-, (b) Laplacian-, and (c) Gaussian-quantizer (variance mismatch).

gamma-quantizer and an optimum quantizer, $\overline{\epsilon_N^2}(c_Q = \text{gamma}, c_X = c)/\overline{\epsilon_N^2}(c_Q = c, c_X = c)$ is shown in Fig. 7(a);[2] Fig. 7(b) and (c) illustrate the corresponding families of curves for the Laplacian- and Gaussian-quantizer. If, for example, a doubling of the mse is acceptable compared with the optimum quantization, a Laplacian-quantizer ($N = 1024$) can process r.v.'s with a shape parameter between 0.75 to 3.5. For the Gaussian-quantizer the corresponding range lies between 1.5 and infinity, whereas for the gamma-quantizer this range is limited to $0.65 \leqslant c \leqslant 2$. (Note that for large c the relative shape modification becomes small, Fig. 3.)

[2]For large N the gamma-quantizer has nearly the same performance as the optimum quantizer at $c = 0.75$. This shows that the corresponding pdf's are similar.

From this one can conclude that the estimated shape parameter c_Q should be a lower bound to c_X in order to avoid the strong degradation arising if $c_Q > c_X$.

IV. Mismatch Relative to the Variance

For optimizing a quantizer, beside the choice of the shape of the pdf model, there must be an estimated value for the variance. In DPCM- or transform coding the variance strongly depends on the statistics of the source signal, especially on the correlation of adjacent samples. Usually when designing the coder the statistics are not known exactly. Thus studies of the performance of quantizers mismatched relative to the variance may be important.

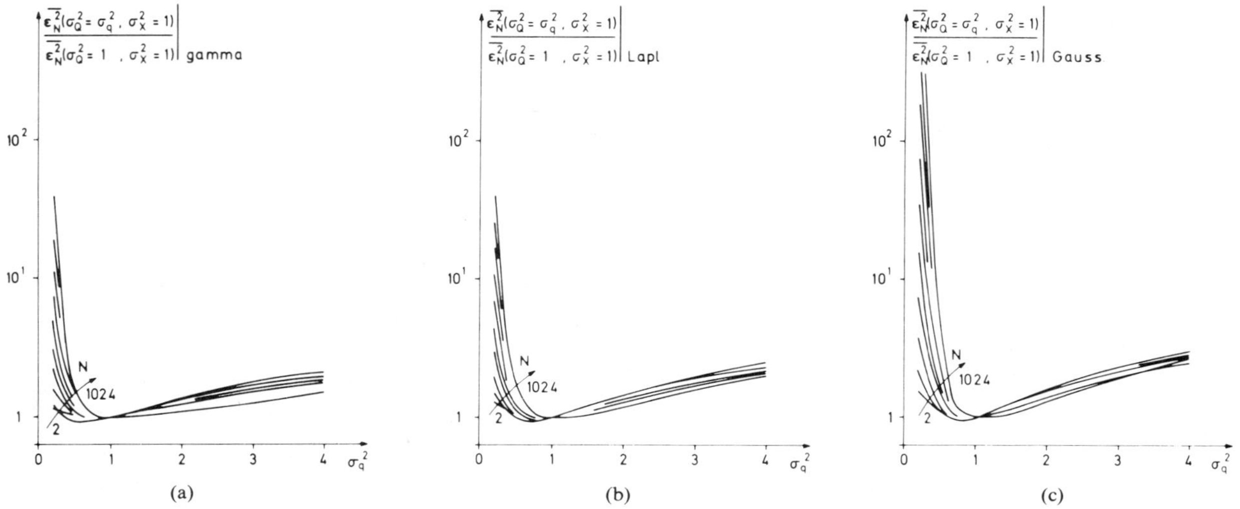

Fig. 9. Ratio of mse of (a) gamma-, (b) Laplacian-, and (c) Gaussian-quantizer and matched quantizer (variance mismatch).

TABLE I
VARIANCE CORRECTION FACTORS FOR LAPLACIAN ($c = 1$) AND GAUSSIAN ($c = 2$) QUANTIZERS

N	$k_c(N)$	
	c = 1	c = 2
2	1,00	1,00
4	0,70	0,81
8	0,73	0,82
16	0,79	0,86
32	0,86	0,90
64	0,92	0,94
128	0,95	0,96
256	0,98	0,98
512	0,99	0,99
1024	1,00	0,99

For this experiment we calculated the mse $\overline{\epsilon_N^2}(\sigma_Q^2 = \sigma_q^2, \sigma_X^2 = 1)$ as a function of σ_q^2 for a gamma-, Laplacian-, and Gaussian-distribution (Fig. 8); this corresponds to the quantization of an r.v. with variance one using a quantizer optimized to the variance σ_q^2. In Fig. 9 the mse is normalized to the mse of the matched quantizer, $\overline{\epsilon_N^2}(\sigma_Q^2 = \sigma_q^2, \sigma_X^2 = 1)/\overline{\epsilon_N^2}(\sigma_Q^2 = 1, \sigma_X^2 = 1)$.

These results can be generalized in the following way: quantizing an r.v. X with the variance σ_ξ^2 using a quantizer optimized to the variance σ_q^2 yields a mse of

$$\overline{\epsilon_N^2}\left(\sigma_Q^2 = \sigma_q^2, \sigma_X^2 = \sigma_\xi^2\right) = \sigma_\xi^2 \cdot \overline{\epsilon_N^2}\left(\sigma_Q^2 = \sigma_q^2/\sigma_\xi^2, \sigma_X^2 = 1\right). \quad (7)$$

From Fig. 9 one can conclude that the Laplacian quantizer is less sensitive to a variance mismatch than the Gaussian quantizer. If the doubling of the mse is acceptable compared with the optimum quantizer, the variance σ_Q^2 may vary between 0.5 and 4. For a Gaussian quantizer the corresponding range is 0.66 to 3. The variance mismatch performance of the gamma-quantizer is similar to that of the Laplacian-quantizer.

There is a property of the mse curves which may be surprising: especially for small numbers of levels, $N \geqslant 2$, the minimum of the mse is not at $\sigma_q^2 = 1$. This means the "exactly matched" quantizer is not the best one, due to the approximation given by (3) and (4). Taking into account a correction factor $k_c(N)$ this can be avoided by optimizing the quantizer with respect to the variance

$$\sigma_{Q,\,\mathrm{opt}}^2 = k_c(N) \cdot \sigma_X^2. \quad (8)$$

The correction factors are listed in Table I for the Laplacian- and Gaussian-distributions.

V. UNIFORMLY PRE-QUANTIZED RANDOM VARIABLES

For the optimization we usually have a continuous pdf model. On the other hand in a digital system an r.v. is stored using a finite number of bits. Thus the r.v. is of the lattice type [8]: the pdf is given by

$$p_{\tilde{X}}(\xi) = \sum_{\ell=-\infty}^{\infty} a_\ell\, \delta(\xi - \ell\Delta). \quad (9)$$

For the weighting factors a_ℓ there is the condition

$$\sum_{\ell=-\infty}^{\infty} a_\ell = 1. \quad (10)$$

(The effects arising from the limited range of values are neglected here.) This is a special case of the mismatch relative to the shape, which we study separately because of its fundamental significance in digital systems.

We started with the following model: a continuously distributed r.v. X with the pdf $p_X(\xi)$ is uniformly quantized, yielding the discretely distributed r.v. $\tilde{X}$. There is an unlimited number of quantizing intervals of width Δ. Thus the weighting factors of (9) are

$$a_\ell = \int_{\Delta(\ell - 1/2)}^{\Delta(\ell + 1/2)} p_X(\xi)\, d\xi, \qquad \ell = \cdots, -1, 0, 1, \cdots. \quad (11)$$

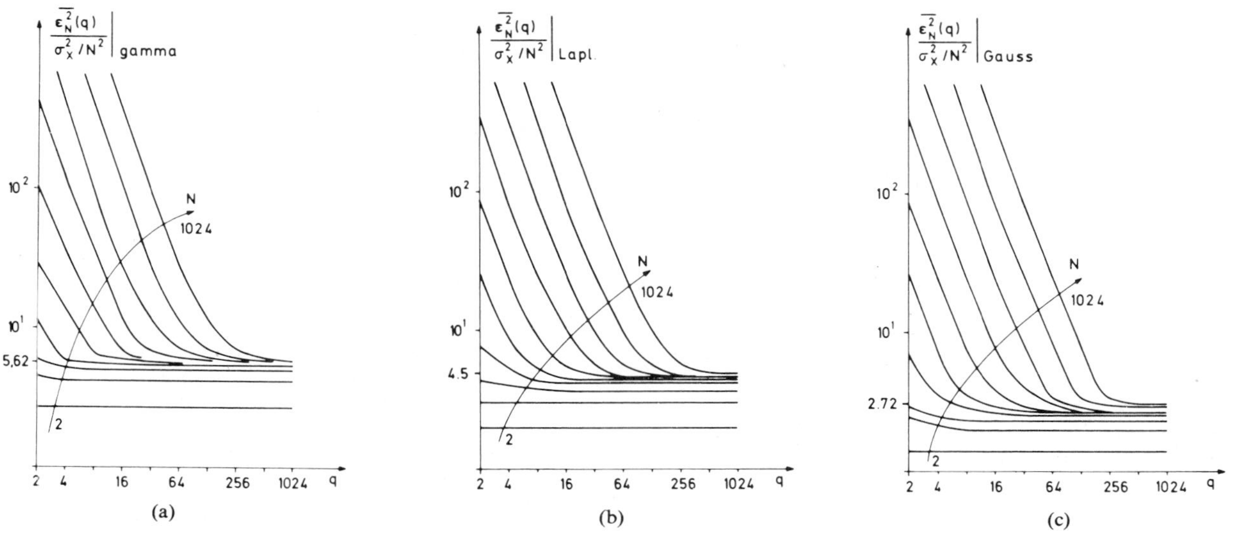

Fig. 10. Normalized mse for (a) gamma-, (b) Laplacian-, and (c) Gaussian-quantizer (uniform pre-quantization).

This procedure we call pre-quantization. We define the ratio

$$q=\sigma_X/\Delta \tag{12}$$

to be the pre-quantizing factor. It denotes the number of discrete levels in the range of the standard deviation of X.

Fig. 10 shows the normalized mse as a function of q for a gamma-, Laplacian-, and Gaussian-quantizer. For all quantizers the mse is nearly independent of q for $q>N/4$. On the other hand there is a distinct degradation for $q<N/4$.

Note that neglecting the limited range of digitally stored numbers has no influence if the minimum value $\tilde{\xi}_-$ and the maximum value $\tilde{\xi}_+$ of $\tilde{X}$ are constrained by

$$\tilde{\xi}_- < g_1 \qquad \tilde{\xi}_+ > g_{N-1}. \tag{13}$$

VI. Conclusion

The generalized Gaussian distribution can be used for source modeling in many cases. It contains the Laplacian-, Gaussian-, and, asymptotically, the uniform-distribution; moreover it includes a good approximation to the gamma-distribution.

Studies of the performance of quantizers mismatched relative to the shape suggest a convenient estimate of the shape parameter: the estimated value c_Q should be a lower bound to c_X.

The matching of a quantizer relative to the variance is not very critical. With a decreasing shape parameter c the quantizers become less sensitive to a mismatch.

Finally a discrete r.v. should assume at least $N/4$ levels in the range of a standard deviation for an N-level quantizer. There is no significant dependence on the shape of the pdf.

References

[1] M. D. Paez and T. H. Glisson, "Minimum mean-squared-error quantization in speech PCM and DPCM systems," *IEEE Trans. Commun.*, vol. COM-20, pp. 225–230, 1972.

[2] A. Habibi and P. A. Wintz, "Image coding by linear transformation and block quantization," *IEEE Trans. Commun.*, vol. COM-19, pp. 50–62, 1971.

[3] R. Zelinski and P. Noll, "Adaptive transform coding of speech signals," *IEEE Trans. Accoust., Speech, Signal Processing*, vol. ASSP-25, pp. 299–309, 1977.

[4] J. Max, "Quantizing for minimum distortion," *IRE Trans. Inform. Theory*, vol. IT-6 pp. 7–12, 1960.

[5] W. Mauersberger, "Optimierung gedächtnisloser Quantisierer," *AEU*, vol. 32, pp. 397–402, 1978.

[6] D. A. Panter and W. Dite, "Quantizing distortion in pulse-count modulation with nonuniform spacing levels," in *Proc. IRE*, vol. 39, pp. 44–48, 1951.

[7] J. H. Miller and J. B. Thomas, "Detectors for discrete-time signals in non-Gaussian noise," *IEEE Trans. Inform. Theory*, vol. IT-18, pp. 241–250, 1972.

[8] A. Papoulis, *Probability, Random Variables and Stochastic Processes.* New York:McGraw-Hill, 1965.

22

Reprinted from *IEEE Trans. Commun.* **COM-23** (4):439–443 (1975)

QUANTIZER MISMATCH

R. M. Gray and L. D. Davisson

***Abstract*—A simple upper bound is derived to the difference in performance obtained from applying a given quantizer to two different sources. This provides a bound on the performance loss or mismatch resulting when applying a quantizer designed for one source to another. The bound is in terms of a generalization of the Vasershtein distance between the source random variables and does not depend on the particular quantizer chosen. In particular, if two sources are sufficiently close in this sense, then any quantizer results in nearly identical performance on either source. Implications for optimal performance bounds are discussed and examples are given.**

I. INTRODUCTION

The majority of papers in the various areas of statistical communication theory begin with the assumption of a source model, i.e., a complete and accurate probabilistic description of the source to be communicated. Most optimal or nearly optimal communications systems depend to some extent on the statistics of the source for which they are designed. Unfortunately, however, in practice such descriptions are rarely known precisely and at best are approximations. This leads to the natural question as to how much performance may be lost if a system designed for a particular source is applied to another source. For example, what is the performance loss if a quantizer designed using Max's tables [5] to be optimal for a Gaussian source with variance σ_1^2 is instead applied to a Gaussian source with variance σ_2^2? As a more extreme example, quantizers are sometimes designed to be optimal for uniformly distributed random variables for simplicity and are subsequently applied to nonuniform random variables better modeled as Gaussian random variables. It would be useful to have a simple bound on the resulting performance mismatch.

One would hope that if two sources are "close" in some sense, e.g., if the mathematical model well approximates the actual source, then the performance difference resulting from a particular quantizer should be small. In this concise paper this intuitive and desirable property is made precise for a large class of quantizers with distortion measured by the average of some power of the magnitude error. The principal tool is a measure of the distance or difference between two probability distributions that is a generalization of the Vasershtein distance [1], [2] and corresponds to a special case of a distance between random processes called the $\bar{\rho}$ ("ρ-bar") distance [3]. These distance measures differ significantly from the more common distance measures such as the Bhattacharyya distance, the

Paper approved by the Associate Editor for Communication Theory of the IEEE Communications Society for publication without oral presentation. Manuscript received August 26, 1974; revised November 14, 1974. This work was supported in part by the National Science Foundation under Grants GK-31630 and GK-14190, and in part by the Joint Services Electronics Program at Stanford Electronics Laboratory under U. S. Navy Contract N00014-67-0112-0044.

R. M. Gray is with the Department of Electrical Engineering, Stanford University, Stanford, Calif. 94305.

L. D. Davisson is with the Department of Electrical Engineering, University of Southern California, Los Angeles, Calif.

variational distance or distribution distance, and the divergence [4]. The generalized Vashershtein distance will be seen to be the appropriate measure for the given problem, yielding simple bounds in several examples. In particular, an easy bound on the mismatch resulting from applying a finite-level uniform quantizer to a Gaussian source is derived and shown to be small for appropriately chosen parameters. This provides a new demonstration of the robustness of uniform quantizers for nonuniform sources.

II. PRELIMINARIES

We assume that the sources to be quantized are discrete-time real-valued random processes. In addition we consider only one-dimensional (single letter) quantizers. Let X denote a real-valued random variable with cumulative distribution function $F_X(\alpha) = \Pr(X \le \alpha)$ and probability density function $f_X(\alpha) = dF_X(\alpha)/d\alpha$. Similarly define Y, $F_Y(\beta)$, and $f_Y(\beta)$. A stationary sequence $\{X_n\}$ of random variables with identical marginal distribution $F_{X_n} = F_X$ is called a source. Similarly define the stationary source $\{Y_n\}$.

A quantizer q consists of a sequence of interval endpoints r_k, $k = 1,\cdots,N+1$, where N is a finite integer and $-\infty = r_1 < r_2 < \cdots < r_N < r_{N+1} = \infty$; and a set of output levels or representative values a_k, $k = 1,\cdots,N$, where $r_k < a_k \le r_{k+1}$. The quantizer q is defined by $q(r) = a_k$ if $r_k < r \le r_{k+1}$. A common measure of the performance of a quantizer q applied to a source X is the average distortion as measured by some integer power of quantization error:

$$\rho_X^{(\nu)}(q) \triangleq E\{|X - q(X)|^\nu\} = \int_{-\infty}^{\infty} |x - q(x)|^\nu f_X(x)\,dx,$$

where $\nu \ge 1$. It is well known [5] that for fixed N and any ν, the quantizer minimizing $\rho_X^{(\nu)}(q)$ must have the endpoints midway between the output levels, i.e.,

$$r_k = (a_{k-1} + a_k)/2, \qquad k = 2,\cdots,N. \tag{1}$$

This relation must hold regardless of the particular source distribution and so we henceforth consider only quantizers satisfying (1).

Much of the literature on quantization has focused on obtaining precise relations for the endpoints and output levels minimizing $\rho_X^{(\nu)}(q)$ for particular source densities [5]–[8]. These relations are usually quite complicated and dependent on the source statistics and require either numerical methods or approximation. These specific relations are not relevant here, however, as we are concerned only with the performance *change* resulting from applying a given q (perhaps designed to be optimal for a specific source) to different sources.

The mismatch bound is in terms of a measure of the distance between random variables which is a generalization of the Vasershtein distance [1], [2] and a special case of the $\bar{\rho}$ distance between random processes [3]. Given two probability density functions $f_X(\alpha)$ and $f_Y(\beta)$ describing random variables X and Y, respectively, define $\mathcal{P}_{XY}$ as the family of all joint probability density functions $f_{XY}(\alpha,\beta)$ with marginals $f_X(\alpha)$ and $f_Y(\beta)$, i.e.,

$$\int_{-\infty}^{\infty} d\alpha f_{XY}(\alpha,\beta) = f_Y(\beta)$$

$$\int_{-\infty}^{\infty} d\beta f_{XY}(\alpha,\beta) = f_X(\alpha).$$

Given a value $\nu \ge 1$, define

$$\bar{\rho}^{(\nu)}(X,Y) \triangleq \inf_{f_{XY}\epsilon\mathcal{P}_{XY}} E\{|X - Y|^\nu\}$$

$$= \inf_{f_{XY}\epsilon\mathcal{P}_{XY}} \int_{-\infty}^{\infty} d\alpha \int_{-\infty}^{\infty} d\beta\, |\alpha - \beta|^\nu f_{XY}(\alpha,\beta).$$

The infimum is required as the minimum may not exist. The distance $\bar{\rho}^{(\nu)}(X,Y)$ measures how well the two random variables can be matched up in an average distortion sense by assigning a consistent joint distribution to them. This distance is different than the distances considered by Kailath [4] in that it measures how well two random variables can be matched up by joint descriptions, but we shall see that it is the appropriate measure for the given problem.

The Vasershtein distance is $\bar{\rho}^{(1)}$. In particular, the Vasershtein distance requires a metric distortion measure, that is, one satisfying the triangle inequality. We note that $|x - y|^\nu$ is not a metric and does not satisfy the triangle inequality unless $\nu = 1$.

III. THE MISMATCH THEOREMS

We consider first the special but important case where distortion is measured by average magnitude error, i.e., $\rho_X^{(1)}(q)$. The fact that $|x - y|$ is a metric yields a particularly simple and intuitive result.

Theorem 1: Given a quantizer q satisfying (1) and two real-valued random variables X and Y, then

$$|\rho_X^{(1)}(q) - \rho_Y^{(1)}(q)| \le \bar{\rho}^{(1)}(X,Y). \tag{2}$$

Proof: Let $f_{XY}(\alpha,\beta)\epsilon\mathcal{P}_{XY}$ approximately yield $\bar{\rho}^{(1)}(X,Y)$, i.e., given $\epsilon > 0$,

$$\int_{-\infty}^{\infty} d\alpha \int_{-\infty}^{\infty} d\beta f_{XY}(\alpha,\beta)\,|\alpha - \beta| \le \bar{\rho}^{(1)}(X,Y) + \epsilon.$$

For a quantizer satisfying (1) we have that $|r - q(r)| = \min_k |r - a_k|$ and hence using the triangle inequality we have that

$$\rho_X^{(1)}(q) = \int_{-\infty}^{\infty} d\alpha f_X(\alpha)\,|\alpha - q(\alpha)|$$

$$= \int_{-\infty}^{\infty} d\beta \int_{-\infty}^{\infty} d\alpha f_{XY}(\alpha,\beta)\cdot\min_k |\alpha - a_k|$$

$$\le \int_{-\infty}^{\infty} d\beta \int_{-\infty}^{\infty} d\alpha f_{XY}(\alpha,\beta) \min_k \{|\alpha - \beta| + |\beta - a_k|\}$$

$$= \int_{-\infty}^{\infty} d\beta \int_{-\infty}^{\infty} d\alpha f_{XY}(\alpha,\beta)\{|\alpha - \beta| + \min_k |\beta - a_k|\}$$

$$= \int_{-\infty}^{\infty} d\beta \int_{-\infty}^{\infty} d\alpha f_{XY}(\alpha,\beta)\,|\alpha - \beta| + \int_{-\infty}^{\infty} d\beta f_Y(\beta)\,|\beta - q(\beta)|$$

$$\le \bar{\rho}^{(1)}(X,Y) + \rho_Y^{(1)}(q) + \epsilon.$$

Since ϵ is arbitrary, $\rho_X^{(1)}(q) \le \bar{\rho}^{(1)}(X,Y) + \rho_Y^{(1)}(q)$. Reversing the roles of X and Y yields the theorem.

Theorem 1 shows that the difference in performance of a single quantizer on separate sources is bound above by their Vasershtein distance—regardless of the form of the quantizer (provided it satisfies (1)). In particular, if a quantizer is designed for X but applied to Y, then little performance will be lost provided the random variables are close in the $\bar{\rho}^{(1)}$ sense. Thus if F_X is fixed, $\rho_Y^{(1)}(q) \to \rho_X^{(1)}(q)$ as $\bar{\rho}^{(1)}(X,Y) \to 0$ so that quantizer performance is a continuous function of the source under the Vasershtein distance.

Corollary 1: For fixed integer N define the optimal attainable performance using any quantizer with N levels (satisfying (1)) by

$$\Delta_X^{(1)}(N) = \inf_q \rho_X^{(1)}(q),$$

then

$$|\Delta_X^{(1)}(N) - \Delta_Y^{(1)}(N)| \le \bar{\rho}^{(1)}(X,Y).$$

Proof: Simply apply Theorem 1 to a quantizer q that is optimal for X, then $\Delta_Y^{(1)}(N) \le \rho_Y^{(1)}(q) \le \rho_X^{(1)}(q) + \bar{\rho}^{(1)}(X,Y) = \Delta_X^{(1)}(N) + \bar{\rho}^{(1)}(X,Y)$. Interchanging X and Y completes the proof.

We next consider general νth law distortion measures for $\nu \ge 1$.

Theorem 2: Given a quantizer q satisfying (1) and two real-valued random variables X and Y, then for any $\nu \geq 1$,

$$|\rho_X^{(\nu)}(q)^{1/\nu} - \rho_Y^{(\nu)}(q)^{1/\nu}| \leq \bar{\rho}^{(\nu)}(X,Y)^{1/\nu} \tag{3}$$

and

$$|\rho_X^{(\nu)}(q) - \rho_Y^{(\nu)}(q)| \leq \nu[\max(\rho_X^{(\nu)}(q),\rho_Y^{(\nu)}(q))]^{1-1/\nu}\bar{\rho}^{(\nu)}(X,Y)^{1/\nu}. \tag{4}$$

Proof: Parallel to the proof of Theorem 1 let $f_{XY}\epsilon\mathcal{P}_{XY}$ approximately yield $\bar{\rho}^{(\nu)}(X,Y)$. We have that

$$\begin{aligned}\rho_X^{(\nu)}(q) &= \int_{-\infty}^{\infty} d\alpha f_X(\alpha)\,|\alpha - q(\alpha)|^\nu \\ &= \int_{-\infty}^{\infty} d\beta \int_{-\infty}^{\infty} d\alpha f_{XY}(\alpha,\beta)\{\min_k |\alpha - a_k|\}^\nu \\ &\leq \int_{-\infty}^{\infty} d\beta \int_{-\infty}^{\infty} d\alpha f_{XY}(\alpha,\beta)\{|\alpha - \beta| + \min_k |\beta - a_k|\}^\nu \\ &= \int_{-\infty}^{\infty} d\beta \int_{-\infty}^{\infty} d\alpha f_{XY}(\alpha,\beta)\{|\alpha - \beta| + |\beta - q(\beta)|\}^\nu.\end{aligned}$$

From Minkowski's inequality we have that

$$\begin{aligned}&\left\{\int_{-\infty}^{\infty} d\beta \int_{-\infty}^{\infty} d\alpha f_{XY}(\alpha,\beta)[|\alpha - \beta| + |\beta - q(\beta)|]^\nu\right\}^{1/\nu} \\ &\leq \left\{\int_{-\infty}^{\infty} d\beta \int_{-\infty}^{\infty} d\alpha f_{XY}(\alpha,\beta)\,|\alpha - \beta|^\nu\right\}^{1/\nu} \\ &\quad + \left\{\int_{-\infty}^{\infty} d\beta \int_{-\infty}^{\infty} d\alpha f_{XY}(\alpha,\beta)\,|\beta - q(\beta)|^\nu\right\}^{1/\nu} \\ &= \{\bar{\rho}^{(\nu)}(X,Y) + \epsilon\}^{1/\nu} + \{\rho_Y^{(\nu)}(q)\}^{1/\nu+}\end{aligned}$$

so that

$$\{\rho_X^{(\nu)}(q)\}^{1/\nu} \leq \{\rho_Y^{(\nu)}(q)\}^{1/\nu} + \{\bar{\rho}^{(\nu)}(X,Y) + \epsilon\}^{1/\nu}.$$

Interchanging the roles of X and Y proves (3). Equation (4) follows (3) by a Taylor series expansion as follows: let $a \geq b \geq 0$ and expand $b^{1/\nu}$ in a Taylor series expansion with remainder term in $(0,a]$ to obtain

$$b^{1/\nu} = a^{1/\nu} + (b - a)\nu^{-1}a^{1/\nu-1} + 1/2(b - a)^2\nu^{-1}(\nu^{-1} - 1)\theta^{1/\nu-2}$$

where $\theta\epsilon(0,1)$. Since the rightmost term is negative, we have that

$$a^{1/\nu} - b^{1/\nu} \geq (a - b)\nu^{-1}a^{1/\nu-1}. \tag{5}$$

Defining a as the maximum and b as the minimum of $\rho_X^{(\nu)}(q)$ and $\rho_Y^{(\nu)}(q)$ in (5) yields

$$\begin{aligned}&|\rho_X^{(\nu)}(q)^{1/\nu} - \rho_Y^{(\nu)}(q)^{1/\nu}| \\ &\quad \geq |\rho_X^{(\nu)}(q) - \rho_Y^{(\nu)}(q)|\,\nu^{-1}[\max(\rho_X^{(\nu)}(q),\rho_Y^{(\nu)}(q))]^{1/\nu-1},\end{aligned}$$

which with (3) yields (4).

Discussion: Theorem 2 is more complicated due to the more general ν. Equation (4) bounds the actual performance difference in terms of the distance and either individual performance (since min (a,b) is less than a or b). Equation (3) does not directly bound the performance difference, but instead bounds the performance of one source given the distance and the performance of the other source—often a more useful relation. For example, if q is designed for Y and applied to X so that $\rho_Y^{(\nu)}(q)$ is known and fixed, then (3) becomes

$$\rho_X^{(\nu)}(q) \leq |\rho_Y^{(\nu)}(q)^{1/\nu} + \bar{\rho}^{(\nu)}(X,Y)^{1/\nu}|^\nu.$$

Thus, for example, $\rho_X^{(\nu)}(q) \to \rho_Y^{(\nu)}(q)$ as $\bar{\rho}^{(\nu)}(X,Y) \to 0$ so that quantizer performance is continuous under the generalized Vasershtein distance for any $\nu \geq 1$.

Choosing optimal quantizers as in Corollary 1, Theorem 2 immediately yields the following.

Corollary 2: For $\nu \geq 1$ and integer N we have that

$$|\Delta_x^{(\nu)}(N)^{1/\nu} - \Delta_Y^{(\nu)}(N)^{1/\nu}| \leq \bar{\rho}^{(\nu)}(X,Y)^{1/\nu}$$

$$\Delta_X^{(\nu)}(N) \leq |\Delta_Y^{(\nu)}(N)^{1/\nu} + \bar{\rho}^{(\nu)}(X,Y)^{1/\nu}|^\nu$$

$$|\Delta_X^{(\nu)}(N) - \Delta_Y^{(\nu)}(N)| \leq \nu\{\max(\Delta_X^{(\nu)}(N),\Delta_Y^{(\nu)}(N))\}^{1-1/\nu} \cdot \bar{\rho}^{(\nu)}(X,Y)^{1/\nu}.$$

As a final comment, metric distortion measures are highly desirable for communications studies since the triangle inequality allows one to show that if the performance loss in each stage of a system is small, then the overall loss is also small. A ν-law distortion measure can be modified to become a metric by simply taking a νth root, that is, when computing time average distortion consider

$$\{n^{-1}\sum_{i=0}^{n-1}|X_i - q(X_i)|^\nu\}^{1/\nu}.$$

For example, consider root-mean-squared-error distortion instead of mean-squared error. For ergodic processes the limit of the above average converges with probability one to $\rho_X^{(\nu)}(q)^{1/\nu}$. That the triangle inequality is satisfied follows from Minkowski's inequality. Thus given a nonmetric law distortion measure, if we "metricize" the distortion measure by defining

$$\begin{aligned}\rho_X(q) &= \lim_{n\to\infty} E\{[n^{-1}\sum_{i=0}^{n-1}|X_i - q(X_i)|^\nu]^{1/\nu}\} \\ &= \rho_X^{(\nu)}(q)^{1/\nu},\end{aligned}$$

then (3) and the corollaries actually give the performance mismatch. Note that the modified distortion measure preserves the ν-law notion of closeness, but regains the desired geometric notion of distance.

IV. EVALUATION AND BOUNDING OF DISTANCE MEASURES

In this section we derive a simple upper bound to the generalized Vasershtein distance and thereby obtain another mismatch bound. Special cases wherein the bound of this section actually yields the Vasershtein distance are described. In the next section these results are coupled with the mismatch theorem to obtain several examples.

Given a cumulative distribution function $F(\alpha)$, let $F^{-1}(\beta)$ denote the inverse distribution function, i.e., $F^{-1}(\beta) = \gamma$ if γ is the smallest value for which $F(\gamma) = \beta$.

Theorem 3: Given two random variables X and Y with distribution functions F_X and F_Y, respectively, then for any $\nu \geq 1$,

$$\bar{\rho}^{(\nu)}(X,Y) \leq \int_0^1 |F_X^{-1}(u) - F_Y^{-1}(u)|^\nu du \tag{6}$$

with equality if $\nu = 1$ in which case

$$\bar{\rho}^{(1)}(X,Y) = \int_{-\infty}^{\infty} |F_X(\alpha) - F_Y(\alpha)|\,d\alpha. \tag{7}$$

Proof: Let U be a uniformly distributed random variable on $[0,1]$, i.e.,

$$f_U(u) = \begin{cases}1, & 0 \leq u \leq 1 \\ 0, & \text{otherwise.}\end{cases}$$

It is well known that the random variables $F_X^{-1}(U)$ and $F_Y^{-1}(U)$ will have distribution functions F_X and F_Y, respectively, i.e., any random variable can be modeled as the inverse distribution function of a uniform random variable. Given a single uniform random variable U, define the joint distribution function

$$F_{XY}(\alpha,\beta) \triangleq \Pr(X \leq \alpha, Y \leq \beta)$$
$$= \Pr(F_X^{-1}(U) \leq \alpha, F_Y^{-1}(U) \leq \beta)$$
$$= \Pr(U \leq F_X(\alpha), U \leq F_Y(\beta)).$$

Since U is a uniform random variable this becomes

$$F_{XY}(\alpha,\beta) \triangleq F_U(\min(F_X(\alpha),F_Y(\beta))) = \min(F_X(\alpha),F_Y(\beta))$$

so that $F_{XY}(\alpha,\infty) = F_X(\alpha)$, i.e., F_{XY} has the correct marginals so that $f_{XY}(\alpha,\beta)\epsilon\mathcal{P}_{XY}$ where

$$f_{XY}(\alpha,\beta) = (d^2/d\alpha\, d\beta)F_{XY}(\alpha,\beta) = f_X(\alpha)f_Y(\beta)\delta(F_X(\alpha) - F_Y(\beta))$$

where δ denotes the Dirac delta or unit impulse function. Since $\bar{\rho}^{(\nu)}(X,Y)$ is the minimum value of $E\{|X - Y|^\nu\}$ over all $f_{XY}\epsilon\mathcal{P}_{XY}$, the above $f_{XY}(\alpha,\beta)$ must yield an upper bound to $\bar{\rho}^{(\nu)}(X,Y)$ and therefore we have from the fundamental theorem of expectation that

$$\bar{\rho}^{(\nu)}(X,Y) \leq E\{|X - Y|^\nu\} = \int_0^1 |F_X^{-1}(u) - F_Y^{-1}(u)|^\nu\, du,$$

proving (6). Note that the guess for f_{XY} relates X and Y in a deterministic fashion since, for example, $X = F_X^{-1}(F_Y(Y))$. Equation (7) is proved by Vallender [2].

We next consider the special but important case of zero-mean Gaussian random variables with different variances.

Corollary 2: Let X and Y be zero-mean Gaussian random variables with variance σ_X^2 and σ_Y^2, respectively. We have that

$$\rho^{(1)}(X,Y) = (2/\pi)^{1/2}|\sigma_x - \sigma_y| \qquad (8)$$

$$\rho^{(2)}(X,Y) = |\sigma_X - \sigma_Y|^2. \qquad (9)$$

Proof: For both $\nu = 1$ and 2 let f_{XY} be as in the proof of the theorem. In this case X and Y are jointly Gaussian perfectly correlated in that the correlation coefficient is one, i.e., $E(XY) = \sigma_X\sigma_Y$. Thus $X - Y$ is zero mean Gaussian with variance $\sigma^2 = E\{(X - Y)^2\} = \sigma_X^2 + \sigma_Y^2 - 2\sigma_X\sigma_Y = |\sigma_X - \sigma_Y|^2$ so that Theorem 3 yields

$$\rho^{(1)}(X,Y) \leq E|X - Y| = (2/\pi)^{1/2}\sigma$$
$$\rho^{(2)}(X,Y) \leq E(X - Y)^2 = \sigma^2.$$

That the above inequalities are actually equalities follows since

$$E|X - Y| \geq |E|X| - E|Y||$$
$$= |(2/\pi)^{1/2}\sigma_X - (2/\pi)^{1/2}\sigma_Y| = (2/\pi)^{1/2}\sigma$$

and from the Cauchy–Schwarz inequality

$$E(X - Y)^2 = \sigma_X^2 + \sigma_Y^2 - 2E(XY)$$
$$\geq \sigma_X^2 + \sigma_Y^2 - 2\sigma_X\sigma_Y$$
$$= \sigma^2.$$

We note that (8) can also be obtained directly from (7) using integration by parts.

It is an open question as to how generally (6) holds with equality.

V. EXAMPLES

Example 1: Let X and Y be zero-mean Gaussian random variables with variances σ_X^2 and σ_Y^2, respectively. Then for any quantizer q satisfying (1)

$$|\rho_X^{(1)}(q) - \rho_Y^{(1)}(q)| \leq (2/\pi)^{1/2}|\sigma_X - \sigma_Y|$$

$$|\rho_X^{(2)}(q)^{1/2} - \rho_Y^{(2)}(q)^{1/2}| \leq |\sigma_X - \sigma_Y|$$

$$|\rho_X^{(2)}(q) - \rho_Y^{(2)}(q)| \leq 2[\max(\rho_X^{(2)}(q),\rho_Y^{(2)}(q))]^{1/2} \cdot |\sigma_X - \sigma_Y|.$$

Thus the performance will be close if the standard deviations are close.

Example 2: Let X and Y be as in Example 1. Say q is designed for X but applied to Y. It is often useful to have an upper bound to $\rho_Y^{(2)}(q)$ in terms of $\bar{\rho}^{(2)}(X,Y)$ and $\rho_X^{(2)}(q)$ rather than a bound on the difference between $\rho_X^{(2)}(q)$ and $\rho_Y^{(2)}(q)$. Using (3) instead of (4) as in Example 1 yields

$$\rho_Y^{(2)}(q) \leq \{\rho_X^{(2)}(q)^{1/2} + \bar{\rho}^{(2)}(X,Y)^{1/2}\}^2$$
$$= \rho_X^{(2)}(q) + \bar{\rho}^{(2)}(X,Y) + 2\{\rho_X^{(2)}(q)\bar{\rho}^{(2)}(X,Y)\}^{1/2}$$
$$= \rho_X^{(2)}(q) + |\sigma_X - \sigma_Y|^2 + 2\rho_X^{(2)}(q)^{1/2}|\sigma_X - \sigma_Y|.$$

Example 3: Quantizers are often designed to be optimal for a uniformly distributed random variable, i.e., uniform quantization over some finite interval. What is the resulting mismatch if such a quantizer is applied to a zero-mean Gaussian source with variance σ^2? For convenience let X be uniformly distributed on $[-1/2,1/2]$, let Y be zero-mean Gaussian with variance σ^2, and let $\nu = 2$. Theorem 3 plus a change of variables implies that

$$\bar{\rho}^{(2)}(X,Y) \leq \int_0^1 |F_X^{-1}(u) - F_Y^{-1}(u)|^2\, du$$
$$= \int_{-\infty}^{\infty} (F_X^{-1}(F_Y(y)) - y)^2 f_y(y)\, dy$$

where $F_X^{-1}(u) = u - 1/2$, $0 \leq u \leq 1$. Define

$$\varphi(y) = (2\pi)^{-1}\exp(-y^2/2)$$
$$\Phi(y) = \int_{-\infty}^{y} \varphi(\alpha)\, d\alpha$$

so that $f_Y(y) = \varphi(y/\sigma)$, $F_Y(y) = \Phi(y/\sigma)$, and $F_X^{-1}(F_Y(y)) = \Phi(y/\sigma) - 1/2$. Integration by parts yields

$$\bar{\rho}^{(2)}(X,Y) \leq \int_{-\infty}^{\infty} [\Phi(y/\sigma) - 1/2 - y]^2\varphi(y/\sigma)\, dy$$
$$= \sigma^2 + 1/12 - \int_{-\infty}^{\infty} 2y\Phi(y/\sigma)\varphi(y/\sigma)\, dy$$
$$= \sigma^2 + 1/12 - 2\sigma\int_{-\infty}^{\infty} t\Phi(t)\varphi(t)\, dt$$
$$= \sigma^2 + 1/12 - 2\sigma\int_{-\infty}^{\infty} \varphi(t)^2\, dt$$
$$= \sigma^2 + 1/12 - \sigma\pi^{-1/2}.$$

This expression is minimized by taking $\sigma^2 = (4\pi)^{-1}$ in which case

$$\bar{\rho}^{(2)}(X,Y) \leq [1 - 3/\pi]/12$$
$$\simeq 0.0038.$$

Thus a Gaussian random variable can be "fitted" surprisingly close in the distance sense to a uniform random variable by the appropriate choice of variance.

An N-level uniform quantizer on $[-1/2,1/2]$ applied to a uniformly distributed random variable Y yields $\rho_Y^{(2)}(q) = (12N^2)^{-1}$ [9, p. 308] so that Theorem 2 yields

$$\rho_X^{(2)}(q) \leq (\rho_Y^{(2)}(q)^{1/2} + \bar{\rho}^{(2)}(X,Y)^{1/2})^2$$
$$\simeq (12N^2)^{-1} + 0.0038 + (0.017)/N.$$

Thus the performance loss is bounded above by $0.0038 + (0.017)/N$. The smallness of the mismatch for moderate N demonstrates the robustness of finite-level uniform quantizers for Gaussian sources with the appropriate choice of parameters.

ACKNOWLEDGMENT

The authors would like to thank Prof. T. Berger, Cornell University, for his helpful comments and suggestions.

REFERENCES

[1] R. L. Dobrushin, "Prescribing a system of random variables by conditional distributions," *Theory Prob. Appl.*, vol. 14, no. 3, pp. 458–486, 1970.

[2] S. S. Vallender, "Computing the Wasserstein distance between probability distributions on the line" (in Russian), *Theory Prob. Appl.*, vol. 18, pp. 824–827, 1973.

[3] R. M. Gray, D. L. Neuhoff, and P. Shields, "A generalization of Ornstein's $\bar{d}$-distance with applications to information theory," *Ann. Prob.*, to be published, Apr. 1975.

[4] T. Kailath, "The divergence and Bhattacharyya distance measures in signal selection," *IEEE Trans. Commun. Technol.*, vol. COM-15, pp. 52–60, Feb. 1967.

[5] J. Max, "Quantizing for minimum distortion," *IRE Trans. Inform. Theory*, vol. IT-6, pp. 7–12, Mar. 1960.

[6] G. M. Roe, "Quantizing for minimum distortion," *IEEE Trans. Inform. Theory* (Corresp.), vol. IT-10, pp. 384–385, Oct. 1964.

[7] R. C. Wood, "On optimum quantization," *IEEE Trans. Inform. Theory*, vol. IT-15, pp. 248–252, Mar. 1969.

[8] V. R. Algazi, "Useful approximations to optimum quantization," *IEEE Trans. Commun. Technol.*, vol. COM-14, pp. 297–301, June 1966.

[9] A. B. Carlson, *Communication Systems*. New York: McGraw-Hill, 1968.

23

Reprinted from *IEEE Trans. Commun.* **COM-22**(12):1897–1901 (1974)

Robust Quantization of Discrete-Time Signals with Independent Samples

JOEL M. MORRIS, STUDENT MEMBER, IEEE, AND V. DAVID VANDELINDE, MEMBER, IEEE

***Abstract*—The N-level uniform quantizer on $[-c,c]$ plus the assignment of $y_0° = -(a_s + c)/2$ and $y_{N+1}° = (a_s + c)/2$ to signal values falling in the saturation regions $[-a_s,-c)$ and $(c,a_s]$, respectively, is shown to be the minimax $(N + 2)$-level quantizer with a nonsaturating input range $[-c,c]$. The performance criterion considered is the mean weighted quantization error and the input signals are only required to be amplitude bounded by $\pm a_s$ where $a_s > c > 0$. The worst case input signal marginal probability distributions are shown to be discrete. From the derivation of this result, the minimax error can be computed. An example is given which illustrates the performance of the minimax quantizer for several input ranges against different input signal probability distributions.**

I. INTRODUCTION

THE input signals to a quantizer in many communications and data systems are described by probability distributions which are generally known or assumed to belong to some very restrictive class H of probability distributions; such as, all distributions with given bounds on certain moments; and/or all distributions of a certain type (e.g., Gaussian, uniform, exponential, or Rayleigh). Quantizers, which may be stationary or adaptive [4], [8], [9], [10], are designed to meet some desired criteria for this *a priori* knowledge and/or assumptions about H. The desired criteria may involve quantization error [2], [6], signal to distortion ratios [7], quantizer output entropy [3], design complexity, and/or cost, etc.

In many cases, the knowledge about the signal distributions may be wrong, the assumptions rash, or the signal distributions may change, causing the quantizer to perform unsatisfactorily with respect to the design criterion. Therefore, a robust quantizer may be advantageous in these cases. Although it would generally be a suboptimal quantizer over the class H, a robust quantizer would guarantee a performance level, i.e., performs as well as possible uniformly, over a wider class A of signal probability distributions within which the class H of the known or assumed distributions is contained.

In this paper, we make the following observation, which allows the consideration of the class A of signal probability distributions as an extremely large class. It is noted that practically all signals in the physical world are amplitude limited or the measuring process itself amplitude limits them. Therefore, most signal probability distributions used in theoretical work (e.g., Gaussian, exponential, etc.) are not really experienced in practice. The signal distributions experienced in practice, i.e., "physical" signal probability distributions, may be similar to the "theoretic" probability distributions on a finite interval but account for the nonzero probability mass in the tails of these "theoretic" distributions by: 1) appending this tail probability mass in the form of step functions at the respective endpoints of that finite interval, or by 2) normalizing the probability mass of the "theoretic" distributions on that finite interval.

The finite interval admitting all "physical" signals for a particular application would be specified by the designer, i.e., $[-a_s,a_s]$, where the assumption of symmetry causes no loss of generality. The criteria considered are mean weighted quantization error criteria encompassing the most common weighting functions such as the squared quantization error and the absolute quantization error functions.

Section II of this paper presents the basic terminology and definitions used in the remaining sections. In Section III, we state the problem to be considered and then summarize the functional analysis method used to obtain the solution. Section IV establishes the existence of worst case marginal distributions and characterizes them. In Section V, the solution to the problem is obtained and the minimax quantizer characterized. We also show that this minimax quantizer is not the maximin quantizer. Section VI shows that this quantizer is also the minimax quantizer if we relax the stationarity requirement on the discrete-time signal $\{S_i : i \in I\}$ but require the random variables S_i to be only independent and confined to the finite range $[-a_s,a_s]$. In Section VII, some examples are presented which show the performance of the minimax quantizer for common signals. Comparisons are made to an optimum quantizer performance. Section VIII contains the concluding remarks and discussion of further research areas.

II. DEFINITIONS AND TERMINOLOGY

It is appropriate at this point to define and discuss terminology before formally stating the problem.

Paper approved by the Associate Editor for Communication Theory of the IEEE Communications Society for publication after presentation at the 8th Annual Princeton Conference on Information Sciences and Systems, Princeton University, Princeton, N. J., March 1974. Manuscript received March 26, 1974; revised July 26, 1974.

The authors are with the Department of Electrical Engineering, The Johns Hopkins University, Baltimore, Md.

Definition 1: $\{S_i : i \in I\}$ is a discrete-time random process where the random variable S_i has the marginal distribution $F_i(s)$.

Definition 2: X is a normed linear space (X^* = dual space of X), where $F(s) = F \in X$ implies:

a) $\| F \| = \mathrm{TV}(F) < \infty$, where $\mathrm{TV}(F)$ is the total variation of $F(s)$; and

b) $F(s)$ is right continuous.

Definition 3: $A = \{F \in X; \| F \| = 1, F$ is nondecreasing$\}$, i.e., A is the set of nondecreasing functions having total variation $F(a_s) - F(-a_s^-) = 1$. Thus, A represents the class of marginal distributions for random variables S_i confined to $[-a_s, a_s]$.

Definition 4: $D = \{q\}$ = the set of all $(N+2)$-level quantizers q with nonsaturating input range $[-c,c]$, characterized by the $2N+5$ parameters $(s_0, \cdots, s_{N+2}; y_0, \cdots, y_{N+1})$ where

a) the output of the quantizer q is,

$$q(s) = y_k \tag{1}$$

with $s, y_k \in (s_k, s_{k+1}]$, $k = 2, \cdots, N$; $s, y_{N+1} \in (s_{N+1}, s_{N+2}]$, $s, y_0 \in [s_0, s_1)$; and $s, y_1 \in [s_1, s_2]$. Hence, the s_k are called the transition values and the y_k are called the representation values [2];

b) $(s_2, \cdots, s_N; y_0, \cdots, y_{N+1})$ are the quantizer parameters to be chosen;

c)
$$s_{N+1} = -s_1 = c \tag{2}$$

are given and we specify

$$s_{N+2} = -s_0 = a_s; \tag{3}$$

and

d) $[-c,c]$ is the nonsaturating input range and $[-a_s, -c)$, $(c, a_s]$ are the saturation regions.

Definition 5:

$$\mathcal{E}(q,F) = \int_{-a_s}^{a_s} g^*(s)\, dF(s) \tag{4}$$

is a Lebesgue–Stieltjes integral where $F \in X$ and $g^*(s)$ is an upper semicontinuous function defined by

$$g^*(s) = \begin{cases} \max[g(s_{k+1} - y_k), g(s_{k+1} - y_{k+1})], & \text{for } s = s_{k+1}, \quad k = 0, \cdots, N \\ g(s - y_k), & \forall s \in (s_k, s_{k+1}), \quad k = 1, \cdots, N \\ g(s - y_0), & \forall s \in [s_0, s_1) = [-a_s, -c) \\ g(s - y_{N+1}), & \forall s \in (s_{N+1}, s_{N+2}] = (c, a_s] \end{cases} \tag{5}$$

where $g(\tau)$ is a continuous function defined by

$$g(\tau) = g(-\tau) \geq 0, \qquad \tau \in [-a_s, a_s]$$
$$g(\tau_2) > g(\tau_1), \qquad |\tau_2| > |\tau_1|, \tag{6}$$

with τ the instantaneous quantization error, i.e., the difference between the quantizer input and its output.

Definition 6: (see [5].) A sequence $\{x_n^*\}$ in X^* is said to converge weak* to the element x^* if for every $x \in X$, $\langle x, x_n^* \rangle \to \langle x, x^* \rangle$.

Definition 7: (see [5].) A set $K \subset X^*$ is said to be weak* compact if every infinite sequence from K contains a weak* convergent sequence.

Definition 8: (see [5].) A functional f defined on a normed space X^* is said to be weak* continuous at x_0^* if given $\in > 0$ there is a $\delta > 0$ and a finite collection $\{x_1, x_2, \cdots, x_n\}$ from X such that $|f(x^*) - f(x_0^*)| < \in$ for all x^* such that $|\langle x^*, x_i \rangle| < \delta$ for $i = 1, 2, \cdots, n$. We say that f is weak* upper semicontinuous at x_0^* if $|f(x^*) - f(x_0^*)| < \in$ is replaced with $f(x^*) - f(x_0^*) < \in$ in the above.

III. PROBLEM STATEMENT

The problem of determining the robust stationary $(N+2)$-level quantizer for generally weak conditions on the class of "physical" signal probability distributions can be formally stated as follows.

Given:

1) A discrete-time signal $\{S_i : i \in I\}$, where the S_i are independent identically distributed (iid) random variables described by a probability distribution function from A.

2) A mean weighted quantization error criterion defined by $\sum \mathcal{E}(q, F_i)$, where $q \in D$ and $F_i \in A$.

Problem 1: Find the $(N+2)$-level quantizer with nonsaturating input range $[-c,c]$, (i.e., the quantizer parameter set $(s_2, \cdots, s_N; y_0, \cdots, y_{N+1}) \sim q \in D$), which minimizes the worst possible value for $\sum \mathcal{E}(q, F_i)$ that results from the given signal having marginal distributions in A.

This problem can be solved using a functional analysis approach. The quantizer parameter set, through the ith weighted quantization error criterion, $\mathcal{E}(q, F_i)$, is shown to define a linear functional on X. Hence, $\mathcal{E}(q, F_i)$ is a linear functional on the subset A which is an admissible class of signal source-marginal distributions. Moreover, because the S_i are iid, it is sufficient to work solely with $\mathcal{E}(q,F) = \mathcal{E}(q, F_i)$, $\forall i \in I$. Hence $\mathcal{E}(q,F)$ is first maximized over the admissible class A for an arbitrary quantizer parameter set corresponding to an arbitrary quantizer $q \in D$, and then minimized over the admissible class D of quantizers to obtain the minimax result. All proofs will only be outlined or summarized. The interested reader is referred to [11].

IV. WORST CASE DISTRIBUTIONS

Before determining the worst case marginal distributions for our problem, we must establish the existence of maximizing marginal distributions in the set A for arbitrary quantizers $q \in D$.

Theorem 1: (See Alaoglu [5].) Let Y be a real-normed linear space. The closed unit sphere in Y^* is weak* compact.

Proof: See [5, p. 128].

Proposition IV.1: There exists a maximizing marginal distribution function for the functional $\mathcal{E}_q(F) = \mathcal{E}(q,F)$ on A.

Outline of Proof: It is straightforward to show that $\mathcal{E}_q(F) = \mathcal{E}(q,F)$, for arbitrary q, is a bounded linear functional on X and A, and hence $\mathcal{E}_q \in X^* =$ dual space of X. X can be also shown to be the dual space of $C =$ space of continuous functions by the Riesz representation theorem [5]. Showing that A is a closed subset of the closed unit ball in X, we have A is weak* compact. Finally, by showing that $\mathcal{E}_q(F)$ is weak* upper semicontinuous on A, we can state that $\mathcal{E}_q(F)$ achieves a maximum on A.

Proposition IV.2: A maximizing marginal distribution function F_q, for the arbitrary quantizer $q \in D$, is any distribution in A having its total probability mass on the points $\{s_i\} \subset \{s_k : k = 0,\cdots,N+1\}$, where $g^*(s)$ achieves its maximum.

Outline of Proof: The proof consists of using the upper semicontinuity of $q^*(s)$ to show that

$$\mathcal{E}_q(F_q) = \max_{F \in A} \mathcal{E}_q(F) = || \mathcal{E}_q || \cdot || F_q ||$$

where

$$|| \mathcal{E}_q || = \sup_{s \in [-a_s,a_s]} | g^*(s) |.$$

In addition, by the definition of g^* there exists a subset of points $\{s_i\}$ from $\{s_k : k = 0,\cdots,N+1\}$ such that $g^*(s_i) = || \mathcal{E}_q ||$. Therefore, one can construct a functional $F_q \in X$ with $|| F_q || = 1$ which is a series of step discontinuities occuring at one or more of the subsets of points $\{s_i\}$.

V. THE MINIMAX QUANTIZER

Now that the existence of the maximizing marginal distribution for an arbitrary quantizer has been established and its characterization presented, we are in a position to determine the minimax quantizer. The method of solution is to minimize $\mathcal{E}(q,F_q)$ with respect to $q \in D$, noting that $\mathcal{E}(q,F_q) = \mathcal{E}_q(F_q) = || \mathcal{E}_q ||$ is a function of q only.

Definition V.1: A N-level uniform quantizer on the interval $[-b,b]$ is a N-level quantizer characterized by the parameters $(s_0,\cdots,s_N;y_0,\cdots,y_{N-1})$ which satisfies $| s_k - y_k | = | s_{k+1} - y_k | = | s_{k+1} - y_{k+1} |$, $\forall s_k \in [-b,b]$, $\forall y_k \in (-b,b)$ and by $s_0 = -b$, $s_N = b$.

Proposition V.1: The N-level uniform quantizer on $[-c,c]$ plus the assignment of $y_0^\circ = -\frac{1}{2}(a_s + c)$ and $y_{N+1}^\circ = \frac{1}{2}(a_s + c)$ is the solution to Problem 1.

Outline of Proof: With the minimizing $q \in D$ satisfying

$$\mathcal{E}_{q^\circ}(F_{q^\circ}) = \min_{q \in D} \mathcal{E}_q(F_q) = \min_{q \in D} || \mathcal{E}_q ||$$

$$= \min_{q \in D} [\max_{s \in [-a_s,a_s]} | g^*(s) |],$$

there are two possible cases to consider:

Case 1: for $| a_s - c | \leq 2c/N$; and

Case 2: for $| a_s - c | > 2c/N$.

The proof consists of showing that the $(N+2)$-level quantizer, which is uniform on $[-c,c]$ and makes the assignment $y_{N+1} = -y_0 = \frac{1}{2}(a_s + c)$, minimizes

$$|| \mathcal{E}_q || = \max_{s \in [-a_s,a_s]} | g^*(s) |$$

by causing the value $|| \mathcal{E}_{q^\circ} ||$ to occur at each of the transition points $s_k \in [-c,c]$, $k = 1,\cdots,N$ if Case 1 is true or at each of the transition points $s_k \in [-a_s,-c) \cup (c,a_s]$, $k = 0,1,N,N+1$ if Case 2 is true. Hence any other value of $|| \mathcal{E}_q ||$, not satisfying the above requirement, is greater than the minimum $|| \mathcal{E}_{q^\circ} ||$.

At this point, we might ask the question, is this minimax quantizer also the maximin quantizer? This is equivalent to asking if the minimax quantizer q°, and the corresponding worst case marginal distribution function F_{q°, are a global saddlepoint pair. We now show that, unfortunately, this is not true.

Proposition V.2: The minimax quantizer q°, and the corresponding worst case marginal distribution function F_{q° are not a global saddlepoint pair.

Outline of Proof: The proof consists of considering Case 2 in the previous proof outline and exhibiting a quantizer $q' \in D$ such that $\mathcal{E}(q',F_{q^\circ}) < \mathcal{E}(q^\circ,F_{q^\circ})$. Hence, $\mathcal{E}(q^\circ,F) \leq \mathcal{E}(q^\circ,F_{q^\circ}) \leq \mathcal{E}(q,F_{q^\circ})$ is not true $\forall F \in A$ and $\forall q \in D$ which is the requirement for (q°,F_{q°) to be a global saddlepoint pair.

VI. THE MINIMAX QUANTIZER FOR NONSTATIONARY SIGNALS

The above results were derived for stationary signal sources. However, this stationary requirement (i.e., the requirement that the S_i are identically distributed) can be relaxed and the minimax quantizer under these conditions is the same quantizer described in Section V. We address ourselves to the following nonstationary signal source problem.

Given:

1) A discrete-time signal $\{S_j : j = 0,\cdots,J-1\}$ such that the random variables S_j are independent but are not necessarily identically distributed on the finite interval $[-a_s,a_s]$.

2) A mean weighted quantization error criterion defined by

$$\mathcal{E}(\bar{q},\bar{F}) = \sum_{j=0}^{J-1} \mathcal{E}(q(j),F(j)),$$

where $q(j) \in D$,

$$\bar{q} \in D^J = \prod_{j=0}^{J-1} D,$$

$F(j) \in A$, and

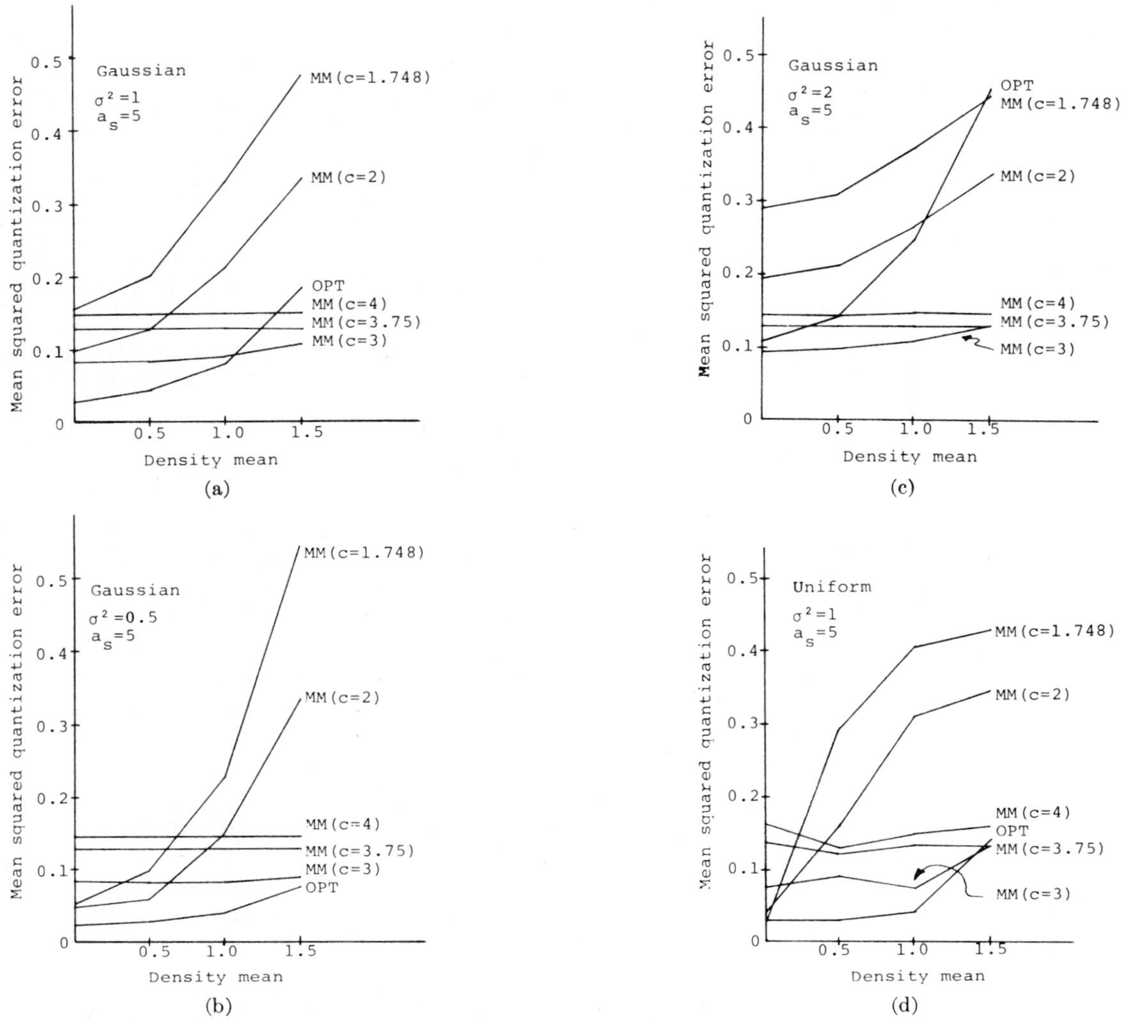

Fig. 1. (a) Mean squared quantization error versus the shift in the mean of a Gaussian signal of $\sigma^2 = 1$ and amplitude bounded to 5. (b) Mean squared quantization error versus the shift in the mean of a Gaussian signal of $\sigma^2 = 0.5$ and amplitude bounded to ± 5. (c) Mean squared quantization error versus the shift in the mean of a Gaussian signal of $\sigma^2 = 2$ and amplitude bounded to ± 5. (d) Mean squared quantization error versus the shift in the mean of a uniform signal of $\sigma^2 = 1$ and amplitude bounded to ± 5.

$$\bar{F} \in A^J = \prod_{j=0}^{J-1} A.$$

Problem 2: Find the discrete time $(N+2)$-level quantizer with nonsaturating input range $[-c,c]$, (i.e., the discrete-time quantizer parameter set $(s_2(j),\cdots,s_N(j);\ y_0(j),\cdots,y_{N+1}(j)) \sim q(j) \in D$), which minimizes the worst possible value for $\mathcal{E}(\bar{q},\bar{F})$ that results from some marginal probability distribution set $\{F(j)\}$ in A^J.

Proposition VI.1: The N-level uniform quantizer on $[-c,c]$ plus the assignment of $y_0{}^\circ(j) = \frac{1}{2}(-a_s - c)$ and $y_{N+1}{}^\circ(j) = \frac{1}{2}(a_s + c)$, $j = 0,\cdots,J-1$ is the solution to Problem 2.

Outline of Proof: The proof here is similar to the proof of Proposition V.1. Because of the independence of the random variables S_j each term $\mathcal{E}(q(j),F(j))$ can be maximized with respect to $F(j) \in A$ for arbitrary $q(j) \in D$ and then minimized over D. The result can be shown to be the same as for Proposition V.1. Therefore the minimax quantizer is stationary and the same as in Proposition V.1, likewise for the worst case marginal distributions.

VII. AN EXAMPLE OF THE MINIMAX QUANTIZER'S PERFORMANCE

We now compare the performance of the minimax quantizer with different input ranges with the performance of a second quantizer in terms of the mean squared quantization error for several signal probability distribution functions. This second quantizer tested was optimized or designed for the zero-mean normal distribution with $\sigma^2 = 1$ [6]. It produces essentially the same error (mean squared) when the signal range is 10 (i.e., $a_s = 5$) and the distribution is adjusted accordingly. This value of the error is 0.0345 for $N = 8$ and corresponds to a percentage error of 0.345 percent with respect to the signal range. Since $s_7 = -s_1 = 1.748$ for this second quantizer, this value was also used to define the input range of one of the minimax quantizers, i.e., $c = 1.748$. The other input ranges used for the minimax quantizer were $c = 2,3,3.75,4$. The distributions used correspond to the Gaussian and uniform density functions with variations in the mean and variance. Fig. 1(a)–(d) display the quantization error of the different quantizers for a particular input signal density shape as the density mean increases. The second quantizer curves are labeled OPT.

For this example, we see that in many instances the minimax quantizer does as well or better when the signal distributions differ from the second quantizer's optimum design distribution. It can also be observed that for the fixed signal range value the error of the minimax quantizer increases rapidly with increasing density mean for the lower values of c, but is fairly constant in error for the higher values of c. This appears to be a result of the squared error weighting terms on the two saturation regions which dominate the total error value for the lower values of c.

VIII. CONCLUSION

In this paper, we have proved that the N-level uniform quantizer on the quantizer nonsaturating input range $[-c,c]$, plus the assignment of $y_0{}^\circ = -\frac{1}{2}(a_s + c)$ and $y_{N+1}{}^\circ = \frac{1}{2}(a_s + c)$ to those input signal values falling in the saturation regions below and above the input range, respectively, is the minimax $(N + 2)$-level quantizer with nonsaturating input range $[-c,c]$ for stationary discrete-time signals defined by marginal probability distributions on the finite interval $[-a_s,a_s]$. It is noted that "physical" signals are amplitude limited, so that, in practical terms, this amplitude bound is not very restrictive. The performance criteria are mean weighted quantization error criteria where the weighting functions are only required to be symmetric in the quantization error and monotonically increasing for positive error values. This result is also shown to apply to the case of nonstationary discrete-time signal sources with independent samples defined on the finite interval $[-a_s,a_s]$.

The results were derived using a functional analysis framework. We guaranteed the existence of and characterized the maximizing marginal distributions for arbitrary quantizers. This led to a minimum norm problem where the minimization was over the quantizer parameters. This minimization gave the minimax result.

An example illustrates the performance of the minimax quantizer compared with a second quantizer for several stationary signal marginal probability distributions including the second quantizer's optimum distribution. For some distributions, the minimax quantizer's performance was comparable or better than the second quantizer's performance. The performance criterion for the example was the mean squared quantization error.

The main contribution of this paper is to provide a step towards the characterization of robust quantizers for various levels of signal information structure. In this paper, the input signal information structure assumed was probably at the lowest level. The design engineer is also provided with a bound on the minimax quantizer performance, since given the quantizer nonsaturating input range $[-c,c]$, the signal range $[-a_s,a_s]$, the quantization error weighting function, and N, one can compute the worst case quantization error.

As a suggestion for further study, one may place more structure on the admissible set of signal distributions and/or consider noise corrupted signals.

It is hoped that this paper will stimulate more interest in the area of quantization since it appears to be of increasing importance in digital communication system performance.

ACKNOWLEDGMENT

The authors are grateful to the reviewers for their helpful comments and Dr. R. M. Gray for his interest in this work.

REFERENCES

[1] R. B. Ash, *Real Analysis and Probability.* New York: Academic, 1972.
[2] J. D. Bruce, "On the optimum quantization of stationary signals," in *Conf. Rec., IEEE Int. Conv.*, 1964, pp. 118–124.
[3] T. J. Goblick, Jr., "Analog source digitization: A comparison of theory and practice," *IEEE Trans. Inform. Theory*, vol. IT-13, pp. 323–326, Apr. 1967.
[4] R. E. Larson, "Optimum quantization in dynamic systems," *IEEE Trans. Automat. Contr.*, vol. AC-12, pp. 162–168, Apr. 1967. R. E. Larson and E. Tse, "Reply," *IEEE Trans. Automat. Contr.*, vol. AC–12, pp. 274–276, Apr. 1972.
[5] D. G. Luenberger, *Optimization by Vector Space Methods.* New York: Wiley, 1969.
[6] J. Max, "Quantizing for minimum distortion," *IRE Trans. Inform. Theory*, vol. IT-6, pp. 7–12, Mar. 1960.
[7] B. Smith, "Instantaneous companding of quantized signals," *Bell Syst. Tech. J.*, pp. 653–709, 1958.
[8] D. J. Goodman and A. Gersho, "Theory of an adaptive quantizer," in *Proc. Symp. Adaptive Processes, Decision, and Control*, Dec. 1973.
[9] P. Commiskey, N. S. Jayant, and J. L. Flanagan, "Adaptive quantization in differential PCM coding of speech," *Bell Syst. Tech. J.*, pp. 1105–1118, Sept. 1973.
[10] N. S. Jayant, "Adaptive quantization with a one-word memory," *Bell Syst. Tech. J.*, pp. 1119–1144, Sept. 1973.
[11] J. M. Morris and V. D. VandeLinde, "Robust quantization of stationary signals," Dep. Elec. Eng., The Johns Hopkins Univ., Baltimore, Md., Tech. Rep. 73–18, Dec. 1973.

Reprinted from *IEEE Trans. Inform. Theory* **IT-28**(2):296–306 (1982)

Robust Memoryless Quantization for Minimum Signal Distortion

WILLIAM G. BATH, MEMBER, IEEE, AND V. DAVID VANDELINDE, SENIOR MEMBER, IEEE

Abstract—Robust quantizers are designed for situations where there is only an incomplete statistical description of the quantizer input. The goal of the design is to closely approximate quantizer inputs by quantizer outputs without using more than a specified number of quantization levels. The exact probability distribution of the input is unknown, but this distribution is known to belong to some set C. The primary set C considered is the set of all unimodal probability distributions which satisfy generalized moment constraint (e.g., mean-square value less than or equal to a constant). A quantizer is derived which minimizes over all quantizers the maximum distortion over all distributions in C. This robust quantizer guarantees a significantly lower worst case distortion than the classical Gaussian-optimal quantizer, while performing nearly as well as the Gaussian-optimal quantizer when the input is, in fact, Gaussian.

I. Introduction

ONE OF THE MOST common operations in a modern communication or control system is analog-to-digital conversion. An analog waveform is first sampled in time, then each of these input samples (or some function of present and past input samples) is approximated by one of a finite number of quantization levels. The set of quantization levels is selected using *a priori* knowledge about the signal to be quantized. Although equally spaced levels are most common, unequally spaced levels can be readily implemented and may produce a lower distortion for a given number of levels. This paper addresses robust quantization; that is, selecting the quantization levels when there is only a limited statistical description of the input signal.

When the input samples are independent, a simple zero-memory (present output depends only on present input) quantizer is usually used. The simplest and most widely used zero-memory quantizers are uniform (quantizers with equally spaced levels). A uniform quantizer is quite reasonable when one believes the input signal is equally likely to be in any portion of its range. If, on the other hand, the signal is more likely to be in some portions of its range than others, it seems prudent to space the quantization levels more finely where the signal is more likely to occur. Techniques for doing this have been developed. Given the probability distribution function of the (independent) input samples, Max [1] has derived a set of necessary conditions for a quantizer to minimize the mean-square quantization error. Optimum quantizers have been found (numerically) for Gaussian [1], gamma, and Laplace [2] distributions, and extensions have been made to quite general distortion measures [3]. However, in each case the probability distribution function of the input signal is needed to select the set of quantization levels.

Manuscript received November 17, 1980.

W. G. Bath is with the Johns Hopkins University Applied Physics Laboratory, Johns Hopkins Rd., Laurel, MD 20810.

V. D. VandeLinde is with the Department of Electrical Engineering, Johns Hopkins University, Baltimore, MD 21218.

In practice, the probability distributions of signals are seldom known precisely. As a result, nonparametric and robust techniques for estimation and detection have been developed which perform well and uniformly for large classes of noise probability distributions. Here, robust quantizers will be developed which perform well and uniformly for large classes of input signal distributions. In particular, a minimax philosophy is taken. If C is the set of all possible input probability distributions having the known characteristics, then the minimax quantizer q^* and the minimax distortion D^* are defined by the following properties.

1) No matter which of the probability distributions in C occurs, the minimax quantizer is guaranteed to produce a distortion no greater than D^*.
2) No other quantizer can guarantee a distortion less than D^*.

This can be formally stated as follows:

$$D^* = \min_{q \in Q_N} \max_{F \in C} D(q, F), \tag{1}$$

where $D(q, F)$ is the distortion of quantizer q given an input with cumulative probability distribution function (cpdf) F, and Q_N is the set of all N-level quantizers.

This problem has been solved [4] for the case where all that is known about the input samples is that their amplitudes lie in the range $[a, b]$, in which case the minimax quantizer is uniform across $[a, b]$, and the worst case distributions consist of atoms halfway between the quantization levels. This result reinforces the intuitive notion that when very little is known about the input signal, a uniform quantizer is a good idea. However, the thought that nature will consistently place the input signal exactly halfway between the quantization levels is indeed pessimistic. Further, there is usually more information about the input signal than its maximum and minimum values. As will be shown below, use of this information can significantly reduce quantizer distortion.

In Section IV, a minimax quantizer is derived for an independent input sequence whose common cpdf belongs to the unimodal generalized moment constrained set (the set of all unimodal cpdf's, with mode zero, which have a generalized moment, such as mean-square value, less than or equal to a constant). This is a reasonable set in practice because 1) many commonly occurring cpdf's are unimodal, and 2) some moment constraint which establishes the scale of the input signal is usually available either from *a priori* knowledge of the input signal or from an automatic gain control (AGC) loop. The minimax quantization levels and minimax distortion are found numerically by a constrained minimization in the Lagrange multiplier space (R^2). The worst case distortion of the minimax quantizer is shown to be significantly lower than the worst case distortion of quantizers designed for a Gaussian input cpdf. This same approach may be used to construct minimax quantizers for other cpdf sets C, including the set of all unimodal distributions and the set of all distributions with a known generalized moment.

II. Statement of Problem

The situation considered is as follows. A sequence of input data $\{X_k\}$ is passed through a zero-memory quantizer q to produce a sequence $\{Y_k\}$ of quantized outputs. The quantizer is then simply a function $q(x)$ mapping the continuous range of an input sample onto a finite number of quantization levels $(y_1, y_2, \cdots, y_N)$. An N-level quantizer can be specified using $2N-1$ numbers $(b_1, \cdots, b_{N-1}, y_1, \cdots, y_N)$:

$$q(x) = \begin{cases} y_1, & \text{if } x \in [-\infty, b_1), \\ y_i, & \text{if } x \in [b_{i-1}, b_i), \quad i = 2, \cdots, N-1, \\ y_N, & \text{if } x \in [b_{N-1}, \infty]. \end{cases} \tag{2}$$

It is assumed that the input sequence $\{X_k\}$ consists of independent random variables with cpdf's $\{F_k\}$, respectively. The sequence of cpdf's may not be known precisely. However, it is known that each F_k is a member of a set C. The set C represents the quantizer designer's *a priori* information about the input sequence.

Quantizer performance is measured by a distortion measure D, which is a function both of the quantizer q and the input cpdf F. The distortion measures considered here have the form

$$D(q, F) = \int e(x, q(x))\, dF(x), \tag{3}$$

where the function $e(x, y)$ will be termed the distortion weighting.

Bounded distortion weightings often most accurately represent the true penalties associated with quantization errors. For example, if the quantization errors in a speech transmission system are so large that the speech is unintelligible, increasing the errors imposes no additional penalty. In addition, if the distortion function is unbounded, particularly "bad" input cpdf's may have the same (possibly infinite) distortion for all quantizers, even though some quantizers are clearly superior to others.

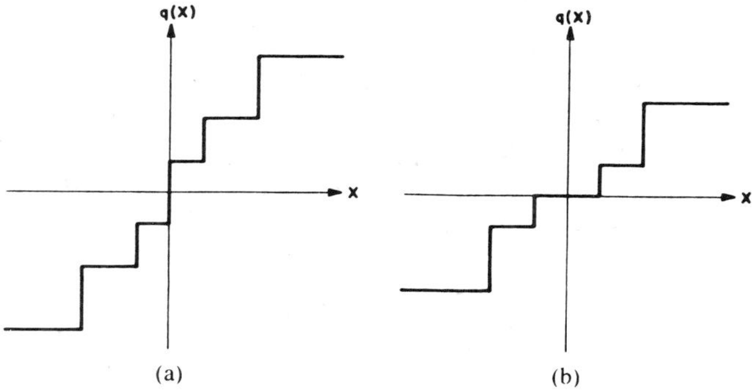

Fig. 1. Quantizer input–output characteristics.

Attention will be restricted to two types of bounded distortion weightings:

$$1)\ e_1(x, y) = \begin{cases} g(x-y), & x \in [b_0, b_N] \\ g(b_N - q(x)), & x > b_N \\ g(b_0 - q(x)), & x < b_0, \end{cases} \tag{4}$$

$$2)\ e_2(x, y) = \begin{cases} g(x-y), & g(x-y) \le L \\ L, & g(x-y) > L, \end{cases} \tag{5}$$

where $g(x)$ is a distortion function satisfying five conditions:

$$\begin{aligned} &1)\quad g(x) = g(-x), \\ &2)\quad g(0) = 0, \\ &3)\quad g(x) \text{ is continuous}, \\ &4)\quad g(x) \text{ is monotonic strictly increasing in } |x|, \\ &5)\quad g(x) \to \infty \text{ as } x \to \infty. \end{aligned} \tag{6}$$

These distortion weightings are bounded versions of the distortion weighting $e(x, y) = g(x-y)$. $e_1(x, y)$ would be used when very large input samples are felt to be of less importance or when the analog system preceding the quantizer has limited dynamic range. $e_2(x, y)$ would be used when no additional penalty accrues for errors larger than a fixed number (L). The design of quantizers for the two distortion weightings is quite similar. Therefore, results will be derived for $e_1(x, y)$. The modifications necessary for $e_2(x, y)$ are discussed in Section VII.

For these distortion measures, no choice of transition levels b_i can be better than

$$b_i = \frac{y_{i+1} + y_i}{2}, \qquad i = 1, \cdots, N-1. \tag{7}$$

This follows from Max's results [1]. Only quantizers satisfying (7) will be considered. A quantizer is then completely specified by the N quantization levels $(y_1, \cdots, y_N)$.

Most results will be derived for one-sided quantization problems (problems where the inputs X_k are nonnegative). Two-sided problems $(-\infty \le X_k \le \infty)$ will be reduced to one-sided problems by assuming symmetric two-sided quantizers. If a two-sided quantizer has an even number of levels, it is called a midriser (MR) quantizer (Fig. 1(a)). If the number of levels is odd, the quantizer is a midstep

(MS) quantizer (Fig. 1(b)). If F is symmetric, an M-level two-sided quantizer has the same distortion as an N-level one-sided quantizer where

1) $M = 2N$ (MR), or
2) $M = 2N - 1$ and y_1 is constrained to be zero (MS).

Since the sets C of possible cpdf's are defined symmetrically ($F_X \in C$ if and only if $F_{-X} \in C$), the worst case distortion of an M-level two-sided quantizer and an N-level one-sided quantizer are equal as well.

The set of possible input cpdf's will be the unimodal generalized moment constrained (UGMC) set C of all cpdf's on $[-\infty, \infty]$ satisfying

1) F is unimodal with mode zero,[1]
2) $\int \rho(s)\, dF(s) \le c$.

The constraint function, $\rho(s)$, satisfies the same properties (6) as the distortion function $g(s)$. Distributions putting mass at $\pm\infty$ are thus not excluded from being unimodal. Note that unimodality precludes atoms at any finite value other than zero.

The robust quantization problem can now be stated. Find a minimax quantizer q^* and a worst case probability distribution F^* satisfying

$$\inf_{q \in Q_N} \sup_{F \in C} D(q, F) = D(q^*, F^*), \tag{8}$$

where Q_N is the set of all N-level one-sided quantizers, C is the unimodal generalized moment class, and D is the distortion measure using weighting $e_1(x, y)$.

III. Outline of Solution

The minimax optimization problem (8) consists of an infinite dimensional inner maximization over F and a finite dimensional outer minimization over q. A functional analysis approach to the inner maximization will be taken as follows. Consider the Banach space NBV$[0, \infty]$ of normalized[2] functions of bounded variation (with the total variation norm). The set of admissible distributions C is a subset of NBV$[0, \infty]$, and $D(q, F)$ is a linear functional on NBV$[0, \infty]$. It is easily shown that C is weak* compact and that $D(q, F)$ is weak* continuous in F. This implies that, for any $q \in Q_N$, a worst case distortion $D^*(q)$ and a worst case distribution $F^* \in C$ exist such that

$$D^*(q) = \max_{F \in C} D(q, F) = D(q, F^*).$$

This maximization can be converted into a two-dimensional constrained minimization by solving the dual minimization problem and exploiting the unimodality of the admissible distributions. The minimization is performed with respect to Lagrange multipliers λ_1 and λ_2 corresponding to the unit mass and generalized moment constraints, respectively.

[1]A cpdf F is defined to be unimodal if it is concave on $(0, \infty)$ and convex on $(-\infty, 0)$.

[2]A distribution F is normalized if it is right continuous and if $F(-\infty) = 0$.

The outer minimization over q is N-dimensional. Q_N is a compact subset of N-dimensional Euclidean space and $D^*(q)$ is lower semicontinuous. Thus a minimax quantizer q^* exists solving (8). This outer minimization can be eliminated by imposing necessary conditions on the Lagrange multipliers λ_1 and λ_2 sufficiently stringent to ensure each pair of multipliers uniquely determines a quantizer. Thus the entire problem (8) is solved by a two-dimensional constrained minimization in the Lagrange multiplier plane.

IV. Construction of the Minimax Quantizer

A procedure for constructing the minimax quantizer q^*, given a distortion function $g(x)$, a constraint function $\rho(x)$, a constraint value c, and the number N of quantization levels, will be derived. The basis of this derivation is the Lagrange duality theorem [5]. Because $D(q, F)$ is a positive linear functional when $F \in C$, $D^*(q)$ is unchanged if C is expanded to include substochastic distributions. Thus an equivalent set of possible cpdf's is

$$C' = \{F \in \text{NBV}[0, \infty]:$$

1) F is nondecreasing,
2) F is unimodal with mode zero,
3) $\int dF \le 1$,
4) $\int \rho\, dF \le c\}$.

If $\bar{C} = \{F \in \text{NBV}[0, \infty]$: F satisfies 1) and 2)$\}$, then $\bar{C}$ is convex (sums of monotonic concave functions are monotonic and concave), and the set C' can be written

$$C' = \{F \in \bar{C}: G(F) \le 0\},$$

where G: NBV$[0, \infty] \to R^2$ and

$$G = \begin{bmatrix} \int dF - 1 \\ \int \rho\, dF - c \end{bmatrix}$$

is a convex functional. A version of the Lagrange duality theorem [5] can then be stated as follows.

Theorem 1: Let $D(q, F)$ be a real-valued convex functional defined on a convex subset $\bar{C}$ of the vector space NBV$[0, \infty]$, and let G be a convex mapping of NBV$[0, \infty]$ into R^2. Suppose an F_1 exists such that $G(F_1) < 0$ and $D^*(q) = \sup\{D(q, F): F \in \bar{C}, G(F) \le 0\}$ is finite. Then

$$\sup_{F \in C'} D(q, F) = \min_{\lambda_1, \lambda_2 \ge 0} \sup_{F \in \bar{C}} \Big[D(q, F) - \lambda_1 \cdot \Big(\int dF - 1 \Big) - \lambda_2 \Big(\int \rho\, dF - c \Big) \Big], \tag{9}$$

and the minimum on the right is achieved by some $\lambda_1^*, \lambda_2^* \ge 0$. If the supremum on the left is achieved by some $F^* \in \bar{C}$, then

$$\lambda_1^* \Big(\int dF^* - 1 \Big) = \lambda_2^* \Big(\int \rho\, dF^* - c \Big) = 0$$

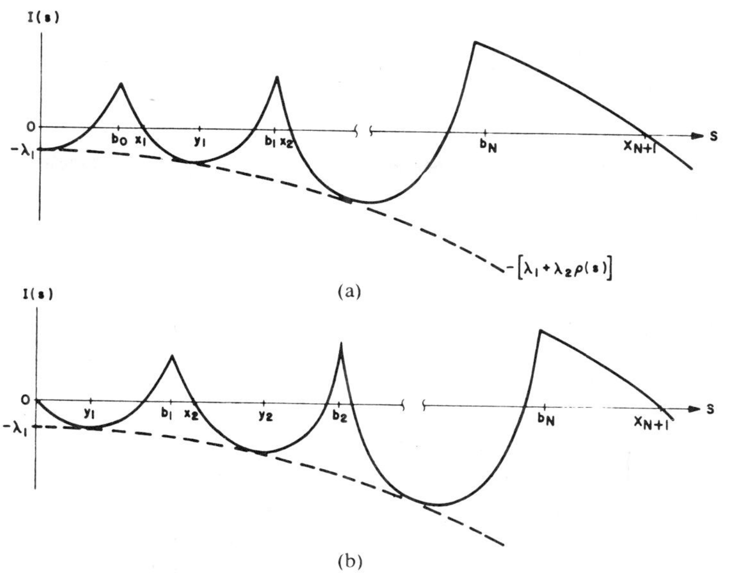

Fig. 2. Form of function $I(s)$ used to construct minimax quantizer. (a) MS quantizer. (b) MR quantizer.

and F^* maximizes $D(q, F) - \lambda_1^*(\int dF - 1) - \lambda_2^*(\int \rho \, dF - c)$ over all $F \in \bar{C}$.

There is such a vector $F_1 \in \bar{C}$ ($F_1(x) = 0.5$, $x \geq 0$), and $D^*(q)$ is finite since the distortion weighting function is bounded. Thus the worst case distortion can be written

$$D^*(q) = \min_{\lambda_1, \lambda_2 \geq 0} [(\lambda_1 + \lambda_2 c) + B(\lambda_1, \lambda_2)],$$

where

$$B(\lambda_1, \lambda_2) = \max_{F \in \bar{C}} \int e(s, q(s)) - \lambda_1 - \lambda_2 \rho(s) \, dF(s).$$

The minimization can certainly be restricted to values of λ_1, λ_2 for which $B(\lambda_1, \lambda_2)$ is finite. Define

$$H(x) = \int_0^x I(s) \, ds,$$

where

$$I(s) = e(s, q(s)) - \lambda_1 - \lambda_2 \rho(s);$$

then $B(\lambda_1, \lambda_2)$ has the following properties:

1) $B(\lambda_1, \lambda_2) < \infty$ if and only if $H(x) \leq 0$ for $x \geq 0$,
2) $B(\lambda_1, \lambda_2) < \infty$ implies $B(\lambda_1, \lambda_2) = 0$.

These can be demonstrated as follows. If $H(x_0) > 0$ for some x_0 in $[0, V]$, then to find a sequence $\{F_n\}$ in C which makes B infinite, pick F_n uniform over $[0, x_0]$ with total mass n and let $n \to \infty$. Conversely, if $H(x) \leq 0$ on $[0, V]$, then the unimodality of F ensures $B < \infty$. Finally, since $F = 0$ on $[0, V]$ is in $\bar{C}$, it follows $H(x) \leq 0$ implies $B = 0$.

The arbitrary quantizer problem can then be written

$$D^*(q) = \min_{\substack{\lambda_1, \lambda_2 \geq 0 \\ H(x) \leq 0}} \lambda_1 + \lambda_2 c. \tag{10}$$

Given a quantizer q, this is a two-dimensional minimization problem subject to an inequality constraint. The minimax quantizer q^* is found by solving

$$D^* = \min_{q \in Q_N} \min_{\substack{\lambda_1, \lambda_2 \geq 0 \\ H(x) \leq 0}} \lambda_1 + \lambda_2 c. \tag{11}$$

In theory, these problems can be solved directly by numerical minimization algorithms. However, more efficient construction procedures can be derived by imposing mild regularity conditions on $g(s)$ and $\rho(s)$.

1) For any $b \in (0, 1)$ and any $a \geq 0$, $g(s - a) - b\rho(s)$ has no maxima.
2) For any $a \geq 0$, there exists a $b \in (0, 1)$ and an $s_0 > 0$ so that $g(s - a) - b\rho(s)$ is monotonic in $|s - s_0|$.

These conditions will be satisfied by most distortion and constraint functions. For example, it is easily verified that they hold for $g(s) = |s|^p$, $\rho(s) = |s|^q$, and $q \geq p \geq 1$.

The above conditions on $g(s)$ and $\rho(s)$ have been chosen to constrain the function $I(s)$ to have the form shown in Fig. 2. Consider the interval $[b_{i-1}, b_i)$. By assumption, $I(s)$ is decreasing at b_{i-1} and achieves at most one relative minimum. Since $I(s)$ is continuous, it has at most two roots in $[b_{i-1}, b_i)$. In the interval $[b_N, \infty)$, $I(s)$ is monotonic so that it has at most one root. Let x_i be the smallest root in each $[b_{i-1}, b_i)$ (if it exists), $i = 1, \cdots, N$, and let x_{N+1} be the root in $[b_N, \infty)$ (if it exists). The x_i are all the downward crossings of zero. Thus $H(x) \leq 0$ for all $x \geq 0$ if and only if $H(x_i) \leq 0$, $i = 1, \cdots, N+1$ for those x_i which exist.

The worst case distortion for a given quantizer $(y_1, \cdots, y_N)$ can now be easily found as follows.

1) Select a pair of multipliers $\lambda_1, \lambda_2 \geq 0$.
2) Compute x_i and $H(x_i)$, $i = 1, \cdots, N+1$.
3) Check that $H(x_i) \leq 0$, $i = 1, \cdots, N+1$ for all those x_i which exist.

4) Repeating steps 1)–3), minimize $\lambda_1 + \lambda_2 c$ over all $\lambda_1, \lambda_2 \geq 0$ for which step 3) is satisfied. The result is the worst case distortion $D^*(q)$.

The required computation grows linearly with the number of quantization levels N.

The minimax quantizer can be constructed by the following procedure.

1) Select a pair of multipliers $\lambda_1, \lambda_2 \geq 0$.
2) For each i ($i = 1,\cdots,N$) choose $y_i \geq x_i$ and b_{i-1} to simultaneously satisfy

$$I(x_i) = 0$$
$$H(x_i) = 0$$
$$b_{i-1} = (y_i + y_{i-1})/2.$$

(In the MS case, y_0 is taken to be zero. In the MR case, b_0 is taken to be zero.)

3) Check that $H(x_{N+1}) \leq 0$.
4) Repeating steps 1)–3), minimize $\lambda_1 + \lambda_2 c$ over all $\lambda_1, \lambda_2 \geq 0$ for which step 3) is satisfied. The result is the worst case distortion D^*. The quantizer $(y_1,\cdots,y_N)$ which achieves the minimum is the minimax quantizer.

Note that the required computation for this procedure also grows linearly with N (as opposed to exponential growth for direct minimization of (11)).

The construction procedure is based upon the following lemma. The conditions on the multipliers (λ_1, λ_2) imposed by the lemma are both necessary for the minimax and sufficiently strict to ensure that the multipliers correspond to exactly one quantizer.

Construction Lemma: A necessary condition for a quantizer $(y_1,\cdots,y_N)$ and a pair of multipliers λ_1, λ_2 to be minimax is that x_i exists and $H(x_i) = 0$ for $i = 1,2,\cdots,N + 1$.

The proof of this lemma is by contradiction [6] and will only be sketched here. Suppose $y_1,\cdots,y_N$ is a minimax quantizer with multipliers λ_1, λ_2. If, for some i, $1 \leq i \leq N$, x_i does not exist or $H(x_i) < 0$, then a small increase in y_i and a small decrease in $y_1,\cdots,y_N$ produce a new quantizer which satisfies $H(x) < \epsilon$ for some $\epsilon > 0$ and for all $x \geq 0$. This new quantizer will have a worst case distortion strictly smaller than that of the original quantizer because λ_1 may now be decreased by ϵ without violating the condition $H(x) \leq 0$, $x \geq 0$. This is a contradiction. A similar argument may be used for $i = N + 1$. The construction theorem can now be presented.

Theorem 2: The construction procedure produces a minimax quantizer.

Proof: The construction procedure solves the problem

$$\min_{\substack{\lambda_1, \lambda_2 \geq 0 \\ H(x_i)=0,\, i=1,\cdots,N \\ H(x_{N+1}) \leq 0}} \lambda_1 + \lambda_2 c.$$

This is the same problem as

$$\min_{\substack{\lambda_1, \lambda_2 \geq 0 \\ H(x) \leq 0}} \lambda_1 + \lambda_2 c \tag{12}$$

since $\{H(x_i) = 0,\ i = 1,\cdots,N;\ H(x_{N+1}) \leq 0\}$ is a necessary condition for optimality (construction lemma) and a sufficient condition for $H(x) \leq 0$ for all $x \geq 0$. However, (12) has the solution (11) which is the minimax quantizer.

V. Characterizing the Worst Case Probability Distributions

At least one worst case distribution $F^*(q)$ exists for any quantizer $q \in Q_N$. This section characterizes any worst case distribution F^* for a minimax quantizer. For large N, finding all the worst case distributions is a tedious numerical problem. However, for any N, a closed-form expression for at least one F^* can be found easily.

Suppose F^* is a worst case distribution for the minimax quantizer problem (8) with multipliers $\lambda_1, \lambda_2 > 0$. Decompose F^* into an atom of intensity p at zero and a density function f on $(0, \infty)$. Then f and p satisfy

1) $p = 0$ in the MS case,
2) $f(x)$ is constant on (x_i, x_{i+1}), $i = i_0,\cdots,N$,
3) $f(x) = 0$ for $x > x_{N+1}$,

where $i_0 = 1$ for MR quantizers, $i_0 = 0$ for MS quantizers, and the x_i are as defined in Section IV.

This may be seen as follows. Theorem 1 implies that

$$\int I(s)\, dF^*(s) = 0. \tag{13}$$

Define

$$S_i = \int_{x_i}^{x_{i+1}} I(s) f(s)\, ds, \qquad i = i_0,\cdots,N,$$
$$S_{N+1} = \int_{x_{N+1}}^{\infty} I(s) f(s)\, ds.$$

The construction procedure ensures that for each i, $i = i_0,\cdots,N$, a z_i exists with $I(s) < 0$ on (x_i, z_i) and $I(s) > 0$ on (z_i, x_{i+1}). This implies that F^* satisfies $[e(0, q(0)) - \lambda_1]p = 0$ and that $S_i = 0$ for $i = 1,\cdots,N + 1$. (To show this, break up each integral S_i into two integrals over (x_i, z_i) and (z_i, x_{i+1}), respectively, and then invoke the unimodality of F.) However, $S_i = 0$ implies $f(s)$ is constant over (x_i, x_{i+1}), $i = i_0,\cdots,N$.

Each worst case distribution F^* then corresponds to a vector $(f_0,\cdots,f_N)$ in R^{N+1} where

$$\begin{aligned} f(x) &= f_i \text{ on } (x_i, x_{i+1}), && i = 1,\cdots,N \\ f(x) &= f_0 \text{ on } (0, x_1), && \text{MS case} \\ p &= f_0, && \text{MR case.} \end{aligned} \tag{14}$$

Necessary and sufficient conditions for a vector $(f_0,\cdots,f_N)$

to be worst case are

$$\sum_{i=0}^{N} \alpha_i f_i = 1 \qquad \sum_{i=0}^{N} \beta_i f_i = c \tag{15}$$

$$\sum_{i=0}^{N} \gamma_i f_i = \lambda_1 + \lambda_2 c \tag{16}$$

$$f_i \geq 0, \qquad i = 0,\cdots,N \tag{17}$$

$$f_{i+1} \leq f_i, \qquad i = i_0,\cdots,N-1, \tag{18}$$

where

$$\alpha_i = x_{i+1} - x_i, \qquad i = i_0,\cdots,N$$

$$\beta_i = \int_{x_i}^{x_{i+1}} \rho(s)\,ds, \qquad i = i_0,\cdots,N$$

$$\gamma_i = \int_{x_i}^{x_{i+1}} e(s,q(s))\,ds, \qquad i = i_0,\cdots,N,$$

where in the MR case $\alpha_0 = 1$, $\beta_0 = 0$, and $\gamma_0 = e(0, q(0))$. Note that (16) is redundant when $\lambda_2 > 0$. This follows from the fact that

$$\gamma_i = \lambda_1 \alpha_i + \lambda_2 \beta_i, \qquad i = 0,\cdots,N,$$

since

$$\int_{x_i}^{x_{i+1}} e(s,q(s)) - \lambda_1 - \lambda_2 \rho(s)\,ds = 0, \qquad i = i_0,\cdots,N,$$

and

$$e(0,q(0)) - \lambda_1 = 0, \qquad \text{MR case},$$

by the construction lemma. These results are summarized as follows.

Theorem 3: Given a minimax quantizer with $\lambda_1, \lambda_2 > 0$, all worst case distributions have the piecewise uniform form (14) and are describable by a vector $(f_0,\cdots,f_N) \in R^{N+1}$. Any unimodal distribution of this form which satisfies the generalized moment constraint with equality is worst case. The set of vectors corresponding to worst case distributions is an $N-1$-dimensional convex polygon in R^{N+1}.

For arbitrary N, finding the set of all worst case distributions reduces to finding the feasible solutions of the $N+1$-dimensional linear programming problem defined by (15)–(18). However, a simpler technique is available to find some of the worst case distributions. Pick F_1 and F_2 to be unimodal distribution functions of the form (4) such that

$$\int_0^{x_{N+1}} \rho(s)\,dF_1(s) = c_1 > c$$

and

$$\int_0^{x_{N+1}} \rho(s)\,dF_2(s) = c_2 < c.$$

Then the distribution

$$F = \left(\frac{c - c_2}{c_1 - c_2}\right) F_1 + \left(\frac{c_1 - c}{c_1 - c_2}\right) F_2 \tag{19}$$

is worst case.

For example, let the uniform density over $(0, x_{N+1})$ be $F_1 = \text{unif}(0, x_{N+1})$ and let $F_2 = 1_{\{(0,\infty)\}}$ be an atom at zero (MR) or $F_2 = \text{unif}(0, x_1)$ (MS). Then

$$F = \begin{cases} \dfrac{c x_{N+1}}{R(x_{N+1})}\,\text{unif}(0, x_{N+1}) + \dfrac{R(x_{N+1}) - c x_{N+1}}{R(x_{N+1})}\,1_{\{(0,\infty)\}}, & \text{MR}, \\[2ex] \dfrac{[c x_1 - R(x_1)]x_{N+1}}{x_1 R(x_{N+1}) - x_{N+1} R(x_1)}\,\text{unif}(0, x_{N+1}) + \dfrac{[R(x_{N+1}) - c x_{N+1}]x_1}{x_1 R(x_{N+1}) - x_{N+1} R(x_1)}\,\text{unif}(0, x_1), & \text{MS} \end{cases} \tag{20}$$

is a worst case distribution for any given N, where

$$R(x) = \int_0^x \rho(s)\,ds.$$

Note that $R(x_{N+1}) \geq c x_{N+1}$ and $R(x_1) < x_1 c$ are guaranteed by the existence of a minimax solution.

The form of the worst case distributions has a geometric interpretation. Consider the vector space of all concave normalized functions of bounded variation on $[0, \infty]$. The positive cone in this space is the set of all unimodal distribution functions. The quantizer distortion is a linear functional on this vector space. The unit mass and generalized moment constraints are linear as well. Thus finding the worst case distortion is a linear programming problem. The worst case distributions correspond then to vertices of an infinite-dimensional polygon. If one takes as a basis for the unimodal distributions the set of step functions $\{F_\alpha\}$ of the form

$$F_\alpha = \begin{cases} x/\alpha, & 0 \leq x \leq \alpha \\ 1, & x > \alpha, \end{cases}$$

then the vertices correspond to linear combinations of a countable number of F_α. The worst case distributions derived above are of this form with the α's assuming the values $x_1,\cdots,x_{N+1}$.

VI. Comparison of Minimax and Classical Quantizers

Minimax quantizers have been constructed for several values of M between 2 and 32 using quadratic distortion and constraint functions ($g(x) = \rho(x) = x^2$), the bounded-distortion weighting $e_1(x, y)$, and a loading factor (defined as $b_N/c^{0.5}$) of four. The quantization levels are listed in Table I. The optimum Gaussian quantization levels derived by Max [1] (for unbounded quadratic distortion) are listed for comparison. The two types of quantizers appear roughly similar for small N. However, the minimax quantizer has a larger dynamic range with levels more closely spaced near zero and with a higher saturation value. For example, the $M = 16$ and $M = 32$ minimax quantizers each have a first quantization level only half as large as the

TABLE I
COMPARISON OF MINIMAX AND GAUSSIAN-OPTIMAL QUANTIZATION LEVELS ($b_N = 4$, $c = 1$, $g(x) = \rho(x) = x^2$)

Number of Levels N	Minimax	Gaussian Optimal
2	.354	.7980
3	0.000	0.000
	1.503	1.224
4	.346	.4979
	1.975	1.494
7	0.000	0.000
	.593	.6508
	1.464	1.302
	2.912	1.952
8	.1750	.2930
	.8196	.8790
	1.693	1.465
	3.066	2.051
15	0.000	0.000
	.2602	.3534
	.5710	.7068
	.9273	1.060
	1.360	1.414
	1.906	1.767
	2.612	2.120
	3.537	2.474
16	.0808	.1676
	.3647	.5028
	.6723	.8380
	1.028	1.173
	1.459	1.508
	1.997	1.844
	2.683	2.179
	3.567	2.514
31	0.000	0.000
	.1197	0.1360
	.2567	.2729
	.3983	.4115
	.5474	.5528
	.7073	.6979
	.8809	.8481
	1.072	1.005
	1.284	1.170
	1.522	1.347
	1.789	1.540
	2.092	1.753
	2.437	1.997
	2.829	2.289
	3.278	2.665
	3.792	3.239
32	.03720	.06590
	.1665	.1981
	.2986	.3314
	.4360	.4668
	.5817	.6050
	.7385	.7473
	.9096	.8947
	1.098	1.049
	1.308	1.212
	1.544	1.387
	1.809	1.577
	2.110	1.788
	2.452	2.029
	2.8411	2.319
	3.286	2.692
	3.795	3.263

corresponding Gaussian-optimal quantizer, while for $7 \leq M \leq 16$ the minimax quantizers each saturate $c^{0.5}$ higher than the corresponding Gaussian-optimal quantizer. The large dynamic range of the minimax quantizer is a natural result of the UGMC set containing both short- and long-tailed distributions.

The most popular measure of quantizer performance is the signal-to-quantization-noise ratio (SNR). This is the ratio of a moment of the input signal to a moment of the quantization error (treated as noise). When the cpdf of the input signal is only partially specified, there is a range of SNR values corresponding to the possible input cpdf's. The worst case SNR is defined as the minimum SNR for any distribution in the UGMC class which satisfies the moment constraint with equality. When $\rho(x) = g(x) = x^p$, the worst case SNR is given by

$$\text{SNR}_c^*(\text{dB}) = \frac{20}{p} \log_{10} \frac{c}{D^*}.$$

Figs. 3 and 4 compare the worst case SNR's of three quantizers: the minimax quantizer as listed in Table I, the Gaussian-optimal quantizer, and a uniform quantizer (with step size optimized for a Gaussian input). These Gaussian-optimal quantizers are designed for a Gaussian distribution with variance c since the linear nature of the distortion functional makes the c worst case value in $[0, c]$. Quadratic distortion and constraint functions, the bounded-distortion weighting $e_1(x, y)$ and $b_N/c^{=0.5} = 4$, are used. Results are shown for two-sided MR and MS quantizers as a function of the number of bits in a quantizer output word. (One level is not used by the MS quantizer.)

The dependence of worst case SNR on the number of bits is roughly linear for the minimax quantizer and increases about 7 dB/bit (i.e., D^* is proportional to $N^{2.2}$) for five bits or less. The worst case SNR for Gaussian-optimal quantizer is initially close to the minimax SNR, but the two curves separate as the number of bits increase. The 5-bit minimax MR quantizer has an 8.4-dB higher worst case SNR than the Gaussian-optimal quantizer (8.0 dB higher for MS). The uniform quantizer's worst case SNR is lower than the Gaussian-optimal quantizer's. Thus the minimax quantizer provides a significantly larger guaranteed SNR than either of the classical quantizers.

Figs. 5 and 6 compare the SNR's of the three quantizers for a Gaussian input signal. The three curves are quite

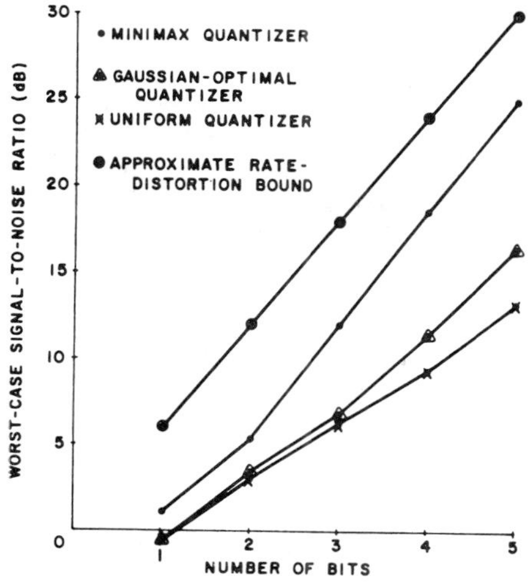

Fig. 3. Comparison of MR quantizers' worst case SNR's across UGMC class of input distributions. $g(x) = \rho(x) = x^2$, $b_N/\sqrt{c} = 4$.

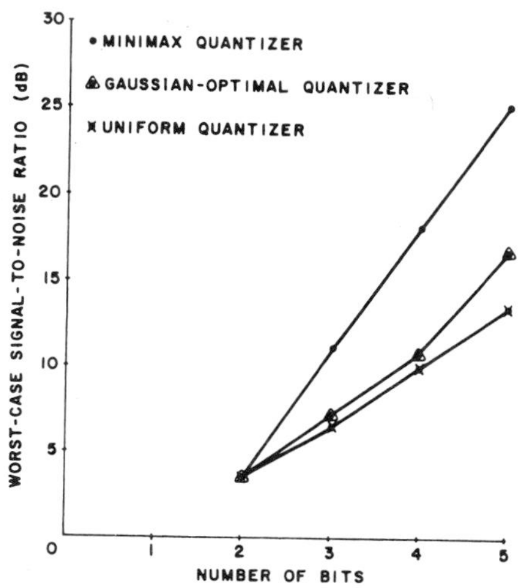

Fig. 4. Comparison of MS quantizers' worst case SNR's across UGMC class of input distributions. $g(x) = \rho(x) = x^2$, $b_N/\sqrt{c} = 4$.

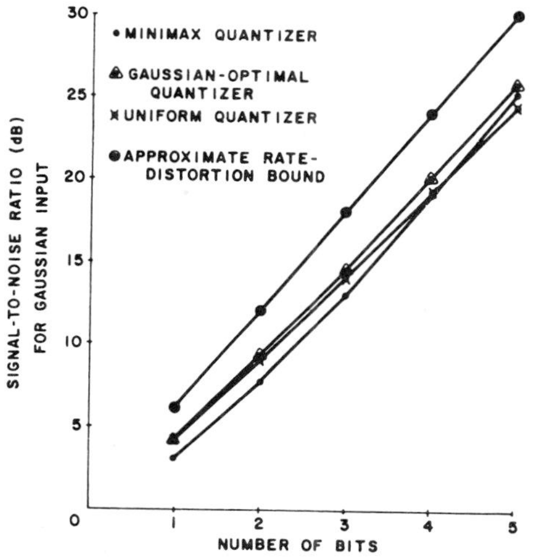

Fig. 5. Comparison of MR quantizers' SNR's for Gaussian input distribution. $g(x) = \rho(x) = x^2$, $b_N/\sqrt{c} = 4$.

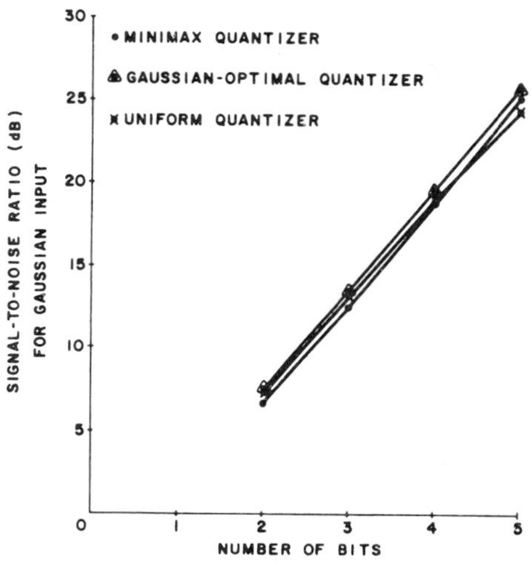

Fig. 6. Comparison of MS quantizers' SNR's for Gaussian input distribution. $g(x) = \rho(x) = x^2$, $b_N/\sqrt{c} = 4$.

similar with a maximum 2-dB difference between the Gaussian-optimal and minimax quantizers. This difference declines, as the number of bits increases, to a 1-dB difference at five bits. (Note that Figs. 5 and 6 apply to both an unbounded quadratic distortion weighting $e(x, y) = (x - y)^2$, and the bounded quadratic distortion weighting $e_1(x, y)$ since the two are within 0.1 dB over the range shown.) Thus only a small penalty in Gaussian performance is paid for achieving robust performance across the entire unimodal generalized moment class.

It is well-known [7] that the Gaussian distribution, F_G, solves the maximin problem

$$D^*_{\mathrm{RD}} = \max_{F \in A} \inf_{b \in \mathrm{BQ}(M)} D_{\mathrm{BQ}}(b, F),$$

where A is the set of continuous cpdf's with zero mean and variance at most c; $\mathrm{BQ}(M)$ is the set of all block quantizers with rate $\log_2 M$ bits/sample and $D_{\mathrm{BQ}}(b, F)$ is the mean-square error between quantizer input and the reconstructed output (averaged over all inputs in the block). The solution is

$$D^*_{\mathrm{RD}} = \inf_{b \in \mathrm{BQ}(M)} D_{\mathrm{BQ}}(b, F_G) = c/M^2.$$

It has also been shown [8] that F_G is worst case for the minimax block quantization problem. That is,

$$D^*_{\mathrm{RD}} = \inf_{b \in \mathrm{BQ}(M)} \sup_{F \in A} D_{\mathrm{BQ}}(b, F) = c/M^2.$$

The infimum is achieved by letting the block length become infinite. However, no procedure for constructing block quantizers arbitrarily close to minimax is known.

Because F_G is unimodal, c/D^*_{RD} is a tight approximate upper bound on the worst case SNR obtainable when block quantizing continuous UGMC distributions. D^*_{RD} is somewhat less than the exact upper bound because D_{BQ} is unbounded quadratic error. The minimax curves in Figs. 3 and 4 represent the largest worst case SNR possible with zero-memory quantization. Since these curves also apply when the UGMC distributions are required to be continu-

TABLE II
COMPARISON OF MINIMAX, GAUSSIAN-OPTIMAL, AND UNIFORM QUANTIZER ENTROPIES FOR GAUSSIAN INPUT

	ENTROPY		
M	Minimax	Gaussian-Optimal	Uniform
2	1.000	1.000	1.000
3	1.446	1.536	1.536
4	1.805	1.911	1.904
7	2.430	2.647	2.598
8	2.616	2.825	2.761
15	3.502	3.677	3.524
16	3.597	3.765	3.602
31	4.557	4.685	4.410
32	4.616	4.730	4.449

ous (Section VII-D) the difference between the approximate upper bound and the minimax curves indicates the improvement theoretically possible by adding memory to the quantizer. This improvement is roughly 5–6 dB (as opposed to 4 dB when the input is known to be Gaussian). The improvement will increase as b_N becomes large. (It is easily shown that, for $\rho(x) = g(x)$, as $b_N \to \infty$, $\lambda_2 \to 1$, so that $\mathrm{SNR}_c^*(\mathrm{dB}) \to 0$ dB for any fixed M.)

One technique for introducing memory into a quantizer is by entropy encoding the quantizer output. This increases considerably the complexity of the quantization system and often requires buffering to produce a constant bit rate output. When the input is Gaussian and the quantizer is Gaussian-optimal, entropy encoding produces about a 1.5-dB SNR improvement for (MR) 5 bits/sample (average). Table II compares the output entropy of the minimax, Gaussian-optimal, and uniform quantizers for a Gaussian input. Since the minimax quantizer has less entropy than the Gaussian-optimal quantizer, entropy encoding will result in a larger gain for the minimax quantizer (for example, 2.2 dB for (MR) 5 bits/sample giving the minimax and Gaussian-optimal quantizers nearly identical performance for a Gaussian input). In order to use entropy encoding when the input signal distribution is only partially specified, either a code must be found which has a guaranteed upper bound on its average rate for any possible input distribution or some form of adaptive entropy encoding must be used.

VII. APPLICATIONS TO SIMILAR PROBLEMS

A. Bounded Difference Distortion Measures

For some applications, the distortion weighting

$$e_2(x, y) = \begin{cases} g(x-y), & g(x-y) \le L \\ L, & g(x-y) > L \end{cases}$$

may be more realistic than the distortion weighting $e_1(x, y)$. The weighting $e_2(x, y)$ is a difference distortion measure and is symmetric in its arguments. Thus a positive difference between quantizer input and output always receives the same weight as an equal negative difference (this is not necessarily true for $e_1(x, y)$). In particular, $e_2(x, y)$ may be applied to DPCM systems where the difference between quantizer input and output equals the difference between DPCM system input and output.

As noted in Section II, the minimax problem for $e_2(x, y)$ is similar to that for $e_1(x, y)$. The distortion functional $D(q, F)$ is still bounded and linear in F so that Theorem 1 holds. The minimax quantizer can be computed using the same construction procedure where now x_{N+1} is computed by

$$x_{N+1} = \rho_+^{-1}\left[\frac{g(L) - \lambda_1}{\lambda_2}\right],$$

where ρ_+^{-1} is the inverse of ρ restricted to the positive axis. To see this, define a quantizer-dependent distortion weighting

$$e_3(x, y) = \begin{cases} g(x-y), & x \in (0, b_{N-1}) \\ \min\left[L, g(x-y)\right], & x \ge b_{N-1}. \end{cases} \tag{21}$$

It is clear that $e_3(x, y) = e_2(x, y)$ if

$$y_i - y_{i-1} \le 2L, \qquad i = 2, \cdots, N,$$

but quantizers satisfying this inequality have uniformly no higher distortion than quantizers in all of Q_N. If $(y_1, \cdots, y_N)$ has $y_i - y_{i-1} = 2L + \Delta$, $\Delta > 0$, then $(y_1, \cdots, y_{i-1}, y_i - \Delta, y_{i+1} - \Delta, \cdots, y_N - \Delta)$ has uniformly no higher distortion because of unimodality. Thus a quantizer is minimax for $e_3(x, y)$, if and only if it is minimax for $e_2(x, y)$. Defining $b_N = y_N + L$ gives $e_1(x, y) = e_2(x, y) = e_3(x, y)$ for $x < b_N$. The construction lemma then holds with minor modifications.

B. Eliminating the Generalized Moment Constraint

If the generalized moment constraint is eliminated, the set C of possible cpdf's consists of all unimodal cpdf's on $[0, \infty]$ with mode zero. This is a very large class of distributions and, as a result, will have a substantially larger minimax distortion than the UGMC class. However, construction of a minimax quantizer is much simpler. The minimax quantizer is unique and nearly uniform (the first level is not the same as for a uniform quantizer). For the special case of a power distortion function ($g(x) = x^p$, $p \ge 1$) the minimax quantizer can be found in closed form.

With the moment constraint eliminated, the dual problem is one-dimensional. An approach similar to that taken is Section IV yields a closed-form solution. In the MR case, the minimax quantizer is

$$y_i = \frac{b_N - (N-1)d}{2} + (i-1)d, \qquad i = 1, \cdots, N, \tag{22}$$

with minimax distortion

$$D^* = \frac{2G(d/2)}{d}, \tag{23}$$

and d satisfies the equation

$$\frac{2G(d/2)}{d} = g\left[\frac{b_N - (N-1)d}{2}\right], \tag{24}$$

where

$$G(x) = \int_0^x g(s)\, ds.$$

For the special case, $g(s) = |s|^p$, $p \geq 1$, (22)–(24) reduce to

$$y_i = \frac{b_N}{\left(\frac{1}{p+1}\right)^{1/p} + (N-1)}\left[0.5\left(\frac{1}{p+1}\right)^{1/p} + (i-1)\right],$$

$$i = 1,\cdots,N$$

$$D^* = \frac{1}{2^p(p+1)}\left[\frac{b_N}{\left(\frac{1}{p+1}\right)^{1/p} + (N-1)}\right]^p$$

$$d = \frac{b_N}{\left(\frac{1}{1+p}\right)^{1/p} + (N-1)}.$$

For the MS case,

$$y_i = id, \qquad i = 1,\cdots,N$$

with minimax distortion still given by (23), but now d satisfies the equation

$$\frac{2G(d/2)}{d} = g(b_N - (N-1)d).$$

For the special case $g(x) = x^p$, $p > 1$, the minimax MS quantizer distortion and step size are

$$D^* = \frac{i}{2^p(p+1)}\left[\frac{b_N}{0.5\left(\frac{1}{p+1}\right)^{1/p} + (N-1)}\right]^p$$

$$d = \frac{b_N}{0.5\left(\frac{1}{p+1}\right)^{1/p} + (N-1)}.$$

The worst case distribution for both MR and MS now include atoms at $\pm\infty$ which correspond to limits of uniform distributions over wider and wider ranges. Note also that using the second distortion weighting $e_2(x, y)$ of (5) results in a degenerate problem where all quantizers have the same distortion ($g(L)$).

C. Eliminating the Unimodal Constraint

There are circumstances where the quantizer input is not known to be unimodal. For example, the input may be a random sample from two or more populations with significantly different modes. If the unimodal constraint is eliminated, the set C of possible cpdf's consists of all cpdf's on $[0, \infty)$ with generalized moment at most c. One disadvantage of eliminating the unimodal constraint is that the worst case distributions are somewhat degenerate (consisting of atoms at the transition levels). For many problems these distributions may be unduly pessimistic.

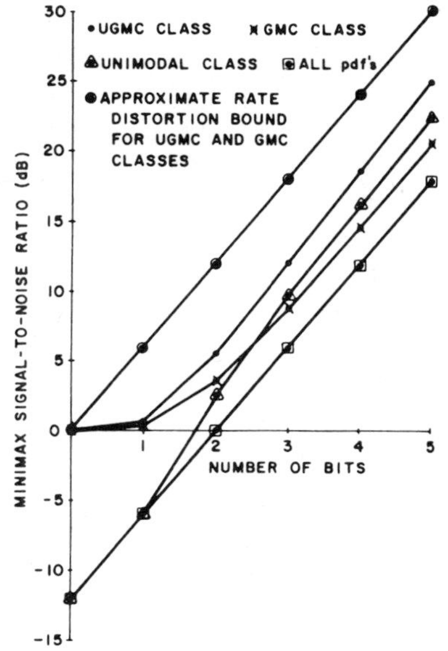

Fig. 7 Minimax MR quantizer SNR's for four classes of input distributions (MR). $g(x) = \rho(x) = x^2$, $b_N/\sqrt{c} = 4$.

Theorem 1 holds without change for the nonunimodal case. The distributions F^* which solve (13) consist of atoms at the maxima of $I(s)$. Thus the problem can be restated:

$$D^*(q) = \min_{\substack{\lambda_1,\lambda_2 \geq 0 \\ I(b_i) \leq 0,\, i=0,\cdots,N}} \lambda_1 + \lambda_2 c \tag{25}$$

and

$$D^* = \min_{q \in Q_N} D^*(q).$$

As in Section IV, these minimizations can be simplified by imposing the regularity conditions on the distortion and constraint functions. The minimax distortion $D^*(q)$ for an arbitrary quantizer can then be found and the minimax quantizer constructed by procedures analogous to those given in Section IV. Define

$$I_i(y_1,\cdots,y_N,\lambda_1,\lambda_2) = e(b_i, q(b_i)) - \lambda_1 - \lambda_2\rho(b_i),$$
$$i = 0,\cdots,N.$$

The construction procedure is based upon $I_i(y_1,\cdots,y_N, \lambda_1,\lambda_2) = 0$, $i = 0,\cdots,N$ being a necessary condition for minimax.

Fig. 7 compares the performance of the three MR minimax quantizers derived so far (the UGMC, GMC, and unimodal classes), as well as the minimax quantizer for the class of all cpdf's derived in [4]. Worst case SNR's are plotted as a function of the number of bits in the quantizer output. The UGMC class has the highest SNR (since it is the smallest class) which is about 7 dB above the SNR for the class of all cpdf's. For two or more bits, the GMC and unimodal classes fall about halfway in between the UGMC and all cpdf classes.

D. Restriction to Continuous Distribution Functions

The unimodal generalized moment class includes cpdf's with atoms at zero. In some applications, however, it may be more realistic to only allow continuous cpdf's. It is shown here that, for most cases of practical interest, the quantizer constructed in Section IV is also minimax for continuous UGMC cpdf's.

It is sufficient to show that the set of worst case distributions contains at least one continuous distribution. This is trivially true for MS quantizers (Theorem 3). For MR quantizers, it suffices to find one continuous unimodal cpdf of the form (14) with

$$\int \rho \, dF \leq c. \tag{26}$$

A continuous worst case distribution can then be constructed using (19). Consider the cpdf F_1 which is uniform over $(0, x_2)$. For F_1, condition (26) becomes

$$R(x_2) \leq cx_2. \tag{27}$$

This condition is both necessary and sufficient for a quantizer which is minimax over the UGMC class to still be minimax when restricted to continuous cpdf's.

It is easily verified that the minimax quantizers listed in Table I satisfy (27) and so are still minimax when the set of input distributions is restricted to be continuous. This condition is not satisfied only when $b_N/c^{0.5}$ is large while N is small (corresponding to very low SNR's of limited practical interest). Even in this case, the minimax quantizer derived in Section IV still minimizes the supremum of the distortion, but there is no continuous distribution which is worst case (i.e., the solution is degenerate). A similar situation occurs when the GMC problem (Section VII-C) is restricted to continuous distributions.

VIII. Conclusions

This paper considers the problem of quantizing data for which there is only an incomplete statistical description. Specifically, instead of assuming that the probability distribution of the input data is known, it is assumed only that it belongs to some set C. The primary set C considered is the UGMC set of all probability distribution which are unimodal with mode zero and which satisfy a generalized moment constraint (e.g., mean-square value less than or equal to a constant).

When the input data are independent, a minimax quantizer exists. This minimax quantizer minimizes, over all possible quantizers, the maximum distortion of each quantizer for any probability distribution in the UGMC set. By examining the dual optimization problem, a numerically efficient procedure for constructing the minimax quantizer is found. Comparison of the minimax quantizer with the Gaussian-optimal quantizer shows that the minimax quantizer guarantees a significantly lower distortion (across the UGMC set) while performing nearly as well as the Gaussian quantizer when the input is, in fact, Gaussian.

When the number of quantization levels is large, the design of a quantizer may be considerably simplified by employing the companding representation. A quantizer is modeled by a uniform quantizer preceeded by an analog nonlinearity (compander), and the quantizer distortion is a function of this nonlinearity and of the input probability distribution. Optimum nonlinearities may be found for a given input probability distribution [9]. A minimax approach to this problem yields both the minimax companding system and a simpler way of finding (approximately) the minimax quantization levels when the number of levels is large. This topic will be considered in a future paper.

References

[1] J. Max, "Quantizing for minimum distortion," *IRE Trans. Inform. Theory*, vol. IT-6, pp. 7–12, Mar. 1960.

[2] M. D. Paez and T. H. Glisson, "Minimum mean-squared-error quantization in speech PCM and DPCM systems," *IEEE Trans. Commun. Technol.*, vol. COM-20, pp. 225–230, Apr. 1972.

[3] D. J. Sharma, "Design of absolutely optimal quantizers for a wide class of distortion measures," *IEEE Trans. Inform. Theory*, vol. IT-24, pp. 693–702, Nov. 1978.

[4] J. M. Morris and V. D. VandeLinde, "Robust quantization of discrete-time signals with independent samples," *IEEE Trans. Commun. Technol.*, vol. COM-22, pp. 1897–1901, Dec. 1974.

[5] D. G. Luenberger, *Optimization by Vector Space Methods.* New York: Wiley, 1969.

[6] W. G. Bath, "Robust quantization of independent data," Ph.D. dissertation, Dept. Elec. Eng., Johns Hopkins Univ., Baltimore, MD, 1980.

[7] T. Berger, *Rate Distortion Theory.* Englewood Cliffs, NJ: Prentice-Hall, 1971.

[8] D. J. Sakrison, "Worst sources and robust codes for difference distortion measures," *IEEE Trans. Inform. Theory*, vol. IT-21, pp. 301–309, May 1975.

[9] P. F. Panter and W. Dite, "Quantization distortion in pulse-count modulation with nonuniform spacing of levels," *Proc. IRE*, vol. 39, pp. 44–48, Jan. 1951.

Reprinted from pages 483-487 of IEEE Conf. Decis. Control, 1979, vol. 1, 620p.

ROBUST QUANTIZERS DESIGNED USING THE COMPANDING APPROXIMATION

W. G. Bath and V. D. VandeLinde

Abstract

This paper considers the design of a quantizer when the probability distribution of the input signal is not known exactly. A robust (minimax) quantizer is derived which produces the largest possible guaranteed signal-to-noise ratio over a sizeable class of input distributions (all unimodal distributions with a known moment.) A closed-form solution for the robust quantization levels is possible by using the companding approximation. In many instances, the robust quantizer guarantees a significantly higher signal-to-noise ratio than quantizers designed for the standard distributions (uniform, Gaussian, Laplace.) A similar approach is used to "robustify" the commonly used μ-law and A-law companders.

1. Introduction

One of the most common operations in a modern communications or control system is analog-to-digital conversion. An analog waveform is first sampled in time, then each of these input samples is approximated by one of a finite number of quantization levels. The set of quantization levels is selected using apriori knowledge about the signal to be quantized. If the probability distribution function (pdf) of the input samples is known precisely, then a quantizer may be used which is optimum for that distribution [1]. However, often the pdf is only partially specified. Here, it is assumed that the pdf is specified only by a generalized moment constraint (such as mean-square value) and by a single, known mode. The design philosophy is to pick a quantizer which minimizes the maximum quantizer distortion over all possible input pdf's. A previous paper [2] has solved this problem by a numerical technique which is valid for any number of quantizer levels. In this paper, an approximate solution is developed for large numbers of levels using the companding approximation.

2. Statement of Problem

The situation considered is as follows. An input datum, X, is passed through a quantizer, q, to produce a quantized output Y. The quantizer has zero memory. An N-level quantizer can be defined in terms of 2N-1 numbers $(b_1,\ldots, b_{N-1}, y_1,\ldots, y_N)$:

$$q(x) = \begin{cases} y_1 & x\varepsilon[-\infty,b_1) \\ y_i & x\varepsilon[b_{i-1},b_i) \quad i=2,\ldots, N-1 \\ y_N & x\varepsilon[b_{N-1},\infty] \end{cases} \tag{1}$$

The goal of quantizer design is to make the quantizer output, Y, statistically close to the quantizer input, X, without using an exorbitant number of levels. The probability distribution function (pdf) of the input, X, is not known precisely but is known to belong to a set C.

Quantizer performance is measured by a distortion measure, D, which is a function both of the quantizer, q, and the input pdf F. The distortion measure has the form

$$D(q,F) = \int e(x,q(x))\, dF(x) \tag{2}$$

For example, if $e(x,y)=(x-y)^2$, then the distribution is the mean-square difference between quantizer input and output. By choosing the distortion weighting, e(x,y), one specifies how the "closeness" of quantizer input and output will be measured.

Bounded distortion weightings often most accurately represent the true penalties associated with quantization errors. For example, if the quantization errors in a speech transmission system are so large that the speech is unintelligible, increasing the errors imposes no additional penalty. In addition, if the distortion function is unbounded, particularly "bad" input pdf's may have the same (possibly infinite) distortion for all quantizers even though some quantizers are clearly better than others.

Consider the bounded distortion weighting:

$$e(x,y) = \begin{cases} g(x-y) & : |x|<V \\ g(|x|-V) : & |x|>V \end{cases} \tag{3}$$

where g(x) is a distortion function satisfying four conditions:

(1) $g(x)=g(-x)$
(2) $g(0)=0$
(3) g(x) is continuous
(4) g(x) is monotonic strictly increasing in $|x|$

This distortion weighting is a bounded version of the distortion weighting $e(x,y) = g(x-y)$ and would be used when very large input samples are felt to be of less importance or when the analog system preceeding the quantizer has limited dynamic range.

For this distortion measure, no choice of transition levels, b_i can be better than

$$b_i = \frac{y_{i+1}+y_i}{2} \qquad i=1,\ldots, N-1 \tag{4}$$

This follows from Max's results [1], and simply corresponds to quantizing an input value, X, to the nearest quantization level. Only quantizers satisfying (4) will be considered. A quantizer is, then, completely specified by the N quantization levels $(y_1,\ldots, y_N)$.

Results will be derived for one-sided quantization problems (problems where the inputs, X, are nonnegative.) Since the sets, C, of possible pdf's are defined symmetrically, if attention is restricted to quantizers with equal numbers of positive and negative levels, there is no loss of generality in solving the one-sided minimax problem.

The set of possible input pdf's considered first will be the unimodal, generalized moment constrained (UGMC) set, C, of all pdf's on $[-\infty,\infty]$ satisfying:

(1) F is unimodal with mode zero
(2) $\int \rho(s)\, dF(s) = c$

The constraint function, $\rho(s)$, satisfies the same four properties as the distortion function g(s). F is defined to be unimodal with mode zero if F(x) is convex in $(-\infty,0)$ and concave in $(0,\infty)$. Note that unimodality precludes atoms at any finite value other than zero.

The robust quantization problem can now be stated explicitly: Find a minimax quantizer, q^* , and a worst-case probability distribution, F^* , satisfying

$$\inf_{q \in Q_N} \sup_{F \in C} D(q,f) = D(q^*,F^*) \tag{5}$$

Q_N is the set of all N-level, one-sided quantizers, C, is the unimodal, generalized moment class and D is the distortion measure using weighting $e(x,y)$.

3. Summary of Results for Quantizers with Arbitrary Numbers of Levels

The robust quantization problem (8) has been solved in [2]. An existence proof plus an application of the Lagrange duality theorem [3] reduces (5) to

$$D^* = \min_{q \in Q_N} \min_{\lambda_1,\lambda_2 \geq 0} [(\lambda_1 + \lambda_2 c) + B] \tag{6}$$

where

$$B = \max_{F \in C} \int e(s,q(s)) - \lambda_1 - \lambda_2 \rho(s)\, dF(s) \tag{7}$$

and λ_1, λ_2 are Lagrange multipliers corresponding to the unit probability and generalized moment constraints respectively. By recognizing that B is either 0 or ∞ and by considering the effect of small perturbations in the quantization levels, (6) becomes [4]

$$D^* = \min_{\lambda_1,\lambda_2 \geq 0} \lambda_1 + \lambda_2 c \qquad H(x_i)=0\ ;\ i=1,\ldots,N+1 \tag{8}$$

where

$$H(s) = \int_0^x e(s,q(s)) - \lambda_1 - \lambda_2 \rho(s)\, ds \tag{8}$$

and the x_i are the nonnegative zero-crossing of the integrand in (9). This is a constrained two-dimensional minimization which is easily solved by numerical minimization in a time linear in N . Several minimax quantizers and their worst-case distortions are tabulated in [2]. In many instances, the minimax quantizer guarantees a significantly lower worst-case distortion than the uniform or Gaussian-optimal quantizers. At the same time, the minimax quantizer performs nearly optimally against a Gaussian input distribution.

4. Distortion Approximation for a Large Number of Levels

A large N approximation for the quantizer distortion has been developed by Bennett [5]. The approximation is based on the companding representation of a quantizer. Any quantizer $q=(y_1,\ldots,y_N)$ can be represented by a continuously differentiable, monotonic increasing function $K(x)$ satisfying $K(0)=0, K(V)=V$;

$$y_i = K^{-1}(u_i);\ i=1,\ldots,N \tag{10}$$

where $(u_1,\ldots,u_N)$ is a uniform quantizer. The quantizer is, then, modeled by (and in fact can be implemented by) Figure 1. Defining

$$D_0(k,F) = \frac{V^p}{2^p(p+1)} \int_0^V \frac{f(s)}{k^p(s)}\, ds \tag{11}$$

where $k(x)$ is the derivative of $K(x)$, the desired approximation (for $g(x)=x^p$) is:

$$D(q,F) \cong \frac{D_0(k,F)}{N^p} \tag{12}$$

The minimax companding problem can now be explicitly stated as follows:

$$D_0(k^*,F^*) = \min_{k \in \Gamma} \max_{F \in C} D_0(k,F) \tag{13}$$

where C is the UGMC set and Γ is the set of admissable companders.

5. The Minimax Compander

A closed-form expression for the minimax compander, k^* , given a distortion function $g(x)=x^p$; $p\geq 1$, a constraint function, $\rho(x)$, and a limiting value V will be derived. The solution to the minimax problem (13) is independent of N so that once k^* is found, an approximation to any N-level minimax quantizer can be found simply by sampling the inverse of K^* . As in the arbitrary N case, the Lagrange duality theorem produces

$$D^* = \min_{k \in \Gamma} \min_{\lambda_1,\lambda_2 \geq 0} \lambda_1 + \lambda_2 c \qquad H(x) \leq 0;\ x \geq 0 \tag{14}$$

where

$$H(x) = \int_0^x I(s)\, ds$$

and

$$I(s) = \frac{V}{2(1+p)^{1/p} k^p(s)} - \lambda_1 - \lambda_2 \rho(s)$$

It can be shown [4] that a necessary condition for optimality is $I(s)=0;\ s \in [0,V]$. This leads to the minimax companding theorem:

Theorem: The compander $k^*(s)$ given by

$$k^*(s) = \frac{V}{2(p+1)^{1/p}[\lambda_1^* + \lambda_2^* \rho(s)]^{1/p}}$$

with λ_1^*, λ_2^* chosen to solve

$$\min_{\lambda_1,\lambda_2 \geq 0} \lambda_1 + \lambda_2 c \qquad \int_0^V k^*(s)\,ds = V \tag{16}$$

solves the minimax companding problem (13).

As an example, consider the quadratic case $p=2$, $\rho(x)=x^2$. If one defines $r=\lambda_2/\lambda_1$, then the minimax distortion can be written

$$D_0^* = \min_{r>0} \frac{1+rc}{12r} \log^2 [Vr^{.5} + (1+rV^2)^{.5}] \tag{17}$$

The minimax multipliers are then found from r^* and D_0^* using (16) and the definition of r . Thus for the quadratic case, the parameters of the minimax compander are found by solving the one-dimensional transcendental equation (17).

6. Worst-Case Distortion for an Arbitrary Compander

Given an arbitrary compander, k, it is useful to know its worst-case distortion across the UGMC class. In the quadratic case, a simple solution is possible for a large class of companders. Suppose $k(s)$ is a compander with

$$l(s) = \frac{V^2}{12k^2(s)} \tag{18}$$

and that $l(s)$ satisfies

(1) $l(s)$ is monotonic increasing
(2) $l(s)$ is three times differentiable.
(3) $\frac{d^3}{ds^3}[l(s)] > 0$ for all $s \geq 0$.
(4) $\lim_{s\to\infty} s^{-2} l(s) = \infty$.

then the worst-case distortion for any distribution in the UGMC set and for $\rho(s)=g(s)=s^2$ is

$$D_0^*(k) = l(0) + \frac{3c}{V^3}\left[\int_0^V l(s)\,ds - Vl(0)\right] \tag{19}$$

For example, the optimum compander for a Gaussian distribution with variance c satisfies these conditions and has a worst-case distortion

$$D_0^* = \frac{c^2}{12}\left[\int_0^{L_F} e^{-s^2/6}ds\right]^2\left[1 + \frac{3}{L_F^3}\int_0^{L_F} e^{s^2/3}ds - \frac{3}{L_F^3}\right]$$

where $L_F=V/c^{.5}$. Similarly, the worst-case distortion for a compander optimized for a Laplacian distribution with variance c is:

$$D_0^* = \frac{3c^2}{8}\left[1 - e^{-2^{.5}L_F/3}\right]^2\left[1+\frac{9}{2^{.5}L_F^3}\left(e^{2L_F 2^{.5}/3}-1\right)-\frac{3}{L_F^3}\right]$$

7. Comparison of Minimax and Classical Companders and Quantizers

The use of a minimax compander (either directly as shown in Figure 1) or to select quantization levels using equation (10) is based on the distortion approximation in section 4. It is desirable to know how large the number of quantization levels, N, must be for the approximation to be good. This can be determined by comparing the worst-case performance of the minimax quantizers derived in [2] with the worst-case performance of the minimax companders derived in section 6. For the comparison to be valid, the distortion of the minimax compander must be computed using the procedure given in [2] for finding the worst-case distortion of an arbitrary quantizer. If the signal-to-noise ratio (SNR) of a quantizer is defined as

$$SNR(q,F) = \frac{c}{D(q,F)} \tag{20}$$

then the resulting SNR penalty for designing a quantizer via the simpler compander approach is tabulated in Table 1 as a function of the number of bits in the quantizer output. The companding loss declines rapidly as the number of bits increases and is less than .1dB for a 5-bit quantizer. Thus for a quantizer with more than a few output bits, the penalty for using the companding approximation is negligible.

Figure 2 compares the worst-case (across the UGMC class) SNR of the minimax compander with the worst-case SNRs of the Gaussian-optimal and Laplace-optimal companders. Quadratic distortion and constraint functions are assumed. Also shown is the worst-case SNR for a uniform quantizer with levels equally spaced over [-V,V]. For small loading factors, all four companders produce nearly uniform quantizers and so have similar signal-to-noise ratios. As the loading factor increases, the SNRs of the three classical companders fall off more rapidly that the minimax SNR.

The Gaussian-optimal compander does quite poorly at large loading factors. This is understandable given the rapid decay of the Gaussian density which leads to a Gaussian-optimal quantizer with very coarsely spaced levels near V. Such a quantizer is easily thwarted by an input distribution with long tails. The uniform quantizer has the reverse problem. Its quantization levels are too coarsely spaced near zero.

The Laplace-optimal compander performs quite well (within 1dB of minimax) for loading factors less than 5. This leads to an interesting parallel. Robust detection and estimation problems (unquantized) are often solved asymptotically by detectors and estimators designed for distributions with exponential tails [6]. Thus it is not surprising that designing a compander for a Laplace distribution produces a compander which is nearly minimax (over a specific range of loading factors.)

8. Maximin Signal-to-Noise Ratio Quantization and Companding

In many applications, the SNR is a better measure of the "closeness" of quantizer input and output than the distortion $D_0(k,F)$ alone. For example, this is generally true when quantizing speech or when quantizing radar video. Clearly, if c is known, then maximizing the signal-to-noise ratio is equivalent to minimizing the distortion. Consider, then, a problem where c is not known exactly but is known to be in an interval $[c_L,c_U]$. One may then pose the maximum SNR problem

$$SNR^* = \max_{q\epsilon Q_N} \min_{c_L\leq c\leq c_U} \min_{F\epsilon C} SNR(q,F) \tag{21}$$

It is straightforward to show that the solution to this problem is the minimax quantizer designed for $c=c_L$.

Thus even when there is a wide uncertainty in the scale of the input signal, a reasonable lower bound on the signal-to-noise ratio can be guaranteed. (This seems to imply that the upper limit, c_U, does not affect the solution of the maximim problem. However in practice, if c is allowed to become arbitrarily large, the choice of the value V to limit the distortion measure may be come unrealistic.)

The problem of guaranteeing a SNR over a range of input signal moments was solved heuristically by Smith [7]. The heuristic solution is the well-known μ-law compander

$$K_\mu(x) = V\frac{\log(1 + \mu x/V)}{\log(1 + \mu)} \tag{22}$$

Another heuristic solution is the A-law compander [8]:

$$K_A(x) = \begin{cases} \dfrac{Ax}{\log(1+A)}; & 0 \leq x \leq V/A \\[2ex] \dfrac{V+V\log(Ax/V)}{1+\log(A)}; & V/A \leq x \leq V \end{cases} \quad (23)$$

Both the heuristic solutions are attempts to approximate an unrealizeable logarithmic compander. Comparison with the minimax compander obtained by integrating (15) shows that all three are essentially logarithmic for large x.

The worst-case distortions of the μ and A law companders computed using an approach similar to that taken in section 5 are:

$$D_0^*(k_\mu) = \frac{V^2}{12}\left[\frac{\log(1+\mu)}{\mu}\right]^2\left[1 + \frac{\mu(3c)^{.5}}{V} + \frac{\mu^2 c}{V^2}\right] \quad (24)$$

and,

$$D^*(k_A) = \frac{(1+\log(A))^2 c}{12A^2}\left[\frac{V^2}{c} + A^2 - 4 + 3/A\right] \quad (25)$$

Equations (24) and (25) may be differentiated to find minimax values of μ and A. These minimax parameters minimize the maximum distortion over the UGMC class subject to the constraint that the compander be μ-law or A-law. Figure 3 shows the minimax values μ^*, A^*, r^* as a function of the worst-case loading factor L_F^*.

Figure 4 compares the worst-case SNR for the minimax compander with the worst-case SNRs for μ and A law companders using the minimax values μ^* and A^*. The three curves are within 1dB over a 70dB range of worst-case loading factors V/c_L. This is not surprising given the similarity in the functional forms of the three companders. It is interesting to note that the μ=255 value used to compand telephone conversations in North America corresponds to a worst-case loading factor of about 35dB while the A=87.6 value used in Europe corresponds to about 32dB.

References

1 Max, J., "Quantizing for Minimum Distortion," IRE Trans. on Information Theory, vol. IT-6, pp. 7-12, March, 1960.

2 Bath, W. G. and VandeLinde, V. D., "Robust Quantizers for Signals with Known Moments and Modes," 22nd Midwest Symposium on Circuits and Systems, June 17-19, 1979, Philadelphia, Pa.

3 Luenberger, D. G., Optimization by Vector Space Methods, New York: Wiley, 1969.

4 Bath, W. G., "Robust Quantization of Independent and Dependent Data," Ph.D. dissertation, Dept. of Elec. Engrg., Johns Hopkins Univ., 1979.

5 Bennett, W. R., "Spectrum of Quantized Signals," Bell Syst. Tech. J., vol. 27, pp. 446-472, July 1948.

6 Huber, P. J., "Robust Estimation of a Location Parameter," Ann. Math. Statist., Vol. 35, pp. 73-101, March 1964.

7 Smith, B., "Instanteous Companding of Quantized Signals," Bell Syst. Tech. J., vol. 27, pp. 446-472, 1948.

8 Cattermole, K. W., "Principles of Pulse Code Modulation," London: Iliffe, 1969.

Number of Bits	Loss (dB)
1	2.5
2	1.8
3	.6
4	.3
5	.1

Table 1 SNR LOSS DUE TO DESIGNING QUANTIZER WITH COMPANDING APPROXIMATION

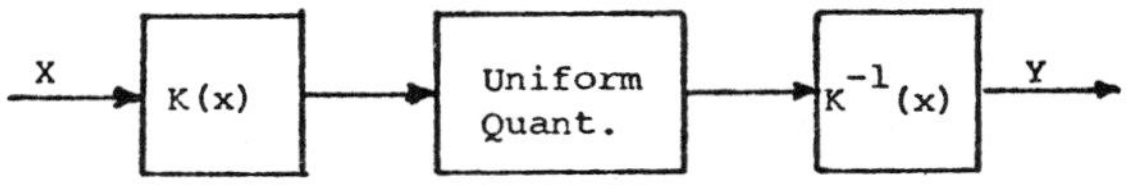

Figure 1 COMPANDING REPRESENTATION OF A QUANTIZER

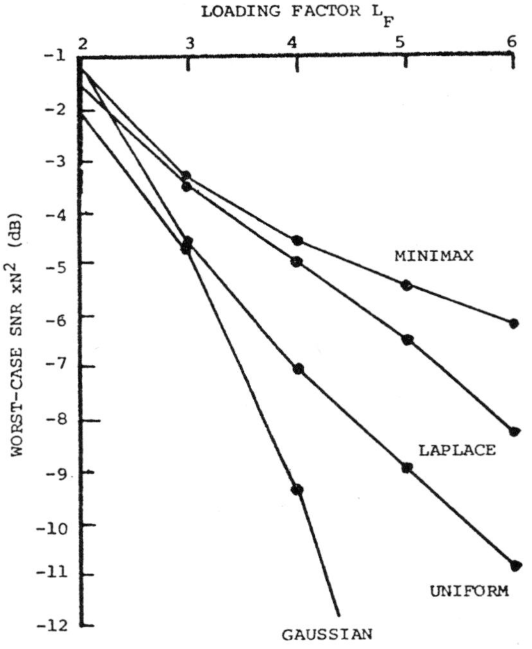

Figure 2 COMPARISON OF MINIMAX AND CLASSICAL COMPANDERS' WORST-CASE DISTORTION

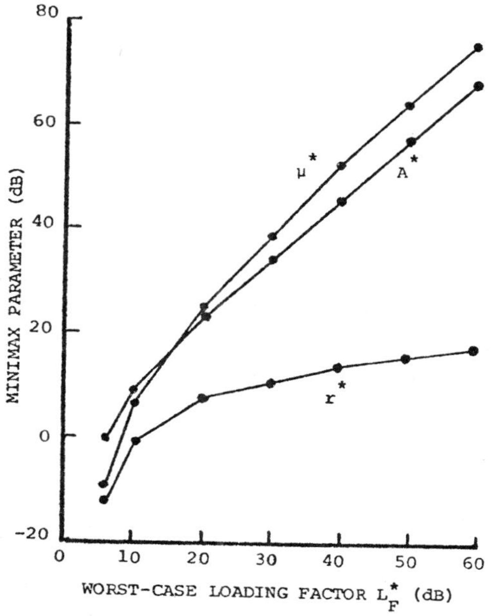

Figure 3 MINIMAX COMPANDING PARAMETERS r^*, μ^*, A^*

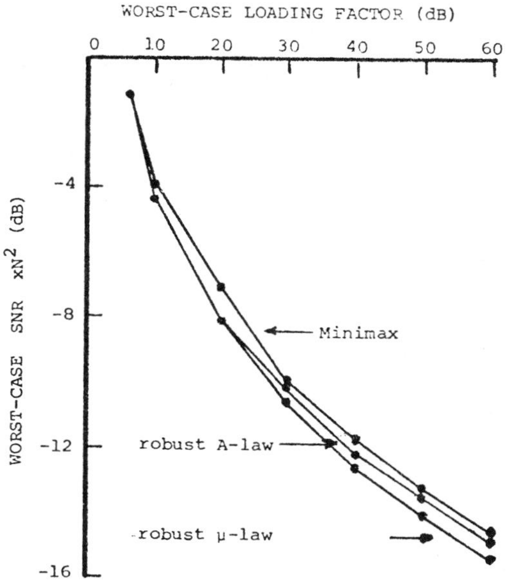

Figure 4 WORST-CASE DISTORTION OF MINIMAX AND "ROBUSTIFIED" μ AND A LAW COMPANDERS

Reprinted from *IEEE Trans. Commun.* **COM-32**(3):240–245 (1984)

Design of Quantizers from Histograms

PETER F. SWASZEK, MEMBER, IEEE, AND JOHN B. THOMAS, FELLOW, IEEE

Abstract—**The design of signal quantizers when the statistical description of the source is a histogram on a finite domain is considered. Minimum mean and minimax error criteria are discussed, both leading to the design of piecewise linear compressor functions. The problem of allocating the histogram intervals in the symmetric source density case is addressed through the use of Chebyshev probability inequalities. Finally, an adaptive implementation of the proposed system is described.**

INTRODUCTION

A signal quantizer is a device which projects a possibly infinitely valued, k-dimensional space onto a finite set of points. Design of an N-level quantizer consists of partitioning the k-space into N disjoint regions and allocating to each region an output point. In the one-dimensional case, the input space is the real line or some subset of it and the N regions are intervals; hence, specifying the $N + 1$ interval endpoints and the N output points uniquely determines the zero-memory quantizer.

In general, the quantizer output will not equal the input signal, the difference being the quantization error. The design procedure should reduce the effect of this error by minimizing some suitable measure of the distortion induced by the error. One common criterion is mean rth error

$$D_r = \int |x - Q(x)|^r \, p(x) \, dx$$

where $p(x)$ is the source probability density function (pdf), $Q(x)$ is the quantizer output for the input x, and the integral is taken over the domain of the source.

In the zero-memory case, this error can be rewritten as

$$D_r = \sum_{i=1}^{N} \int_{x_i}^{x_{i+1}} |x - y_i|^r \, p(x) \, dx \qquad (1)$$

where the x_i are the interval endpoints and the y_i are the associated output points. Max [1] considered this distortion measure for $r = 2$ and $p(x)$ the Gaussian density and found necessary conditions upon the x_i and y_i to minimize D_2. One condition, which holds for most mean error criteria, is that the breakpoints should be Dirichlet partitions of the

Paper approved by the Editor for Communication Theory of the IEEE Communications Society for publication after presentation at the 15th Annual Johns Hopkins Conference on Information Sciences and Systems, Baltimore, MD, March 1981. Manuscript received August 4, 1982; revised July 10, 1983. This work was supported by the National Science Foundation under Grants ECS-79-18915 and ECS-8206237 and by the U.S. Army Research Office under Grant DAAG29-82-K-0095.

P. F. Swaszek was with the Department of Electrical Engineering and Computer Science, Princeton University, Princeton, NJ 08544. He is now with the Department of Electrical Engineering, University of Washington, Seattle, WA 98195.

J. B. Thomas is with the Department of Electrical Engineering and Computer Science, Princeton University, Princeton, NJ 08544.

output points

$$x_i = \tfrac{1}{2}(y_{i-1} + y_i); \quad i = 2, 3, \cdots, N, \quad x_1 = -\infty,$$

$$x_{N+1} = \infty.$$

This condition states that the optimal quantizer should map each input point to the nearest output point. All schemes considered herein will have this property which reduces the specification of an N-level quantizer to the allocation of the N output points.

For a large number of levels, a nonuniform quantizer has been modeled as a three-part system [2]-[4]: a compressor nonlinearity g, a uniform quantizer Q_U, and an expandor nonlinearity $h = g^{-1}$ (see Fig. 1). The compressor function g maps the domain of x onto $[-1, 1]$ and the quantizer Q_U projects $[-1, 1]$ onto N equally spaced output points. Selection of the compressor function g determines the system performance. For the mean rth error criterion, the asymptotic error ($N \to \infty$) for a compressor g with signal pdf $p(x)$ is

$$D_r \approx \frac{1}{(r+1)N^r} \int_{-\infty}^{\infty} \frac{p(x)}{|g'(x)|^r}\, dx. \tag{2}$$

The calculus of variations can be used to find the best compressor function for the particular pdf:

$$g_r(x) = -1 + \frac{2\int_{-\infty}^{x} p(y)^{1/(r+1)}\, dy}{\int_{-\infty}^{\infty} p(y)^{1/(r+1)}\, dy}. \tag{3}$$

For this compressor, the associated asymptotic mean rth distortion is

$$D_r \approx \frac{1}{2^r N^r (r+1)} \left[\int_{-\infty}^{\infty} p(x)^{1/(r+1)}\, dx\right]^{r+1}. \tag{4}$$

Robust Quantizers

Minimization of this distortion measure requires exact knowledge of the signal pdf; most of the available literature on quantizer design assumes this pdf is known. It is well known that when the source and quantizer are not matched, severe degradation can occur. The performance of mismatched quantizers was considered in detail by Mauersberger [5]. He evaluated numerically mean square error rates for variance and density shape mismatches of generalized Gaussian density quantizers. He also presented suggestions for design under this known density functional form with unknown parameters.

Robust quantizers, defined as those that perform well over a range of inputs, are desirable. For the situation in which the only available statistical information is that the source has a finite domain (the interval $[-c, c]$), Morris and VandeLinde [6] solved the minimax problem

$$\min_{q \in Q} \max_{p \in P} D(p, q)$$

where Q is the set of N-level quantizers, P is the class of density functions on $[-c, c]$, and $D(p, q)$ is the distortion measure for density p and quantizer q. They investigated mean error distortion measures and characterized the worst case density and the resulting minimax quantizer. The worst case pdf consisted of atoms located at the quantizer breakpoints (the points of maximum error) and the minimax quantizer

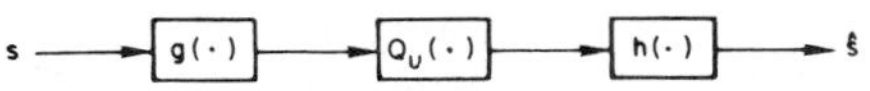

Fig. 1. Compandor system model.

was a uniform (equal step size) quantizer on $[-c, c]$. Bath and VandeLinde [7] later investigated the minimax quantizer for a unimodal source which satisfies an integral (moment) constraint. In this case, the worst case pdf was piecewise uniform and a numerical solution procedure was described.

The specific problem considered herein is the design of quantizers when the available statistical information consists of a source histogram. An M-region histogram is characterized by the division of the real line into M disjoint regions $[h_i, h_{i+1})$, $i = 1, \cdots, M$, and associating with each region the probability p_i that the source takes a value in that interval. If $p(x)$ is the underlying source density, then

$$p_i = \int_{h_i}^{h_{i+1}} p(x)\, dx; \quad -\infty \leqslant h_1 \leqslant \cdots \leqslant h_M \leqslant \infty.$$

It will be assumed that $p(x)$ has finite support of $[-L, L]$. Finite support is necessary for the histogram quantizer design and relaxation of this condition will be mentioned later.

Quantizer Design from a Source Histogram

Given an M-region histogram with regions $[h_i, h_{i+1})$ and probabilities p_i, define the region widths

$$\Delta_i = h_{i+1} - h_i; \quad i = 1, \cdots, M.$$

A simple approach to the quantizer design would be to assume that the density is piecewise constant of value p_i/Δ_i on the region $[h_i, h_{i+1})$. With this assumption, the optimal compressor characteristic from (3) is

$$g_r(x) = s_j x + b_j; \quad x \in [h_j, h_{j+1}), \quad j = 1, 2, \cdots, M \tag{5}$$

where s_j and b_j are defined by

$$s_j = \frac{2\left(\frac{p_j}{\Delta_j}\right)^{1/(r+1)}}{\sum_{i=1}^{M} (p_i \Delta_i^r)^{1/(r+1)}} \tag{6}$$

and

$$b_j = \frac{2\sum_{i=1}^{j-1} (p_i \Delta_i^r)^{1/(r+1)} - 2\left(\frac{p_j}{\Delta_j}\right)^{1/(r+1)} h_j}{\sum_{i=1}^{M} (p_i \Delta_i^r)^{1/(r+1)}} - 1. \tag{7}$$

This compressor is piecewise linear and asymptotically has mean rth error

$$D_r \approx \frac{1}{2^r N^r (r+1)} \left[\sum_{i=1}^{M} (p_i \Delta_i^r)^{1/(r+1)}\right]^{r+1}. \tag{8}$$

A somewhat more conservative approach would be to consider a minimax-type problem. Direct minimization of the maximum error leads to a uniform quantizer on $[-L, L]$. Instead, consider a single histogram region $[h_j, h_{j+1})$. Generalizing Morris and VandeLinde's result, the quantizer on this region should be uniform. On this region, $g(x)$ is linear and the overall compressor is again piecewise linear:

$$g_m(x) = \alpha_j x + \beta_j; \quad x \in [h_j, h_{j+1}).$$

Continuity of the compressor function requires that

$$\beta_j = -1 + \sum_{i=1}^{j-1} \alpha_i \Delta_i - \alpha_j h_j.$$

The uniform quantizer Q_U has N equispaced outputs on $[-1, 1]$. Region j of the histogram, $[h_j, h_{j+1})$, with width Δ_j and compressor slope α_j, maps onto an interval of width $\Delta_j\alpha_j$ in $[-1, 1]$. For large N, the number of outputs covered by $[h_j, h_{j+1})$ is $N_j = N\alpha_j\Delta_j/2$. The maximum error in region j (since the spacing of levels in $[h_j, h_{j+1})$ is uniform) is

$$d_j = \frac{\Delta_j}{2N_j} = \frac{1}{N\alpha_j}.$$

For a general error functional $e(\ )$ (monotonically increasing in the absolute value of its argument), the maximum error on region j is $e(1/N\alpha_j)$. This error would be due to a point mass located within $[h_j, h_{j+1})$ with error $1/N\alpha_j$. Since error occurs in each region where $p_i > 0$, taking a mean maximum error measure yields

$$D_M = \sum_{i=1}^{M} p_i e(1/N\alpha_i).$$

The constraint on the α_j's is

$$\sum_{i=1}^{M} \alpha_i \Delta_i = 2; \qquad \alpha_i \geqslant 0.$$

Minimizing this sum with respect to the constraint gives the condition

$$\alpha_j^2 \Delta_j \propto p_j e'(1/N\alpha_j).$$

For the error measure $e(\ ) = |\ |^r$, this condition simplifies to $\alpha_j = s_j$ for s_j as defined by (7). Using this value of α_j yields $\beta_j = b_j$ for b_j from (8). The resulting mean rth maximum error is

$$D_M = \frac{1}{N^r}\left[\sum_{i=1}^{M} (p_i\Delta_i^r)^{1/(r+1)}\right]^{r+1} = 2^r(r+1)D_r. \tag{9}$$

Both the piecewise constant density approach and the mean maximum error method produce the same solution when the error functional is rth power. Also, the error measures D_r and D_M differ only by the multiplicative constant $2^r(r + 1)$.

Histogram Selection

If the histogram data are not prespecified, the designer may have control over the allocation of the histogram regions. Both of the error measures in (8) and (9) resulted in increasing functions of the sum

$$S = \sum_{i=1}^{M} (p_i\Delta_i^r)^{1/(r+1)}. \tag{10}$$

Bounding this term will bound the error. Let us assume that the underlying density and resulting histogram are symmetric about zero. Chebyshev type probability inequalities will be used to provide upper bounds on the histogram region probabilities, the p_j's. Then the above sum will be minimized over the region widths, the Δ_j's. We note that the use of the Chebyshev inequality holds for all symmetric densities, including multimodal and discrete forms.

Consider the four region histogram for a symmetric density on $[-L, L]$ with unit variance. Denote the two regions on $[0, L]$ by $[0, \alpha]$ and $[\alpha, L]$. The Chebyshev inequality [8] (σ^2 is the source power)

$$\text{Prob}(x \geqslant k) \leqslant \begin{cases} 1/2; & 0 \leqslant k \leqslant \sigma \\ \sigma^2/2k^2; & \sigma \leqslant k \end{cases}$$

bounds the p_i

$$p_3 = \text{Prob}(0 \leqslant x < \alpha) \leqslant \text{Prob}(0 \leqslant x) = 1/2$$

$$p_4 = \text{Prob}(\alpha \leqslant x \leqslant L) \leqslant \frac{\sigma^2}{2\alpha^2} = \frac{1}{2\alpha^2}.$$

The sum in (10) is then bounded (since $p_1 = p_4$, $\Delta_1 = \Delta_4$, $p_2 = p_3$, and $\Delta_2 = \Delta_3$ by the symmetry assumption)

$$S \leqslant 2\left[\frac{\alpha^2}{2}\right]^{1/(r+1)} + 2\left[\frac{(L-\alpha)^r}{2\alpha^2}\right]^{1/(r+1)}$$

and can be minimized for $\alpha \in [0, L]$. Optimal region placement is a function of the actual underlying distribution and, hence, does not result; however, suboptimal allocations do occur and an understanding of region placement develops. Larger values of M are analyzed in a similar manner.

Plots of the resulting region placements for $M = 4$, 6, and 8 region histograms using the Chebyshev inequality are displayed in Fig. 2 for the squared error functional ($r = 2$). The graphs indicate the locations of the histogram breakpoints. Notice that as σ/L increases, the solution for $M = 8$ degenerates to the $M = 6$ selection. Similarly, the $M = 6$ lines collapse to the four region case. The set of permissible solutions ($-L \leqslant h_1 < h_2 < \cdots < h_M \leqslant L$) is convex and the degeneration signifies that the minimum is achieved on the set's boundary (one or more of the Δ_i going to zero). For larger M, the ratio of σ/L must be small to obtain nonzero Δ_i.

As more information about the underlying density becomes known, tighter bounds on the region probabilities may be found. For example, for symmetric, unimodal densities, the Gauss inequality [8]

$$\text{Prob}(x \geqslant k) \leqslant \begin{cases} \frac{1}{2}(1 - k/\sigma\sqrt{3}); & 0 \leqslant k \leqslant 2\sigma/\sqrt{3} \\ 2\sigma^2/9k^2; & 2\sigma/\sqrt{3} \leqslant k \end{cases}$$

may be employed to solve for the histogram boundaries. Similar plots to those of Fig. 2 are presented in Fig. 3 for the Gauss inequality.

Numerical Comparisons

The following examples compare the piecewise linear compressors to the optimal and μ-law compressors under the mean square error criterion. Due to the symmetry of the example density functions, only graphs of the positive half of each compressor function will be provided. For the example pdf's, the optimal compressor is found from (3) with the asymptotic error given in (4). The μ-law quantizer (with $\mu = 255$) has a compressor function on $[0, L]$ of

$$g_\mu(x) = \frac{\ln(1 + \mu x/L)}{\ln(1 + \mu)}.$$

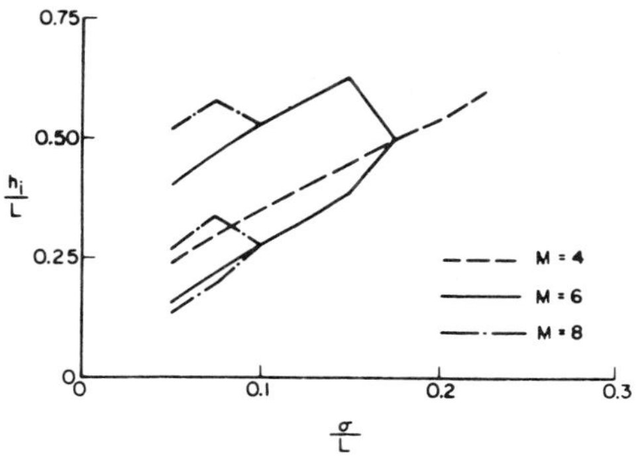

Fig. 2. Chebyshev inequality region placement.

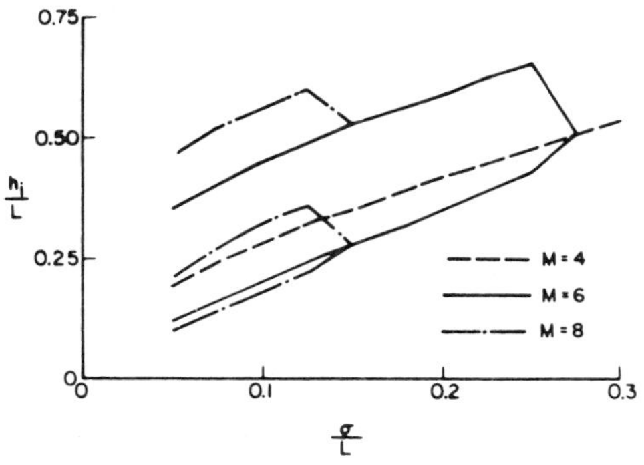

Fig. 3. Gauss inequality region placement.

Substituting into (2) with $r = 2$ yields the μ-law compressor's asymptotic mean square error

$$D_\mu \approx \frac{L^2 \ln^2 (1+\mu)}{3\mu^2 N^2} \left[1 + \frac{2\mu}{L} E\{|x|\} + \frac{\mu^2 \sigma^2}{L^2} \right].$$

The comparison of performance for small N ($N = 16$) quantizers is also tabulated. The outputs are found by an inverse mapping through each compressor function of 16 points equally spaced in $[-1, 1]$. Dirichlet partitions define the quantizer breakpoints (the x_i) and the mean square error is found from (1).

Gaussian Source: The unit normal source is the canonical choice for the comparison of quantization schemes. The pdf for $x \in [-L, L]$ is

$$p(x) = Ke^{-x^2/2}$$

with K chosen for unit mass on $[-L, L]$. The following results are for $L = 5$ (5σ loading). Piecewise linear compressors for four, six, and eight region histograms are compared (corresponding to two, three, and four regions on $[0, 5]$). The Gauss inequality bound with $\sigma/L = 0.2$ yields suboptimal region placement for M equal to 4 and 6. For $M = 8$, equispacing and a modified Gauss placement are tabulated. Figs. 4 and 5 display the compressor functions. Table I lists the histogram region endpoints and the associated asymptotic error rates. The $M = 2$ quantizer is the uniform quantizer on $[-5, 5]$.

For small N ($N = 16$), Table II lists the positive output values for Max's optimal, μ-law, and piecewise linear compressors. The piecewise linear examples are the $M = 4$ (Gauss bound) and $M = 8$ (equispaced) versions. Values of mean square error

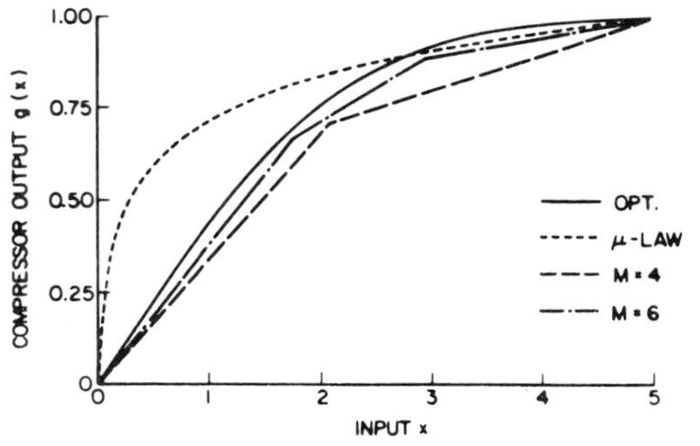

Fig. 4. Gaussian source compressors.

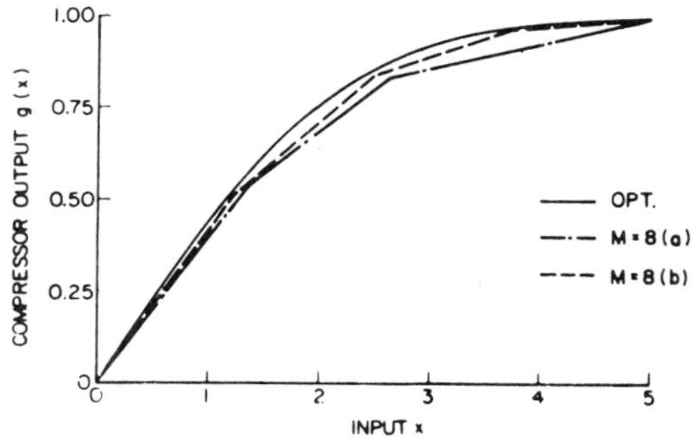

Fig. 5. Gaussian source compressors.

and the associated signal-to-noise ratio are tabulated where

$$\text{SNR} = 10 \log_{10} \frac{\sigma^2}{\text{MSE}} \text{ dB}.$$

Laplacian Source: When modeling signal sources, the Laplace source on $[-L, L]$ is sometimes considered

$$p(x) = Ke^{-\sqrt{2}|x|}$$

where K is chosen for unit mass on $[-L, L]$. For this pdf, we will take $L = 8$. Again, the Gauss inequality bound yields suboptimal region selection for $\sigma/L = 0.125$. Figs. 6 and 7 depict the optimal, μ-law, and piecewise linear compressors for $M = 4$, 6, and 8. Table III lists the histogram breakpoints and the asymptotic error rates.

Adams and Geisler [9] tabulated optimal output values for the Laplace source when $N = 16$ and $r = 2$. Table IV lists these values along with the outputs for the μ-law device and the $M = 4$ and 8 histogram quantizers. As in the previous example, MSE and SNR are tabulated for each scheme.

Adaptive Implementation

It is often of interest to be able to design a quantizer from a training sequence rather than from a preassumed statistical characterization of the source. This idea has been explored in [10] for the design of optimum quantizers, the resulting design requiring an iterative solution. Below we consider the design of the piecewise linear quantizers given a training sequence. Since the resulting design procedure is simple, it can be employed adaptively to keep the system "tuned" to the source.

Consider a training sequence consisting of n independent samples of the input source sorted to the various histogram regions, n_i samples falling with the ith region ($\Sigma_{i=1}^{M} n_i = n$).

TABLE I
GAUSSIAN SOURCE ASYMPTOTIC ERROR RATES

M	Histogram Divisions				N^2D
2	5.				8.33
4	2.1	5.			4.00
6	1.75	2.95	5.		3.22
8(a)	0.7	1.4	2.65	5.	3.31
8(b)	1.25	2.5	3.75	5.	3.01
μ-law compressor					10.6
optimal compressor					2.69

TABLE II
N = 16 GAUSSIAN SOURCE QUANTIZERS

	Max Opt.	μ-law	4 region	8 region
y_1	0.1284	0.008122	0.1855	0.1517
y_2	0.3881	0.03585	0.5565	0.4551
y_3	0.6568	0.09131	0.9276	0.7586
y_4	0.9424	0.2022	1.299	1.062
y_5	1.256	0.4241	1.670	1.433
y_6	1.618	0.8677	2.041	1.913
y_7	2.069	1.755	3.141	2.393
y_8	2.733	3.530	4.380	3.446
MSE	0.009513	0.04098	0.01452	0.01090
SNR(dB)	20.2	13.9	18.4	19.6

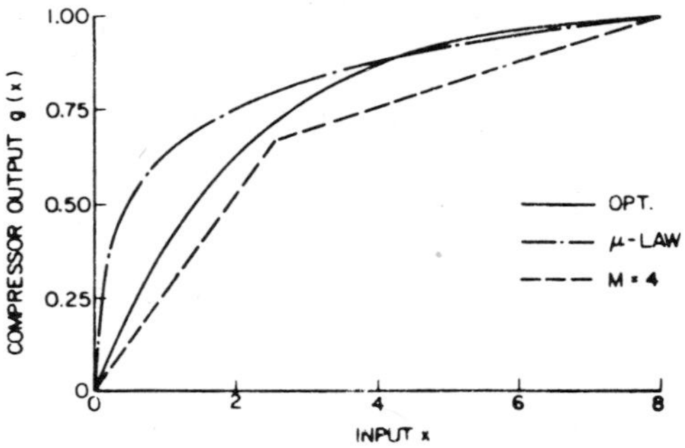

Fig. 6. Laplacian source compressors.

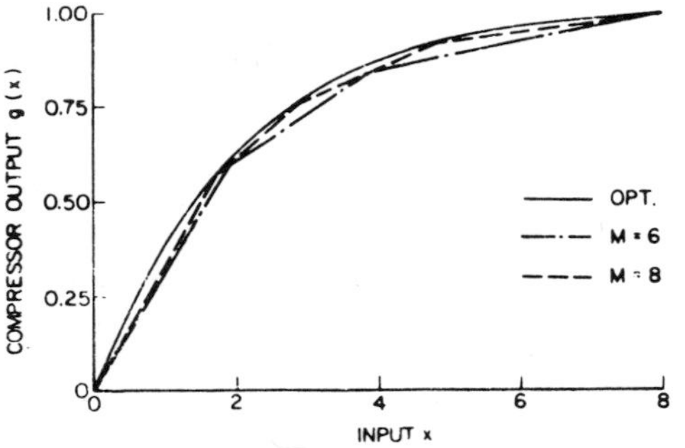

Fig. 7. Laplacian source compressors.

TABLE III
LAPLACIAN SOURCE ASYMPTOTIC ERROR RATES

M	Histogram Divisions				N^2D
2	8				21.3
4	2.56	8.			7.16
6	1.92	3.92	8.		5.47
8	1.76	2.88	4.8	8.	5.00
μ-law compressor					10.7
optimal compressor					4.20

TABLE IV
N = 16 LAPLACIAN SOURCE QUANTIZERS

	A&G Opt.	μ-law	4 region	8 region
y_1	0.1240	0.01299	0.2398	0.1915
y_2	0.4048	0.05736	0.7195	0.5745
y_3	0.7287	0.1461	1.199	0.9575
y_4	1.111	0.3236	1.679	1.341
y_5	1.578	0.6785	2.158	1.724
y_6	2.177	1.388	2.892	2.477
y_7	3.017	2.808	4.935	3.628
y_8	4.431	5.648	6.978	5.810
MSE	0.0161	0.0408	0.0255	0.0192
SNR(dB)	17.9	13.9	15.9	17.2

This event is multinomially distributed

$$p(n_1, n_2, \cdots, n_M) = \frac{n!}{n_1! n_2! \cdots n_M!} \, p_1^{n_1} p_2^{n_2} \cdots p_M^{n_M}.$$

Maximum likelihood estimates (MLE's) of the parameters are those values which maximize the probability of occurrence. In this case, the MLE's are the frequencies of occurrence

$$\hat{p}_i = \frac{n_i}{n}.$$

For the piecewise linear compressor functions developed herein, specification requires the slopes and intercepts defined in (6) and (7). The only unknowns in these expressions are the p_i; hence, due to the functional invariance of MLE's, the maximum likelihood estimates of the slopes and intercepts are those values given by (6) and (7) evaluated at $p_i = \hat{p}_i$. In summary, a training sequence design of this scheme would perform the simple tasks of estimating the p_i from a sample input sequence and evaluating the slopes and intercepts of the piecewise linear compressor from (6) and (7). Similarly, in adaptive mode, the system could occasionally reestimate the p_i and recompute the compressor parameters.

CONCLUSIONS

Robust quantizer design depends entirely upon the amount of information assumed about the class of permissible input distributions. As previously mentioned, the uniform quantizer is the minimax quantizer when only finite support of the density is known and the μ-law quantizer performs well for most input statistics. From the examples, the proposed method of piecewise linear compressor quantizer design is seen to present a viable alternative to uniform quantization or optimal quantization of the "known" pdf. A few other points need to be considered.

1) Initially, the input was assumed to have finite support of $[-L, L]$. Without this constraint, some of the histogram region widths, the Δ_i, would be infinite and the resulting piecewise linear compressor would have sections of zero slope. For nonfinite support, select L such that the probability of overload (x outside $[-L, L]$) is small and map the overload regions to the nearest output, y_1 or y_N.

2) The use of probability inequalities in histogram region selection provided degenerate solutions for $M > 8$ unless σ/L approached zero. The results for $M = 8$ from Table I indicate that equal subdivisions of $[-L, L]$ is a reasonable procedure for larger M. For Δ_i of order $1/M$, the histogram converges uniformly to the underlying density [11] and the piecewise linear compressor converges to the optimal compressor.

References

[1] J. Max, "Quantizing for minimum distortion," *IRE Trans. Inform. Theory*, vol. IT-6, pp. 7–12, Mar. 1960.

[2] W. R. Bennett, "Spectra of quantized signals," *Bell Syst. Tech. J.*, vol. 27, pp. 446–472, July 1948.

[3] V. R. Algazi, "Useful approximations to optimal quantization," *IEEE Trans. Commun. Technol.*, vol. COM-14, pp. 297–301, June 1966.

[4] A. Gersho, "Principles of quantization," *IEEE Trans. Circuits Syst.*, vol. CAS-25, pp. 427–437, July 1978; also *IEEE Commun. Soc. Mag.*, pp. 16–29, Sept. 1977.

[5] W. Mauersberger, "Experimental results on the performance of mismatched quantizers," *IEEE Trans. Inform. Theory*, vol. IT-25, pp. 381–386, July 1979.

[6] J. M. Morris and V. D. VandeLinde, "Robust quantization of discrete-time signals with independent samples," *IEEE Trans. Commun.*, vol. COM-22, pp. 1897–1902, Dec. 1974.

[7] W. G. Bath and V. D. VandeLinde, "Robust memoryless quantization for minimum signal distortion," *IEEE Trans. Inform. Theory*, vol. IT-28, pp. 296–306, Mar. 1982.

[8] J. K. Patel, C. H. Kapadia, and D. B. Owen, *Handbook of Statistical Distributions*. New York: Dekker, 1976.

[9] W. C. Adams, Jr. and C. E. Giesler, "Quantization characteristics for signals having Laplacian amplitude probability density functions," *IEEE Trans. Commun.*, vol. COM-26, pp. 1295–1297, Aug. 1978.

[10] Y. Linde, A. Buzo, and R. M. Gray, "An algorithm for vector quantization design," *IEEE Trans. Commun.*, vol. COM-28, pp. 84–95, Jan. 1980.

[11] W. Wertz, *Statistical Density Estimation: A Survey*. Vandehoech & Ruprecht, 1978.

Part V

SUBOPTIMUM VECTOR SIGNAL QUANTIZERS

Editor's Comments on Papers 27 Through 33

In the papers of Part III, the performance benefits of optimum vector quantization were developed. The papers in this section all consider vector quantization schemes that are designed under some previously assumed constraint upon the system and are thus suboptimum in performance. Several reasons may exist for constraining the system implementation. For example, complex valued signals (e.g., image data or DFT sequences) may be obtained more easily in the polar coordinate form of magnitude and phase rather than the rectangular abscissa and ordinate form; hence, polar coordinate quantization may be a more natural operation (see Papers 28 and 29). A second reason might be that the system implementation is required to employ currently available (off the shelf) items such as scalar quantizers. In this case, the system might apply some transformation to the data for redundancy removal and quantize the resulting data with scalar devices. Such a scheme is considered in Paper 27.

A final, and perhaps the major, reason for specifying such a design constraint is that optimum vector quantizers are, in general, difficult to design. In this context, "difficult" means that the design process requires a large number of mathematical computations, much larger than is required for designing scalar quantizers. For example, consider the necessary conditions for optimum vector quantizers of centroids and nearest-neighbor regions that are used in the optimum vector quantizer design algorithm described in several papers in Part III. The centroid of one of the N quantization regions, S_i, is given by:

$$\mathbf{y}_i = \int_{S_i} \mathbf{x}\, p(\mathbf{x})\, d\mathbf{x} \Big/ \int_{S_i} p(\mathbf{x})\, d\mathbf{x}.$$

Specifying the k coordinates of this point involves the computation of $k + 1$ k-dimensional integrations and k scalar divisions. For the N distinct regions, this results in a total of $O(kN)$ k-dimensional integrations and $O(kN)$ scalar mathematical operations. The other step of the iterative technique is the assignment of the S_i as nearest-neighbor regions about each output value. This partitioning of $\mathcal{R}^k$ is most simply accomplished by computing the intersections of half spaces:

$$S_i = \bigcap_{\substack{j \neq i \\ j = 1}}^{N} \{\mathbf{x}: |\mathbf{x}-\mathbf{x}_i| < |\mathbf{x}-\mathbf{x}_j|\}.$$

With this definition, specification of each S_i requires computing the intersections of $N-1$ k-dimensional halfspaces for an overall total of $O(N^2)$ k-dimensional intersections. Further, remember that both of these sets of operations must be performed for every iteration of the design algorithm. So far, the only examples available in the literature involve $k \leq 5$ and $N \leq 32$ (1 bit per dimension).

For comparison, consider those same iterative design computations for scalar quantizers. To make things comparable, assume that the system employs k separate scalar quantizers for the k coordinates. Further, let the quantizer for the i th coordinate have M_i levels for a total of $\prod_{i=1}^{k} M_i = N$ parallelepiped regions in $\mathcal{R}^k$. Since the quantizers are scalar, the designs are done independently. For one of these quantizers, each design iteration evaluates $M_i + 1$ scalar integrations and M_i scalar divisions for the centroid computation. As to the regions, the intersections of semi-infinite lines (one-dimensional halfspaces) are intervals that require only $M_i - 1$ scalar additions and $M_i - 1$ scalar divisions for their specification. To design all k scalar quantizers requires $\sum_{i=1}^{k} M_i + k$ scalar integrations and $3 \sum_{i=1}^{k} M_i - 2k$ scalar operations (additions or divisions). If the levels are factored equally among the rectangular coordinates, $M_i = N^{1/k}$ (optimum if the source has identical marginal densities), then $\sum_{i=1}^{k} M_i = kN^{1/k}$ and the k-scalar quantizers require $O(kN^{1/k})$ scalar integrations and $O(kN^{1/k})$ scalar operations per cycle of the iterative design algorithm. For moderate values of k and N, the difference between the number of operations for scalar and vector design is substantial.

A similar argument permits a demonstration that the vector quantizers also require many more operations for implementation. For the optimum vector quantizer, one simple implementation computes the distance from **x** to each of the N output points and selects the smallest. Since squaring a set of positive values doesn't effect their ordering, an equivalent technique is to compute and compare the distances squared. With this simplification, the implementation requires N distances squared, each of which involves N subtractions, N multiplications (squaring), and $N - 1$ additions for a total of $O(N^2)$ scalar operations. Also, sorting these N values requires $O(N)$ scalar comparisons. In a recent paper, Gersho (1982) suggests an alternative implementation for optimum vector quantizers. Unfortunately, it is no simpler in terms of the number of computations required. For the scalar case, again let N be factored to $N = \prod_{i=1}^{k} M_i$. To implement each scalar quantizer, no k-dimensional distance need be calculated: each scalar input must be sorted to one of M_i bins, requiring $O(\log_2 M_i)$ comparisons. For the k separate scalar quantizers this is a total of $\sum_{i=1}^{k} \log_2 M_i = \log_2 N$ scalar comparisons. The results of this discussion are summarized in Table I. In conclusion, note that with optimum vector quantizers there is a tradeoff of a large increase in complexity of design and implementation versus gain in performance over that of scalar quantizers.

The papers in this section have all attempted to reap some of the performance gain of vector quantizers without the inherent increase in complexity described above. The first paper, by Huang and Schultheiss (Paper 27), considers

Table 1. Comparison of Complexity of Vector and Scalar Quantizers.

	Vector Quantizer	*Scalar Quantizer*
Design		
centroids	$O(kN)$ k-dim integrations	$O(kN^{1/k})$ scalar integrations
	$O(kN)$ scalar divisions	$O(kN^{1/k})$ scalar divisions
regions	$O(N^2)$ k-dim intersections	$O(kN^{1/k})$ scalar operations
Implementation	$O(N^2)$ scalar operations	none
	$O(N)$ scalar comparisons	$O(\log_2 N)$ scalar comparisons

the correlated Gaussian source with the quantizer constrained to be a series combination of a linear (matrix) operator, a bank of k scalar Max-Lloyd quantizers, and another linear operator. With this system, Huang and Schultheiss remove the redundancy inherent in the source yet keep the implementation simple by employing scalar Gaussian quantizers. In Paper 27 the matrix operators are selected and the N levels are allocated to the scalar quantizers so as to minimize the MSE. It is interesting to note that the optimum choice of the linear operator transforms the correlated input into independent variables. A discussion of the nonstationarity of the resulting noise vector appears in Tasto and Wintz (1973). Segall (1976) further examines the bit allocation problem for all encoding schemes.

Bucklew and Gallagher, in Paper 28, and Pearlman, in Paper 29, consider the minimum MSE quantization of the independent bivariate Gaussian source through scalar quantization of the polar coordinate representation of the input. For this system the quantization regions are constrained to be sections of annuli. Both papers develop the fact that a uniform quantizer is the optimum choice for the phase quantizer. Necessary conditions upon the scalar magnitude quantizer's parameters are derived and are shown to be satisfied by a scaled version of the minimum MSE quantizer for the magnitude density function. Allocation of the number of levels to the phase and magnitude quantizers is discussed, acknowledging that the phase contains more of the information content of the signal and, hence, should be more finely quantized. In a later paper, Bucklew and Gallagher (1979) present a similar analysis for the polar coordinate quantization of other circularly symmetric sources. Also, it is seen in both papers that, for fine quantization (large N), optimum polar coordinate quantization has smaller MSE than optimum rectangular coordinate quantization. Bucklew and Gallagher show this through an asymptotic analysis, Pearlman by tabulating results for N up to 1024. Pearlman (Paper 29) continues by tabulating similar results for the case of a pair of uniform scalar quantizers for the polar coordinates. He also notes that the above performance relation reverses if the output is required to have fixed entropy.

Papers 30 and 31 consider unrestricted polar quantizers (UPQs), a generalization of the polar quantizers above. The idea of the UPQs, as developed by Wilson in Paper 30, is that the polar form discussed in Papers 28 and 29 is too restrictive because it fixes the number of phase quantization levels independent of the magnitude value. By allowing the number of phase divisions to vary from magnitude ring to magnitude ring (but always summing to N), Wilson's quantizers were able to outperform the strictly polar and rectangular coordinate quantizers for the bivariate Gaussian source in the small N range ($N<32$). It was in this range that the polar coordinate quantizers did not outperform rectangular coordinate quantizers (see Papers 28 and 29). In Paper 31 Swaszek and Ku extend Wilson's results on the UPQs to larger values of N and other circularly symmetric sources. The technique involves employing a compander system for the magnitude quantization and a compressorlike argument for dividing the angle divisions to each magnitude ring.

In Paper 32 Swaszek and Thomas generalize polar quantizers to more than two dimensions by considering scalar quantization of the spherical coordinate representation of a *k*-dimensional source. In this paper the source is assumed to be spherically symmetric; the multivariate density function has contours of equal probability that are hyperspheres centered at the origin. This source restriction includes the important independent multivariate Gaussian source. The topics discussed in Paper 32 are similar to those of Papers 28 and 29: conditions upon the scalar quantizers' parameters and results on the allocation of levels to each quantizer. The Gaussian example included shows that three-dimensional spherical coordinate quantization outperforms polar coordinate quantization for that source. In an earlier paper, Dunn (1965) considered quantizing the *k*-dimensional Gaussian source to one magnitude value and various angle combinations.

In the last paper of this section (Paper 33), Swaszek and Thomas examine polar coordinate quantizers for the bivariate Gaussian source. In particular, they note that the conditions of centroidal output values and nearest-neighbor regions are not satisfied by the polar quantizers. (The necessity of these conditions is discussed in Paper 11). In Paper 33 the vector version of Lloyd's Method I design algorithm is applied to the polar quantizers, producing quantizers that display the necessary conditions as well as maintain the circular symmetry of the source. Unfortunately, these quantizers do not outperform polar quantizers by very much. An earlier development of a similar bivariate scheme can be found in the discussion by Dallas (Paper 35). Also developed in Paper 33 are Dirichlet rotated polar quantizers (DRPQs). These DRPQs display circular symmetry, satisfy the necessary conditions of optimum quantizers, and, in examples for the bivariate Gaussian source, have performance very near that of the optimum bivariate quantizer (see also Swaszek, 1983).

REFERENCES

Bucklew, J. A., and N. C. Gallagher Jr., 1979, Two-Dimensional Quantization of Bivariate Circularly Symmetric Densities, *IEEE Trans. Inform. Theory* **IT-25**(6):667–671.

Dunn, J. G., 1965, The Performance of a Class of n Dimensional Quantizers for a Gaussian Source, *Columbia Symp. Signal Tran. Proc.*, pp. 76–81.

Gersho, A., 1982, On the Structure of Vector Quantizers, *IEEE Trans. Inform. Theory* **IT-28**(2):157–166.

Segall, A., 1976, Bit Allocation and Encoding for Vector Sources, *IEEE Trans. Inform. Theory* **IT-22**(2):162–169.

Swaszek, P. F., 1983, Further Notes on Circularly Symmetric Quantizers, *Johns Hopkins Conf. Inform. Sci. & Sys. Proc.*, pp. 794–800.

Tasto, M., and P. A. Wintz, 1973, Note on the Error Signal of Block Quantizers, *IEEE Trans. Commun.* **COM-21**(3):216–219.

27

Reprinted from *IEEE Trans. Commun. Syst.* **CS-11**(3):289–296 (1963)

Block Quantization of Correlated Gaussian Random Variables*

J. J. Y. HUANG†, MEMBER, IEEE, AND P. M. SCHULTHEISS‡, MEMBER, IEEE

Summary—**The paper analyzes a procedure for quantizing blocks of N correlated Gaussian random variables. A linear transformation (P) first converts the N dependent random variables into N independent random variables. These are then quantized, one at a time, in optimal fashion. The output of each quantizer is transmitted by a binary code. The total number of binary digits available for the block of N symbols is fixed. Finally, a second $N \times N$ linear transformation (R) constructs from the quantized values the best estimate (in a mean-square sense) of the original variables. It is shown that the best choice of R is $R = P^{-1}$, regardless of other considerations. If $R = P^{-1}$, the best choice for P is the transpose of the orthogonal matrix wich diagonalizes the moment matrix of the original (correlated) random variables. An approximate expression is obtained for the manner in which the available binary digits should be assigned to the N quantized variables, i.e., the manner in which the number of levels for each quantizer should be chosen. The final selection of the optimal set of quantizers then becomes a matter of a few simple trials. A number of examples are worked out and substantial improvements over single sample quantizing are attained with blocks of relatively short length.**

I. Introduction

For the transmission of information from a continuous source over a digital channel the output from the source must be represented by a discrete set of variables that can be coded into the channel. The usual procedure is to sample the continuous source output at regular time intervals and then quantize the samples in amplitude. The errors arising from these two operations are linked only through the fact that the digital channel has a fixed number of digits available in any given time interval so that a high sampling rate implies coarse quantization and vice versa. Much useful insight can therefore be gained by separate studies of sampling and quantizing. This paper is concerned only with the quantization process.

Lloyd [1] and Max [2] have independently solved the problem of designing an optimum quantizer for operating on single samples of a random process with specified probability density. If successive samples are statistically dependent it should clearly be possible to reduce the average quantization error without increasing the total number of digits used for representing a set of N samples. Unfortunately the procedure developed by Lloyd and Max requires fairly time-consuming digital computation even for the one-dimensional (single sample) case. A straightforward generalization to even two dimensions, while offering no difficulty in principle, does not therefore appear feasible from a practical point of view. Other general approaches to the multidimensional quantization problem have been made but have yielded to date only asymptotic results valid for extremely fine quantizers. [3].

The present paper deals with a more restricted procedure. Its merits are the absence of any limitation on fineness of quantization and the relative ease of designing quantizers of moderate block length. Fig. 1 illustrates

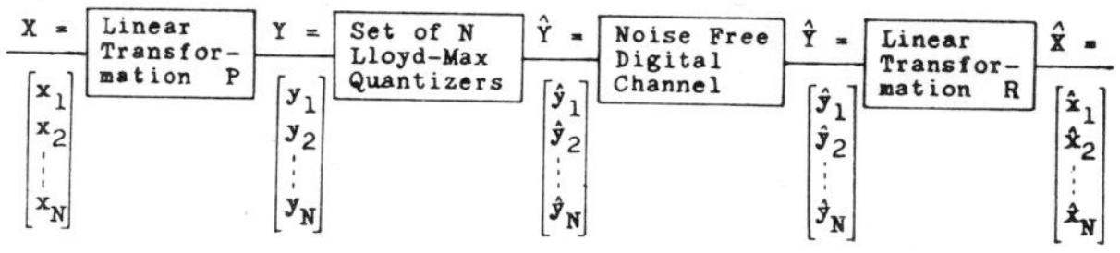

Fig. 1—Schematic diagram of block quantizer.

the proposed system in block diagram form. The input X is a column matrix of N samples $\{x_i\}$ of the output from a continuous source. Throughout most of this paper, and unless an explicit statement to the contrary is made, X will be assumed to be an N-dimensional Gaussian random variable with zero mean. A nonsingular transformation matrix P operates on X to yield a column matrix Y of *uncorrelated* random variables. If X is Gaussian, Y also has a Gaussian distribution. Hence its components $\{y_i\}$ are not only uncorrelated but actually independent. The transformation matrix P is clearly not unique. Its adjustment will be one of the problems to be considered. Y is next quantized by a set of N Lloyd-Max quantizers, *i.e.*, y_i is processed by a quantizer adjusted so as to minimize the quantization error $(\hat{y}_i - y_i)$ in some appropriate average sense. A mean-square error criterion will be used throughout this paper. The quantizer output $\hat{Y}$ is transmitted over a noise-free digital channel and then operated upon by a linear transformation matrix R which yields the column matrix $\hat{X}$, an approximation of X. The intuitive idea behind the proposed instrumentation is obviously to exploit correlation between input samples $\{x_i\}$ by first generating a set of uncorrelated variables $\{y_i\}$ and then quantizing these sample by sample.

Coding into the channel is assumed to take place on a symbol by symbol basis. If y_i is quantized into 2^{b_i} levels, then b_i binary digits are required to transmit y_i over the channel. While the various values of $\hat{y}_i$ are not precisely equiprobable in general, Max' results indicate that the differences in probability are always quite small for the Gaussian case so that any benefits derived from block coding would be trivial compared to the increased complexity of instrumentation. In any case, similar improvements could be made in transmitting the output of a single sample quantizer and are therefore without interest here. Under these assumptions the only significant

* Received April 17, 1963.
† San Jose State College, San Jose, Calif.
‡ Yale University, New Haven, Conn.

effect of the noise-free channel is to fix the number of quantization digits (quantum levels) available per source sample.

The problem may now be stated as follows. A fixed number of channel digits is available to represent the quantized version of N samples from the source. It is desired to minimize the mean-square error between input and output, *i.e.*, to minimize $E\{(X-\hat{X})^T(X-\hat{X})\}$,[1] by adjustment of the matrices P and R and by assignment of an appropriate number of quantization digits to each of the N quantizers.

II. Results

The analysis leads to the following general results:

1) The best choice for the matrix R is the inverse of P.

2) The best choice for the matrix P is the transpose of the orthogonal matrix which diagonalizes the moment matrix M_x of the input process. In other words, $P = U^T$, where U is the orthogonal matrix which satisfies

$$U^T M_x U = \begin{bmatrix} \lambda_1 & & & 0 \\ & \cdot & & \\ & & \cdot & \\ 0 & & & \lambda_N \end{bmatrix} \tag{1}$$

$\lambda_1 \geq \lambda_2 \geq \cdots \geq \lambda_N$ are the eigenvalues of M_x arranged in descending order. This choice is optimal regardless of the number of digits b_i assigned to each of the N quantizers as long as $b_1 \geq b_2 \geq \cdots \geq b_N$.

3) The best choice of the number of binary digits b_i for the ith quantizer is given *approximately* by the expression

$$b_i = \theta + 1/2 \log_2 \frac{\lambda_i}{[\text{Det } M_x]^{1/N}} \tag{2}$$

where θ is the (fixed) average number of binary digits per sample available in the channel and Det M_x is the determinant of the moment matrix M_x. The values of b_i resulting from this expression need not be integers and may even be negative in special cases. As a consequence some trial and error is required to find the optimum digit assignment. However, with (2) as a guide only a limited number of fairly simple computations need be made. A series of examples are worked out in Section IV to indicate the type of situations that might arise in practice.

4) The proposed block quantizing scheme leads to a mean-square error smaller than that attainable through single sample quantization by a factor of approximately $(\text{Det } M_x)^{1/N}/Q$, where Q is the average source power. In all cases investigated the computational error due to this approximation was less (and often substantially less) than 15 per cent. Hence the choice of an appropriate block length N can be made on the basis of a very simple computation. A lower bound on the mean-square error with block length N is $2^{-2\theta}$ $(\text{Det } M_x)^{1/N}$.

[1] The symbol T denotes the transpose of a matrix and $E\{\ \}$ designates the expectation of the bracketed quantity.

III. Derivation

It is convenient to derive the results in the order in which they were listed in Section II, because the conclusion $R = P^{-1}$ is independent of the choice of P and $\{b_i\}$ and the conclusion $P = U^T$ does not depend on the choice of $\{b_i\}$.

Theorem I:

The mean-square error $D = (1/NE)\{(X-\hat{X})^T(X-\hat{X})\}$ is minimized by choosing $R = P^{-1}$.

Proof:

$$Y = PX, \qquad \hat{X} = R\hat{Y}$$

hence

$$\begin{aligned} D &= (1/N)E\{(P^{-1}Y - R\hat{Y})^T(P^{-1}Y - R\hat{Y})\} \\ &= (1/N)E\{[P^{-1}(Y - \hat{Y}) + (P^{-1} - R)\hat{Y}]^T \\ &\qquad \cdot [P^{-1}(Y - \hat{Y}) + (P^{-1} - R)\hat{Y}]\} \\ &= (1/N)E\{(Y - \hat{Y})^T(P^{-1})^TP^{-1}(Y - \hat{Y})\} \\ &\qquad + (1/N)E\{\hat{Y}^T(P^{-1} - R)^T(P^{-1} - R)\hat{Y}\} \\ &\qquad + (1/N)E\{\hat{Y}^T(P^{-1} - R)^TP^{-1}(Y - \hat{Y})\} \\ &\qquad + (1/N)E\{(Y - \hat{Y})^T(P^{-1})^T(P^{-1} - R)\hat{Y}\}. \end{aligned} \tag{3}$$

For convenience in further manipulation let

$$(P^{-1} - R)^TP^{-1} \equiv C \equiv [c_{ij}]$$

and consider the last two terms of (3),

$$\begin{aligned} &(1/N)E\{\hat{Y}^T(P^{-1} - R)^TP^{-1}(Y - \hat{Y})\} \\ &\qquad + (1/N)E\{(Y - \hat{Y})^T(P^{-1})^T(P^{-1} - R)\hat{Y}\} \\ &= (1/N)E\{Y^TC(Y - \hat{Y})\} + (1/NE)\{(Y - \hat{Y})^TC^T\hat{Y}\} \\ &= (2/N) \sum_{i=1}^{N} \sum_{j=1}^{N} c_{ij}[E(\hat{y}_iy_j) - E(\hat{y}_i\hat{y}_j)] \\ &= (2/N) \sum_{i=1}^{N} c_{ii}[E(\hat{y}_iy_i) - E(\hat{y}_i^2)] \\ &\qquad + (2/N) \sum_{\substack{i=1 \\ i\neq j}}^{N} \sum_{j=1}^{N} c_{ij}[E(\hat{y}_iy_j) - E(\hat{y}_i\hat{y}_j)]. \end{aligned} \tag{4}$$

Defining

$P(S_i^k)$ ≡ the probability that y_i falls into S_i^k, the kth interval of the ith quantizer

$P(y_i/S_i^k)$ ≡ the conditional probability density function of y_i knowing that the variable falls into S_i^k

M_i ≡ the number of quantization intervals of the ith quantizer

$\hat{y}_i^{(k)}$ = the output of the ith quantizer when y_i falls into S_i^k

and recognizing that the Lloyd-Max quantizer assigns the values of $\hat{y}_i$ according to the expression

$$\hat{y}_i^{(k)} = \int_{S_i^k} y_iP(y_i/S_i^k)\, dy_i, \tag{5}$$

one obtains

$$E[\hat{y}_i y_i] = \sum_{k=1}^{M_i} P(S_i^k)\hat{y}_i^{(k)} \int_{S_i{}^k} y_i P(y_i/S_i^k)\, dy_i = E[\hat{y}_i^2]. \quad (6)$$

Thus the next to last term in (4) vanishes. In the last term, because y_i and y_j are zero-mean uncorrelated Gaussian variables

$$E(\hat{y}_i y_j) - E(\hat{y}_i \hat{y}_j) = E(\hat{y}_i)E(y_j) - E(\hat{y}_i)E(\hat{y}_j), \quad (7)$$

but

$$E(\hat{y}_i) = \sum_{k=1}^{M_i} P(S_i^k) \int_{S_i{}^k} y_i P(y_i/S_i^k)\, dy_i = E(y_i). \quad (8)$$

Hence, the last term of (4) vanishes also and from (3)

$$D = (1/N)E\{(Y - \hat{Y})(P^{-1})^T P^{-1}(Y - \hat{Y})\} + (1/N)E\{\hat{Y}^T(P^{-1} - R)^T(P^{-1} - R)\hat{Y}\}. \quad (9)$$

Both terms of (9) are non-negative definite quadratic forms. Therefore, D is minimized by choosing $R = P^{-1}$ and the resulting mean-square error is

$$D = (1/N)E\{(Y - \hat{Y})(P^{-1})^T P^{-1}(Y - \hat{Y})\}. \quad (10)$$

This completes the proof.

For ease in further manipulation, define

$$(P^{-1})^T P^{-1} \equiv W \equiv [w_{ij}]. \quad (11)$$

Since $E\{(y_i - \hat{y}_i)(y_j - \hat{y}_j)\} = 0$ for $i \neq j$, (10) becomes

$$D = (1/N) \sum_{i=1}^{N} w_{ii} E[(y_i - \hat{y}_i)^2] = (1/N) \sum_{i=1}^{N} w_{ii}\sigma_i^2 E\left[\left(\frac{y_i}{\sigma_i} - \frac{\hat{y}_i}{\sigma_i}\right)^2\right] \quad (12)$$

where

$$\sigma_i^2 \equiv E(y_i^2). \quad (13)$$

Thus the term

$$D_i \equiv E\left[\left(\frac{y_i}{\sigma_i} - \frac{\hat{y}_i}{\sigma_i}\right)^2\right] \quad (14)$$

is the mean-square error of a Lloyd-Max quantizer designed to operate in optimal fashion on a random variable y_i/σ_i with unit variance. Since the input process $\{x_i\}$ is Gaussian the y_i/σ_i are Gaussian and the corresponding quantizer has been discussed in detail by Lloyd and Max. Thus, the overall mean-square error per sample of the proposed instrumentations can be written in terms of the mean-square errors of the separate Lloyd-Max quantizers as follows:

$$D = (1/N) \sum_{i=1}^{N} w_{ii}\sigma_i^2 D_i. \quad (15)$$

Eq. (15) is to be minimized by adjustment of the w_{ii}, σ_i^2 and D_i. The D_i are under the designer's control only to the extent that he can assign the total number of quantization digits available for the representation of N source samples in arbitrary fashion among the N quantizers. Thus if the number of binary digits assigned to the ith quantizer is b_i(*i.e.*, the quantizer has 2^{b_i} output levels) and the channel provides θ binary digits per source sample, then the $\{b_i\}$ can be any set of integers satisfying

$$\sum_{i=1}^{N} b_i = N\theta. \quad (16)$$

The parameters $\{w_{ii}\sigma_i^2\}$ are determined by the transformation matrix P. Their adjustment for minimum D is clearly subject to the constraint that P must be a realizable linear transformation converting the moment matrix M_x to a diagonal moment matrix with elements $\sigma_1^2, \cdots, \sigma_N^2$ along the principal diagonal. Thus the $\{w_{ii}\sigma_i^2\}$ are obviously constrained by the equation

$$\text{Trace}(M_x) = E(X^T X) = E[Y^T(P^{-1})^T P^{-1} Y] = \sum_{i=1}^{N} w_{ii}\sigma_i^2. \quad (17)$$

Let the b_i (and hence the D_i) be fixed and ordered such that

$$b_1 \geq b_2 \geq \cdots \geq b_N. \quad (18)$$

The use of a larger number of levels clearly decreases the quantizing error. Hence

$$D_1 \leq D_2 \leq \cdots \leq D_N. \quad (19)$$

In the light of (15) and (17) it therefore appears desirable to choose $w_{11}\sigma_1^2 = \text{Trace}(M_x)$, $w_{ii}\sigma_i^2 = 0$, $i > 1$. Unfortunately, this choice does not lead to a realizable transformation P in most cases. Since one can clearly think of the $w_{ii}\sigma_i^2$ as weights on a known, monotone nondecreasing sequence $\{D_i\}$, the crux of the design problem becomes the determination of the maximum values which successive $w_{ii}\sigma_i^2$ can assume without making the transformation P unrealizable.

The required insight into this problem is given by the following theorem.

Theorem 2:

Let $\lambda_1 \geq \lambda_2 \geq \cdots \geq \lambda_N$ be the eigenvalues of the moment matrix M_x. Let P be any nonsingular $N \times N$ matrix satisfying

$$PM_xP^T = \begin{bmatrix} \sigma_1^2 & & 0 \\ & \ddots & \\ 0 & & \sigma_N^2 \end{bmatrix} \equiv \Sigma \quad (20)$$

where the σ_i are arbitrary constants. Define [as in (11)]

$$(PP^T)^{-1} \equiv W \equiv [w_{ij}]. \quad (21)$$

Then the mean-square error D (15) satisfies the inequality

$$D = (1/N) \sum_{i=1}^{N} w_{ii}\sigma_i^2 D_i \geq 1/N \sum_{i=1}^{N} \lambda_i D_i. \quad (22)$$

Proof:

Using the definitions

$$\mathfrak{D} \equiv \begin{bmatrix} D_1 & & O \\ & \ddots & \\ O & & D_N \end{bmatrix} \qquad \Lambda \equiv \begin{bmatrix} \lambda_1 & & O \\ & \ddots & \\ O & & \lambda_N \end{bmatrix}, \tag{23}$$

the quantity ND can be written in the form

$$ND = Tr\{(PP^T)^{-1} \sum \mathfrak{D}\}. \tag{24}$$ [2]

Let U be the orthogonal matrix satisfying

$$U^T M_x U = \Lambda. \tag{25}$$

Then from (20) and (25)

$$\sum = PM_xP^T = P(U\Lambda U^T)P^T = (PU\Lambda^{1/2})(\Lambda^{1/2}U^TP^T). \tag{26}$$

Premultiplying and postmultiplying by $\Lambda^{-\frac{1}{2}}$, one obtains

$$(\sum^{-1/2}PU\Lambda^{1/2})(\Lambda^{1/2}U^TP^T\sum^{-1/2})$$
$$= (\sum^{-1/2}PU\Lambda^{1/2})(\sum^{-1/2}PU\Lambda^{1/2})^T = I. \tag{27}$$

Hence,

$$A \equiv \sum^{-1/2}PU\Lambda^{1/2} \tag{28}$$

is an orthogonal matrix. In terms of A

$$P = \sum^{+1/2}A\Lambda^{-1/2}U^T. \tag{29}$$

Hence,

$$(PP^T)^{-1} = [\sum^{1/2}A\Lambda^{-1/2}U^TU\Lambda^{-1/2}A^T\sum^{1/2}]^{-1}$$
$$= [\sum^{1/2}A\Lambda^{-1}A^T\sum^{1/2}]^{-1}$$
$$= \sum^{-1/2}A\Lambda A^T\sum^{-1/2}. \tag{30}$$

Eq. (24) can now be rewritten as follows:

$$ND = Tr\{\sum^{-1/2}A\Lambda A^T\sum^{1/2}\mathfrak{D}\}$$
$$= Tr\{A\Lambda A^T\sum^{1/2}\mathfrak{D}\sum^{-1/2}\} = Tr\{A\Lambda A^T\mathfrak{D}\}, \tag{31}$$

since the trace of a matrix product is invariant under cyclic permutations of the factors and since the diagonal matrices $\mathfrak{D}$ and $\sum^{-\frac{1}{2}}$ commute.

If the elements of A are written as $\{a_{ij}\}$, (31) becomes

$$ND = \sum_{i=1}^{N}\sum_{j=1}^{N} a_{ij}^2 D_i \lambda_j. \tag{32}$$

Let

$$D_j' = \sum_{i=1}^{N} a_{ij}^2 D_i, \tag{33}$$

so that

$$ND = \sum_{j=1}^{N} \lambda_j D_j'. \tag{34}$$

A is an orthogonal matrix so that the diagonal elements of $AA^T = A^TA$ are all unity. Hence,

$$\sum_{i=1}^{N} a_{ij}^2 = \sum_{j=1}^{N} a_{ij}^2 = 1.$$

[2] $Tr\{\ \}$ designates the trace of the indicated matrix.

Therefore the $\{D_i'\}$ are averages of the $\{D_i\}$ in the sense of Hardy, Littlewood and Pólya.[3] Hence, if the $\{D_i'\}$ are arranged in ascending order,

$$\sum_{j=1}^{k} D_j' \geq \sum_{j=1}^{k} D_j \qquad k \leq N. \tag{35}$$

If the $\{D_j'\}$ are not arranged in ascending order this becomes a strict inequality. Now consider

$$\sum_{j=1}^{N} \lambda_j D_j' = \sum_{j=2}^{N} \lambda_j \left[\sum_{i=1}^{j} (D_i') - \sum_{i=1}^{j-1} (D_i') \right] + \lambda_1 D_1'$$
$$= \sum_{j=2}^{N} \lambda_j \sum_{i=1}^{j} D_i' - \sum_{j=1}^{N-1} \lambda_{j+1} \sum_{i=1}^{j} D_i' + \lambda_1 D_1'$$
$$= \sum_{j=2}^{N-1} (\lambda_j - \lambda_{j+1}) \sum_{i=1}^{j} D_i' + \lambda_N \sum_{i=1}^{N} D_i' + (\lambda_1 - \lambda_2) D_1'. \tag{36}$$

Since $\lambda_1 \geq \lambda_2 \geq \cdots \geq \lambda_N > 0$, use of (35) yields immediately

$$\sum_{j=1}^{N} \lambda_j D_j' \geq \sum_{j=2}^{N-1} (\lambda_j - \lambda_{j+1}) \sum_{i=1}^{j} D_i$$
$$+ \lambda_N \sum_{i=1}^{N} D_i + (\lambda_1 - \lambda_2) D_1. \tag{37}$$

Reversing the sequence of operations performed in (36), one obtains

$$\sum_{j=2}^{N-1} (\lambda_j - \lambda_{j+1}) \sum_{i=1}^{j} D_i$$
$$+ \lambda_N \sum_{i=1}^{N} D_i + (\lambda_1 - \lambda_2) D_1 = \sum_{j=1}^{N} \lambda_j D_j. \tag{38}$$

Hence,

$$\sum_{j=1}^{N} \lambda_j D_j' \geq \sum_{j=1}^{N} \lambda_j D_j. \tag{39}$$

Substitution of (39) into (34) completes the proof.

It is now a simple matter to show that the lower bound of (22) is indeed attainable.

Theorem 3:

Let U be the orthogonal matrix satisfying

$$U^T M_x U = \Lambda. \tag{40}$$

Then for any given set of $\{b_i\}$ satisfying $b_1 \geq b_2 \geq \cdots \geq b_N$, the mean-square error D is minimized by the linear transformation

$$P = U^T \tag{41}$$

and the minimum is

$$D = (1/N) \sum_{i=1}^{N} \lambda_i D_i. \tag{42}$$

[3] G. H. Hardy, J. E. Littlewood, G. Pólya, "Inequalities," Cambridge University Press, London, England, Section 2.20; 1959.

Proof:

The transformation $P = U^T$ leads to the moment matrix

$$M_y = E[YY^T] = E[U^TXX^TU] = U^TM_xU = \Lambda. \quad (43)$$

Thus $\sigma_i^2 = \lambda_i$. Furthermore $U^TU = I$, so that [from (11)] $w_{ii} = 1$, $i = 1, 2, \cdots N$. Hence $\sigma_i^2 w_{ii} = \lambda_i$ and (22) becomes an equality. Note also that the matrix A (28) is now the identity matrix so that (33) yields $D_j' = D_j$.

Theorem 3 demonstrates that the optimum transformations P and R are U^T and U respectively, regardless of the assignments of binary digits b_i to the various quantizers.[4] On the other hand, it is obvious from (42) that the choice of the $\{b_i\}$ (and hence $\{D_i\}$) affects the mean-square error D. The remaining problem is therefore the determination of the optimum set of $\{b_i\}$. The following approximate computation provides a guide to the optimum choice that reduces the number of trials required for the final selection to a relatively small number in most cases of practical interest. The argument is based on the observation [2] that D_i, the mean-square error of a single Lloyd-Max quantizer (with input of unit variance), varies almost linearly with 2^{-2b_i}. Formally one can write

$$D_i = K(b_i)2^{-2b_i} = K(b_i)e^{-2b_i \ln 2}, \quad (44)$$

where $\ln 2 = \log_e 2$ and $K(b_i)$ is a function varying only slowly with b_i. From the computational results given by Lloyd [1] and Max [2] it is a simple matter to plot $K(b_i)$ for the case of a Gaussian variable. The result is shown in Fig. 2.[5] For $b_i \geq 3$, $K(b_i)$ varies only slightly and even for $b_i < 3$ the variation over any given range is quite small compared to the variation of 2^{-2b_i}. The assumption $K(b_i) = K$, a constant, therefore suggests itself as a first approximation. Using this approximation as well as (44) one can rewrite (42) as follows:

$$D \cong (1/N) \sum_{i=1}^{N} \lambda_i K e^{-2b_i \ln 2}. \quad (45)$$

This expression must be minimized by adjustment of the b_i under the constraint of (16). Treating b_i as a continuous variable and using an undetermined multiplier β, one obtains

$$\frac{\partial}{\partial b_j}\left[(1/N) \sum_{i=1}^{N} \lambda_i K e^{-2b_i \ln 2} + \beta \sum_{i=1}^{N} b_i\right]$$
$$= -(1/N)\lambda_j K 2 \ln 2 e^{-2b_j \ln 2} + \beta = 0. \quad (46)$$

It follows that

$$\lambda_j e^{-2b_j \ln 2} = \frac{N\beta}{K2 \ln 2} = C, \text{ a constant} \quad (47)$$

$$\text{for } j = 1, 2, \cdots N.$$

[4] The condition $b_1 \geq b_2 \geq \cdots \geq b_N$ would always be satisfied in practice because one would clearly not assign a smaller number of quantization digits to a variable with a larger mean-square value.

[5] Physical meaning is obviously attached only to integral values of b_i or, at most, to integral numbers of levels $N_i = 2^{b_i}$.

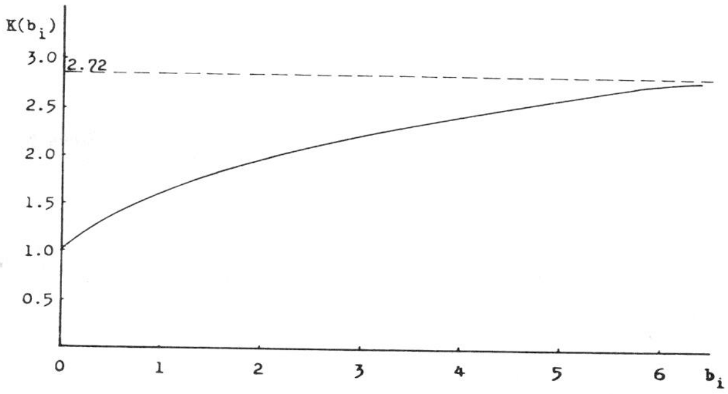

Fig. 2—The function $K(b_i)$.

Solving (47) for b_j and using (16) to evaluate the constant C, one obtains

$$b_j = \theta + \frac{1}{2 \ln 2}\left[\ln \lambda_j - (1/N) \ln\left(\prod_{j=1}^{N} \lambda_j\right)\right]$$
$$= \theta + \frac{1}{2 \ln 2} [\ln \lambda_j - (1/N) \ln (\text{Det } M_x)]$$
$$= \theta + 1/2 \log_2 \frac{\lambda_j}{(\text{Det } M_x)} \quad (48)$$

where Det M_x is the determinant of the moment matrix M_x. Formal substitution of (48) into (45) leads to

$$D \cong K2^{-2\theta}(\text{Det } M_x)^{1/N}. \quad (49)$$

Recognizing that $K(b_i) \geq 1$ for all i, one obtains the following easily computed lower bound ($D_{l.b.}$) on the minimum mean-square error.

$$D_{l.b.} = 2^{-2\theta}(\text{Det } M_x)^{1/N}. \quad (50)$$

An interesting equivalent expression for $D_{l.b.}$ can be obtained by observing that the entropy per sample (H') of N samples from a Gaussian process with moment matrix M_x is given by

$$H' = (1/N) \ln [(2\pi e)^{N/2}(\text{Det } M_x)^{1/2}]. \quad (51)^6$$

Hence,

$$(\text{Det } M_x)^{1/N} = \frac{1}{2\pi e} e^{2H'}. \quad (52)$$

When $N \to \infty$, H' becomes the entropy per degree of freedom defined by Shannon[7] and $(1/2\pi e)e^{2H'}$ becomes the entropy power (Q_1) of the process, *i.e.*, the power of a set of independent Gaussian samples having the same entropy as the sequence of samples under study. Thus, for large blocks

$$D_{l.b.} \to Q_1 2^{-2\theta}. \quad (53)$$

If $N = 1$, Det $M_x = Q$, the average power of the process whose samples are being quantized. Thus Q_1/Q, or more

[6] In this definition H' is measured in natural units.

[7] C. E. Shannon, "A mathematical theory of communication," *Bell Sys. Tech. J.*, p. 633, October, 1948.

generally $(\text{Det } M_x)^{1/N}/Q$, can be used as an approximate measure of the improvement attainable through the proposed block quantization scheme.

The lower bound given by (50) is not in general realizable for two reasons. 1) $K(b_i) > 1$ for $b_i > 0$. 2) The theoretically optimal b_i given by (32) need not be integers. If the samples are very highly correlated it is even possible that the formal solution will yield negative values of b_i. One would expect the increase in error due to 2) to be fairly small in most practical cases and would therefore anticipate that a reasonably accurate estimate of the minimum attainable error would be given by

$$D \cong K(\theta)2^{-2\theta}(\text{Det } M_x)^{1/N}. \qquad (54)$$

Since sample by sample quantization with θ binary digits leads to a mean-square error of $K(\theta)2^{-2\theta}Q$, (54) is equivalent to the previous observation that the ratio $(\text{Det } M_x)^{1/N}/Q$ should furnish a measure of the improvement attainable through block quantization.

The actual selection of integral b_i is a matter of some trial and error. Having the formal result of (48), one need only consider combinations of integers close to the values given by that equation. With the $w_{ii}\sigma_i^2 = \lambda_i$ known and the D_i in known relation to the b_i (44) the necessary trial computations can be made rapidly by use of (42). In selecting appropriate integral combinations of b_i it is worth keeping in mind that the assumption $K(b_i)$ = constant assigns too much relative error to quantizers with small b_i. When (48) yields a value roughly midway between two small integers, one would therefore expect the lower choice to lead to a smaller mean-square error.

IV. Examples

As a first example, consider a continuous Gaussian process with the autocorrelation function $\phi(\tau) = e^{-|\tau|}$ (a Markov process). Two sampling rates will be considered. 1) A sampling rate such that the correlation coefficient ρ between successive samples is 0.8. 2) A sampling rate such that the correlation coefficient ρ between successive samples is 0.9.

For case 1) Figs. 3–5 give the normalized mean-square error

$$V = \frac{D}{QK(\theta)2^{-2\theta}} \qquad (55)$$

as a function of N. Values of θ equal to 2 (coarse quantizing), 4(moderately fine quantizing) and ∞ (extremely fine quantizing) have been used. As an illustration of the computational procedure, consider the combination $\theta = 2$, $N = 4$. The eigenvalues of M_x are $\lambda_1 = 3.103$, $\lambda_2 = 0.559$, $\lambda_3 = 0.209$, $\lambda_4 = 0.129$. Eq. (48) yields the following estimates of optimum digit assignments:

$$b_1 = 3.34,\ b_2 = 2.11,\ b_3 = 1.48,\ b_4 = 1.06. \qquad (56)$$

From (42) one obtains

$$D = 0.776D_1 + 0.140D_2 + 0.052D_3 + 0.032D_4. \qquad (57)$$

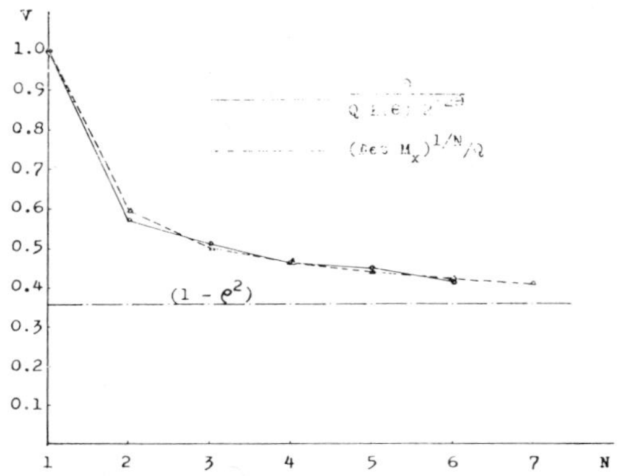

Fig. 3—Block quantizer performance. Markov input, $\rho = 0.8$, $\theta = 2$.

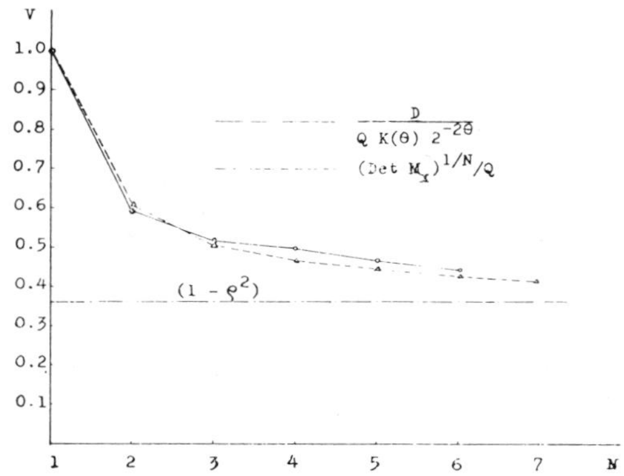

Fig. 4—Block quantizer performance. Markov input, $\rho = 0.8$, $\theta = 4$.

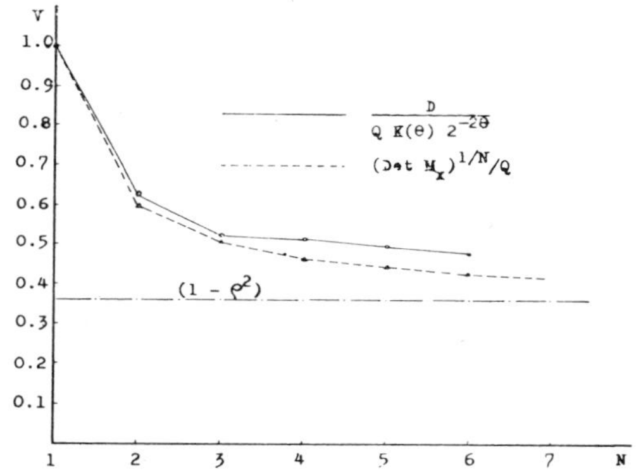

Fig. 5—Block quantizer performance. Markov input, $\rho = 0.8$, $\theta = \infty$.

Table I shows combinations of integral values of $\{b_i\}$ sufficiently close to (56) to merit investigation, and the resulting values of $\{D_i\}$ and D. Since an increase (decrease) by one unit in b_i reduces (increases) D_i by a factor somewhat smaller than 4, it is obvious that all other combinations will lead to larger values of D. No more than three trials were necessary to find the optimum combination in any of the examples investigated and in most instances one or two trials were sufficient. In addition to the actual value of V, Figs. 3–5 show the approximate value $(\text{Det } M_x)^{1/N}/Q$. For $\theta = 2$ the actual values are scattered on both sides of the approximate curve; for large values of θ they lie consistently above it. This is due to the circumstance that the use of nonintegral b_i in the approximation decreases the error whereas the assumption of constant $K(b_i)$ tends to increase it. Thus, the two approximations introduce inaccuracies that tend to cancel. For large θ, $K(b_i)$ is in fact very nearly constant, hence only the first factor is operative. Aside from this rather minor effect the improvement ratio V is evidently independent of θ.

TABLE I

b_1	b_2	b_3	b_4	D_1	D_2	D_3	D_4	D
3	2	2	1	0.0268	0.0164	0.0061	0.0116	0.0609
4	2	1	1	0.0074	0.0164	0.0189	0.0116	0.0543
4	2	2	0	0.0074	0.0164	0.0061	0.0320	0.0619

Shown also in Figs. 3–5 is the line $1 - \rho^2 = 0.36$, a lower bound on V. For $N \geq 2$ it is the conditional variance of a sample knowing $N - 1$ past samples. Since the process under study has the Markov property, this conditional variance is constant for $N \geq 2$. It is a simple matter to show that

$$\lim_{N\to\infty} \frac{(\text{Det } M_x)^{1/N}}{Q} = 1 - \rho^2, \tag{58}$$

so that at least the approximate value of V asymptotically approaches the minimum attainable with the available information. From a practical point of view, V is sufficiently close to the asymptotic value at $N = 3$ or 4 so that it would be difficult to justify the use of larger blocks.

Fig. 6 presents similar information for case 2) ($\rho = 0.9$) with $\theta = 2$. Again the actual values of V are scattered about the approximate curve which approaches the lower bound $1 - \rho^2 = 0.19$ for large N. As in the first case, little improvement is attained by going to values of N larger than 3 or 4.

The first example is rather special in one sense, *i.e.*, because of the Markov property one expects (and finds) relatively little gain from the use of blocks of length greater than 2. The second example is therefore chosen from a category characterized by strong dependence of the conditional distribution of a given sample on more than the immediately preceding sample. Specifically, the autocorrelation function of the continuous input process will be taken as

$$\phi(\tau) = e^{-\tau^2}. \tag{59}$$[8]

[8] This corresponds to a spectrum of the form $e^{-\omega^2}$ which is nonfactorable in the sense of Wiener ("Extrapolation, Interpolation and Smoothing of Stationary Time Series," John Wiley and Sons, Inc., New York, N. Y., § 2.2; 1949.) It may be regarded as the limiting version of a factorable spectrum.

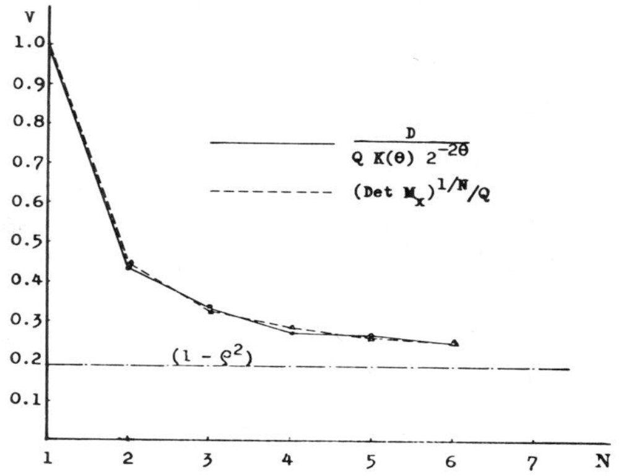

Fig. 6—Block quantizer performance. Markov input, $\rho = 0.9$, $\theta = 2$.

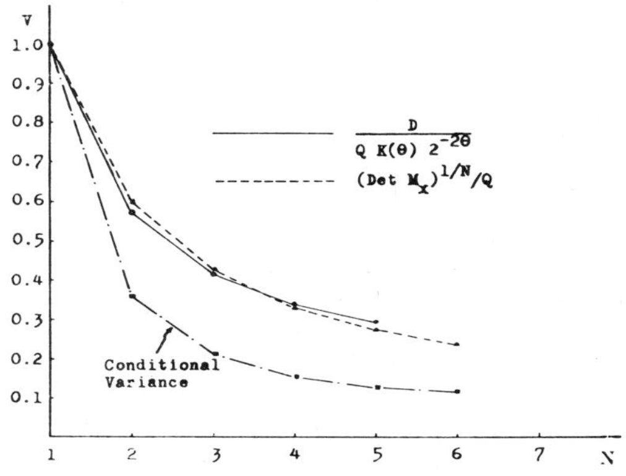

Fig. 7—Block quantizer performance. Input autocorrelation $e^{-\tau^2}$, $\rho = 0.8$, $\theta = 2$.

Sampling is assumed to take place at a rate such that the correlation coefficient between adjacent samples is 0.8. Fig. 7 shows V and $(\text{Det } M_x)^{1/N}/Q$ for $\theta = 2$. As anticipated, the use of larger blocks yields considerably more improvement in the second example than in the first. The lower bound shown in Fig. 7 is again the conditional variance of a sample given the $N - 1$ preceding samples. It is not obvious in this case whether $(\text{Det } M_x)^{1/N}/Q$ approaches the lower bound for large N.

V. Concluding Remarks

The procedure for block quantizing developed in this article is optimal only if the input samples have a jointly Gaussian probability distribution. One would expect, of course, that the same technique would also lead to substantial error reductions (relative to single sample quantization) in many non-Gaussian situations. The difficulty in the non-Gaussian case hinges on the fact that the linear transformation P in Fig. 1 affects the probability density of Y. The $\{y_i\}$ will not, in general, have probability density functions of the same form, nor will y_i have a probability density of the same form as x_i. Since the quantity D_i (14) depends on the form of the probability density of y_i as well as on b_i, one may be able to select a transformation P other than U^T

which lowers the D_i sufficiently to reduce D (15), even though the $w_{ii}\sigma_i^2$ no longer satisfy $w_{ii}\sigma_i^2 = \lambda_i$. In the Gaussian case this problem does not arise because the $\{y_i\}$ are Gaussian and hence the D_i are independent of the transformation P. Formal optimization in the non-Gaussian case appears to be a difficult problem and it is not clear whether in situations of practical interest the attainable improvement would differ greatly from that achieved in the Gaussian case.

VI. Acknowledgement

The contents of this paper are derived in essence from a dissertation submitted by J. J. Y. Huang to the faculty of the Yale University School of Engineering in partial fulfillment of the requirements for the degree of Doctor of Engineering. The research on which it is based was supported by Bell Telephone Laboratories Inc., Murray Hill, N. J. The authors are grateful to various staff members of Bell Telephone Laboratories for critical comments on the manuscript and wish to acknowledge in particular detailed suggestions for a concise proof of Theorem 2 made by E. Kimme and A. J. Goldstein.

References

[1] S. P. Lloyd, "Least Square Quantization in PCM," Bell Telephone Lab. Memo., Murray Hill, N. J. (unpublished).
[2] J. Max, "Quantizing for minimum distortion," IRE Trans. on Information Theory, vol. IT-6, pp. 7–12; March, 1960.
[3] P. Zador, Bell Telephone Lab., Murray Hill, N. J. (unpublished).
[4] J. Y. Huang, "Quantization of Correlated Random Variables," Ph.D. dissertation, Yale University, New Haven, Conn.; 1962.

28

Reprinted from *IEEE Trans. Inform. Theory* **IT-25**(5):537-543 (1979)

Quantization Schemes for Bivariate Gaussian Random Variables

JAMES A. BUCKLEW AND NEAL C. GALLAGHER, JR., MEMBER, IEEE

***Abstract*—The problem of quantizing two-dimensional Gaussian random variables is considered. It is shown that, for all but a finite number of cases, a polar representation gives a smaller mean square quantization error than a Cartesian representation. Applications of the results to a transform coding scheme known as spectral phase coding are discussed.**

I. Introduction

CONSIDER a two-dimensional Gaussian random variable X with independent components. For many applications in signal processing and digital communications it is necessary to represent this quantity by a finite set of values. One possible representation of X is in Cartesian coordinates, obtained by individually quantizing the two rectangular components of X. An alternative representation, in polar coordinates, is obtained by quantizing the magnitude and phase angle of X.

In [1] experimental data are put forward to show that, in all of the cases treated, polar formatting is better than rectangular. The purpose of this paper is to give a more rigorous treatment of the problem and to ascertain which of the representations leads to a smaller mean square quantization error.

In the first section we will derive the exact error expression for the polar format. The second and third sections deal with computer simulations of the expression and compare the polar and rectangular formats. It is shown that, in almost all cases, the polar format gives a smaller quantization error.

If the polar format is to be used, the question arises as to the best ratio of the number of phase quantizer levels to the number of magnitude quantizer levels. Pearlman [2] used distortion rate theory to derive a bound for this expression. In the fourth section we derive an asymptotic expression that agrees with the Pearlman result and perform computer simulations showing the validity of this bound.

In the fifth section we apply the above results to a transform coding scheme, spectral phase coding (SPC). Theoretical arguments are given for the observed robustness of SPC, and an exact error expression is derived. Computer simulations are then made demonstrating the robustness of SPC.

Manuscript received November 18, 1977; revised December 18, 1978. This work was supported by the Air Force Office of Scientific Research, Air System Command, USAF, under grant AFOSR-78-3605.
The authors are with the School of Electrical Engineering, Purdue University, West Lafayette, IN 47907.

II. Development

Consider the mean square quantization error E_p of a polar format representation:

$$E_p=\sum_{j=1}^{N_\theta}\sum_{i=1}^{N_r}\int_{c_{j-1}}^{c_j}\int_{a_{i-1}}^{a_i}|r\exp(j\theta)-b_i\exp(jd_j)|^2 \cdot\frac{f_r(r)\,dr\,d\theta}{2\pi}, \quad (1)$$

where N_θ and N_r are the number of levels in the phase and magnitude quantizers, respectively. The b_i and d_j are the output levels of the magnitude and phase quantizers corresponding to input levels lying in the intervals $(a_{i-1},a_i]$ and $(c_{j-1},c_j]$, respectively. The function $f_r(r)$ is the input density of the magnitude which is Rayleigh distributed and independent of the random phase θ which is uniformly distributed over $[-\pi,\pi]$.

After squaring out the integrand and integrating over θ from c_{j-1} to c_j, we obtain

$$E_p=\sum_{j=1}^{N_\theta}\sum_{i=1}^{N_r}\int_{a_{i-1}}^{a_i}\left[(c_j-c_{j-1})\left[r^2+b_i^2\right]-2rb_i\left[\sin(c_j-d_j)-\sin(c_{j-1}-d_j)\right]\frac{f(r)\,dr}{2\pi}\right]. \quad (2)$$

Setting $\partial E_p/\partial d_j=0$ leads to the equations

$$c_j-d_j=\frac{\pi}{N_\theta}=d_j-c_{j-1} \quad (3a)$$

$$c_j-c_{j-1}=\frac{2\pi}{N_\theta}, \quad (3b)$$

for $j=1,\cdots,N_\theta$. It should be noted that these are simply the equations for a uniform quantizer. Consequently, the expression for mean square error becomes

$$E_p=\sum_{i=1}^{N_r}\int_{a_{i-1}}^{a_i}\left[r^2+b_i^2-2rb_i\,\mathrm{sinc}(1/N_\theta)\right]f(r)\,dr, \quad (4)$$

where $\mathrm{sinc}(\cdot)=\sin\pi(\cdot)/\pi(\cdot)$. A differentiation with respect to b_i yields the optimum b_i as

$$b_i=\mathrm{sinc}(1/N_\theta)\frac{\int_{a_{i-1}}^{a_i} rf(r)\,dr}{\int_{a_{i-1}}^{a_i} f(r)\,dr}. \quad (5)$$

Substituting this value back into (4), we find

$$E_p = \overline{r^2} - \sum_{i=1}^{N_r} \operatorname{sinc}^2(1/N_\theta) \frac{\left[\int_{a_{i-1}}^{a_i} rf(r)\,dr\right]^2}{\int_{a_{i-1}}^{a_i} f(r)\,dr}, \tag{6}$$

where the upper bar indicates the statistical expectation operator. Let $E(N_r, r)$ denote the mean square quantization error produced by an optimal, one-dimensional, N_r output level, Rayleigh quantizer. It is shown in [3] that $E(N_r, r)$ is given by the difference between the variance of the quantizer input and the variance of the output. Hence $E(N_r, r)$ may be written as

$$E(N_r, r) = \overline{r^2} - \sum_{i=1}^{N_r} b_i'^2 \left[\int_{a'_{i-1}}^{a'_i} f(r)\,dr\right], \tag{7}$$

where the $\{a'_i\}$ are the quantizer input interval endpoints and the $\{b'_i\}$ are the quantizer output levels. Max [4] shows that the $\{b'_i\}$ and $\{a'_i\}$ satisfy

$$a'_i = \frac{b'_i + b'_{i+1}}{2} \tag{8a}$$

$$b'_i = \frac{\int_{a'_{i-1}}^{a'_i} rf(r)\,dr}{\int_{a'_{i-1}}^{a'_i} f(r)\,dr}. \tag{8b}$$

These equations may be written as

$$a'_i = \frac{\int_{a'_{i-1}}^{a'_i} rf(r)\,dr}{2\int_{a'_{i-1}}^{a'_i} f(r)\,dr} + \frac{\int_{a'_i}^{a'_{i+1}} rf(r)\,dr}{2\int_{a'_i}^{a'_{i+1}} f(r)\,dr}. \tag{9}$$

Minimizing (4) with respect to the a_i yields

$$a_i = \frac{b_i + b_{i+1}}{2\operatorname{sinc}(1/N_\theta)}, \tag{10}$$

and substituting (5) into the above gives

$$a_i = \frac{\int_{a_{i-1}}^{a_i} rf(r)}{2\int_{a_{i-1}}^{a_i} f(r)\,dr} + \frac{\int_{a_i}^{a_{i+1}} rf(r)\,dr}{2\int_{a_i}^{a_{i+1}} f(r)\,dr}, \tag{11}$$

which is identical to (9). Fleisher [5] shows that Max's conditions (i.e., (8a) and (8b)) are necessary and sufficient for the optimality of the Rayleigh quantizer. Thus we are assured that the solutions to (11) are unique, leading us to the conclusion that

$$a_i = a'_i.$$

The polar format error expression then becomes

$$E_p = \operatorname{sinc}^2(1/N_\theta) E(N_r, r) + \left(1 - \operatorname{sinc}^2(1/N_\theta)\right)\overline{r^2}. \tag{12}$$

If we assume bit rate limited signal transmission, then we must constrain the product of N_r and N_θ to be less than or equal to some constant, let us say N. To compare the rectangular and polar formats, it is assumed that the product of N_x and N_y, the number of output levels of the rectangular format quantizers, must also equal N. By use of symmetry arguments it may be shown that, for optimal rectangular format operation, N_x must equal N_y. Therefore,

$$N_x = N_y = N^{1/2}. \tag{13}$$

Let $E(N_x, g)$ denote the mean square quantization error produced by an optimal N_x output level Gaussian quantizer. The rectangular format error E_{rect} is given by

$$E_{\text{rect}} = 2E(N_x, g) = 2E(\sqrt{N}, g). \tag{14}$$

The problem is now to compare (12) with (14).

III. Exact Computer Simulation

In this section we make use of Max's [4] tabulated results for $E(N_x, g)$. Max gives values of this function from $N_x = 1$ to $N_x = 36$. We duplicate Max's work for the Rayleigh quantizer and obtain values for $E(N_r, r)$. Using an exhaustive search, we compute the smallest values of error obtainable for (12) and (14) for values of N from 1 to 2000. For all of these cases, there are only 31 values of N for which the rectangular format is better. These values

TABLE I
Values of N where Rectangular Format Is Superior to Polar Format

Based upon exact expressions	Based upon approximate expressions
	1, 2, 3, 4,
6	6
8	8
9	9
12	12
13	13, 15
16	16
17	17
20	20
21	21
25	24, 25
26	26, 27, 28, 29, 30, 31, 32
35	35
36	36
37	37
38	38
42	42
43	43, 44, 48
49	49
50	50
51	51
56	56
57	57
58	58
59	59
63	63
64	64, 65, 66, 67
72	72
73	73
74	74, 81, 82, 83, 84, 99
100	
101	
	110, 111, 112, 113

TABLE II
A TABULATION OF THE RELATIVE EFFICIENCY $\eta=(E_p-E_r)/E_p$ OF POLAR QUANTIZATION OVER THAT OF RECTANGULAR QUANTIZATION, THE BEST NUMBER OF MAGNITUDE LEVELS N_r, AND THE BEST NUMBER OF RECTANGULAR FORMAT LEVELS N_x, AS A FUNCTION OF N

N	η	N_r	N_x	N	η	N_r	N_x	N	η	N_r	N_x
1	.000	1	1	51	3.801	4	7	101	.578	5	10
2	-.001	1	1	52	-3.544	4	7	102	-4.424	6	10
3	-28.572	1	1	53	-3.544	4	7	103	-4.424	6	10
4	-.005	1	2	54	-.990	4	6	104	-4.424	6	10
5	-16.226	1	2	55	-3.459	5	6	105	-4.424	6	10
6	2.468	1	2	56	1.613	4	7	106	-4.424	6	10
7	-4.084	1	2	57	1.613	4	7	107	-4.424	6	10
8	3.325	2	2	58	1.613	4	7	108	-6.469	6	9
9	22.679	1	3	59	1.613	4	7	109	-6.469	6	9
10	-.703	2	3	60	-5.316	5	7	110	-1.554	6	10
11	-.703	2	3	61	-5.316	5	7	111	-1.554	6	10
12	.620	2	3	62	-5.316	5	7	112	-1.554	6	10
13	.620	2	3	63	3.655	5	7	113	-1.554	6	10
14	-15.048	2	3	64	4.369	4	8	114	-6.812	6	10
15	-1.004	2	3	65	-1.544	5	8	115	-6.812	6	10
16	1.936	2	4	66	-1.544	5	8	116	-6.812	6	10
17	1.936	2	4	67	-1.544	5	8	117	-6.204	6	9
18	-6.640	2	4	68	-1.544	5	8	118	-6.204	6	9
19	-6.640	2	4	69	-1.544	5	8	119	-8.422	7	9
20	4.377	2	4	70	-6.538	5	7	120	-4.120	6	10
21	1.220	3	4	71	-6.538	5	7	121	-1.919	6	11
22	-.660	2	4	72	.689	5	8	122	-1.919	6	11
23	-.660	2	4	73	.689	5	8	123	-1.919	6	11
24	-2.631	3	4	74	.689	5	8	124	-1.919	6	11
25	6.493	3	5	75	-6.442	5	8	125	-1.919	6	11
26	6.493	3	5	76	-6.442	5	8	126	-6.151	6	11
27	-5.911	3	5	77	-6.442	5	8	127	-6.151	6	11
28	-5.911	3	5	78	-6.442	5	8	128	-6.151	6	11
29	-5.911	3	5	79	-6.442	5	8	129	-6.151	6	11
30	-1.022	3	5	80	-4.180	5	8	130	-2.146	6	10
31	-1.022	3	5	81	-.972	5	9	131	-2.146	6	10
32	-1.022	3	5	82	-.972	5	9	132	-1.865	6	11
33	-9.643	3	5	83	-.972	5	9	133	-4.015	7	11
34	-9.643	3	5	84	-2.232	6	9	134	-4.015	7	11
35	1.471	3	5	85	-6.499	5	9	135	-4.015	7	11
36	1.388	3	6	86	-6.499	5	9	136	-4.015	7	11
37	1.388	3	6	87	-6.499	5	9	137	-4.015	7	11
38	1.388	3	6	88	-2.789	5	8	138	-5.298	6	11
39	-4.288	3	6	89	-2.789	5	8	139	-5.298	6	11
40	-3.440	4	5	90	-1.765	5	9	140	-8.395	7	10
41	-3.440	4	5	91	-1.765	5	9	141	-8.395	7	10
42	3.885	3	6	92	-1.765	5	9	142	-8.395	7	10
43	3.885	3	6	93	-1.765	5	9	143	-2.599	7	11
44	-2.196	4	6	94	-1.765	5	9	144	-.750	7	12
45	-2.196	4	6	95	-6.090	5	9	145	-.750	7	12
46	-2.196	4	6	96	-8.522	6	9	146	-.750	7	12
47	-2.196	4	6	97	-8.522	6	9	147	-5.660	7	12
48	-1.140	4	6	98	-8.522	6	9	148	-5.660	7	12
49	3.801	4	7	99	-.593	6	9	149	-5.660	7	12
50	3.801	4	7	100	.578	5	10	150	-5.660	7	12

for N correspond in general to regions where N is a perfect square. Apparently, for values of N greater than 101, polar formatting is always the better of the two methods. The left column of Table I contains a listing of the 31 values of N for which rectangular format gives smaller error. Table II gives an indication of the relative efficiency of polar and rectangular formatting by tabulating $(E_p-E_r)/E_p$ for values of N from 1 to 150. Also in Table II may be found the best number of magnitude levels N_r (with N_θ = greatest integer less than N/N_r) and the best number of rectangular format levels N_x (with N_y = greatest integer less than N/N_x) for each value of N from 1 to 150. For values of N larger than 2000, we may make use of approximation methods.

IV. Approximate Computer Simulation

Wood [6] describes a technique whereby one can approximate the mean square error of an optimal quantizer for large N. He then gives an expression for the error of an N level Gaussian quantizer which agrees to within about one percent with the actual computed mean square error given by Max [4]. This error expression is

$$E(N_x,g)=\frac{2.73N_x\sigma^2}{(N_x+0.853)^3}. \tag{15}$$

Using Wood's approximations, we obtain for the Rayleigh density a similar error expression which also agrees well with the actual computed error. This error expression is

$$E(N_r,r)=\frac{0.9287N_r\sigma^2}{(0.596+N_r)^3}. \tag{16}$$

By use of these approximate error expressions, we again find the values of N where rectangular format gives smaller error than polar format. Computer simulations are run up to a value of $N=10^6$. We find that for values of N greater than 113, polar format is always better.

Table I summarizes the results of the last two sections. In the first column we find the values of N for which the

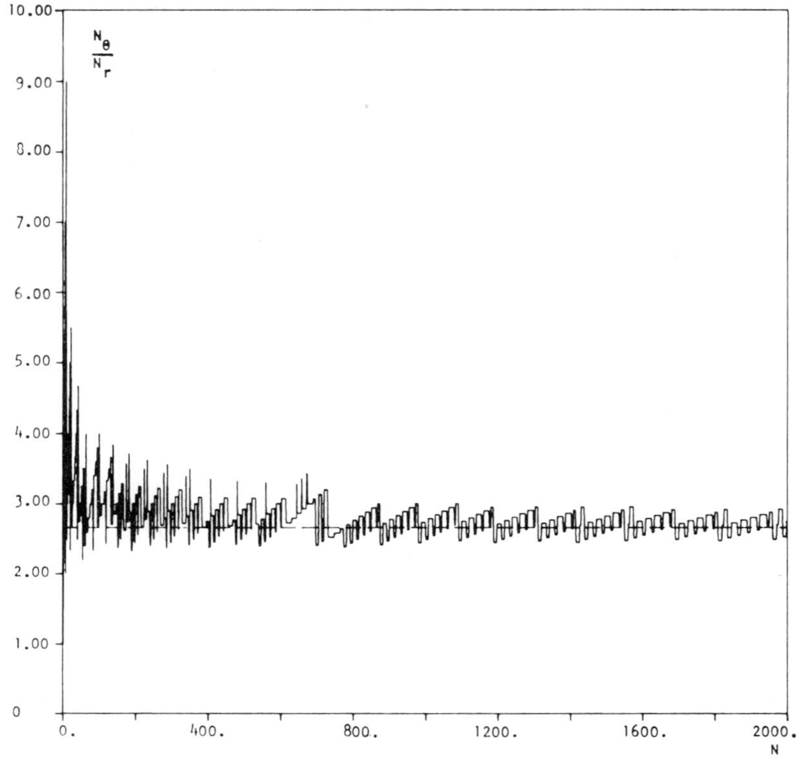

Fig. 1. Ratio of optimum number of phase quantizer levels to magnitude quantizer levels as a function of N.

Cartesian format error is smaller than the polar format error according to the exact error expressions. In the second column we find the values of N for which rectangular format is better than polar format according to the approximate error expressions. It can be seen that, in general, the approximate expressions are more pessimistic than the exact quantities.

V. Magnitude-Phase Information Comparison

An interesting problem that arises in using the polar representation is to find the best choice for the ratio of phase quantization levels to magnitude quantization levels. Pearlman [2] used distortion rate theory to obtain the ratio $N_\theta / N_r = 2.596$. We now give a somewhat different derivation.

We minimize (12), assuming N is large. We note that

$$\frac{\sin x}{x} \cong 1 - \frac{x^2}{6} \tag{17}$$

and

$$\left(1 - \frac{x^2}{6}\right)^2 \cong 1 - \frac{x^2}{3}. \tag{18}$$

Using these approximations, and (16) together with (12), we obtain

$$E_p \cong \left(1 - (\pi / N_\theta)^2 / 3\right) \frac{0.9287 N_r}{(0.5965 + N_r)^3} + \tfrac{2}{3}(\pi / N_\theta)^2. \tag{19}$$

Assuming $(0.5965 + N_r)^3 \cong N_r^3$, we substitute $N_r = N / N_\theta$ into (19), differentiate with respect to N_θ, and set the resulting expression equal to zero. Solving for N_θ, we find

$$N_\theta = 1.63 N^{1/2} \tag{20a}$$

or

$$\frac{N_\theta}{N_r} = 2.662, \tag{20b}$$

which agrees closely with the Pearlman bound. Fig. 1 shows a computer plot of the actual ratio plotted as a function of N. The dotted line is the value 2.662. Using this value in (19), it is a simple matter to show that, for large N, the polar format error is smaller than the rectangular format error.

VI. Applications to Spectral Phase Coding

From the preceeding sections, we know that if 3.33 bits or more per sample is to be used to quantize a white Gaussian sequence, it is better to pair the members of the sequence and quantize them in a polar format rather than simply quantizing the samples individually. We also know that the phase information is much more important than the magnitude information for minimizing the mean square quantization error.

Spectral phase coding (SPC) [1], [7] is one way in which we may make use of the above two properties. Consider some arbitrary data sequence $x_0, x_1, \cdots, x_L$, where in our examples we let $L = 4096$. The message sequence is divided into blocks of N samples; we consider the case

$N=32$. Each block of N terms is then divided in half, with the first $N/2$ terms forming the sequence $\{a_{1,n}\}_{n=0}^{N/2-1}$ and the second group of $N/2$ terms forming $\{a_{2,n}\}_{n=0}^{N/2-1}$. The complex-valued sequence $\{a_n\}_{n=0}^{N/2-1}$ is formed from

$$a_n = a_{1,n} + ia_{2,n}. \tag{21}$$

We then form the spectral sequence $\{A_p \exp(i\theta_p)\}$ from

$$A_p \exp(i\theta_p) = \sum_{n=0}^{(N/2)-1} a_n \exp(-i4\theta np/N), \qquad p=0,\cdots,\frac{N}{2}-1. \tag{22}$$

The SPC sequence $\{\psi_p\}_{p=0}^{N-1}$ is described by the following equations:

$$a_n = \frac{2}{N}\sum_{p=0}^{(N/2)-1} \frac{S}{2}\left[\exp(i\psi_p)+\exp(i\psi_{p+(N/2)})\right] \cdot \exp(i4\pi np/N), \qquad n=0,\cdots,\frac{N}{2}-1, \tag{23}$$

where

$$S = \max_p \{A_p\}, \tag{24a}$$

$$\psi_p = \theta_p + \phi_p, \tag{24b}$$

$$\psi_{p+(N/2)} = \theta_p - \phi_p, \tag{24c}$$

and $\phi_p = \cos^{-1}(A_p/S)$. Equation (24) describes the coding procedure and (23) the decoding procedure.

SPC is essentially a polar format representation of the discrete Fourier transform (DFT) of a random phase time series. In [8] the conditions under which the real and imaginary parts of the samples from the DFT tend to independent normal random variables are discussed. This is an asymptotic result, and it tells us that the magnitude of the DFT is Rayleigh and independent of the uniformly distributed phase. The uniform $(-\pi,\pi)$ distribution of the phase makes it a simple matter to quantize this quantity in an optimum fashion. Because of the relatively high phase information content, this case of quantization is important. Indeed, as is shown in Section IV, as long as the phase is optimally quantized, the quantizer characteristics for the magnitude component are much less important. In addition to the uniform phase property for the asymptotic case, we can show that in some special cases the phase has this property for small as well as large N.

Consider (22). We assume a_n can be represented as $r_n \exp(i\theta_n)$ where θ_n is uniform and independent of r_n for all θ_i, $i \neq n$. Under these assumptions, we have the following theorem.

Theorem: A_p is independent of θ_p, and θ_p is uniformly distributed for any arbitrary block size N.

Proof:

$$\mathrm{Re}\{A_p\} = \sum_{k=0}^{(N/2)-1} r_k \cos\phi_k \tag{25a}$$

$$\mathrm{Im}\{A_p\} = \sum_{k=0}^{(N/2)-1} r_k \sin\phi_k, \tag{25b}$$

where

$$\phi_k = \theta_k - \frac{4\pi}{N}kp. \tag{25c}$$

Consider the joint characteristic function of these two random variables:

$$\begin{aligned}
\Psi_N(\omega_1,\omega_2) &= E_{r_k}E_{\phi_k}\{\exp(j(\omega_1 \mathrm{Re}\{A_p\}+\omega_2 \mathrm{Im}\{A_p\}))\} \\
&= E_{r_k}E_{\phi_k}\left\{\exp\left(j\left[\omega_1 \sum_{k=0}^{(N/2)-1} r_k\cos\phi_k + \omega_2\sum_{k=0}^{(N/2)-1} r_k \sin\phi_k\right]\right)\right\} \\
&= \frac{1}{(2\pi)^{(N/2)-1}} E_{r_k}\left\{\int_{-\pi}^{\pi} \exp\left(j\left(\sum_{k=0}^{(N/2)-1} r_k \cdot[\omega_1\cos\phi_k+\omega_2\sin\phi_k]\right)\right) d\phi_1\, d\phi_2 \cdots d\phi_{(N/2)-1}\right\} \\
&= \frac{1}{(2\pi)^{(N/2)-1}} E_{r_k}\left\{\int_{-\pi}^{\pi} \exp\left(j\sum_{k=0}^{(N/2)-1}\sqrt{\omega_1^2 r_k^2+\omega_2^2 r_k^2} \cdot\cos\left(\phi_k+\tan^{-1}\frac{\omega_2}{\omega_1}\right)\right) d\phi_1\cdots d\phi_{(N/2)-1}\right\} \\
&= E_{r_k}\left\{\prod_{k=0}^{(N/2)-1} J_0\left((\omega_1^2+\omega_2^2)^{1/2} r_k\right)\right\}
\end{aligned} \tag{26}$$

where E_{r_k} and E_{ϕ_k} are the expectation operators over the subscripted random variables. However, this is circularly symmetric. Using the properties of the two-dimensional Fourier transform, we know that the bivariate density must also be circularly symmetric. However, this can happen if and only if the magnitude is independent of the phase and the phase is uniformly distributed over a region of support 2π.

This theorem tells us that with the given assumptions, we can guarantee that the optimal transform phase quantizer is the uniform quantizer. In many cases, experimental data indicate that we are not far from the optimum result even when the conditions for the theorem do not hold for a particular sequence.

We now derive an expression for the quantizing error of the SPC representation. The ideal unquantized SPC representation is

$$A_p \exp(i\theta_p) = \frac{S}{2}\left[\exp(i\psi_p)+\exp(i\psi_{p+(N/2)})\right]. \tag{27}$$

To begin with, we assume that the phase terms $\{\psi_p\}$ are quantized to M equal step size quantization levels. From [2] we have

$$e^{i\hat{\psi}_p} = \sum_{m=-\infty}^{\infty} \mathrm{sinc}(m+1/M)\exp(i(mM+1)\psi_p), \tag{28}$$

where $\hat{\psi}_p$ is the quantized version of $\hat{\psi}_p$. From experimental results it is found that quantization of the S parameter is negligible and will henceforth be ignored. The quantization error E can now be expressed as

$$E=\hat{A}_p\exp(i\hat{\theta}_p)-A_p\exp(i\theta_p) \tag{29}$$

where $\hat{A}_p\exp(i\hat{\theta}_p)$ represents $A_p\exp(i\theta_p)$ using the quantized parameter $\hat{\psi}_p$. Using (28) in (29) we have

$$\begin{aligned}E=\frac{S}{2}\sum_{m\neq 0}&\operatorname{sinc}(m+1/M)\\&\left[\exp(i(mM+1)\psi_p)+\exp(i(mM+1)\psi_{p+(N/2)})\right]\\&+(1-\operatorname{sinc}(1/M))\frac{S}{2}(\exp(i\psi_p)+\exp(i\psi_{p+(N/2)})).\end{aligned} \tag{30}$$

We square this quantity and take its expectation, using the following expressions derived in Appendix A:

$$\begin{aligned}\Bigg|\frac{S}{2}\sum_{m\neq 0}&\operatorname{sinc}(m+1/M)\\&\cdot\left[\exp(i(mM+1)\psi_p)+\exp(i(mM+1)\psi_{p+(N/2)})\right]\Bigg|^2\\&=\frac{S^2}{2}\sum_{l\neq 0}\operatorname{sinc}^2(l+1/M)\\&\cdot\left[1+\cos(2\phi_p(Ml+1))\right],\end{aligned} \tag{31}$$

$$\begin{aligned}\boldsymbol{E}\Bigg\{&(\exp(i\psi_p)+\exp(i\psi_{p+(N/2)}))\sum_{m\neq 0}\operatorname{sinc}(m+1/M)\\&\cdot\left[\exp(-i(mM+1)\psi_p)+\exp(i(mM+1)\psi_{p+(N/2)})\right]\Bigg\}=0,\end{aligned} \tag{32}$$

and

$$\begin{aligned}E\left\{(1-\operatorname{sinc}(1/M))^2\left|\frac{S}{2}(\exp(i\psi_p)+\exp(i\psi_{p+(N/2)}))\right|^2\right\}\\=(1-\operatorname{sinc}(1/M))^2E\{A_p^2\},\end{aligned} \tag{33}$$

where $E\{\cdot\}$ is the statistical expectation operator. Then

$$\begin{aligned}E_r=\boldsymbol{E}\{(A_p^2)\}&(1-\operatorname{sinc}(1/M))^2\\&+\frac{S^2}{2}\sum_{l\neq 0}\operatorname{sinc}(l+1/M)\\&\cdot\boldsymbol{E}\{1+\cos[2(lM+1)\phi_p]\}.\end{aligned} \tag{34}$$

From the Riemann–Lebesgue lemma [9] we know that, for large M, $E[\cos 2(lM+1)\phi_p]\ll 1$. Also,

$$\begin{aligned}\sum_{l\neq 0}\operatorname{sinc}^2(l+1/M)&=1-\operatorname{sinc}^2(1/M)\\&=(1-\operatorname{sinc}(1/M))(1+\operatorname{sinc}(1/M))\end{aligned} \tag{35}$$

$$\cong 2(1-\operatorname{sinc}(1/M)), \tag{36}$$

so that

$$E_r=E\{A_p^2\}(1-\operatorname{sinc}(1/M))^2+S^2(1-\operatorname{sinc}(1/M)). \tag{37}$$

TABLE III
A COMPARISON OF NORMALIZED QUANTIZATION ERROR FOR AN SPC SEQUENCE AND AN OPTIMAL UNIT VARIANCE GAUSSIAN QUANTIZER FOR DIFFERENT PROBABILITY DENSITIES

Density	Error (Gaussian)	Error SPC
N(0, 1)	0.91 E-2	2.38 E-2
N(0, 2)	1.90 E-2	2.40 E-2
N(0, 4)	7.00 E-2	2.38 E-2
$U(-\frac{\sqrt{12}}{2}, \frac{\sqrt{12}}{2})$	0.73 E-2	8.58 E-2
$U(-\sqrt{12}, \sqrt{12})$	1.34 E-2	8.50 E-2
U(-4, 4)	3.50 E-2	8.43 E-2
U(-5, 5)	9.84 E-2	8.58 E-2
X(2)	3.50 E-2	12.40 E-2
X(1)	18.60 E-2	12.35 E-2
X(.5)	62.70 E-2	12.39 E-2

This error expression agrees extremely closely with computer simulations and with the error expression found in [1, eq. (22)] which is derived by a different method. The second term contributes the most to E_r.

We now present examples that make use of a sequence of 4096 zero mean, unit variance Gaussian random variables. We first form the SPC version of this sequence allowing four bits per SPC sample. The error expression in (37) predicts a mean square error of 2.2×10^{-2} per sample. The actual computed average error per sample for SPC block sizes of 32 is 2.3×10^{-2}. An optimal Max [4] quantizer would give an error of 0.91×10^{-2} per sample. By using SPC we create only a little over twice the minimum achievable error for this signal and this number of quantization levels. However, if the signal statistics change and the same quantizers are employed, what is the expected result?

Table III summarizes a number of computer simulations for Gaussian, double sided exponential, and uniform random variables coded using both the optimal unit variance Gaussian–Max quantizer and SPC. $N(0,A)$ is the zero mean, variance of A, Gaussian density; $U(-A/2, A/2)$ is the zero mean, variance of $A^2/12$, uniform density; and $X(A)$ is the zero mean, variance of $1/A^2$, double sided exponential density.

For this example, one can see that the large variance signals have lower quantizing error if coded with SPC. Because the Max–Gaussian quantizer has very small step sizes near the origin, we expect that it will produce small errors for those signals that have a large amount of probability in that region. The most striking characteristic of these results is the way the normalized SPC mean square error remains virtually constant for each particular distribution. SPC tracks variations in signal power very well.

VII. Conclusion

In this paper we have investigated in detail the optimum quantization of two-dimensional Gaussian random variables. Results are put forward to prove that, in general, polar format is superior to rectangular format. Applications of this to a coding scheme (SPC) are studied in order to explain why SPC seems to exhibit robustness with respect to variations in signal statistics and signal power.

Appendix A

We will now derive (31). Taking the square of the expression and moving the expectation operator through the sum leaves

$$\frac{S^2}{4}\sum_{m\neq 0}\sum_{l\neq 0}\mathrm{sinc}(m+1/M)\,\mathrm{sinc}(l+1/M)$$
$$\cdot E\{\exp(iM(m-l)\psi_p)+\exp(iM(m\psi_p-l\psi_{p+(N/2)}))$$
$$\cdot\exp(i(\psi_p-\psi_{p+(N/2)}))$$
$$+\exp(-iM(l\psi_p-m\psi_{p+(N/2)}))\exp(-i(\psi_p-\psi_{p+(N/2)}))$$
$$+\exp(iM(m-l)\psi_{p+(N/2)})\}. \quad \text{(A1)}$$

Assume that A_p is independent of θ_p. This means that the θ_p and ϕ_p used in the expressions for ψ_p and $\psi_{p+(N/2)}$ are also independent. Therefore, the expectation in (A1) is zero except for those terms where $l=m$. Consequently, this expression is equivalent to

$$\frac{S^2}{4}\sum_{l\neq 0}\mathrm{sinc}^2(l+1/M)$$
$$E\{2+2\cos[(Ml+1)(\psi_p-\psi_{p+(N/2)})]\}. \quad \text{(A2)}$$

Because

$$\psi_p-\psi_{p+(N/2)}=2\phi_p, \quad \text{(A3)}$$

we have

$$\frac{S^2}{2}\sum_{l\neq 0}\mathrm{sinc}^2(l+1/M)[1+\cos 2\psi_p(Ml+1)]. \quad \text{(A4)}$$

Equation (32) is obtained by a similar argument. For (33), we recognize that

$$\frac{S}{2}(\exp(i\psi_p)+\exp(i\psi_{p+(N/2)}))=A_p\exp(i\theta_p). \quad \text{(A5)}$$

Therefore,

$$E\{(1-\mathrm{sinc}(1/M))^2|A_p\exp(i\theta_p)|^2\}$$
$$=(1-\mathrm{sinc}(1/M))^2E\{A_p^2\}. \quad \text{(A6)}$$

References

[1] N. C. Gallagher, "Quatizing schemes for the discrete Fourier transform of a random time series," *IEEE Trans. Inform. Theory*, vol. IT-24, pp. 156–163, Mar. 1978.

[2] W. A. Pearlman, "Quantization error bounds for computer generated holograms," Stanford Univ. Inform. Syst. Lab., Stanford, CA, Tech. Rep #6503-1, Aug. 1974.

[3] J. A. Bucklew and N. C. Gallagher, "A note on optimum quantization," *IEEE Trans. Inform. Theory*, vol. IT-25, pp. 365–366, May, 1979.

[4] J. Max, "Quantization for minimum distortion," *IRE Trans. Inform. Theory*, vol. IT-6, pp. 7–12, Mar. 1960.

[5] P. E. Fleischer, "Sufficient conditions for achieving minimum distortion in a quantizer," *IEEE Int. Conv. Rec.*, Part I, pp. 104–111, 1964.

[6] R. C. Wood, "On optimum quantization," *IEEE Trans. Inform. Theory*, vol. IT-5, pp. 248–252, Mar. 1969.

[7] N. C. Gallagher, "Discrete spectral phase coding," *IEEE Trans. Inform. Theory*, vol. IT-22, pp. 622–624, Sept. 1976.

[8] N. C. Gallagher and B. Liu, "Statistical properties of the Fourier transform on random phase diffusers," *Optik*, vol. 42, pp. 65–86, Feb. 1975.

[9] H. L. Royden, *Real Analysis*. Toronto: MacMillan, 1968, pg. 90.

Reprinted from *IEEE Trans. Commun.* **COM-27**(6):892–899 (1979)

Polar Quantization of a Complex Gaussian Random Variable

WILLIAM A. PEARLMAN, MEMBER, IEEE

Abstract—We solve numerically the optimum fixed-level non-uniform and uniform quantization of a circularly symmetric complex (or bivariate) Gaussian random variable for the mean absolute squared error criterion. For a given number of total levels, we determine its factorization into the product of numbers of magnitude and phase levels that produces the minimum distortion. We tabulate the results for numbers of "useful" output levels up to 1024, giving their optimal factorizations, minimum distortion, and entropy. For uncoded quantizer outputs, we find that the optimal splitting of rate between magnitude and phase, averaging to 1.52 and 1.47 bits more in the phase angle than magnitude for optimum and uniform quantization, respectively, compares well with the optimal polar coding formula of 1.376 bits of Pearlman and Gray [3]. We also compare the performance of polar to rectangular quantization by real and imaginary parts for both uncoded and coded output levels. We find that, for coded outputs, both polar quantizers are outperformed by the rectangular ones, whose distortion-rate curves nearly coincide with Pearlman and Gray's polar coding bound. For uncoded outputs, however, we determine that the polar quantizers surpass in performance their rectangular counterparts for all useful rates above 6.0 bits for both optimum and uniform quantization. Below this rate, the respective polar quantizers are either slightly inferior or comparable.

I. INTRODUCTION

Our purpose is to present comprehensive and exact numerical solutions of the optimum uniform and nonuniform polar quantizations of a circularly symmetric (bivariate) complex Gaussian random variable using the error criterion of mean absolute squared difference. This random variable has independent and identically distributed real and imaginary parts. Although a special case of the general bivariate Gaussian, it is one that arises quite often. The discrete Fourier transforms of complex and real stationary N-sequences (aside from the real-valued zero and $N/2$ terms for the real sequence) consist of elements with identically distributed and uncorrelated real and imaginary parts which approach Gaussian in distribution as N becomes large [1]. Part of an optimal procedure for encoding a general bivariate (or multivariate) random variable is first to change to a co-ordinate system where the components are uncorrelated [1, 2]. The only real restriction in this case then is the identical distribution of the components, which is by far the more common occurrence. By polar quantization we mean quantization by magnitude and phase angle in contrast to the more common rectangular quantization by real and imaginary parts. Polar quantization has had applications in computer holography, DFT encoding, image processing, and communications [3, 4, 5]. We consider both optimum non-uniform and optimum uniform quantizations, which, for the sake of brevity, we call "optimum" and "uniform," respectively. We shall subsequently show that fixed-level optimum or uniform polar quantization surpasses the performance of the corresponding optimal rectangular quantization (pair of Max [6] quantizers) for nearly all numbers of quantization output levels. This result has never before been corroborated, but has been hinted at by other workers. Recently, Gallagher [4] calculated, for the one case of the total number of levels about 400, that the optimum (non-uniform) and (optimum) uniform polar quantizers are both superior to their optimum rectangular counterparts using a pair of identical Max quantizers [6]. Pearlman and Gray [1] have derived optimal performance bounds for polar encoding and conclude that polar quantizing schemes may exist which surpass in performance the corresponding rectangular Max schemes. Powers [3] has used a convenient suboptimal polar quantizing scheme whose performance approached, but did not surpass optimal rectangular quantization.

We address the cases of polar and rectangular quantization through no claims of optimality for either or both, but for the reason that they are certainly the most convenient. Their coordinate systems are the natural ones for expressing a complex variable; and the coordinates in each of these systems are statistically independent for the random variable under consideration. The optimal quantization scheme may involve decision regions in the complex plane which are not expressible as constant values of the coordinates in either of these coordinate systems [7]. We do not touch upon this issue here.

A comprehensive solution to the polar quantization problem is the specification of the decision and output levels for every value of magnitude and phase angle, the minimum distortion (mean absolute squared error), and the entropy. We are able to provide these quantities for any number of levels in optimal fixed-level quantization for both optimum and uniform quantizations. Furthermore, we must solve the optimal allocation of levels between magnitude and phase angle for any given number of total levels. Gallagher [4] provided the optimal allocation of 12 magnitude and 33 phase levels for the one case of 396 total levels in both optimum and uniform polar quantization. Pearlman and Gray [1] have found that the phase angle should receive 1.376 bits more rate than the magnitude in an optimal polar encoding scheme. Gallagher's result for fixed level quantization (non-entropy coded) corresponds to 1.46 bits more in the phase than the magnitude. Here we have provided the optimal numbers of magnitude and phase levels for total numbers of useful levels from 1 to 1024 in both optimum non-uniform and uniform polar quantization. The definition of useful levels involves acceptable values of distortion and will be explained in Section IV. For every useful level in optimum and uniform quantization, we have also calculated the distortion and the entropy. From these numbers we have generated graphs which show the superiority of pclar over rectangular quantization in fixed-level schemes, but its inferiority in entropy coded schemes. Furthermore, we are able to conclude by comparison with the polar coding bound of Pearlman and Gray [1] that if you desire nearly optimum performance in the distortion-rate sense (minimum distortion for a fixed entropy) then you can obtain it much more easily with rectangular quantization.

II. OPTIMUM POLAR QUANTIZATION

We turn now to the mathematical formulation of optimum polar quantization under the distortion criterion of minimum mean absolute squared error. We divide the complex plane into

Paper approved by the Editor for Communication Theory of the IEEE Communications Society for publication without oral presentation. Manuscript received June 30, 1978; revised November 20, 1978. This work was supported by the Graduate School, University of Wisconsin-Madison, Madison, WI.

The author is with the Department of Electrical and Computer Engineering, University of Wisconsin-Madison, Madison, WI 53706.

N regions $R_{mp} = \{z = re^{i\theta} : r_{m-1} \leqslant r < r_m, \theta_{p-1} \leqslant \theta < \theta_p\}$ for $m = 1, 2, \cdots, M$ and $p = 1, 2, \cdots, P$. Here r and θ are polar coordinates in the plane, $r_o = 0, r_M = \infty, \theta_o = 0; \theta_P = 2\pi$, and $N = MP$. A fixed N-level polar quantizer of a complex random variable Z with magnitude R and phase angle Θ and probability density $p(z) = g(r)h(\theta)$ is a mapping of $z = re^{i\theta}$ to $\hat{z}_{mp} = \hat{r}_m e^{i\hat{\theta}_p}$, whenever z belongs to R_{mp}. The sets $\{(r_m, \theta_p)\}_{m=0,p=0}^{M,P}$ and $\{(\hat{r}_m, \hat{\theta}_p)\}_{m=1,p=1}^{M,P}$ are, respectively, the decision and output levels of the quantizer. We choose the mapping to minimize

$$D_z(N) = \sum_{m=1}^{M} \sum_{p=1}^{P} \int_{r_{m-1}}^{r_m} \int_{\theta_{p-1}}^{\theta_p} |re^{i\theta} - \hat{r}_m e^{i\hat{\theta}_p}|^2 g(r)h(\theta)\, dr d\theta \tag{1}$$

where

$$g(r) = \frac{r}{\sigma^2} e^{-r^2/2\sigma^2}, \qquad 0 \leqslant r < \infty$$

$$h(\theta) = \frac{1}{2\pi}, \qquad 0 \leqslant \theta < 2\pi$$

and $N = MP$. The variance of Z is $2\sigma^2$. The necessary conditions for a solution are that

$$\frac{\partial D_z}{\partial \theta_p} = 0 \qquad p = 1, 2, \cdots, P-1 \tag{2a}$$

$$\frac{\partial D_z}{\partial \hat{\theta}_p} = 0 \qquad p = 1, 2, \cdots, P \tag{2b}$$

$$\frac{\partial D_z}{\partial r_m} = 0 \qquad m = 1, 2, \cdots, M-1 \tag{2c}$$

$$\frac{\partial D_z}{\partial \hat{r}_m} = 0 \qquad m = 1, 2, \cdots, M. \tag{2d}$$

The conditions (2a) and (2b) on the phase angle's decision and output levels yield

$$\theta_p = \frac{p}{P}(2\pi), \qquad p = 0, 1, \cdots, P \tag{3a}$$

$$\hat{\theta}_p = \frac{(p - \frac{1}{2})}{P}(2\pi), \qquad p = 1, 2, \cdots, P \tag{3b}$$

which is the anticipated uniform quantization. The conditions (2c) and (2d) on the magnitude's decision and output levels at the stationary phase angle levels of (3a) and (3b) are

$$r_m = \frac{1}{S(P)} \left(\frac{\hat{r}_m + \hat{r}_{m+1}}{2} \right) \qquad m = 1, 2, \cdots, M-1 \tag{3c}$$

$$\hat{r}_m = S(P) \frac{\int_{r_{m-1}}^{r_m} rg(r)\, dr}{\int_{r_{m-1}}^{r_m} g(r)\, dr} \qquad m = 1, 2, \cdots, M \tag{3d}$$

where

$$S(P) \triangleq \text{sinc}\left(\frac{1}{P}\right) = \frac{\sin \pi/P}{\pi/P}.$$

These solutions have been obtained previously by Powers [3], Gallagher [4] and Senge [8] among others. If we define

$$\bar{r}_m = \frac{1}{\text{sinc}(1/P)} \hat{r}_m \qquad m = 0, 1, 2, \cdots, M \tag{4}$$

we can replace the conditions (3c) and (3d) by

$$r_m = \tfrac{1}{2}(\bar{r}_m + \bar{r}_{m+1}) \qquad m = 1, 2, \cdots, M-1 \tag{5a}$$

$$\bar{r}_m = \frac{\int_{r_{m-1}}^{r_m} rg(r)\, dr}{\int_{r_{m-1}}^{r_m} g(r)\, dr} \qquad m = 1, 2, \cdots, M. \tag{5b}$$

The scaled output levels $\bar{r}_m$ are the centroids of the Rayleigh probability density in their respective decision ranges $r_{m-1} \leqslant r < r_m$. For a given number M of magnitude output levels, we can solve for the decision levels r_m and the scaled output levels $\bar{r}_m$ without regard to the number P of phase output levels through Equations (5a) and (5b). Then for any given P we can determine the output magnitude levels through Equation (4). The decision levels r_m are independent of the number of phase levels. In fact, Equations (5) (along with $r_o = 0$ and $r_M = \infty$) are the same conditions for solving the Rayleigh magnitude quantization alone. This problem has already been solved by Pearlman and Senge [9], who derived a new algorithm for the solution and presented a complete tabulation of decision levels, output levels, minimum distortions, and entropies for $M = 1$ to 64. So these results can be used directly for r_m and $\bar{r}_m$. It turns out that the minimum distortions can be put to use here as well. The magnitude distortion corresponding to the solution in (5) is

$$D_r(M) = \sum_{m=1}^{M} \int_{r_{m-1}}^{r_m} (r - \bar{r}_m)^2 g(r)\, dr. \tag{6}$$

When we calculate the distortion in Equation (1) at the stationary points of Equations (3), we find that

$$D_z(N) = D_r(M) + (1 - (S(P))^2)C(M) \tag{7}$$

with

$$C(M) = \sum_{m=1}^{M} \bar{r}_m^{\,2} \int_{r_{m-1}}^{r_m} g(r)\, dr.$$

It is now evident that for a given M it is only necessary to calculate the decision levels, output levels, the magnitude distortion, and the quantity $C(M)$ for the Rayleigh density $g(r)$ only. (In the limit as M approaches infinity, we have $\lim_{M\to\infty} C(M) = \int_0^\infty r^2 g(r)dr = 2\sigma^2$.) Then for any P such that $P = N/M$ we can substitute into Equation (7) to find the total distortion for complex Gaussian polar quantization. The minimality of $D_r(M)$ is guaranteed by Fleischer's sufficient condition [10] that ln $g(r)$ is convex downward. Minimality of the total distortion $D_z(N)$ is assured by the fact that the (Hessian) matrix of all second-order partial derivatives of $D_z(N)$ with respect to the decision and output variables r_m, $\bar{r}_m$, at the stationary points in Equations (3-5) is positive definite.

III. UNIFORM POLAR QUANTIZATION

For uniform polar quantization we demand that, apart from a possibly semi-infinite end interval, both the decision and output levels of the magnitude be equally spaced. The optimal phase levels are already uniformly spaced as given in Equations (3a) and (3b). If we substitute that solution directly into the total mean-squared error distortion of (1) and integrate over θ, we obtain:

$$D_z(N) = \sum_{m=1}^{M} \int_{r_{m-1}}^{r_m} (r^2 + \hat{r}_m{}^2 - 2S(P)r\hat{r}_m)g(r)\, dr. \tag{8}$$

We must minimize $D_z(N)$ under the constraints that

$$\begin{aligned} r_m &= mh & m &= 1, 2, \cdots, M-1 \\ \hat{r}_m &= (m - \tfrac{1}{2})h & m &= 1, 2, \cdots, M \end{aligned} \tag{9}$$

where

$r_0 = 0$, $r_M = \infty$, and $N = MP$.

According to these constraints, we need only determine the interval size h which minimizes $D_z(N)$ in (8). The necessary condition is that

$$\frac{\partial D_z(N)}{\partial h} = \sum_{m=1}^{M} Q(m)(2m-1)[(m-\tfrac{1}{2})h - S(P)\bar{r}_m] = 0,$$

where (10)

$$Q(m) = \int_{(m-1)h}^{mh} g(r)\, dr$$

is the probability of the mth magnitude decision interval and $\bar{r}_m$ is its centroid as defined in (5b) with $r_{m-1} = (m-1)h$ and $r_m = mh$. The sufficient condition for the solution in (10) to yield a minimum distortion is that the second derivative, which we calculate as

$$\frac{\partial^2 D_z(N)}{\partial h^2} = 2\sum_{m=1}^{M} (m-\tfrac{1}{2})^2 Q(m) - (2 - S(P))h \sum_{m=1}^{M-1} m^2 g(mh), \tag{11}$$

be positive. The distortion in (8) at the stationary points of (10) is

$$\begin{aligned} D_z(N) &= 2\sigma^2 - \sum_{m=1}^{M} Q(m)[(2m-1)hS(P)\bar{r}_m - (m-\tfrac{1}{2})^2 h^2] \\ &= 2\sigma^2 - h^2\left(\sum_{m=1}^{M} (m-\tfrac{1}{2})^2 Q(m)\right). \end{aligned} \tag{12}$$

The optimal interval size h in Equation (10) can easily be solved by computer through a one-dimensional Newton-Raphson technique. Such computations reveal that the second derivative in (11) is always positive, so that a minimum of $D_z(N)$ is always obtained. Although this uniform quantization is easy to implement since the interval size specifies all the decision and output levels, it is more difficult than the optimum (non-uniform) quantization in one aspect. Not only is the optimal h a function of the number M of magnitude levels, it is also a function of the number P of phase levels, as seen in Equation (10) where the value of P in $S(P)$ must be substituted in order to solve for h. For larger values of P one is tempted to approximate $S(P)$ by one with little loss of accuracy in the resulting h. In fact, Gallagher [4] used this approximation in his calculations. Then, we might use the values of h given by Pearlman and Senge [9] for Rayleigh uniform quantization. Nevertheless, the best approximation at $P = 64$ gave errors in h up to .005, where we require our numerical errors to be less than 5×10^{-7}. We therefore did not use any approximations, so that we could obtain solutions as accurate as possible. For non-uniform quantization, we could solve for all the decision and output levels for a given M without knowledge of P. For any given M in uniform quantization, a different optimal interval size must be specified for each given value of P. Since it necessarily involves such a lengthy listing, we do not present here a table of the optimal interval sizes for every value of M and P. We shall present a table (Table 2) of optimal interval sizes for useful values of M and P, as explained in the next section. Once an optimal value of h is known for given M and P, the distortion $D_z(N)$ is calculated through the second expression in (12), which does not depend explicitly on P.

IV. OPTIMAL DIVISION OF MAGNITUDE AND PHASE LEVELS

The previous solutions for optimum and uniform magnitude-phase angle quantization do not reveal how to select the numbers of magnitude and phase angle levels, M and P, respectively, that produce the smallest distortion for any given total number of levels $N = MP$. Our purpose now is to find these optimal choices of M and P, the resulting minimum distortion, and the entropy for any given N. The only exact results reported previously have been for the single case of $N = 396$ by Gallagher [4], who found that $M = 12$ and $P = 33$ ($P/M = 2.75$) yield a minimum normalized distortion of 5.95×10^{-3} for optimum quantization and 6.67×10^{-3} for uniform quantization. These distortion values are each less than that obtained by their corresponding optimal (Max [6]) quantization of real and imaginary parts. The result of quantizing the phase angle more finely than the magnitude has been noted previously by several workers in holography and image processing [3, 5, 11, 12, 13]. Among them is Powers [3], who obtained suboptimal solutions for the decision and output magnitude levels for some values of N less than 400, as he split the magnitude into equiprobable decision intervals. In all cases, consistent with Gallagher, the number of phase levels exceeded the number of magnitude levels, the minimum ratio being 2.46. Pearlman and Gray [1] have derived through distortion-rate theory formulas for optimal encoding of the logarithm of the magnitude and the phase angle. This scheme yields a mean-squared error distortion D_z of $z = re^{i\theta}$ which is nearly minimal. It provides a lower bound to this distortion of

$$\frac{D_z(R)}{2\sigma^2} \geqslant 1 - \exp\{-(1.781)2^{-R}\} \text{ for } R > 1.376 \text{ bits} \tag{13}$$

with R the information rate for the complex variable. The bound is quite tight for rates above 4.2 bits. The optimum choices for the information rates of the phase angle R_θ and the magnitude R_r must satisfy

$$R_\theta - R_r = 1.376 \text{ bits.} \tag{14}$$

For fixed level quantization, this splitting of rates corresponds to a P/M ratio of 2.60. We shall compare our subsequent quantization results with these optimal formulas both in the entropy-coded output level case and in the uncoded case, where we can make the association

$$N = 2^R, \qquad M = 2^{R_r}, \qquad \text{and } P = 2^{R_\theta}.$$

Here N, M, and P may not be integers, whereas in actual quantization they must be.

The first task is to find the values of M and P that give the minimum distortion of $D_z(N)$ for any given value of N in both optimum and uniform polar quantization. Let us consider optimum quantization first. For any given value of N, we calculate $D_z(N)$ by Equation (7) for all integers M and P such that $N = MP$. (Determining all the MP factorizations of N requires little calculation and search on the computer since we need to try only integer values of M no larger than $\sqrt{N}$.) The magnitude distortion $D_r(M)$ and the quantity $C(M)$ are provided through the Rayleigh quantization program. We then note the minimum distortion and the factorization for the given value of N. We determined the minimum distortion and optimal (integral) factorizations in this manner for all N from 1 to 1024. We can now construct a table of optimal factorizations and minimum distortions for each value of N. Such a table reveals that the distortion varies with N rather erratically. Certainly one finds the smaller distortions at some of the larger values of N, but also rather large distortions at other large values of N. The reason is that many values of N do not allow the degree of flexibility needed for factoring into a product of integers close to the optimum. For example, a prime number N allows only two factorizations, $M = N$, $P = 1$ and $M = 1$, $P = N$. Inevitably, the smaller distortion, which results from the larger number of phase levels ($M = 1$, $P = N$), is larger than minimum distortions obtained with a smaller N. There are many values of N having relatively few factorizations that produce larger minimum distortions than those belonging to lower N numbers. Such values of N are not considered to be useful for quantization purposes and are expunged from our table. In fact, based on previous work [1, 3], it is safe to assume that a distortion for a value of N greater than 2 is neither minimal for that N or useful whenever the number of phase levels does not exceed the number of magnitude levels by at least a factor of 2. We used this information to avoid the calculation of $D_r(M)$ for all values of M up to 1024. If the factorization routine for a given N called for an M greater than 64, we automatically set the associated $D_z(N)$ to an arbitrarily high value which was subsequently eliminated in seeking either the minimum $D_z(N)$ for that N or the useful factorizations for all N.

Factorization inflexibility causes other values of N to have relatively low utility compared to others and hence are not considered to be useful, either. These values are also expunged from the table. There are two utility criteria. The first is that a unit increment in N does not produce enough of a decrease in distortion and the second is that it produces too much. In the first case we eliminate the higher value because it does not pay off in enough of a distortion decrease. The decrease is considered sufficient if the ratio of the absolute difference in logarithmic distortion to the increase in rate is greater than 0.5. In the second case, the payoff in distortion decrease is so large for incrementing the number of levels by just one that we consider it mandatory to use the next level and to eliminate the initial level. When the previous ratio exceeds 2.0, the initial level is eliminated and the incremental level is retained. Such tests for usefulness are applied at every value of N. We call this entire procedure our level sorting algorithm. The final table of factorizations and minimum distortions for each "useful" value of N appears in Table 1. The entropies of the quantization levels for each of these values of N also have been calculated and are presented in the table.

The same philosophy for computing minimum distortions for each N and sorting out the useful values of N applies to uniform polar quantization. We first must solve for h in Equation (10) for all combinations of M and P that allow us to find the minimum distortions $D_z(N)$ in Equation (12) for all N from 1 to 1024. At first, it may seem that we have to find h for all M and P from 1 to 1024. Based on previous experience, we know that a minimum distortion is always obtained for at least twice as many phases as magnitude levels. Therefore, we solved for h only for M from 1 to 32 and P from 1 to 64. When we obtained a factorization of N that called for M or P greater than 32 or 64, respectively, we set its distortion to some high value which would be automatically eliminated by the level sorting algorithm. For every M and P where a solution for h is found, we substitute into $D_z(N)$ in Equation (12) to find the associated quantization distortion. We then find the minimum distortions for each N by searching the distortions for each factorization and sort out the useful values of N by the same procedure described for optimum quantization. In Table 2 are presented the optimum factorizations, minimum distortions, and entropies of the quantization levels for each useful value of N. Listed also in Table 2 is the optimal interval size for the magnitude quantization at every useful value of N. The minimum distortions for optimum and uniform polar quantization in Tables 1 and 2 are plotted versus the total number of quantization levels in Figure 1. As expected nonuniform quantization yields smaller distortions for a given number N of levels, but the difference is rather small for all N and indistinguishable at small N, where the larger number of phase levels, which are uniformly quantized for both cases, dominates the quantization effects.

V. COMPARISON TO OTHER SCHEMES

We wish now to compare our polar quantization results with corresponding ones for rectangular (real and imaginary part) quantization and with optimal encoding bounds. The minimum mean squared error fixed-level optimum and uniform quantizations of a real Gaussian random variable have been solved by Max [5]. Since the real and imaginary parts of a complex Gaussian variable are independent and identically distributed real Gaussian variables, the optimal quantization procedure is identical and independent optimal quantizations on the real and imaginary parts. For fixed level schemes, we would use Max quantizers with the same number of levels. Only when the number N of total levels is a perfect square is it a product of identical integral factors. We guess that we should only accept integral factorizations that give as close to identical factors as possible, but we really do not know how close

TABLE 1
USEFUL LEVELS IN OPTIMUM POLAR QUANTIZATION

NUMBER OF LEVELS TOTAL	MAGNITUDE	PHASE	NORMALIZED DISTORTION	ENTROPY IN BITS
1	1	1	.100000+01	.723672-08
2	1	2	.681690+00	.100000+01
3	1	3	.462852+00	.158496+01
4	1	4	.363380+00	.200000+01
5	1	5	.312666+00	.232193+01
6	1	6	.283803+00	.258496+01
10	2	5	.188873+00	.328586+01
12	2	6	.154811+00	.354889+01
14	2	7	.133728+00	.377129+01
16	2	8	.119822+00	.396393+01
18	2	9	.110185+00	.413386+01
20	2	10	.103240+00	.428586+01
21	3	7	.999403-01	.431430+01
24	3	8	.854914-01	.450695+01
27	3	9	.754788-01	.467687+01
30	3	10	.682632-01	.482888+01
33	3	11	.628957-01	.496638+01
36	3	12	.587967-01	.509191+01
39	3	13	.555968-01	.520739+01
40	4	10	.541061-01	.521427+01
44	4	11	.486570-01	.535177+01
48	4	12	.444957-01	.547730+01
52	4	13	.412472-01	.559278+01
56	4	14	.386633-01	.569970+01
60	4	15	.365747-01	.579923+01
64	4	16	.348627-01	.589234+01
65	5	13	.340146-01	.589341+01
70	5	14	.314112-01	.600032+01
75	5	15	.293069-01	.609986+01
80	5	16	.275819-01	.619297+01
85	5	17	.261504-01	.628043+01
90	5	18	.249495-01	.636289+01
96	6	16	.233966-01	.644009+01
102	6	17	.219590-01	.652755+01
108	6	18	.207530-01	.661001+01
114	6	19	.197314-01	.668802+01
120	6	20	.188585-01	.676202+01
126	6	21	.181068-01	.683241+01
133	7	19	.170936-01	.689810+01
140	7	20	.162184-01	.697210+01
147	7	21	.154647-01	.704248+01
154	7	22	.148110-01	.710960+01
161	7	23	.142404-01	.717373+01
168	8	21	.136967-01	.722533+01
176	8	22	.130419-01	.729244+01
184	8	23	.124703-01	.735657+01
192	8	24	.119684-01	.741797+01
200	8	25	.115254-01	.747687+01
207	9	23	.112274-01	.751851+01
208	8	26	.111324-01	.753345+01
216	9	24	.107249-01	.757991+01
225	9	25	.102813-01	.763881+01
234	9	26	.988785-02	.769539+01
24[illegible]	9	27	.953713-02	.774984+01
25[illegible]	9	28	.922322-02	.780231+01
260	10	26	.898041-02	.784076+01
261	9	29	.894119-02	.785293+01
270	10	27	.862937-02	.789521+01
280	10	28	.831517-02	.794768+01
290	10	29	.803288-02	.799830+01
300	10	30	.777827-02	.804721+01
308	11	28	.763265-02	.807958+01
310	10	31	.754786-02	.809452+01
319	11	29	.735017-02	.813021+01
320	10	32	.733865-02	.814032+01
330	11	30	.709538-02	.817912+01
341	11	31	.686481-02	.822642+01
352	11	32	.665546-02	.827223+01
360	12	30	.656910-02	.829986+01
363	11	33	.646485-02	.831662+01
372	12	31	.633840-02	.834717+01
374	11	34	.629078-02	.835969+01
384	12	32	.612894-02	.839297+01
396	12	33	.593823-02	.843737+01
408	12	34	.576407-02	.848043+01
420	12	35	.560459-02	.852225+01
429	13	33	.552378-02	.854870+01
432	12	36	.545820-02	.856290+01
442	13	34	.534955-02	.859177+01
444	12	37	.532350-02	.860243+01
455	13	35	.519001-02	.863359+01
468	13	36	.504355-02	.867423+01
481	13	37	.490880-02	.871376+01
490	14	35	.485786-02	.873688+01
494	13	38	.478452-02	.875223+01
504	14	36	.471136-02	.877753+01
507	13	39	.466967-02	.878971+01
518	14	37	.457656-02	.881706+01
520	13	40	.456334-02	.882623+01
532	14	38	.445224-02	.885553+01
546	14	39	.433735-02	.889300+01
560	14	40	.423098-02	.892953+01
570	15	38	.418192-02	.895187+01
574	14	41	.413227-02	.896515+01
585	15	39	.406700-02	.898935+01
588	14	42	.404053-02	.899992+01
600	15	40	.396060-02	.902587+01
615	15	41	.386186-02	.906150+01
630	15	42	.377009-02	.909626+01
640	16	40	.373767-02	.911615+01
645	15	43	.368463-02	.913021+01
656	16	41	.363891-02	.915177+01
660	15	44	.360495-02	.916338+01
672	16	42	.354713-02	.918654+01
675	15	45	.353049-02	.919580+01
688	16	43	.346165-02	.922049+01
704	16	44	.338194-02	.925365+01
720	16	45	.330747-02	.928607+01
731	17	43	.327564-02	.930541+01
736	16	46	.323782-02	.931778+01
748	17	44	.319592-02	.933858+01
752	16	47	.317256-02	.934881+01
765	17	45	.312143-02	.937100+01
768	16	48	.311133-02	.937918+01
782	17	46	.305177-02	.940271+01
799	17	47	.298650-02	.943374+01
810	18	45	.296462-02	.945118+01
816	17	48	.292526-02	.946411+01
828	18	46	.289495-02	.948289+01
833	17	49	.286771-02	.949386+01
846	18	47	.282967-02	.951392+01
850	17	50	.281361-02	.952301+01
864	18	48	.276841-02	.954429+01
867	17	51	.276263-02	.955157+01
882	18	49	.271085-02	.957404+01
900	18	50	.265675-02	.960319+01
912	19	48	.263499-02	.962023+01
918	18	51	.260576-02	.963176+01
931	19	49	.257742-02	.964998+01
936	18	52	.255768-02	.965977+01
950	19	50	.252330-02	.967913+01
954	18	53	.251231-02	.968725+01
969	19	51	.247231-02	.970769+01
988	19	52	.242423-02	.973571+01
1000	20	50	.240884-02	.975125+01
1007	19	53	.237885-02	.976319+01
1020	20	51	.235784-02	.977982+01

TABLE 2
USEFUL LEVELS IN UNIFORM POLAR QUANTIZATION

NUMBER OF LEVELS TOTAL	MAGNITUDE	PHASE	MAGNITUDE INTERVAL SIZE	NORMALIZED DISTORTION	ENTROPY IN BITS
1	1	1	.000000	.100000+01	.000000
2	1	2	.797885+00	.681690+00	.100000+01
3	1	3	.103648+01	.462852+00	.158496+01
4	1	4	.112838+01	.363380+00	.200000+01
5	1	5	.117246+01	.312666+00	.232193+01
6	1	6	.119683+01	.283803+00	.258496+01
10	2	5	.108455+01	.199719+00	.331306+01
12	2	6	.113268+01	.164137+00	.358293+01
14	2	7	.116424+01	.142309+00	.380718+01
16	2	8	.118594+01	.128019+00	.399993+01
18	2	9	.120144+01	.118175+00	.416935+01
20	2	10	.121286+01	.111115+00	.432069+01
21	3	7	.791119+00	.106063+00	.435302+01
24	3	8	.811847+00	.907864-01	.454138+01
27	3	9	.827065+00	.802930-01	.470690+01
30	3	10	.838519+00	.727869-01	.485492+01
33	3	11	.847330+00	.672372-01	.498899+01
36	3	12	.854236+00	.630203-01	.511162+01
39	3	13	.859741+00	.597421-01	.522465+01
40	4	10	.642623+00	.576605-01	.523746+01
44	4	11	.651399+00	.519335-01	.536848+01
48	4	12	.658390+00	.475875-01	.548848+01
52	4	13	.664036+00	.442131-01	.559926+01
56	4	14	.668652+00	.415414-01	.570220+01
60	4	15	.672470+00	.393903-01	.579835+01
65	5	13	.541417+00	.364832-01	.589357+01
70	5	14	.546072+00	.337617-01	.599464+01
75	5	15	.549960+00	.315720-01	.608916+01
80	5	16	.553235+00	.297844-01	.617795+01
85	5	17	.556018+00	.283061-01	.626167+01
90	5	18	.558399+00	.270699-01	.634089+01
96	6	16	.468655+00	.252876-01	.641725+01
102	6	17	.471475+00	.237910-01	.649977+01
108	6	18	.473904+00	.225400-01	.657792+01
114	6	19	.476010+00	.214836-01	.655214+01
120	6	20	.477845+00	.205836-01	.672281+01
126	7	18	.411749+00	.196925-01	.678058+01
133	7	19	.413876+00	.186263-01	.685388+01
140	7	20	.415740+00	.177181-01	.692373+01
147	7	21	.417382+00	.169383-01	.699043+01
154	7	22	.418833+00	.162637-01	.705426+01
160	8	20	.367991+00	.157938-01	.709954+01
161		23	.420123+00	.156762-01	.711545+01
168	8	21	.369646+00	.150083-01	.716553+01
176	8	22	.371116+00	.143288-01	.722871+01
184	8	23	.372427+00	.137373-01	.728932+01
192	8	24	.373600+00	.132192-01	.734754+01
198	9	22	.333201+00	.129634-01	.738400+01
200	8	25	.374654+00	.127627-01	.740358+01
207	9	23	.334522+00	.123683-01	.744404+01
216	9	24	.335707+00	.118471-01	.750175+01
225	9	25	.336775+00	.113880-01	.755730+01
234	9	26	.337740+00	.109817-01	.761085+01
240	10	24	.304820+00	.108407-01	.764082+01
243	9	27	.338615+00	.106202-01	.766255+01
250	10	25	.305895+00	.103793-01	.769592+01
252	9	28	.339410+00	.102973-01	.771250+01
260	10	26	.306869+00	.997094-02	.774905+01
270	10	27	.307754+00	.960775-02	.780034+01
280	10	28	.308560+00	.928329-02	.784993+01
290	10	29	.309296+00	.899229-02	.789793+01
297	11	27	.282077+00	.884153-02	.792591+01
300	10	30	.309970+00	.873025-02	.794442+01
308	11	28	.282890+00	.851570-02	.797515+01
319	11	29	.283634+00	.822350-02	.802281+01
330	11	30	.284317+00	.796041-02	.806899+01
341	11	31	.284944+00	.772273-02	.811379+01
348	12	29	.261925+00	.762684-02	.813758+01
352	11	32	.285522+00	.750726-02	.815728+01
360	12	30	.262613+00	.736277-02	.818346+01
363	11	33	.286055+00	.731134-02	.819954+01
372	12	31	.263246+00	.712424-02	.822798+01
384	12	32	.263830+00	.690800-02	.827121+01
396	12	33	.264370+00	.671141-02	.831322+01
403	13	31	.244634+00	.664989-02	.833367+01
408	12	34	.264870+00	.653215-02	.835408+01
416	13	32	.245223+00	.643295-02	.837664+01
420	12	35	.265334+00	.636821-02	.839384+01
429	13	33	.245767+00	.623572-02	.841842+01
442	13	34	.246272+00	.605589-02	.845905+01
455	13	35	.246742+00	.589146-02	.849860+01
468	13	36	.247178+00	.574073-02	.853714+01
476	14	34	.230130+00	.567175-02	.855678+01
481	13	37	.247585+00	.560219-02	.857470+01
490	14	35	.230604+00	.550666-02	.859612+01
494	13	38	.247964+00	.547459-02	.861134+01
504	14	36	.231044+00	.535551-02	.863446+01
518	14	37	.231455+00	.521659-02	.867184+01
532	14	38	.231839+00	.508866-02	.870830+01
540	15	36	.216900+00	.503999-02	.872554+01
546	14	39	.232199+00	.497054-02	.874389+01
555	15	37	.217314+00	.490071-02	.876273+01
560	14	40	.232535+00	.486131-02	.877865+01
570	15	38	.217701+00	.477246-02	.879903+01
585	15	39	.218064+00	.465405-02	.883446+01
600	15	40	.218404+00	.454456-02	.886906+01
608	16	38	.205199+00	.450980-02	.888429+01
615	15	41	.218724+00	.444307-02	.890289+01
624	16	39	.205564+00	.439111-02	.891957+01
630	15	42	.219024+00	.434885-02	.893596+01
640	16	40	.205907+00	.428136-02	.895402+01
645	15	43	.219306+00	.426119-02	.896831+01
656	16	41	.206229+00	.417964-02	.898770+01
672	16	42	.206532+00	.408520-02	.902064+01
680	17	40	.194771+00	.406032-02	.903419+01
688	16	43	.206818+00	.399735-02	.905286+01
697	17	41	.195095+00	.395838-02	.906773+01
704	16	44	.207087+00	.39[illegible]-02	.908440+01
714	17	42	.195401+00	.38[illegible]3-02	.910053+01
720	16	45	.207341+00	.383912-02	.911529+01
731	17	43	.195688+00	.377570-02	.913263+01
748	17	44	.195960+00	.369368-02	.916405+01
765	17	45	.196217+00	.361714-02	.919482+01
774	18	43	.185704+00	.358762-02	.920814+01
782	17	46	.196459+00	.354560-02	.922496+01
792	18	44	.185977+00	.350543-02	.923944+01
799	17	47	.196689+00	.347865-02	.925451+01
810	18	45	.186236+00	.342874-02	.927010+01
816	17	48	.196906+00	.341588-02	.928349+01
828	18	46	.186481+00	.335707-02	.930014+01
846	18	47	.186712+00	.328999-02	.932958+01
855	19	45	.177228+00	.326742-02	.934158+01
864	18	48	.186932+00	.322711-02	.935846+01
874	19	46	.177475+00	.319562-02	.937152+01
882	18	49	.187141+00	.316806-02	.938678+01
893	19	47	.177708+00	.312842-02	.940086+01
900	18	50	.187338+00	.311261-02	.941458+01
912	19	48	.177930+00	.306542-02	.942964+01
931	19	49	.178140+00	.300627-02	.945787+01
940	20	47	.169538+00	.298894-02	.946873+01
950	19	50	.178339+00	.295072-02	.948558+01
960	20	48	.169761+00	.292583-02	.949741+01
969	19	51	.178529+00	.289841-02	.951278+01
980	20	49	.169973+00	.286658-02	.952555+01
988	19	52	.178710+00	.284912-02	.953949+01
1000	20	50	.170174+00	.281094-02	.955317+01
1020	20	51	.170365+00	.275854-02	.958028+01

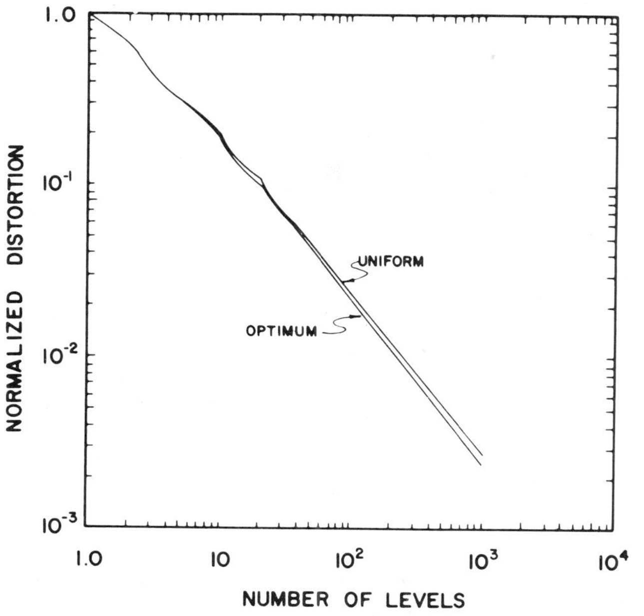

Figure 1 Distortion versus number of levels for optimum and uniform polar quantizers.

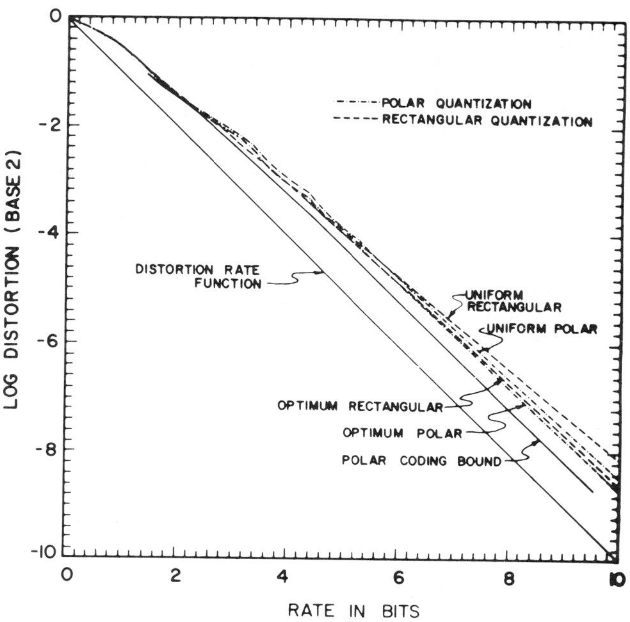

Figure 2 Fixed-level complex quantization–logarithmic distortion versus rate (logarithm of the number of levels) for optimum and uniform polar and rectangular quantizers.

they need to be. We therefore adapted our computer programs to treat the real-imaginary part cases. We used Max's tables for uniform quantization and our own program for optimum quantization [8, 9] to provide the required level number, distortion, and entropy data. We calculated the distortions for every factorization of a given N from 1 to 1024, found the minimum distortion and associated factorization for each N, and ran our level sorting program to identify the useful level numbers, their minimum distortions, and factorizations. Not surprisingly, the useful values of N for both nonuniform and uniform quantization turned out to be only those where the numbers of real and imaginary part levels differed by zero or one.

In order to compare the performance of fixed-level rectangular and polar quantization, we have plotted in Figure 2 the logarithm (base 2) of the minimum distortion for the useful levels versus the rate in bits (base 2 logarithm of the number of levels) for optimum and uniform rectangular and polar quantizations. The significant result depicted in this plot is that both optimum and uniform polar quantizations outperform their rectangular counterparts at the higher rates and are either comparable or insignificantly inferior at smaller rates. The performance crossover point for both types of quantization is at 6.0 bits, approximately. This means that for any N above 64, the distortion of rectangular quantization exceeds that of a polar quantization for the same or a lower value of N.* As the rates grow larger, the improvement of polar over rectangular for the uniform case becomes more significant than that for the nonuniform case. For example, at 9.8 bits (900 levels), the ratio of rectangular to polar distortion for the uniform and nonuniform cases are 1.27 and 1.07, respectively. We remark that nonuniform quantization is superior to uniform for every rate and for both quantization methods. As the rates grow larger, the curves appear to approach straight lines. We have therefore run a least-squares, straight line fit to each of the four described quantization curves for rates exceeding 6.5 bits per complex variable. The resulting empirical formulas for the distortion D_z versus the rate R in bits are as follows in decreasing order of asymptotic performance:

Optimum Polar:

$$D_z/2\sigma^2 = (2.056)2^{-.977R}, \qquad 7.69 \leqslant R \leqslant 10$$

Optimum Rectangular:

$$D_z/2\sigma^2 = (1.887)2^{-.955R}, \qquad 6.64 \leqslant R \leqslant 10$$

Uniform Polar:

$$D_z/2\sigma^2 = (1.849)2^{-.939R}, \qquad 7.26 \leqslant R \leqslant 10$$

Uniform Rectangular:

$$D_z/2\sigma^2 = (1.358)2^{-.861R}, \qquad 6.64 \leqslant R \leqslant 10$$

where $R \triangleq \log_2 N$. (The standard errors of the weighted residuals of the logarithmic (base 2) distortion are, with the exception of the optimum rectangular's 6.1×10^{-3}, less than 3.5×10^{-3}.) The formulas for the polar coding bound of Pearlman and Gray [1] and the ultimate performance bound of the distortion-rate function are, respectively,

$$D_{PG}/2\sigma^2 = 1 - \exp\{-(1.781)2^{-R}\}, \qquad R > 1.376$$

$$\approx (1.781)2^{-R}, \qquad R > 4.12$$

$$D(R)/2\sigma^2 = 2^{-R}, \qquad R > 0$$

* In an independent effort [16], a performance crossover point of N equal to 101 is reported for optimum quantization. The method there is equivalent to the retention of all values of N that produce a monotonically decreasing distortion versus N characteristic without the subsequent deletion of values failing our rate of decrease utility criteria.

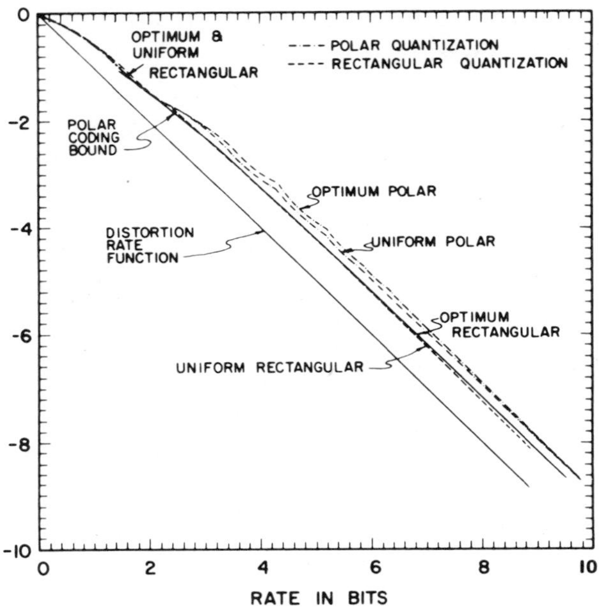

Figure 3 Entropy-coded complex quantization–logarithmic distortion versus rate (entropy) for optimum and uniform, polar and rectangular quantizers.

for R expressed in bits. These bounds are also plotted in Figure 2 for the purposes of comparison.

Some interesting conclusions can be reached when we consider entropy (Huffman) coding of the complex quantizers' output levels. As the minimum achievable output rate of a quantizer is now its entropy, we plot in Figure 3 the distortion versus entropy (in bits) of the rectangular and polar, optimum and uniform, quantizers.

For the larger rates, where all the curves appear to approach straight lines, we ran a least squares straight line curve fit just as we did for the curves in the previous figure. (The standard errors of the weighted residuals of the logarithmic (base 2) distortion are all less than 3.2×10^{-3}.) The results in decreasing order of asymptotic performance are:

Optimum Polar:

$$D_z/2\sigma^2 = (2.00)2^{-.995R}, \qquad 7.52 \leqslant R \leqslant 9.78$$

Uniform Polar:

$$D_z/2\sigma^2 = (2.20)2^{-1.012R}, \qquad 7.11 \leqslant R \leqslant 9.58$$

Optimum Rectangular:

$$D_z/2\sigma^2 = (1.786)2^{-.996R}, \qquad 6.25 \leqslant R \leqslant 9.46$$

Uniform Rectangular:

$$D_z/2\sigma^2 = (1.900)2^{-1.018R}, \qquad 6.06 \leqslant R \leqslant 8.90.$$

The rate R equals the entropy in bits. The magnitudes of the coefficients of R in the exponents of the uniform quantization formulas exceed one for the range of rates given. Clearly, one cannot simply extrapolate these formulas to all higher rates, because their curves would eventually intersect the distortion-rate function, which is the absolute lower bound on performance.

A comparison of the curves in Figure 3 shows that both coded rectangular quantizations are superior to any coded polar quantization. In fact, the curve of uniform rectangular is slightly below and the nonuniform rectangular nearly coincident with the polar coding bound. For the polar quantizers, the uniform one provides slightly lower distortion than the nonuniform one for any given rate, but as the rate grows larger their distortions become nearly equal. At these larger rates, the rate difference between the polar quantizations and the polar coding bound is about .30 bits. Closing this gap in performance by encoding polar coordinates very likely involves the rather difficult and computationally complex procedure of searching a tree or trellis for good codes. The best quantizer in the distortion versus output entropy sense will only close the gap marginally because of the dominant effect of the uniform quantization of the phase angle [14]. It seems rather wasteful of resources to apply tree or trellis coding to polar coordinates when better performance can be obtained much more easily by buffer-instrumented variable-length (Huffman) codes or permutation codes [15] on the outputs of fixed-level uniform real and imaginary part (Max) quantizers. Our conclusion is that, unless there is some other practical reason for polar quantization, one should use complex rectangular coding if optimum distortion-rate performance is desired. For the more convenient fixed-level complex quantizations, whose performances are plotted in Figure 2, the polar schemes are preferable for nearly all rates.

We wish now to draw some conclusions about the optimal splitting of rate between magnitude and phase angle for polar quantization. From Tables 1 and 2 which give the optimal number of magnitude and phase levels, we can easily calculate the corresponding rate differences for all of these useful total level numbers. Since we have decided that polar quantization is advantageous for fixed-level schemes, we wish to compare the optimal rate divisions in these cases to the prescription of $R_\theta - R_r = 1.376$ bits for optimal polar encoding in Equation (14). The constraint of integral factorization does not allow, except perhaps in rare circumstances, the direct testing of this prescription. Our calculations of rate differences (or level ratios) from the tables show that the factorizations may come as close to the prescription as the integral factorization constraint would permit. The average rate difference may be a reliable indicator of the optimum for quantization, since it smoothes out the constraint effects. The averages of all the rate differences for total rates above 1.376 bits are 11.52 and 11.47 bits ($P/M = 2.89$ and $P/M = 2.77$) for optimum and uniform quantization, respectively. Considering that the prescription of 1.376 bits applies to optimal encoding and yields a lower encoding bound which is an accurate approximation to ultimate performance for rates exceeding 4.12 bits, it is a reasonably good indicator of how to divide the rate between magnitude and phase in a polar quantization scheme. The average rate differences of 1.52 bits and 1.47 bits can now be regarded as guidelines for splitting the rate between phase angle and magnitude for fixed-level optimum and uniform polar quantization, respectively.

VI. CONCLUSION

We have presented comprehensive and accurate numerical solutions to the fixed-level polar quantization of a complex Gaussian random variable with minimum mean absolute squared error. Through our level sorting algorithm we have identified those levels judged useful for quantization purposes, as they exhibit the steepest decrease in distortion as rate increases. When we consider only the useful levels from $N = 1$ to 1024, we conclude that fixed-level optimum and uniform polar quantizations are superior to their rectangular counterparts at rates higher than 6.0 bits and either comparable or only slightly inferior at smaller rates. We have given the optimal division of rate between magnitude and phase angle for all useful rates up to 10 bits. We conclude that entropy coded rectangular quantizers are not only superior to the entropy coded polar ones, but also attain the performance of the polar coding bound of Pearlman and Gray [1]. Therefore, optimum polar quantization or coding in the sense of minimizing distortion for a given entropy, is not likely to be rewarding given the computational complexity involved, especially since the same performance can be obtained by entropy coding the outputs of optimum or uniform rectangular quantizers. Polar quantization is definitely advantageous for fixed level applications beyond rates of 6.0 bits per complex variable.

REFERENCES

1. W. A. Pearlman and R. M. Gray, "Source Coding of the Discrete Fourier Transform," *IEEE Trans. Inform. Theory,* vol. IT-24, pp. 683-692, Nov. 1978.
2. J. J. Y. Huang and P. M. Schultheiss, "Block Quantization of Correlated Gaussian Random Variables," *IEEE Trans. Commun. Syst.,* vol. CS-11, pp. 289-296, Sept. 1963.
3. R. S. Powers and J. W. Goodman, "Error Rates in Computer-Generated Holographic Memories," *Applied Optics,* vol. 14, pp. 1690-1701, July 1975.
4. N. C. Gallagher, Jr., "Quantizing Schemes for the Discrete Fourier Transform of a Random Time Series," *IEEE Trans. Inform. Theory,* vol. IT-24, pp. 156-163, March 1978.
5. A. G. Tescher, *The Role of Phase in Adaptive Image Coding,* Technical Report 510, Image Processing Institute, Electronic Sciences Laboratory, University of Southern California, Los Angeles, CA, Dec. 1973.
6. J. Max, "Quantizing for Minimum Distortion," *IRE Trans. Inform. Theory,* vol. IT-6, pp. 7-12, March 1960.
7. A. Gersho, "Asymptotically Optimal Block Quantization," *IEEE Trans. Inform. Theory,* vol. IT-25, July 1979.
8. G. H. Senge, *Quantization of Image Transforms with Minimum Distortion,* Technical Report No. ECE-77-8, Dept. of Electrical and Computer Engineering, University of Wisconsin, Madison, WI, June 1977.
9. W. A. Pearlman and G. H. Senge, "Optimal Quantization of the Rayleigh Probability Distribution," *IEEE Trans. Commun.,* vol. COM-27, pp. 101-112, Jan. 1979.
10. P. E. Fleischer, "Sufficient Conditions for Achieving Minimum Distortion in a Quantizer," *IEEE Int. Conv. Rec., Part I,* pp. 104-111, 1964.
11. L. B. Lesem, P. M. Hirsch, and J. A. Jordan, Jr., "The Kinoform: A New Wavefront Reconstruction Device," *IBM J. Res. and Dev.,* vol. 13, pp. 150-155, March 1969.
12. D. Kermisch, "Image Reconstruction from Phase Information Only," *J. Opt. Soc. Am.,* vol. 60, pp. 15-17, Jan. 1970.
13. N. C. Gallagher and B. Liu, "Method for Computing Kinoforms that Reduces Image Reconstruction Error," *Applied Optics,* vol. 12, pp. 2328-2335, Oct. 1973.
14. P. Noll and R. Zelinski, "Bounds on Quantizer Performance in the Low Bit-Rate Region," *IEEE Trans. Commun.,* vol. COM-26, pp. 300-304, Feb. 1978.
15. T. Berger, "Optimum Quantizers and Permutation Codes," *IEEE Trans. Inform. Theory,* vol. IT-18, pp. 759-765, Nov. 1972.
16. J. A. Bucklew and N. C. Gallagher, "Quantization Schemes for Bivariate Gaussian Random Variables," *IEEE Trans. Inform. Theory,* to be published.

Reprinted from *IEEE Trans. Commun.* **COM-28**(11):1924–1929 (1980)

MAGNITUDE/PHASE QUANTIZATION OF INDEPENDENT GAUSSIAN VARIATES

S. G. Wilson

Abstract—**The performance of two-dimensional polar quantization of independent Gaussian variates is evaluated as a possible improvement over one-dimensional quantizers. The distortion measure is assumed to be squared error. The quantizers are a generalization of more restrictive polar quantizers analyzed previously, admitting a differing number of phase positions in each magnitude sector. Results are provided for several bit rates up to 2.5 bits/sample. The optimal design always performs as well or better than the restricted polar quantizer, and, in the cases analyzed, slightly outperforms the Max (single-sample) quantizers for a fixed number of levels.**

I. INTRODUCTION

Single-sample quantization of analog samples is a simple source encoding scheme which is widely practiced, and which has received wide theoretical attention in the literature. Familiar applications include speech transmission using PCM or DPCM, A/D conversion for digital signal processing, and quantization of transform coefficients produced by a unitary transform. Often these devices are made to have acceptably small distortion, or error, by choosing a sufficiently large number of approximating values. It is known, however, that for a variety of distortion criteria and memoryless source

Paper approved by the Editor for Communication Theory of the IEEE Communications Society for publication without oral presentation. Manuscript received September 20, 1979; revised May 4, 1980.

The author is with the Department of Electrical Engineering, University of Virginia, Charlottesville, VA 22901.

distributions, that single-sample quantizers are rather inefficient in terms of rate-distortion theoretic limits.

This paper addresses potential improvements achievable by jointly quantizing two samples, referred to as *2-D* quantization. *2-D* quantization represents the first step toward general *M*-dimensional quantizers. The rate-distortion bound will only be approached as *M* becomes large; however, it is intuitive that joint bivariate quantization offers potentially better use of resources (information rate) than does one-at-a time processing. We consider a restrictive but important quantization problem: represent a bivariate Gaussian independent pair (X_1, X_2) using N quantization regions so that the mean square error is minimized. This problem has the usual analytical appeal: the joint density function is circularly-symmetric which suggests polar quantizers, and rate-distortion results are well known which provide a standard of comparison. Beyond this, however, the problem is practically motivated. Various processing techniques, when applied to non-Gaussian sources with memory, produce sequences which are "approximately" independent and Gaussian. As an example, we cite the transform coefficients produced by the DFT of a stationary sequence, provided the blocklength is "long." On the other hand, we could think of this as a blocklength-2 strategy for quantizing an explicitly Gaussian i.i.d. sequence.

The optimal univariate quantizer was found numerically by Max [1] for the Gaussian case, under the constraint of minimizing mean-squared error (MSE) for a given number of quantization regions N. Both optimal uniform and nonuniform designs were obtained for $N \leqslant 36$. Of course, these induce a trivial *2-D* quantizer having rectangular regions by simply using an optimal Max quantizer on each coordinate by itself. This procedure provides a basis of comparison for more sophisticated *2-D* techniques.

A slightly different approach is to minimize MSE subject to the quantizer output maintaining a given entropy, H bits/sample, alluding to the possible use of buffer-instrumented entropy coder on the quantizer output. This was solved for the univariate case by Goblick and Holsinger [2], whose essential result is that quantizers exist whose entropy lies 0.25 bits/sample above the MSE $R(D)$ bound at all rates. This corresponds to a distortion inefficiency of about 41 percent, or about 1.5 dB. This is a significant gain over the Max quantizer, particularly at high rates; however, entropy coding becomes progressively more complicated.

More recent work has focused on optimization in two dimensions, as an attempt to do better in the $R(D)$ sense. Some of this work has been directed at asymptotic results, i.e., design and performance as N becomes large. One key result is that regular hexagons form the region boundaries for large N [3], which follows from the geometric result that among admissible polygons which tile the plane, hexagons have the extremal property of maximizing the area covered, subject to being inside a circle of given error radius. Hexagonal partitions are rather difficult to implement, however, and for small-to-moderate values of N are not as flexible as the polar quantizers we describe below. Further, no optimality claim can be made for hexagonal quantizers for moderate N. Gersho [3] goes on to develop optimal asymptotic strategies in higher dimensions.

In a more practical vein, several workers have considered polar quantizers, meaning the quantizer regions are bounded by curves of constant radius and by curves of constant phase. This assumption, while having itself no optimality claims, has great intuitive appeal for circularly-symmetric dentities we consider here. Further, the implementation of the quantizer is straightforward. Pearlman and Gray [4], in analyzing polar quantization of DFT coefficients, obtain a MSE lower bound of $D/2\sigma^2 \cong 1.78(2^{-R})$ for $R > 4.1$ bits/complex variable. Further, optimal source coding calls for assigning about 1.37 bits/complex variable more rate to the phase variable than the amplitude variable. In other words, for large N, $P \cong 2.6\ M$, where M is the number of magnitude levels and P is the number of phase positions.

Gallagher [5] attains similar conclusions for large N and shows that for $N = 396$ (about 4.3 bits/variable) the MSE, $D/2\sigma^2$, for the optimal polar quantizer is 5.93×10^{-3}. Somewhat disappointing is the fact that this is only slightly better than using a $N = 20$ Max quantizer on each coordinate. The latter induces a 400-cell rectangular quantizer in two dimensions. If entropy-coding is allowed, the ranking actually reverses.

Pearlman [6] recently obtained more exhaustive results on polar quantizers for $N = 1$ to 1024. For uncoded quantizers, the conclusion is that for $N \gtrsim 64$, optimal polar quantizers outperform optimum rectangular (Max) quantizers. However, the difference is never very dramatic; for $N = 900$, the optimal polar quantizer achieves $D/2\sigma^2 = 2.65 \times 10^{-3}$, while use of a 30-cell Max quantizer twice achieves $D/2\sigma^2 = 2.8 \times 10^{-3}$. The gain is rather miniscule when one notes that the rate distortion bound for any type of coding is a $D/2\sigma^2 = 1.1 \times 10^{-3}$. Further, if entropy-coding is anticipated, Pearlman observes that rectangular quantizing is the superior alternative. Bucklew and Gallagher [7] also provide an extensive analysis for general N, finding superiority of polar designs for $N > 101$.

Part of the difficulty with polar quantizers is the factorization requirement: $P = N/M$. One assigns a fixed number of phase positions at each magnitude level. We shall refer to these as strictly polar quantizers (SPQ's). One anticipates that this is not optimal use of resources (number of levels), particularly noting the relative density of quantizer regions at the origin, and their departure from more desirable circular regions. Our approach here is to allow freedom in the number of phases assigned at each magnitude level. This extra freedom will allow still smaller MSE, at the expense of a slightly more involved quantization process.

Our interest is in the region of low N, or small rate. This is the region for which SPQ's were inferior to rectangular quantizers [6], [7]. Also, the general optimization for any N with the extra freedom is time consuming. Finally, the small N region is of interest in source coding applications involving low data rate transform coding of speech and images.

II. POLAR QUANTIZATION: DESCRIPTION AND OPTIMIZATION

We assume the data to be encoded are a pair of zero-mean, unit-variance, independent Gaussian variates. The data vector (X_1, X_2) has a joint density given by

$$p_{x_1 x_2}(x_1, x_2) = \frac{1}{2\pi} \exp\left[-(x_1^2 + x_2^2)/2\right],$$

$$-\infty < x_1, x_2 < \infty. \tag{1}$$

By a rectangular-to-polar change of coordinates, we have

$$P_{R_{j\theta}}(r,\theta) = \frac{r}{2\pi}\exp[-r^2/2] \qquad 0 \leqslant r < \infty,\ 0 \leqslant \theta < 2\pi \tag{2}$$

where

$$R = (X_2{}^2 + X_2{}^2)^{1/2} \tag{3}$$

$$\theta = \tan^{-1}\left(\frac{X_2}{X_1}\right). \tag{4}$$

Prompted by the circular symmetry of (2), and by the desire for straightforward implementation, we focus on 2-*D* quantizers with N regions which are equiangular sectors of concentric annuli, as shown in Fig. 1. The region boundaries are formed by curves of constant radius and of constant angle.

We configure an N region quantizer by first partitioning the magnitude range $[0, \infty)$ into M levels. The radii of these annuli are denoted $0 < R_1 < R_2 < \cdots < R_M = \infty$. The spacings of these radii are not required to be uniform. Within each magnitude annulus, indexed by $m = 1, 2, \cdots, M$, we allow P_m equal-sized phase regions, whose boundaries are defined by

$$(p-1)\frac{2\pi}{P_m} \leqslant \theta < p\left(\frac{2\pi}{P_m}\right) \qquad p = 1, 2, \cdots, P_m. \tag{5}$$

A constraint is then

$$\sum_{m=1}^{M} P_m = N. \tag{6}$$

Within the constraint (6), complete freedom is allowed in assigning the available number of regions among phases on different annuli. We dub these general designs *unrestricted polar quantizers* (UPQ's) in distinction to the *strictly polar quantizers* (SPQ's) analyzed previously in the literature [6], [7]. The latter designs enforce $P_m = P$ at all levels such that $N = MP$. This, of course, severely limits the design of the polar quantizer. For example, $N = 16$ factors as $M = 1, 2, 4, 8, 16$, with respective $P = 16, 8, 4, 2, 1$; however, as we shall see, a UPQ with three magnitude levels proves optimum.

We remark that these designs, while allowing more freedom, are not claimed to be optimal, for any specific N, among all possible 2-*D* quantizers. Nonetheless, they are easily configurable for any N of interest and are rather simply implemented. In this regard, determination of the region index n is conceptually performed as follows. Given (X_1, X_2) perform the polar conversion. Scan a table of (R_n, θ_n) pairs ordered according to increasing R_n, then for fixed R_n, ordered by increasing θ_n. (R_n, θ_n) represent the upper limits on magnitude and phase for region n. As soon as $R_n \geqslant R$, we have located the proper annulus. Within this section of the table, scan in θ_n until $\theta \geqslant \theta_n$, whereupon, the right phase region is located. The associated position in the overall table is the region index n.

The following compact notation is used to describe a particular configuration. $(N, M, P_1, \cdots, P_M)$ denotes an N-level, M magnitude quantizer, with $P_1, \cdots, P_M$ phase regions at the various levels. Thus, a (16, 3, 1, 6, 9) UPQ has 16 regions on the plane, with one circular region centered at the origin, six regions in the next annulus, and nine regions in the outer level.

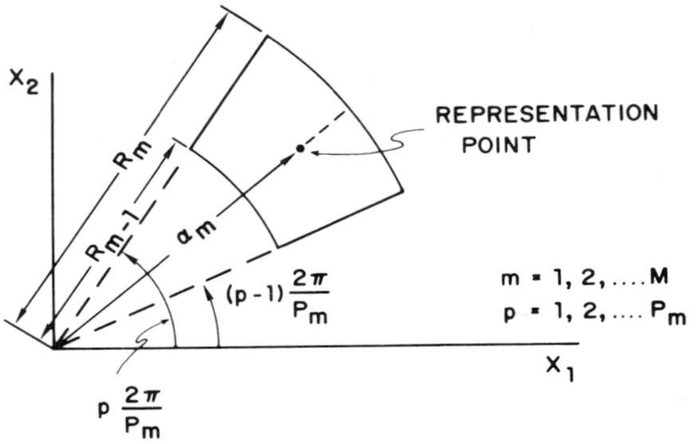

Fig. 1. Notation for unrestricted polar quantizers.

For a given N, there are clearly many possible integer selections for M and the corresponding P_m, such that $\Sigma P_m = N$, and the problem becomes more cumbersome with increasing N. Many of the possibilities can be dismissed outright as having no optimality potential. For example, the (5, 2, 3, 2) quantizer is a poor choice among $N = 5$ designs.

The optimization strategy is simple. We *assume* a set $(N, M, P_1, \cdots, P_M)$ meeting the constraint. We then need to select the $M - 1$ radii (recall $R_M = \infty$), the radii of the quantization points, α_m, $m = 1, 2, \cdots, M$, as in Fig. 1, the phase boundaries, and the phases of the quantization point. The selection is such that the mean-square error, denoted D, is minimized for the particular configuration. A new configuration is then selected and optimized. By exhaustive search, or by at least searching all reasonable configurations, the globally optimal UPQ is determined.

We emphasize the objective is minimum D for a fixed N, referred to as the uncoded quantizer case. Since, in general, the N regions will not be equiprobable, entropy coding of the region indices at the quantizer output can reduce the bit rate below $\log_2 N$. We shall report the associated entropies of the various designs, if such coding is of interest. However, another optimization philosophy which minimizes D for a given R (entropy), or vice versa, will yield somewhat improved performance in this case.

Now consider a specific parameter set $(N, M, P_1, \cdots, P_M)$. The MSE may be expressed as

$$D = \sum_{n=1}^{N} \int_\theta \int_r\limits_{\mathcal{R}_n} |re^{j\theta} - \alpha_n e^{j\beta_n}|^2 p(r,\theta)\, dr\, d\theta \tag{7}$$

where (α_n, β_n) are the magnitude and phase of the representation point for region $\mathcal{R}_n$. Choice of the phase limits is fixed by specification of P_m, i.e., (5). It is also clear that selection of β_n should be midway between the two phase limits. The number of free variables, thus, easily reduces to the $M - 1$ R_m and the M values of α_m.

Necessary conditions for a stationary point are given by differentiation with respect to R_m and α_m as in [6]. The results are easily generalized to the UPQ case:

$$\alpha_m = \frac{P_m}{\pi}\left(\sin\frac{\pi}{P_m}\right)\frac{\int_{R_{m-1}}^{R_m} rp(r)\,dr}{\int_{R_{m-1}}^{R_m} p(r)\,dr} \qquad m = 1, 2, \cdots, M \tag{8a}$$

$$\frac{\partial D}{\partial R_m} = 2R_m\left[\left(\frac{P_{m+1}}{\pi}\sin\frac{\pi}{P_{m+1}}\right)\alpha_{m+1} - \left(\frac{P_m}{\pi}\sin\frac{\pi}{P_m}\right)\alpha_m\right] - \alpha_{m+1}^2 + \alpha_m^2 = 0 \qquad m = 1, 2, \cdots, M-1. \tag{8b}$$

Note that, as often observed, α_m is dependent on the centroid of the probability density for the magnitude on the mth annulus. Simultaneous satisfaction of the $2M - 1$ equations (8) defines the desired minimum. Pearlman [6] in optimizing SPQ's, where $P_m = P$, notes that the problem is essentially determined by the solution of the optimal quantizer for the Rayleigh distribution [8]. That appeal fails here since P_m varies with m.

Optimization of a given quantizer is accomplished numerically as follows. Given a trial vector $(R_1, \cdots, R_{M-1})$, we use (8a) to compute the radii of the quantization points which are optimal for that vector. The integrals are calculated with a Gauss-quadrature procedure chosen to give four decimal-place numerical accuracy. Next, the partial derivatives $\partial D/\partial R_m$ (8b) are computed and used in a gradient corrector procedure to adjust the vector $(R_1, \cdots, R_{M-1})$:

$$\begin{bmatrix} R_1 \\ \vdots \\ R_{M-1} \end{bmatrix}_{k+1} = \begin{bmatrix} R_1 \\ \vdots \\ R_{M-1} \end{bmatrix}_k + C \cdot \begin{bmatrix} \partial D/\partial R_1 \\ \vdots \\ \partial D/\partial R_{M-1} \end{bmatrix} \tag{9}$$

where C is a constant empirically selected. After convergence, the MSE and entropy are calculated.

After some manipulation, the MSE may be written as

$$D = 2 + \alpha_1^2[1 - e^{-R_1^2/2}] + \alpha_2^2[e^{-R_1^2/2} - e^{-R_2^2/2}] + \cdots + \alpha_M^2[e^{-R_{M-1}^2/2}] \cdot \frac{-2P_1\alpha_1}{\pi}\sin\left(\frac{\pi}{P_1}\right)\int_0^{R_1} r^2 e^{-r^2/2}\,dr + \cdots \frac{-2P_M\alpha_M}{\pi}\sin\left(\frac{\pi}{P_M}\right)\int_{R_{M-1}}^{\infty} r^2 e^{-r^2/2}\,dr. \tag{10}$$

This provides the procedure for optimization of a given configuration; a new $(N, M, P_1, \cdots, P_M)$ is selected, and the procedure repeated. We can readily determine if that perturbation of the integer set improved upon the minimum found previously; if not, integer perturbations in another direction are tried.

III. RESULTS AND COMPARISONS

Optimization has been performed for N-level UPQ's for $N = 1, 2, \cdots, 16, 25, 32$, and 36. The small N were of interest for low bit rate transform coding applications, as well as to discern whether better performance could be obtained relative to rectangular quantizers as reported by Pearlman. The larger N values are either perfect squares, admitting ready comparison with optimal rectangular designs, or are a power of two, a convenient operational choice.

Table I provides the quantizer data versus N. In the interest of conserving space, only the optimal UPQ is listed at a given N, along with the R_m, α_m, D', and entropy in bits. Other designs which may be more easily instrumented are often close in performance. It is noted that D' listed in the table is average distortion per coordinate, i.e., $D/2$ where D is given by (10). In this way, we may directly compare the results with univariate encoding, or any other multidimensional scheme.

The performance of certain SPQ's is also shown, taken from [6] and used as a check on our numerical procedure. For N, a perfect square, we also list the performance of the optimal rectangular design, which again is formed by a Max quantizer acting independently on each coordinate.

Several conclusions may be drawn about the performance of 2-D UPQ's. First, relative to optimal 2-D rectangular quantizers, UPQ's are as good or better for all cases where N is a perfect square. 2-D rectangular quantizers can be configured for any N, of course, and it is our conjecture that UPQ's are as good or better for all N, since for N, not a square, less symmetry exists. The gains are not dramatic: 0.06 dB for $N = 9$, 0.3 dB for $N = 16$, 0.34 dB for $N = 25$, and 0.4 dB for $N = 36$. Thus, UPQ quantizers improve on the finding of Pearlman [6] that polar uncoded quantizers (SPQ's) only outperform rectangular designs for $N \gtrsim 64$.

A second conclusion is that UPQ's always outperform (or are equivalent with for $N \leqslant 4$) SPQ quantizers. This is not a surprising conclusion when one understands SPQ's as a subclass of UPQ's. However, the improvement is sometimes quite large, where the best factorization of $N = MP$ still is poor. This is always true for N, a prime number, but for other cases of interest as well. For example, with $N = 8$, (assigning 3 bits to two samples), the optimal SPQ (8, 2, 4, 4) achieves $D = 0.125$ while the optimum UPQ (8, 2, 1, 7) achieves $D = 0.102$. One might say that by virtue of the sum constraint instead of the factoring constraint, UPQ's are more flexible, yielding good designs for any N.

Fig. 2 presents the distortion-rate performance for uncoded optimal UPQ quantizers and the performance for optimal rectangular designs for N, a perfect square. Rate here is defined as $(\log_2 N)/2$. Shown for comparison is the MSE rate-distortion function for the Gaussian memoryless source

$$R(D') = \tfrac{1}{2}\log_2\left(\frac{1}{D'}\right) \qquad 0 < D' \leqslant 1. \tag{11}$$

Although some improvement has been gained in going to 2-D quantization, the distance to the rate-distortion bound

TABLE I
QUANTIZER PERFORMANCE SUMMARY

N	Type	R_1	R_2	α_1	α_2	α_3	D	H*	comment**
1	(1,1,1)	--	--	0.000	--	--	1.000	0.000	1,2,3
2	(2,1,2)	--	--	0.798	--	--	0.682	1.000	1,2,3
3	(3,1,3)	--	--	1.037	--	--	0.463	1.585	1,2
4	(4,1,4)	--	--	1.128	--	--	0.363	2.000	1,2,3
5	(5,1,5)	--	--	1.173	--	--	0.313	2.322	1
	(5,2,1,4)	0.752	--	0.000	1.354	--	0.309	2.313	2
6	(6,1,6)	--	--	1.197	--	--	0.284	2.585	1
	(6,2,1,5)	0.752	--	0.000	1.407	--	0.254	2.556	2
7	(7,2,1,6)	0.752	--	0.000	1.436	--	0.223	2.754	2
8	(8,2,1,7)	0.752	--	0.000	1.454	--	0.204	2.922	2
	(8,2,2,6)	0.889	--	0.362	1.513	--	0.207	2.979	
	(8,2,4,4)	1.375	--	0.747	1.729	--	0.249	2.964	1
9	(9,2,2,7)	0.884	--	0.361	1.529	--	0.188	3.131	
	(9,2,3,6)	1.066	--	0.554	1.622	--	0.187	3.139	2
							0.190	3.072	3
10	(10,2,3,7)	1.051	--	0.547	1.634	--	0.169	3.272	2
	(10,2,5,5)	1.375	--	0.776	1.796	--	0.189	3.286	1
11	(11,2,4,7)	1.179	--	0.658	1.718	--	0.155	3.403	2
12	(12,2,4,8)	1.163	--	0.651	1.722	--	0.143	3.508	2
	(12,2,6,6)	1.375	--	0.792	1.834	--	0.155	3.549	1
13	(13,2,5,8)	1.247	--	0.717	1.780	--	0.133	3.629	2
14	(14,2,5,9)	1.234	--	0.710	1.780	--	0.126	3.715	
	(14,3,1,5,8)	0.458	1.340	0.000	0.856	1.846	0.125	3.729	2
15	(15,3,1,6,8)	0.482	1.424	0.000	0.920	1.908	0.117	3.819	2
16	(16,2,8,8)	1.375	--	0.808	1.871	--	0.120	3.964	1
	(16,3,1,6,9)	0.475	1.400	0.000	0.907	1.900	0.109	3.895	2
							0.118	3.822	3
25	(25,3,4,10,11)	0.798	1.674	0.464	1.192	2.124	0.0739	4.512	2
							0.0799	4.406	3
32	(32,4,1,7,12,12)	0.363	1.031	0.000	0.701	1.384	0.0584	4.844	2
		(also R_3 = 1.846, α_4 = 2.267)							
36	(36,4,1,8,13,14)	0.369	1.051	0.000	0.719	1.397	0.0528	4.990	2
		(also R_3 = 1.848, α_4 = 2.275)							
							0.0580	4.886	3

*entropy in bits per symbol pair

**1: optimal SPQ; 2: optimal UPQ; 3: optimal rectangular

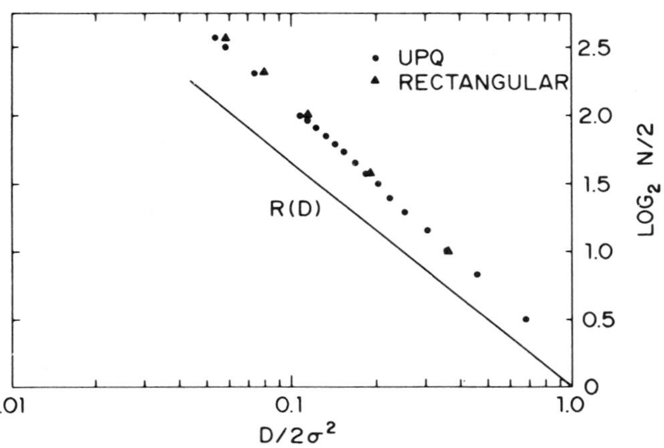

Fig. 2. Rate $[(\log_2 N)/2]$ versus distortion for optimal quantizers.

is still relatively large. Slightly better 2-D designs may exist, although likely not with the same implementation simplicity. However, it is apparent that attainment of performance near $R(D)$ will require much longer blocklengths: either multi-dimensional quantizers analyzed by Gersho [3], or another form of long-block quantizing, tree/trellis coding (see, e.g., Wilson and Lytle [91]).

Some of this gap can be closed, of course, by entropy-coding the output of the quantizer. Fig. 3 shows distortion versus entropy for the UPQ designs and certain rectangular designs. We now observe that the margin of UPQ improvement decreases, and, in some cases, the rectangular quantizer does better. Entropy-coding is again a "great equalizer." Similar conclusions were noted by Pearlman, who observed that the utility of polar quantizing was small if entropy coding were planned. One caveat is, however, that both polar and rectangular designs may be improved if we explicitly optimize entropy versus distortion [2].

Also interesting is the evolution of optimal UPQ designs as N grows. Fig. 4 depicts schematically the geometry of the optimal quantizer. For $N < 5$, best performance is achieved

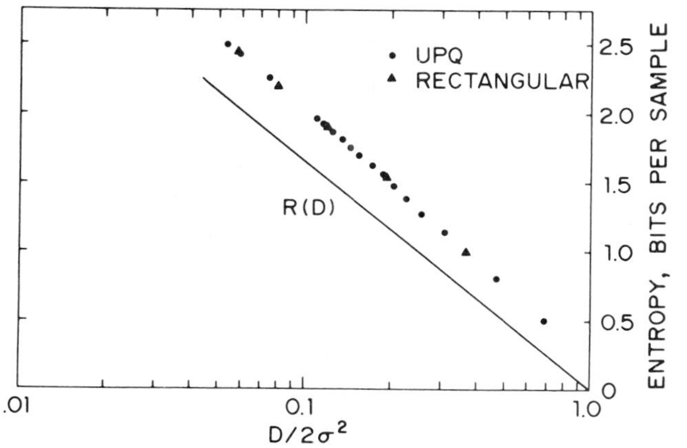

Fig. 3. Rate (entropy) versus distortion for optimal quantizers.

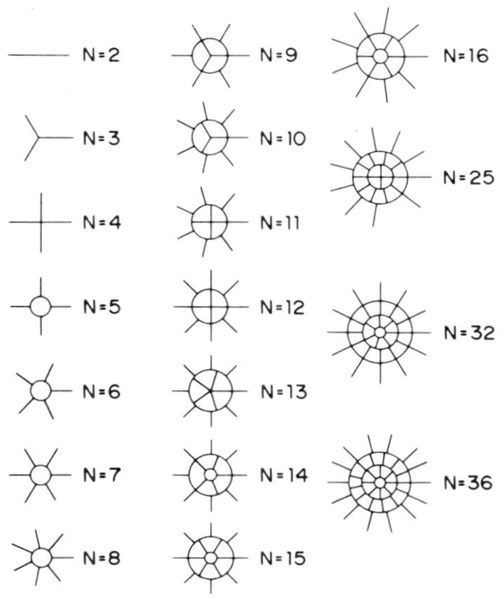

Fig. 4. Geometry of optimal UPQ's (roughly to scale).

by adding more phase regions at the only magnitude level. For $N = 5$, instead of adding another phase, it becomes better to add a single region centered at the origin. As N increases further through $N = 13$, more regions are added to both inner and outer magnitude levels, the former number growing faster percentagewise. At $N = 14$, the same phenomenon repeats that occured for $N = 5$, namely another magnitude region is added. Currently, quantitative descriptions of this process are difficult to supply; perhaps analytical results for large N could show the best distribution of phases and magnitudes.

The evolution, however, is not always "continuous," as illustrated in the transition from $N = 8$ to $N = 9$. (8, 2, 1, 7) is optimal at $N = 8$, and it might be expected that the best $N = 9$ design would be (9, 2, 1, 8) or (9, 2, 2, 7), i.e., addition of one region to either magnitude. However, (9, 2, 3, 6) proves best. This behavior is also observed for larger N as well.

IV. CONCLUSION

Two-dimensional polar quantizers have been described which are as good or better than optimal rectangular quantizers for any fixed N, assuming a bivariate Gaussian distribution having circular symmetry. Improved performance over strictly polar quantization is afforded by allowing a variable number of phases at each magnitude level. The distortion reduction turns out to be rather small, although the designs provide a great deal of flexibility versus N, and are readily implemented.

If entropy coding of the quantizer outputs is allowed, then the advantages of polar over rectangular quantization disappear for the designs reported here. If entropy coding is planned, however, a slightly different optimization should be pursued, perhaps yielding better performance. This would appear to be an interesting research problem, as would extension of these results to larger N.

A further interesting question is the robustness of these quantizers to gain changes, i.e., as the variance of the data becomes larger or smaller than one. It is our feeling that polar designs are more robust in this regard than are rectangular designs.

REFERENCES

[1] J. Max, "Quantizing for minimum distortion," *IRE Trans. Inform. Theory*, pp. 7–12, Mar. 1960.
[2] T. J. Goblick and J. L. Holsinger, "Analog source digitization: Comparison of theory and practice," *IEEE Trans. Inform. Theory*, vol. IT-13, Apr. 1967.
[3] A. Gersho, "Asymptotically optimal block quantization," *IEEE Trans. Inform. Theory*, vol. IT-25, pp. 373–380, vol. IT-25, July 1979.
[4] W. A. Pearlman and R. M. Gray, "Source coding of the discrete Fourier transform," *IEEE Trans. Inform. Theory*, vol. IT-24, pp. 683–692, Nov. 1978.
[5] N. Gallagher, Jr., "Quantization schemes for the discrete Fourier transform of a random time series," *IEEE Trans. Inform. Theory*, Mar. 1978.
[6] W. A. Pearlman, "Polar quantization of a complex Gaussian random variable," *IEEE Trans. Commun.*, vol. COM-27, pp. 892–898, June 1979.
[7] J. A. Bucklew and N. C. Gallagher, "Quantization schemes for bivariate Gaussian random variables," *IEEE Trans. Inform. Theory*, vol. IT-25, pp. 537–543, Sept. 1979.
[8] W. A. Pearlman and G. H. Senge, "Optimal quantization of the Rayleigh probability distribution," *IEEE Trans. Commun.*, vol. COM-27, pp. 101–112, Jan. 1979.
[9] S. G. Wilson and D. W. Lytle, "Trellis encoding of continuous-amplitude memoryless sources," *IEEE Trans. Inform. Theory*, vol. IT-23, pp. 404–408, May 1977.

Reprinted from pages 581-586 of *Proceedings of the 1984 Conference on Information Sciences and Systems,* March 14-16, 1984, Princeton University, 693p.

ASYMPTOTIC PERFORMANCE OF UNRESTRICTED POLAR QUANTIZERS

Peter F. Swaszek & TsuWei Ku

ABSTRACT

For the mean square error criterion, polar coordinate quantizers have been analyzed for a variety of circularly symmetric sources. To obtain further gains in performance for the independent bivariate Gaussian source, a somewhat more complex scheme called unrestricted polar quantizers (UPQs) was introduced. These UPQs have the same form as the polar quantizers except that the number of phase divisions is allowed to vary from magnitude ring to magnitude ring. Previously, they were designed for the small number of levels case (N<36); a range in which the traditional rectangular coordinate quantizers match or beat the performance of polar coordinate quantizers. This paper contains an asymptotic analysis of the MSE performance of the UPQs with a circularly symmetric source.

I - INTRODUCTION

Previously, several researchers have considered polar coordinate quantization with the goal of improving the system performance over that of rectangular coordinate (scalar) quantization without the design and implementation complexity of optimum vector quantizers [1-4]. In essence, polar quantization to N levels refers to the following operations:

1 - the rectangular coordinate representation of the source, (x,y), is transformed to the polar coordinates of magnitude r and phase ϕ through the relations

$$r = (x^2 + y^2)^{1/2} \quad ; \quad r \in [0,\infty)$$

$$\phi = \tan^{-1} \frac{y}{x} \quad ; \quad \phi \in [0,2\pi)$$

2 - the polar variables are quantized by separate scalar Lloyd-Max type quantizers, $Q_r(r)$ and $Q_\phi(\phi)$ with N_r and N_ϕ levels respectively, where $N_r x N_\phi = N$ is the total number of quantization levels.

With this scheme, the bivariate quantization pattern is seen to be a partitioning of the plane into N_r rings concentric about the origin with each ring divided angularly into N_ϕ pieces. All inputs within one of these regions are mapped to a particular output value, usually also within the region.

Three of the referenced papers [1-3] employed the mean square error (MSE) criterion for the quantization of the independent, bivariate Gaussian source; minimizing the MSE by identifying necessary conditions upon the quantizers' parameters and specifying the factorization of N into $N_r x N_\phi$. In all three cases, the gain in performance over rectangular coordinate quantization was small with the worst performance being for small values of N (in fact, for some N < 100, rectangular quantization outperformed polar quantization). This concept of changing coordinates and performing scalar quantization has received continued research interest and has been analyzed for several other situations: the bivariate non-Gaussian circularly symmetric source [4], the k-dimensional spherically symmetric source [5], and the spherically symmetric source with the quantizers constrained to be uniform scalar quantizers [6]; the discussions in [4] and [5] containing details on both the large (asymptotic) and small N cases.

In [7], Wilson overcame the lack of performance of small N polar quantizers for the bivariate Gaussian source by defining and designing unrestricted polar quantizers (UPQs). Whereas the strictly polar quantizers (SPQs) described above employed separate fixed quantizers for both the magnitude and the phase, the UPQs employ a fixed quantizer for the magnitude and a variable quantizer for the phase. In the polar quantization analyses mentioned above, the phase random variable was always uniformly distributed and the optimum phase quantizer was found to be a uniform quantizer on $[0,2\pi)$ with N_ϕ levels (N_ϕ a fixed constant). For the UPQs, Wilson let the number of regions in the uniform phase quantizer vary depending upon the result of the magnitude quantization. To have a total of N regions for the quantizer, the UPQs must satisfy the constraint

$$\sum_{i=1}^{N_r} N_i = N$$

where N_i is the number of phase divisions in the phase quantizer when r is in the i-th magnitude ring. We see that the UPQs are a more general form of polar quantizer and hence they demonstrate better performance that the SPQs. Figure 1 depicts both a strictly polar quantizer and a UPQ for the N=10 case.

Unfortunately, the bivariate Gaussian source results in [7] are purely numerical results for small N (N<36). The larger N case was not considered do to the numerical complexity of evaluation. In particular, the design of the UPQs requires specification of the parameters of the N_r level magnitude quantizer and the number of phase levels per magnitude ring, N_i, $i=1,2,\ldots N_r$. In [7], for each value of N, Wilson evaluated the MSE performance of all combinations of the N_i which summed to N, using a steepest descent approach for finding the magnitude quantizer's parameters. For small N, this is a lengthy task; for large N this approach is computationally prohibitive.

The goal of this paper then, is an asymptotic analysis of the performance of the UPQs when the number of levels N is large. The solution includes both the specification of the magnitude quantizer and the N_i as well as an evaluation of the resulting MSE. Section II contains the derivation of these results. Section III describes several examples of UPQs with exact calculations of the MSE to validate the analytical results. Section IV closes this paper with a discussion of the results and additional comments.

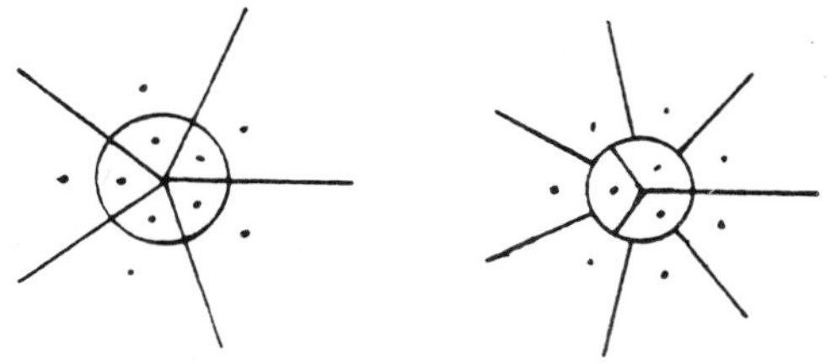

Fig. 1 - Comparison of SPQ and UPQ.

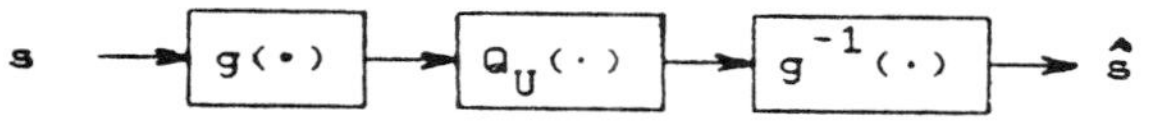

Fig. 2 - Scalar quantizer compandor model.

II - DERIVATION

For this analysis, we will assume that the input is from a continuous, circularly symmetric source with unit variance rectangular coordinate marginals. Such a source has its density function of the form

$$p(x,y) = p(\sqrt{x^2 + y^2})$$

Transforming to polar coordinates, the phase ϕ is uniformly distributed on $[0,2\pi)$ and the magnitude is distributed on $[0,\infty)$ with density function $f(r) = r\,p(r)$. Although this form is restrictive, the important Gaussian case is included.

For the input pair (x,y), the mean square error is defined by

$$MSE = \int_0^\infty \int_0^\infty [(x - \hat{x})^2 + (y - \hat{y})^2]\, p(x,y)\, dx\, dy$$

where hats indicate quantized values. Converting to polar coordinates, the MSE becomes

$$MSE = \int_0^\infty \int_0^{2\pi} [r^2 + \hat{r}^2 - 2 r \hat{r} \cos(\phi - \hat{\phi})]\, \frac{f(r)}{2\pi}\, dr\, d\phi$$

By definition, the UPQs employ separate scalar quantizers for r and ϕ. For the derivation, we will employ the compandor model to implement the scalar quantizers. With this model, we replace each quantizer with a series connection of three elements (Fig. 2): a compandor g, a uniform quantizer Q_U, and an expandor g^{-1}. The magnitude expandor will be labelled g; there will be no phase compandor since the the phase quantizer is already assumed to be uniform. Furthermore, we will employ the additive noise approximation for uniform quantizers with large numbers of levels. With this model, each uniform quantizer Q_U is replaced by an additive noise source, independent of the quantized variable and uniformly distributed on $[-\Delta/2,\Delta/2]$ where Δ is the step size of Q_U. The magnitude quantizer will have an additive source α with $\Delta=1/N_r$. With this additive noise source model, the output of the compandor implementation of the magnitude quantizer can approximated for large N_r (small α) by

$$\hat{r} = g^{-1}[\ g(r)\ +\ \alpha\ ]$$

$$\approx g^{-1}[g(r)] + \alpha\ g^{-1'}[g(r)]$$

$$\approx r + \frac{\alpha}{|g'(r)|}$$

For the phase quantizer the number of levels changes with each magnitude ring. For the asymptotic analysis with N_r large, the number of phase levels will be represented by a function of r, $N_\phi(r)$, and the step size of Q_ϕ, and hence the range of the additive noise γ, will be $\Delta_\phi = 2\pi/N_\phi(r)$. With this noise variable, the phase quantizer's output is $\hat{\phi} \approx \phi + \gamma$. Although the additive noise model is strictly false, its application will provide results which are shown in Section III to be consistent with exact UPQ MSE calculations.

Applying the compandor and additive noise models to the expression for the MSE yields

$$MSE \approx \int_0^{2\pi} \int_0^\infty \int_{-\Delta_\phi/2}^{\Delta_\phi/2} \int_{-\Delta_r/2}^{\Delta_r/2} \left[r^2 + (r + \frac{\alpha}{|g'(r)|})^2 - 2r (r + \frac{\alpha}{|g'(r)|}) \cos\gamma \right] \frac{f(r)}{2\pi \Delta_r \Delta_\phi} d\alpha\, d\gamma\, dr\, d\phi$$

To simplify this expression, we use several facts. First, α, γ, r, and ϕ are all independent random variables (r and ϕ by definition, α and γ by assumption). Next, since α is uniformly distributed on $[-\Delta_r/2, \Delta_r/2]$ then $E(\alpha^2) = 1/12N_r^2$ and $E(\alpha) = 0$. For γ uniform on $[-N_\phi(r)/2, N_\phi(r)/2]$ and $N_\phi(r)$ large, we have

$$E(\cos\gamma) = \frac{N_\phi(r)}{\pi} \sin\frac{\pi}{N_\phi(r)}$$

$$\approx 1 - \frac{\pi^2}{6\ N_\phi^2(r)}$$

Finally, since we previously assumed that the source had unit variance marginals, then $E(r^2) = 2$. Putting these facts all into the previous expression for the MSE yields

$$MSE \approx \frac{1}{12\ N_r^2} \int_0^\infty \frac{f(r)}{|g'(r)|^2} dr$$

$$+ \frac{\pi^2}{3} \int_0^\infty \frac{r^2\ f(r)}{N_\phi^2(r)} dr$$

For any given magnitude density $f(r)$ and magnitude compressor $g(r)$, we wish to establish the functional form of $N_\phi(r)$ to minimize the above expression for the MSE. This function $N_\phi(r)$ evaluated at the i-th magnitude ring's output, $\hat{r}_i$, represents the number of phase divisions for the magnitude ring about $\hat{r}_i$. For a fixed N quantizer, the total number of levels in all of the magnitude rings must sum to N

$$\sum_{i=1}^{N_r} N_\phi(\hat{r}_i) = N$$

Dividing by N_r, this becomes

$$\sum_{i=1}^{N_r} \frac{N_\phi(\hat{r}_i)}{N_r} = \frac{N}{N_r}$$

Let Δ_i equal the width of the i-th magnitude quantization interval. From the definition of the magnitude compressor, this region maps through g into a interval of width $1/N_r$ on the range of the uniform quantizer Q_U. For r^* the midpoint of this magnitude region, we can write this as

$$|g(r^* + \Delta_i/2) - g(r^* - \Delta_i/2)| = 1/N_r$$

We can approximate this difference by a derivative $1/N_r \approx |g'(r^*)|\ \Delta_i$ to yield

$$\sum_{i=1}^{N_r} N_\phi(r^*)\ g'(r^*)\ \Delta_i \approx \frac{N}{N_r}$$

For large N_r (small Δ_i), approximate the sum by an integral to establish the constraint

$$\int_0^\infty N_\phi(r)\ |g'(r)|\ dr = \frac{N}{N_r}$$

With this constraint, the above expression for the MSE can be optimized over the choice of $N_\phi(r)$. Through a calculus of variations argument, the function $N_\phi(r)$ which minimizes the MSE is

$$N_\phi(r) = \lambda \frac{r^{2/3} f^{1/3}(r)}{|g'(r)|^{1/3}} \quad (1)$$

where λ is a constant so that $N_\phi(r)$ satisfies its constraint. Substituting this into the expression for the MSE yields

$$MSE \approx A N_r^{-2} + B N_r^{2} N^{-2}$$

where

$$A = \frac{1}{12} \int_0^\infty \frac{f(r)}{|g'(r)|^2} dr \quad ;$$

$$B = \frac{\pi^2}{3} \left[\int_0^\infty r^{2/3} |g'(r)|^{2/3} f^{1/3}(r) dr \right]^3$$

Optimizing this expression over the choice of N_r yields

$$N_r = \left[\frac{A}{B} \right]^{1/4} N^{1/2} \quad (2)$$

and

$$MSE \approx 2 A^{1/2} B^{1/2} N^{-1} \quad (3)$$

Unfortunately, we don't know how to optimize this expression over the choice of the magnitude compressor $g(r)$. If we assume that $g(r)$ is the optimum MSE compressor for the magnitude density function

$$g(r) = \int_0^r f^{1/3}(r) dr \Big/ \int_0^\infty f^{1/3}(r) dr$$

then the MSE in (3) reduces to

$$MSE = \frac{\pi}{3} C^{1/2} D^{3/2} N^{-1} \quad (4)$$

where

$$C = \int_0^\infty f^{1/3}(r) dr \quad ;$$

$$D = \int_0^\infty r^{2/3} f^{5/9}(r) dr$$

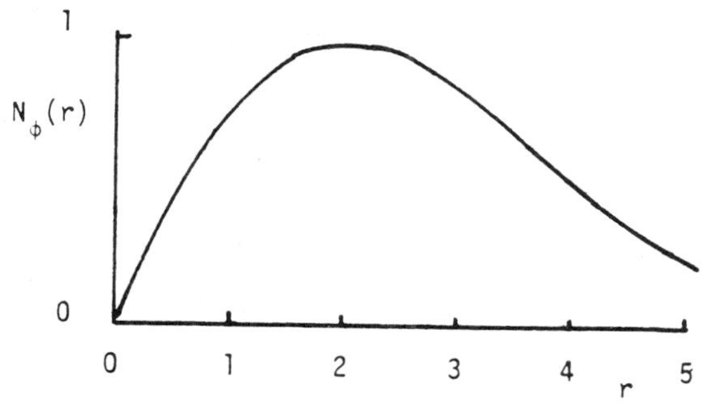

Fig. 3 - $N_\phi(r)$ for the bivariate Gaussian source (normalized).

III - EXAMPLES

As a measure of the usefulness of this result, let us consider the UPQs for the bivariate Gaussian source:

$$p(x,y) = \frac{1}{2\pi} e^{-(x^2 + y^2)/2}$$

The magnitude compressor used in the derivation of (4) is then the Rayleigh compressor. With this source and compressor, the phase level distribution function, $N_\phi(r)$, can be computed. A normalized plot of this function for the Gaussian source appears in Figure 3. As to the MSE, evaluating (4) with this source and compressor yields $MSE \approx 4.30 N^{-1}$. For comparison, equivalent expressions for the MSE of rectangular, polar (SPQ), and optimum quantizers are

$$MSE_{rect} \approx 5.44 N^{-1}; \quad MSE_{SPQ} \approx 4.95 N^{-1};$$

$$MSE_{opt} \approx 4.02 N^{-1}$$

Note that the UPQs outperform the SPQs by quite a large margin; hence, this technique could be very valuable to the quantization of the bivariate Gaussian and other circularly symmetric sources.

To validate the assumptions of the additive noise model made in the derivation of (3) and (4), exact calculations for several UPQs with the bivariate Gaussian source were made. After assuming that $g(r)$ was the Rayleigh density compressor and selecting a value of N, the value of N_r was evaluated through (2); actually, the closest integer value was chosen. Next, the optimum breakpoint and output values for an N_r level Rayleigh quantizer were found from the tables in [8]. Finally, the numbers of phase levels in the rings were found by evaluating $N_\phi(r)$ at the N_r output values, again taking the nearest integer values. Notice that the rounding of the calculated values of N_r and $N_\phi(r)$ yields a total number of levels slightly different than the original value of N. However, if a fixed value of N is desired, the $N_\phi(r)$ values can be adjusted slightly (rounded up or down) to sum to N exactly. Experience with the calculations show that the performance is relatively insensitive to this rounding.

Tables I, II, and III contain data on examples with approximately 90, 400, and 900 levels respectively. The data includes the output and breakpoint values for the Rayleigh quantizers along with the number of phase levels in each magnitude ring. Also listed are the exact MSE values calculated. For all three examples, the exact value of the MSE was very close to the expected value of 4.30 N^{-1}. Figure 4 displays the quantization pattern of the N=90 UPQ on the $x,y \in [-5,5]$ section of the bivariate plane.

Finally, small N applications of the design technique discribed above was considered. In particular, the number of magnitude rings and phase divisions per ring were calculated for N equal to 10, 13, 25, and 32 and are compared in Table IV to the equivalent UPQ results from [7]. Wilson's number of phase regions are listed under UPQ while the phase region numbers from (2) and the sampling of $N_\phi(\mathbf{r})$ are listed under AUPQ. We note that the values are very close to the optimum UPQs.

breakpoint	output	N_ϕ
0.0000	0.4016	7
0.6140	0.8264	12
1.0273	1.2281	16
1.4407	1.6534	19
1.9038	2.1542	18
2.5076	2.8610	16

Total number of regions = 88

exact MSE = 0.051548 = 4.092 N^{-1}

Table I - N = 88 UPQ example.

breakpoint	output	N_ϕ
0.0000	0.2326	9
0.3511	0.4696	17
0.5740	0.6783	23
0.7772	0.8760	28
0.9731	1.0702	32
1.1681	1.2661	36
1.3672	1.4682	38
1.5750	1.6818	40
1.7976	1.9133	41
2.0431	2.1728	41
2.3253	2.4779	39
2.6718	2.8657	35
3.1573	3.4489	28

Total number of regions = 407

exact MSE = 0.010386 = 4.227 N^{-1}

Table II - N = 407 UPQ example.

breakpoint	output	N_ϕ
0.0000	0.1703	11
0.2563	0.3423	20
0.4169	0.4916	27
0.5608	0.6301	33
0.6964	0.7627	38
0.8274	0.8922	43
0.9563	1.0204	47
1.0846	1.1488	51
1.2138	1.2787	54
1.3452	1.4116	57
1.4801	1.5487	59
1.6201	1.6916	60
1.7671	1.8426	61
1.9233	2.0041	62
2.0920	2.1799	61
2.2777	2.3754	60
2.4876	2.5997	57
2.7347	2.8689	53
3.0433	3.2178	46
3.4854	3.7529	35

Total number of regions = 935

exact MSE = 0.004560 = 4.264 N^{-1}

Table III - N = 935 UPQ example.

N	UPQ	AUPQ
10	(3,7)	(3,6)
13	(5,8)	(4,7)
25	(4,10,11)	(4,9,10)
32	(1,7,12,12)	(1,7,11,11)

Table IV - Comparison of the large N design and Wilson's solutions for small N.

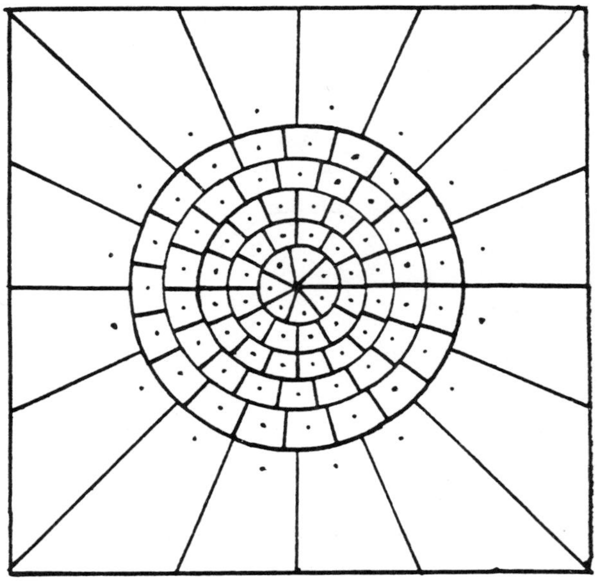

Fig. 4 - N=88 Gaussian UPQ.

IV – COMMENTS

In this paper, an asymptotic expression for the MSE performance of the unrestricted polar quantizers was derived. The optimum choice of the number of magnitude rings was presented. Also included was a derivation of the optimum distribution of the levels to the phase quantizers through the function $N_\varphi(r)$. These results were seen to be dependent only upon the density function of the source and the magnitude compressor function employed. Finally, the analytical results were confirmed through several exact examples with the Gaussian source.

One additional note is required. During the derivation, the performance expression in (3) was not optimized over the choice of the magnitude compressor g(r) to yield (4). In particular, (3) is a function of g(r) in both terms, something that we did not know how to optimize. Hence, the performance of the UPQs could be even better given that g(r) was selected optimally.

ACKNOWLEDGEMENT

This research has been supported by the National Science Foundation NEFRI Grant ECS-8206237 and by a Physio-Control Corporation Career Development Award.

POSTSCRIPT

Much to our chagrin, after the final preparation of this paper, we were able to optimize the expression for the MSE in (3) over the choice of the magnitude compressor. In particular, a Schwarz inequality argument shows that the minimum MSE compressor function is of the form

$$g(r) = \frac{\int_0^r r^{-1/4} f^{1/4}(r)\, dr}{\int_0^\infty r^{-1/4} f^{1/4}(r)\, dr}$$

and the resulting minimum MSE is

$$MSE \approx \frac{\pi}{3N}\left[\int_0^\infty r^{1/2} f^{1/2}(r)\, dr\right]^2$$

For the Gaussian example of Section III, this result is MSE ≈ 4.18 N^{-1}.

REFERENCES

1- N.C. Gallagher Jr., "Quantizing Schemes for the Discrete Fourier Transform of a Random Time Series," IEEE Trans. Inform. Theory, Vol. IT-24, March 1978, pp.156-163.

2- W.A. Pearlman, "Polar Quantization of a Complex Gaussian Random Variable," IEEE Trans. Comm., Vol. COM-27, June 1979, pp.892-899.

3- J.A. Bucklew & N.C. Gallagher Jr., "Quantization Schemes for Bivariate Gaussian Random Variables," IEEE Trans. Inform. Theory, Vol. IT-25, Sept. 1979, pp.537-543.

4- J.A. Bucklew & N.C. Gallagher Jr., "Two-Dimensional Quantization of Bivariate Circularly Symmetric Densities," IEEE Trans. Inform. Theory, Vol. IT-25, Nov. 1979, pp.667-671.

5- P.F. Swaszek & J.B. Thomas, "Multidimensional Spherical Coordinates Quantization," IEEE Trans. Inform. Theory, July 1983, pp.570-576.

6- P.F. Swaszek, "Uniform Spherical Coordinate Quantizers," Proc. Allerton Conf. Comm. Cont. & Comp., Oct. 1983.

7- S.G. Wilson, "Magnitude/Phase Quantization of Independent Gaussian Variates," IEEE Trans. Comm., Vol.COM-28, Nov. 1980, pp.1924-1929.

8- W.A. Pearlman & G.H. Senge, "Optimal Quantization of the Rayleigh Probability Density," IEEE Trans. Comm., Vol. COM-27, Jan. 1979, pp.101-112.

Reprinted from *IEEE Trans. Inform. Theory* **IT-29**(4):570–576 (1983)

Multidimensional Spherical Coordinates Quantization

PETER F. SWASZEK, MEMBER, IEEE, AND JOHN B. THOMAS, FELLOW, IEEE

***Abstract*—Several investigators have considered polar coordinates quantization of a circularly symmetric source; in particular, the independent bivariate Gaussian source. Their schemes quantize the polar coordinates independently in an attempt to reduce the mean-square error below that of an analogous rectangular coordinates quantizer yet retain an implementation simpler than that of the optimal bivariate quantizer. The design of a spherical coordinates quantizer in k dimensions with $k > 2$ ($k = 2$ matches published results) is considered. Examples are presented along with comparisons to the rectangular (one-dimensional) and optimal schemes.**

I. INTRODUCTION

A DATA quantizer is a mapping of a vector-valued source onto a finite number of points. A general multidimensional characterization of an N-level quantizer consists of a partition of the input space into N disjoint regions and the assignment of a particular output value to each region. Implementation requires deciding which of the regions contains the input, often a time-consuming task. In one dimension, a scalar or zero-memory quantizer has regions which are intervals on the real line; hence its implementation is simple. Both uniform and nonuniform interval width quantizers have been designed for a variety of fidelity criteria and source statistics. The canonical example is Max's unit power, Gaussian probability density function quantizer [1] for the mean-square error (MSE) performance criterion. The MSE criterion has universal appeal in its tractability and the relationship to the intuitive notions of noise power and signal-to-noise ratio. Rate distortion theory, however, suggests that multidimensional quantizers may be more efficient.

Research in multidimensional quantization began with the works of Huang and Schultheiss [2] and Zador [3]. Huang and Schultheiss considered the quantization of a correlated Gaussian source. Their system transformed the input vector by a linear device to a set of uncorrelated (hence independent) coordinates and quantized each new coordinate separately. This scheme, although suboptimal, retained a simple implementation and reduced the MSE below that of separate quantizers for the correlated rectangular coordinates. The solution included the factorization of the levels for each quantizer since the product of the number of levels in each quantizer must equal the total number of levels N. Zador considered asymptotic error rates for optimal multidimensional quantization in k dimensions. He derived bounds on the minimally attainable distortion but did not present the actual quantizer design. Later, Gersho [4] and Conway and Sloane [5] discussed optimal quantizer designs of particular dimensionalities.

A major area of interest in suboptimal multidimensional quantizer design involves the use of polar coordinates ($k = 2$). After effecting a change from rectangular to polar coordinates, the resulting magnitude and phase are quantized using separate scalar, Max-type quantizers. Pearlman [6] and Bucklew and Gallagher [7] considered the quantizing of independent bivariate Gaussian random variables in this manner. DFT coefficients, holographic data, or a pair of inputs from an independent and identically distributed

Manuscript received October 29, 1981; revised December 1, 1982. This research was supported in part by the Office of Naval research under Contract N00014-81-K0146 and in part by the National Science Foundation under Grant ECS-79-18915 and NEFRI Grant ECS-8206237. This work was partially presented at the Allerton Conference on Communication, Control, and Computing, Montecello, IL, September 1981.

P. F. Swaszek is with the Department of Electrical Engineering, FT-10, University of Washington, Seattle, WA 98195.

J. B. Thomas is with the Department of Electrical Engineering and computer Science, B-330 Engineering Quadrangle, Princeton University, Princeton, NJ 08544.

Gaussian source can be considered as the output of a bivariate Gaussian source. Published results show that the polar form almost always outperforms the rectangular format, a pair of Max quantizers, for Gaussian variates. Bucklew and Gallagher [8] later extended the polar form to any circularly symmetric density of which the bivariate Gaussian is one example. Noting that rectangular quantizers outperform the polar quantizers when N is small. Wilson [9] defined unrestricted polar quantizers which have lower MSE values. Swaszek and Thomas [10] investigated the optimality of the polar schemes and introduced a scheme which resembles the optimal quantizer for the bivariate Gaussian source, a tesselation of distorted hexagons, but has a simpler implementation.

This paper describes an extension of the polar quantizers mentioned above to the quantization of a k-dimensional spherically symmetric random source. Section II considers the data vector $\boldsymbol{x}$ of length k with rectangular coordinate elements x_j, $j = 1, 2, \cdots, k$. The source statistics are contained in its k-dimensional density function $f(\boldsymbol{x})$. Using a transformation to k-dimensional spherical coordinates, the resulting magnitude and $k - 1$ angles will be separately quantized. In Section III, the magnitude and angles quantizers are derived for the MSE performance criterion. Asymptotic results and allocation of the number of levels to each separate quantizer are the topics addressed in Section IV. Section V contains several examples.

II. Spherical Coordinates Quantizers

Spherically symmetric sources are characterized by contours of constant height which are hyperspheres in the k-dimensional space. These spherically symmetric densities can be generated by replacing the independent variable of a zero-mean unit power univariate density, say $p(x)$, with the square root of a quadratic form [11]. The resulting density has the form:

$$f(\boldsymbol{x}) = \gamma p\left(\sqrt{\boldsymbol{x}^T\boldsymbol{x}}\right); \qquad \boldsymbol{x} \in R^k,$$

where γ is a scaling constant so that the density has unit mass. The resulting multivariate density, however, does not in general have as its marginals the univariate $p(x)$. We also note that in general, the above spherically symmetric density does not factor into the product of the marginal densities of the rectangular coordinates. Hence, the rectangular coordinates are uncorrelated but not independent random variables. The only exception to this property is the multivariate Gaussian distribution.

In one-dimensional or rectangular quantization, each coordinate x_j is quantized independently by a Max-type scalar quantizer Q_j. The resulting quantization regions, being the cross product of k intervals, are k-dimensional rectangular parallelepipeds. Each coordinate has a MSE term $E_1(N_j)$ which is the error associated with a scalar quantizer for the marginal density with N_j levels ($N_1 \times N_2 \cdots \times N_k = N$). The errors, being independent and orthogonal, sum to the total error. A symmetry argument shows that this rectangular quantization error is minimized when each $N_j = N^{1/k}$. The total error is then $k \times E_1(N^{1/k})$.

Another set of coordinates with which to describe an input $\boldsymbol{x}$ of R^k is the magnitude r and $k - 1$ angles φ_j of the k-dimensional spherical coordinates system [12]. The following transformations produce these coordinates:

$$r = \left(\sum_{j=1}^{k} x_j^2\right)^{1/2};$$

$$\varphi_j = \tan^{-1}\left[\frac{x_{j+1}}{\left(\sum_{i=1}^{j} x_i^2\right)^{1/2}}\right], \qquad j = 1, 2, \cdots, k-1.$$

This transformation is a straightforward generalization of the rectangular to polar change of coordinates which has already received much theoretical attention. A change of variables produces the source density in the spherical coordinates

$$f(\boldsymbol{x}) \to f(r, \boldsymbol{\varphi}) = f_k(r) \prod_{j=1}^{k-1} f_j(\varphi_j),$$

where

$$f_k(r) = \frac{2\pi^{k/2}}{\Gamma(k/2)} r^{k-1}\gamma p(r), \qquad r \in [0, \infty)$$

is the magnitude density with $p(\cdot)$ as defined above. The $k - 1$ angle densities are

$$f_1(\varphi_1) = \frac{1}{2\pi}, \qquad \varphi_1 \in [0, 2\pi),$$

and

$$f_j(\varphi_j) = \frac{\Gamma[(j+1)/2]}{\Gamma(1/2)\Gamma(j/2)} \cos^{j-1}\varphi_j,$$

$$\varphi_j \in [-\pi/2, \pi/2],\ j = 2, 3, \cdots, k-1.$$

The resulting spherical coordinates are statistically independent.

Generalizing the concept of polar coordinate quantizing to the spherical coordinates representation, we will employ separate scalar quantizers ($Q_r, Q_1, Q_2, \cdots, Q_{k-1}$) for this spherical coordinates vector. Doing so defines the typical quantization region as the intersection of a nonzero width spherical shell centered at zero with a pyramid of apex zero (see Fig. 1 for $k = 2$ and 3 examples). The spherical coordinates MSE expression is not as simple as that of the rectangular quantizer and is derived below. Another problem of interest is the factorization of the number of levels to each quantizer ($N = N_r \times N_1 \times N_2 \cdots \times N_{k-1}$). In the bivariate case ($k = 2$), the ratio of N_φ to N_r which minimizes the MSE has already been found [8].

For a quantizer with input $\boldsymbol{x}$ and output $\hat{\boldsymbol{x}}$ the MSE is

$$\text{MSE} = \int_{R^k} |\boldsymbol{x} - \hat{\boldsymbol{x}}|^2 f(\boldsymbol{x})\, d\boldsymbol{x},$$

where hats indicate quantized values. Transforming to spherical coordinates and simplifying yields the spherical

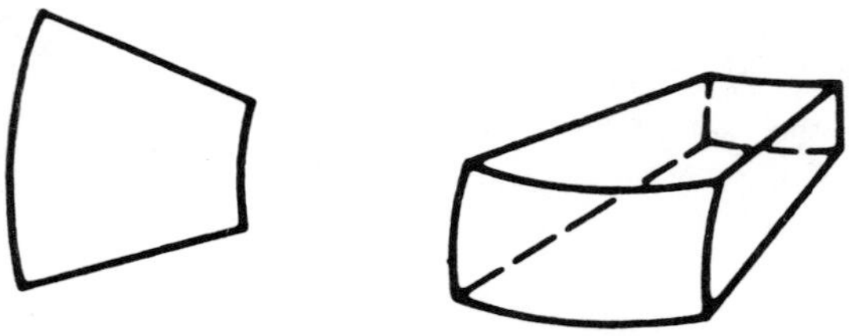

Fig. 1. Examples of $k = 2$ and $k = 3$ spherical coordinates quantizer regions.

coordinates quantizer's MSE D

$$D = \int_0^\infty r^2 f_k(r)\,dr + \int_0^\infty \hat{r}^2 f_k(r)\,dr - 2M_{k-1}\int_0^\infty r\hat{r} f_k(r)\,dr, \quad (1)$$

where M_{k-1} is defined sequentially by ($M_0 = 1$)

$$M_j = \frac{\Gamma[(j+1)/2]}{\Gamma(1/2)\Gamma(j/2)} \int_{-\pi/2}^{\pi/2} (\sin\varphi_j \sin\hat{\varphi}_j + M_{j-1}\cos\varphi_j\cos\hat{\varphi}_j)\cos^{j-1}\varphi_j\,d\varphi_j. \quad (2)$$

Assuming that M_{k-1} is fixed, (1) can be minimized over the magnitude quantizer Q_r. The three integrals present in that expression are all positive since the magnitude r is always positive; hence, D can also be minimized independently of r by maximizing the value of M_{k-1}. It can be shown that maximizing each $M_j(\varphi_j)$ term over the φ_j quantizer sequentially maximizes M_{k-1}.

III. Quantizer Optimization

The quantizers designed in this paper are all scalar processors. Their specification requires the computation of the output values and the endpoints of the quantization intervals. For a quantizer Q_s with N_s levels operating upon an input s, adopt the notation $\hat{s}_i$ as the ith output value and $[s_i, s_{i+1})$ as the ith interval, $i = 1, 2, \cdots, N_s$, with s_i as the ith breakpoint, $i = 1, 2, \cdots, N_s + 1$. When the number of quantization levels is large, the compandor model of Bennett [13] for a nonuniform quantizer will be employed (see Fig. 2). Under this model, the quantizer Q_s is a three-part system: an invertible, differentiable compressor nonlinearity g_s mapping the range of the input to [0, 1], a uniform quantizer Q_U with N_s levels on [0, 1], and an expandor $h_s = g_s^{-1}$ mapping [0, 1] back to the range of the input signal. For this model, specification of the compressor g_s and the number of levels N_s completely determines the quantizer.

The minimization of D in (1) or the maximization of each M_j in (2) can be accomplished in two ways depending upon the value of N_s. Partial derivatives with respect to the quantizer's parameters will yield necessary conditions for the extremum similar to those found by Max. A second derivative test similar to Fleischer's analysis [14] demonstrates sufficiency. These two sets of vector necessary conditions are useful when N_s is small and may be employed iteratively, as suggested by Fleischer, to solve numerically for the optimal quantizer's parameters.

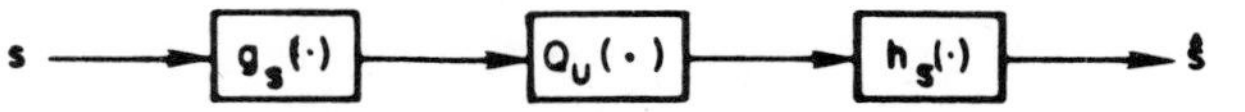

Fig. 2. Compandor model for Q_s.

When N_s is large, the output of a compandor system for an input s can be approximated by

$$\hat{s} \approx s + \epsilon h_s'[g_s(s)] = s + \frac{\epsilon}{g_s'(s)},$$

where ϵ is an independent noise source uniformly distributed on $[-\Delta/2, \Delta/2]$ ($\Delta = 1/N_s$ = the step size of the quantizer Q_U). After substituting for $\hat{s}$, the calculus of variations may be employed to yield the best compressor function. In both cases (maximizing M_j or minimizing D), the sign of the second variation exhibits the sufficiency of the proposed compressor function solution.

Magnitude Quantizer: Taking partial derivatives of D in (1) with respect to the magnitude quantizer's parameters yield:

$$r_i = \frac{1}{2M_{k-1}}(\hat{r}_{i-1} + \hat{r}_i); \qquad i = 2, 3, \cdots, N_r,$$
$$r_1 = 0, \; r_{N_r+1} = \infty, \quad (3)$$

$$\hat{r}_i = \frac{M_{k-1}\int_{r_i}^{r_{i+1}} r f_k(r)\,dr}{\int_{r_i}^{r_{i+1}} f_k(r)\,dr}, \qquad i = 1, 2, \cdots, N_r. \quad (4)$$

These expressions, except for the M_{k-1} terms, are equivalent to the equations defining the minimum MSE quantizer derived by Max. His optimal, N_r-level quantizer is defined by

$$t_i = \tfrac{1}{2}(\hat{t}_{i-1} + \hat{t}_i), \qquad i = 2, 3, \cdots, N_r, \; t_1 = 0, \; t_{N_r+1} = \infty,$$

$$\hat{t}_i = \frac{\int_{t_i}^{t_{i+1}} t f_k(t)\,dt}{\int_{t_i}^{t_{i+1}} f_k(t)\,dt}, \qquad i = 1, 2, \cdots, N_r.$$

From the Max quantizer, define a new quantizer with the same breakpoints and the outputs scaled by M_{k-1}

$$r_i = t_i; \qquad \hat{r}_i = M_{k-1}\hat{t}_i.$$

This new quantizer can be shown to satisfy the necessary conditions imposed on the magnitude quantizer by (3) and (4). It can also be shown that $0 \leqslant M_{k-1} \leqslant 1$ [15] and is usually approximately unity so that the scaling does not remove the output points from their respective regions.

Since the optimal magnitude quantizer is a scaled version of the Max quantizer for the magnitude density, for large N_r we employ the minimum MSE compressor for the magnitude density

$$g_r(r) = K_r \int_0^r f_k^{1/3}(t)\,dt, \quad (5)$$

where K_r is a constant such that g_r maps onto [0, 1]. The actual compandor system has its expandor scaled by M_{k-1}.

Employing the notation E_k as the unscaled quantizer's MSE, the use of the optimal quantizer simplifies the MSE in (1) to

$$D = M_{k-1}^2 E_k(N_r) + k(1 - M_{k-1}^2).$$

Maximizing M_j over the φ_j Quantizer $j = 1, 2, \cdots, k-1$: Partial derivatives of M_j from (2) with respect to the angle quantizer Q_j's parameters ϑ_i and $\hat{\vartheta}_i$ yield the following necessary conditions ($M_0 = 1$):

$$\vartheta_i = \tan^{-1}\left[\frac{M_{j-1}(\cos\hat{\vartheta}_i - \cos\hat{\vartheta}_{i-1})}{(\sin\hat{\vartheta}_{i-1} - \sin\hat{\vartheta}_i)}\right], \qquad i = 2, 3, \cdots, N_j, \tag{6}$$

$$\hat{\vartheta}_i = \tan^{-1}\left[\frac{\int_{\vartheta_i}^{\vartheta_{i+1}} \cos^{j-1}\vartheta \sin\vartheta \, d\vartheta}{M_{j-1}\int_{\vartheta_i}^{\vartheta_{i+1}} \cos^j \vartheta \, d\vartheta}\right], \qquad i = 1, 2, \cdots, N_j, \tag{7}$$

where ϑ_1 and ϑ_{N_j+1} are the endpoints of the interval of definition of φ_j. For the first angle quantizer, these expressions yield a uniform quantizer:

$$\vartheta_i = 2\pi(i-1)/N_1, \qquad i = 1, 2, \cdots, N_1 + 1,$$

$$\hat{\vartheta}_i = \pi(2i-1)/N_1, \qquad i = 1, 2, \cdots, N_1,$$

and the resulting value of M_1 is

$$M_1 = \frac{\sin(\pi/N_1)}{(\pi/N_1)}.$$

The other angle quantizers Q_j, $j \geq 2$, are nonuniform.

For large N_j we cannot immediately employ the minimum MSE compressor for the angle densities since we are trying to maximize M_j in (2) for each j, not minimize the mean-square error between φ_j and $\hat{\varphi}_j$. Assuming that $M_{j-1} \approx 1$ in (2) for M_j (since N_{j-1} is also large, M_{j-1} will be close to unity), the term in parenthesis simplifies to $\cos(\varphi_j - \hat{\varphi}_j)$. Expanding this in a Taylor series about zero, since for large N_j the region widths will be small and keeping only the first two terms, yields

$$\cos(\varphi_j - \hat{\varphi}_j) \approx 1 - \frac{(\varphi_j - \hat{\varphi}_j)^2}{2}.$$

Applying the compandor approximation and the calculus of variations yields the compressor function for the jth angle:

$$g_j(\varphi_j) = K_j \int_{-\pi/2}^{\varphi_j} \cos^{(j-1)/3}\vartheta \, d\vartheta, \tag{8}$$

where again K_j is a constant so that g_j maps onto [0, 1]. This resulting compressor is seen to be the minimum MSE compressor for the angle density and is proportional to an incomplete beta function [16]. For the φ_1 quantizer, the lower limit in (8) is zero, $g_1(\varphi_1)$ is linear and the quantizer is uniform.

The overall result is that to minimize the MSE of a k-dimensional spherical coordinates quantizer, a factorization of the total number of levels N must be selected: $N = N_r \times N_1 \cdots \times N_{k-1}$. For small N, the N_r-level magnitude density quantizer is found by (3) and (4), the φ_1 quantizer is uniform with N_1 levels and the M_j, $j = 2, \cdots, k-1$ are maximized sequentially, each maximization in turn specifying the φ_j quantizer Q_j by (6) and (7). When N_j is large, the factorization is again performed and the compressor functions are computed from (5) and (8).

IV. Asymptotic Results

This section considers asymptotic MSE rates and the solution to the problem of factoring the number of levels N to each of the spherical coordinates quantizers. We assume that the number of levels in each quantizer is large so that the compandor approximation is appropriate. The levels are factored to each quantizer by

$$N = N_r \times N_1 \times N_2 \cdots \times N_{k-1} = N_r \times N_\varphi,$$

where N_φ is denoted as the product of the number of levels in all of the angle quantizers. Previously, we developed the expression

$$D = M_{k-1}^2(N_\varphi) \times E_k(N_r) + k(1 + M_{k-1}^2(N_\varphi)). \tag{9}$$

Hence, we require expressions for the magnitude error E_k as a function of N_r and for M_{k-1} as a function of N_φ. Previous asymptotic results [17] yield the E_k term

$$E_k(N_r) \approx \frac{1}{12N_r^2}\left\{\int_0^\infty f^{1/3}(r)\, dr\right\}^3 = \frac{E_r}{N_r^2},$$

where $f_k(r)$ is the magnitude density.

To solve for M_{k-1} as a function of N_φ, we begin by referring to the discussion of Section II. There, we saw that the M_j terms were defined sequentially in (2). For large N_1, we can make the following approximation

$$M_1 = \frac{\sin(\pi/N_1)}{(\pi/N_1)} \approx 1 - \frac{\pi^2}{6}\frac{1}{N_1^2}.$$

Using this approximation, employing trigonometric identities and small angle approximations, integrating, ignoring higher power terms in the N_j, and maximizing over the factorization of N_φ yield

$$M_j \approx 1 - \frac{C_j}{\prod_{i=1}^{j} N_i^{2/j}}$$

with

$$C_j = \frac{j^2 C_{j-1}}{j^2 - 1} T_j, \qquad C_1 = \frac{\pi^2}{6},$$

and

$$T_j = \left\{\frac{\pi\Gamma\left(\frac{j+1}{2}\right)\Gamma^3\left(\frac{j+2}{6}\right)}{24\Gamma\left(\frac{j}{2}\right)\Gamma^3\left(\frac{j+5}{6}\right)} \frac{(j^2-1)}{jC_{j-1}}\right\}^{1/j}.$$

This result also yields the solution to the factorization of the number of levels in the angle quantizers:

$$N_j \approx \frac{T_j^{(j-1)/2}}{\prod_{i=j+1}^{k-1} T_i^{1/2}} N_\varphi^{1/(k-1)}.$$

A second-derivative test shows that this factorization of N_φ maximizes M_{k-1} for any spherically symmetric density. From the value of C_{k-1}, (9) is minimized by

$$N_r \approx \left\{\frac{E_r(k-1)}{2kC_{k-1}}\right\}^{(k-1)/2k} N^{1/k}.$$

Remembering that $N_\varphi = N/N_r$, the minimum spherical MSE is

$$D \approx k\left\{\frac{2kC_{k-1}}{k-1}\right\}^{(k-1)/k} E_r^{1/k} N^{-2/k}. \qquad (10)$$

This result for general values of k can be compared to the result for the polar case ($k = 2$) which has already appeared in the literature. In particular, with $N = N_r \times N_\varphi$ and $C_1 = \pi^2/6$ then

$$N_r \approx \left(\frac{3E_r}{2\pi^2}\right)^{1/4} N^{1/2}; \qquad N_\varphi \approx \left(\frac{2\pi^2}{3E_r}\right)^{1/4} N^{1/2},$$

and

$$D_2 \approx 2\pi\sqrt{\frac{2E_r}{3}}\, N^{-1},$$

which matches the result in [8] precisely.

V. Examples

The compared quantization schemes are the rectangular coordinates quantizer (k Max-type quantizers), the above described spherical coordinates quantizer and the optimal k-dimensional quantizer discussed by Zador. In order to compare error rates of schemes for different numbers of dimensions, we divide the MSE by k yielding MSE per dimension. Since all of the presented schemes have MSE proportional to $N^{-2/k}$, only the coefficient of the MSE will be compared. Rectangular quantizers yield orthogonal errors making the coefficient a constant, independent of dimension. The spherical coordinates quantizer's MSE is found by evaluating (10) with the appropriate E_r term. The minimally achievable MSE is presented by the upper and lower bounds derived by Zador. This value is

$$\mathrm{MSE}_{\min} \approx \frac{C(k,2)}{N^{2/k}}\left\{\int_{R^k} p^{k/(k+2)}(x)\, dx\right\}^{(k+2)/k},$$

where $C(k,2)$ is a constant dependent upon the optimal uniform quantizer. Zador provided bounds on this constant. More recently, Conway and Sloane found tighter upper limits on $C(k,2)$ for k between 3 and 10. Note, however, that the optimal scheme requires the implementation of the optimal k-dimensional uniform quantizer, usually a vector input device, while the rectangular and spherical schemes require only scalar processors.

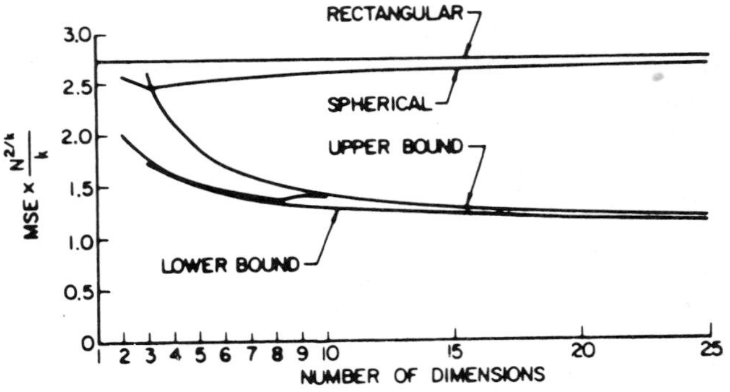

Fig. 3. Coefficient of MSE for Gaussian source.

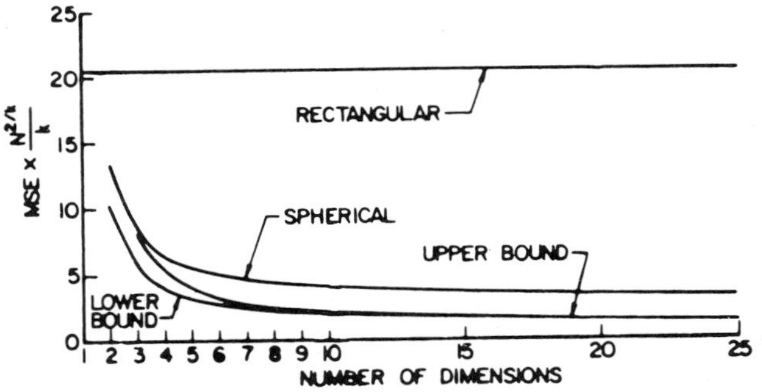

Fig. 4. Coefficient of MSE for Pearson VII source, $\nu = 1.25$.

The first source we consider is the independent Gaussian source with probability density function

$$f(x) = \frac{1}{(2\pi)^{k/2}} e^{-x^T x/2}.$$

Fig. 3 is a plot of the coefficient of the MSE for rectangular, spherical and optimal quantizers versus the number of dimensions. Notice that $k = 3$ yields the best of the spherical error rates and that this value is only slightly below that of the polar ($k = 2$) and rectangular ($k = 1$) quantizers.

Another spherically symmetric source is the Pearson Type II source ($\nu > 0$) with pdf:

$$f(x) = \frac{\Gamma(\nu+1)\left[2(\nu+1) - x^T x\right]^{\nu - k/2}}{\pi^{k/2} 2^\nu (\nu+1)^\nu \Gamma(\nu + 1 - k/2)} \cdot U\left[2(\nu+1) - x^T x\right],$$

where $U(\cdot)$ is the unit step function. This source has finite range and a Pearson II marginal density with parameter ν. Results for this source are not presented because the two-dimensional case performed best in all of the examples attempted (valid polar results can be found in [8]).

Another source with infinite range is the Pearson Type VII source ($\nu > 1$) with Pearson VII marginals:

$$f(x) = \frac{2^\nu(\nu-1)^\nu \Gamma(\nu + k/2)}{\pi^{k/2}\Gamma(\nu)\left[2(\nu-1) + x^T x\right]^{\nu + k/2}}.$$

These Pearson sources have restrictions on the value of the parameter ν in order to assure unit power marginals [18]. Plots of the coefficient of the MSE rate versus dimension for the Pearson VII source with various values of the

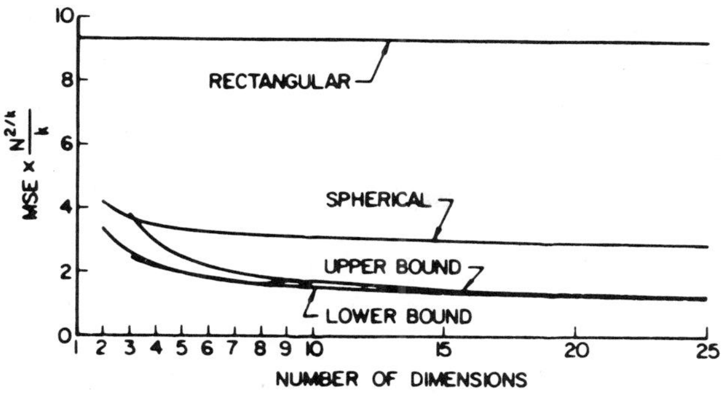

Fig. 5. Coefficient of MSE for Pearson VII source, $\nu = 2.5$.

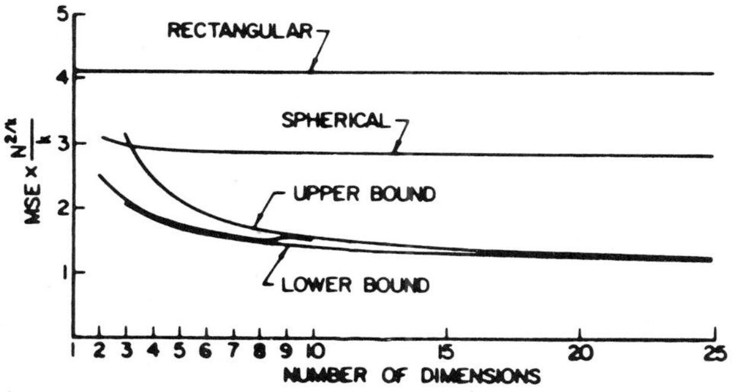

Fig. 6. Coefficient of MSE for Pearson VII source, $\nu = 5$.

parameter ν are shown in Figs. 4–6. The greatest performance gains by the spherical quantizers were for those Pearson VII sources which are furthest from the Gaussian (ν approaches unity). These sources are more peaked at the origin and heavier tailed. Fig. 7 compares the marginal densities of the considered sources to illustrate this point. Note that as $\nu \to \infty$, the Pearson VII source approaches the Gaussian source.

VI. Conclusion

This paper presents the generalization of polar quantizers to greater than two dimensions for all spherically symmetric densities. In comparison, the spherical scheme is applicable to any number of dimensions k and has a scalar processor implementation while the optimal quantizers are available only when the k-dimensional uniform quantizer can be implemented. The derived performance expressions may be used to decide if spherical schemes are of value in a particular application.

The MSE rates presented were for large values of N only. Research in one-dimensional compandor approximations suggest that these error rates are also approximately valid for smaller N. Several examples of the three-dimensional Gaussian quantizers were compared for $N \in [8^3, 18^3]$. It was seen, as was also noted in the published polar results, that rectangular quantizers perform better than spherical quantizers for small values of N, but as N increases the asymptotic rates become valid and the three-dimensional quantizers performed best. Of course, the optimal MSE is always lower.

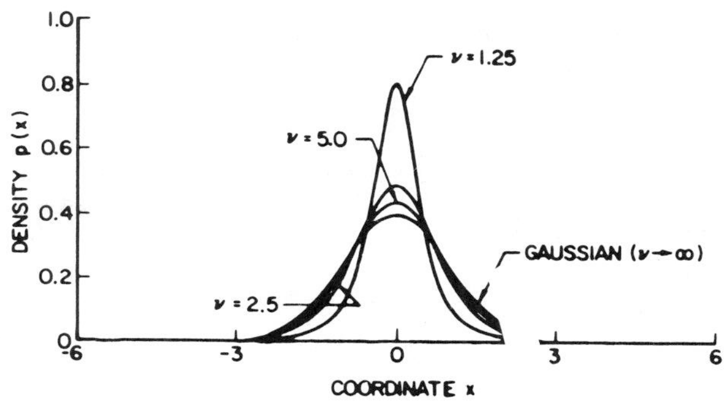

Fig. 7. Marginal densities of compared sources.

The results presented for the multidimensional spherical quantizers show that spherical coordinates encoding of spherically symmetric sources can be more efficient in a MSE sense than one-dimensional rectangular coordinates quantizing. For the Pearson VII source with $\nu = 1.25$, polar quantizing ($k = 2$) has a gain of 1.9 dB in signal-to-noise ratio (SNR) over rectangular quantization, while spherical quantization in six dimensions showed an increase of 6 dB. The optimal gain for $k = 6$ is approximately 9 dB over the rectangular rate. For the Gaussian example, the $k = 3$ design was shown to have the best performance, slightly better than the polar ($k = 2$) quantizers. Unfortunately, these schemes do not perform appreciably better than the rectangular quantizers of this Gaussian source. For instances when their performance is much better than rectangular quantizers, an intuitive explanation is that they preserve the spherical symmetry inherent in the multidimensional densities.

References

[1] J. Max, "Quantizing for minimum distortion," *IRE Trans. Inform. Theory*, vol IT-6, pp. 7–12, Mar. 1960.

[2] J. J. Y. Huang and P. M. Schultheiss, "Block quantization of correlated Gaussian random variables." *IEEE Trans. Comm. Syst.*, vol CS-11, pp 289–296, Sept. 1963.

[3] P. Zador, "Development and evaluation of procedures for quantizing multivariate distributions," Stanford Univ. Ph.D. dissertation, Stat. Dept., Dec. 1963.

[4] A. Gersho, "Asymptotically optimal block quantizers," *IEEE Trans. Inform. Theory*, vol. IT-25, pp. 373–380, July 1979.

[5] J. H. Conway and N. J. A. Sloane, "Voronoi regions of lattices, second moments of polytopes and quantization," *IEEE Trans. Inform. Theory*, vol. IT-28, pp. 211–226, Mar. 1982.

[6] W. A. Pearlman, "Polar quantization of a complex Gaussian random variable," *IEEE Trans. Comm.* vol. COM-27, pp. 892–899, June 1979.

[7] J. A. Bucklew and N. C. Gallagher, Jr., "Quantization schemes for bivariate Gaussian random variables," *IEEE Trans. Inform. Theory*, vol. IT-25, pp. 537–543, Sept. 1979.

[8] ——, "Two-dimensional quantization of bivariate circularly symmetric densities," *IEEE Trans. Inform. Theory*, vol. IT-25, pp. 667–671, Nov. 1979.

[9] S. G. Wilson, "Magnitude phase quantization of independent Gaussian variates," *IEEE Trans. Comm.*, vol. COM-28, pp. 1924–1929, Nov. 1980.

[10] P. F. Swaszek and J. B. Thomas, "Optimal circularly symmetric quantizers," *J. Franklin Inst.*, vol. 313, pp. 272–384, June 1982.

[11] R. D. Lord, "The use of the Hankel transform in statistics," *Biometrika*, vol. 41, pp. 44–55, June 1954.

[12] M. Kendall, *The Geometry of n Dimensions*. London: Griffin, 1962.

[13] W. R. Bennett, "Spectra of quantized signals," *Bell Syst. Tech. J.*, pp. 446–472, July 1948.

[14] P. E. Fleischer, "Sufficient conditions for achieving minimum distortion in a quantizer," *IEEE Int'l. Conv. Rec., Part 1*, pp. 104–111, 1964.

[15] P. F. Swaszek, "Robust quantization, vector quantization and detection," Ph.D. dissertation, Princeton Univ., Princeton, NJ, Dept. of EECS, Aug. 1982.

[16] H. Abramowitz and I. A. Stegun, *Handbook of Mathematical Functions.* National Bureau of Standards Applied Mathematics Series #55, Dec. 1972.

[17] V. R. Algazi, "Useful approximations to optimal quantization," *IEEE Trans. Comm. Tech.*, vol. COM-14, pp. 297–301, June 1966.

[18] D. K. McGraw and J. F. Wagner, "Elliptically symmetric distributions," *IEEE Trans. Inform. Theory*, vol. IT-14, pp. 110–120, Jan. 1968.

Reprinted from *Franklin Inst. J.* **313**(6):373–384 (1982)

Optimal Circularly Symmetric Quantizers†

by PETER F. SWASZEK AND JOHN B. THOMAS

Department of Electrical Engineering and Computer Science, Princeton University, Princeton, NJ 08544, U.S.A.

ABSTRACT: *Polar coordinates quantization of the bivariate Gaussian and other circularly symmetric sources has already been investigated. The schemes quantize the polar coordinates representation of the random variables independently in an attempt to reduce the mean square error below that of an analogous rectangular coordinates quantizer. It has been shown for the Gaussian case that the polar quantizer outperforms the rectangular quantizer when the number of levels* N *is large, while for small* N, *the rectangular form is often better than the polar form. This paper is an investigation of the optimality of polar quantizers with the subsequent development of optimal circularly symmetric quantizers (labelled Dirichlet polar quantizers).*

I. Introduction

The canonical example of a one-dimensional or zero-memory quantizer is Max's Gaussian probability density function quantizer (**1**) for the performance criterion Mean Square Error (MSE). The MSE criterion has universal appeal in its tractability and its intuitive relationship to noise power, hence signal-to-noise ratio (SNR). Rate distortion theory, however, suggests that multidimensional or block quantizers may be more efficient. Research interest in multidimensional quantization began with the work of Huang and Schultheiss (**2**) who considered the problem of quantizing a correlated Gaussian source efficiently. Their solution was to uncorrelate the source by an appropriate linear filter, thereby changing the set of coordinates, and to quantize the resulting independent Gaussian random variables with separate Max-type quantizers.

Zador (**3**) examined the more general problem of quantizing a multidimensional source under the assumption of a large number of levels. He employed Bennett's compandor model and derived error rates depending upon

†This research is supported by the National Science Foundation under Grant ECS-79-18915 and by the Office of Naval Research under Contract N00014-81-K-0146.

the compressor function and uniform quantizer used. His expressions showed that the problem of optimal quantization could be divided into two separate problems: finding the best compressor function on the multidimensional input space and implementing the optimal multidimensional uniform quantizer on the unit hypercube. In two dimensions, the optimal uniform quantizer is a honeycomb-like tesselation of hexagons. When mapped by the inverse of the compandor function, the quantizer becomes a pattern of distorted hexagons on the plane (**4**).

Another major area of interest in multidimensional quantizer design rests in the use of polar coordinates for the independent, bivariate case. Specifically, rather than separately quantizing the abscissa and ordinate as in Fig. 1, a change of variables to polar coordinates is effected. The resulting magnitude and phase are quantized separately by real-time one-dimensional quantizers. Of particular interest is the quantizing of independent, bivariate Gaussian random variables with density

$$p(x, y) = \frac{1}{2\pi} e^{-(x^2+y^2)/2}.$$

For example DFT coefficients, holographic data or pairs of inputs from an iid Gaussian source can be considered as the output of a bivariate Gaussian source.

Independent, unit-power Gaussian variates in rectangular coordinates transform to independent magnitude and phase on the polar coordinates plane by the transformations

$$r = \sqrt{(x^2 + y^2)}; \quad \phi = \tan^{-1}\frac{y}{x}.$$

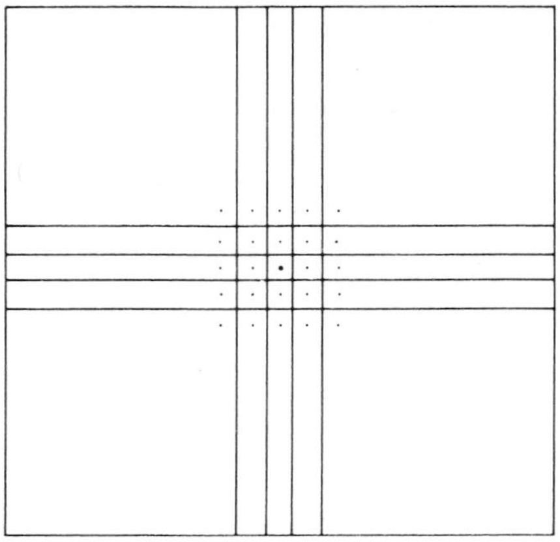

FIG. 1. Typical bivariate rectangular coordinate quantizer.

The resulting source density expressed as a function of the polar coordinates is

$$p(r, \phi) = \frac{1}{2\pi} r\, e^{-r^2/2}.$$

The magnitude r is Rayleigh distributed on $[0, \infty)$ and the phase ϕ is uniformly distributed on $[0, 2\pi)$. Minimizing MSE results in a uniform quantizer for the phase angle and a scaled Max-type Rayleigh quantizer for the magnitude. It has been shown (**5, 6**) that polar coordinates quantizers for a bivariate Gaussian source almost always have smaller MSE than rectangular quantizers.

This paper examines the optimality of the polar quantizers developed by Pearlman (**5**) and Bucklew and Gallagher (**6**) for the MSE criterion. It is well known (**4**) that two conditions are necessary for a *local* minimum of MSE: centroidal output points and Dirichlet partition boundaries. Polar quantizers do not conform to these conditions. Permutations which do conform (labeled Dirichlet polar quantizers) will be developed and compared to other available two-dimensional quantization schemes. Wilson's technique (**7**) will be mentioned and considered as an input for the Dirichlet form. Although this paper will pursue in depth only the bivariate Gaussian case, the extensions to higher dimensions (**8**) and other circularly symmetric densities (**9**) will be outlined.

II. Optimal Two-Dimensional Quantizers

Define the minimum MSE, N-level quantizer Q_N on the plane by $\{S_i, \hat{\mathbf{x}}_i;\ i = 1, 2, \ldots, N\}$ where the S_i are disjoint regions such that their union covers the plane and the $\hat{\mathbf{x}}_i$ are the output points associated by the quantizer to these regions. The quantizer's operation for the input vector $\mathbf{x}$ is

$$Q_N(\mathbf{x}) = \hat{\mathbf{x}}_i; \quad \mathbf{x} \epsilon S_i.$$

For an input $\mathbf{x}$ with bivariate pdf $p(\mathbf{x})$, the MSE D is

$$D = \int_{-\infty}^{\infty}\int_{-\infty}^{\infty} |\mathbf{x} - Q_N(\mathbf{x})|^2 p(\mathbf{x})\, d\mathbf{x}.$$

Minimizing D over the choice of S_i and $\hat{\mathbf{x}}_i$, the following are necessary conditions:

$$\hat{\mathbf{x}}_i = \int_{S_i}\int \mathbf{x} p(\mathbf{x})\, d\mathbf{x} \div \int_{S_i}\int p(\mathbf{x})\, d\mathbf{x} \qquad (1)$$

which states that the output $\hat{\mathbf{x}}_i$ is the centroid of the region S_i with density $p(\mathbf{x})$ and

$$S_i = \bigcap_{j=1, j\neq i}^{N} \{\mathbf{x} : |\mathbf{x} - \hat{\mathbf{x}}_i| < |\mathbf{x} - \hat{\mathbf{x}}_j|\} \tag{2}$$

which states that S_i is formed by taking the intersection of nearest neighbor or Dirichlet partitions of $\hat{\mathbf{x}}_i$ and the other output points. The points of equality in (2) are the region boundaries which are assigned to either region and contribute equivalent error either way. A Dirichlet partition is the perpendicular bisector of the line segment connecting a pair of output points. From (2), it can be shown that the resulting S_i are all convex, simply connected regions. This partitioning holds for most mean error measures while centroidal outputs holds only for MSE. The resulting MSE for this optimal quantizer is

$$D = \sigma_{\mathbf{x}}^2 - \sum_{i=1}^{N} |\hat{\mathbf{x}}_i|^2 \int_{S_i}\int p(\mathbf{x})\, d\mathbf{x}$$

where $\sigma_{\mathbf{x}}^2$ is the signal power.

For the uniform density on the unit square, the optimal region pattern for fine quantization, ignoring edge effects, is known to be a tesselation of regular hexagons. For other densities, (1) and (2) may be used iteratively to converge to a local minimum of MSE. Note that if the regions are fixed, (1) is necessary and sufficient to minimize D. When the output points are fixed, (2) is necessary and sufficient to minimize D. The iterative design method, as previously proposed for one-dimensional problem solutions (**10**), is to select a set of outputs $\{\hat{\mathbf{x}}_i\}$ and to employ (2) to select the S_i optimally. This set of outputs and regions has a distortion measure D_1. Usually, (1) is not satisfied, the $\{\hat{\mathbf{x}}_i\}$ not being optimal for the generated regions, so redefining the outputs by (1) will decrease D to a value smaller than D_1, say D_2. Similarly, now (2) is probably not satisfied, so redefining the regions will again decrease the error. This iterative scheme converges to a local minimum of D due to the fact that D is reduced by each application of (1) or (2) and that D is lower bounded by zero being the integral of a positive quantity.

III. Dirichlet Polar Quantizers

Wilson (**7**) classified two types of polar quantizers: Strictly Polar (SPQ) and Unrestricted Polar (UPQ) Quantizers. For the SPQs, the total number of outputs N is factored into $N_r \times N_\phi$, the number of magnitude and phase levels respectively. The UPQs, a larger class of quantizers, require only that the number of outputs sum up to N ($P_1 + P_2 + \ldots P_M = N$); hence different radii levels can have different numbers of phase levels. For small N, UPQs have been shown to substantially reduce the MSE. All polar quantization regions are partial annuli, delimited by rays of constant angle and arcs of constant radius as in Fig. 2. Unfortunately, these quantizers do not satisfy (1) and (2). In particular, the magnitude boundaries are not Dirichlet partitions. From (2), each S_i is a convex polygon which partial annuli are not.

The iterative technique, as explained above using (1) and (2), may be employed to the strictly polar quantizers to reduce MSE and converge to a local minimum. After selecting a factorization of $N = N_r \times N_\phi$, applying (2) yields a pattern as in Fig. 3. The inherent symmetry of this pattern allows the analysis to focus on one slice of angle $2\pi/N_\phi$. The iterative use of (1) and (2) will not change the phase boundaries; only the magnitude boundaries will move. Similarly, the output points will vary along the ray bisecting the phase boundaries. Hence, a one-dimensional iteration will yield these Dirichlet Polar Quantizers (DPQs) from the SPQs.

From Fig. 3, it is seen that the DPQs can be implemented as follows. First, quantize the phase to one of N_ϕ levels with a uniform quantizer on $[0, 2\pi)$. The second coordinate used to specify the output is its distance s along the quantized phase ray

$$s = r \cos (\phi - \hat{\phi})$$

for $\hat{\phi}$, the quantized version of ϕ. The univariate probability distribution function of this distance coordinate can be found to be

$$f(s) = \frac{2N_\phi}{\sqrt{(2\pi)}} e^{-s^2/2} [\Phi(s \tan \pi/N_\phi) - 1/2]$$

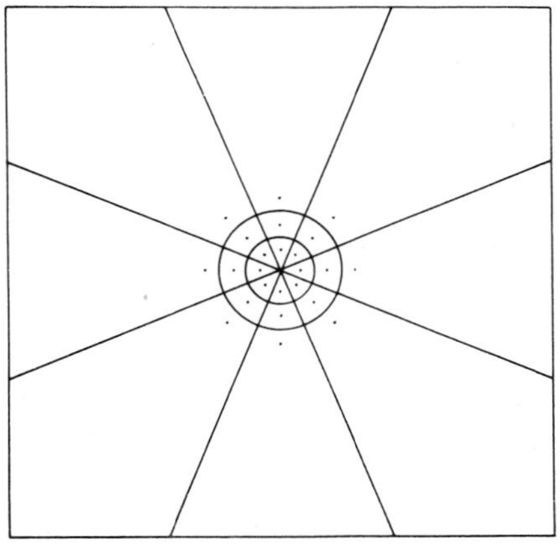

FIG. 2. Polar quantizer (SPQ).

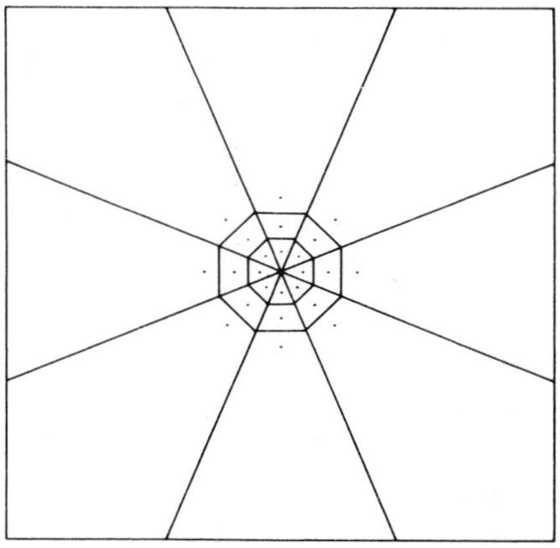

FIG. 3. Dirichlet Polar Quantizer (DPQ) pattern.

for Φ the error function integral

$$\Phi(y) = \int_{-\infty}^{y} \frac{1}{\sqrt{(2\pi)}} e^{-x^2/2} \, dx.$$

As $N_\phi \to \infty$, $f(s)$ approaches the Rayleigh density.

Standard Max-type expressions may be used to define the minimum MSE, N_r-level quantizer for s

$$s_i = \frac{\hat{s}_{i-1} + \hat{s}_i}{2}; \; \hat{s}_i = \frac{\int_{s_i}^{s_{i+1}} t f(t) \, dt}{\int_{s_i}^{s_{i+1}} f(t) \, dt}$$

where the $\hat{s}_i$ are the quantizer outputs and the s_i are the region endpoints. Uniqueness of this quantizer is shown by applying Fleischer's test (**10**) to the distance density

$$\frac{\partial^2}{\partial s^2} \log f(s) < 0.$$

Wilson's solutions of the UPQs for $N = 1, \ldots, 32$ may also be considered with the iterative technique. His $N = 1, 2, 3$ and 4 cases are already optimum. The $N = 5$, 6, 7 and 8 are easily extended. Unfortunately, for $N > 8$, the boundaries are no longer easy to compute and the resulting analysis is not included here. He only considered small N since the number of factorizations grows quickly with N and because the small N region is of importance since it is here that rectangular formats outperform polar forms.

IV. Dirichlet Rotated Polar Quantizers

The previous section showed that (1) and (2) can be applied to a set of outputs to iterate toward a local minimum of MSE. The resulting quantizer will vary depending upon the initial output point pattern. A rectangular starting grid produces a rectangular quantizer, a pair of Max Gaussian quantizers, since the partitions will always move perpendicularly. A polar initial pattern produces the Dirichlet Polar Quantizer already introduced.

Consider the polar quantizer (SPQ) where the magnitude and phase are independently quantized. A rotation of every other magnitude ring, as in Fig. 4, does not change the associated MSE. This new pattern, when applied as a starting point for the iterative method with (1) and (2), will yield a quantization pattern as in Fig. 5, quite different from the Dirichlet Polar Quantizer. This Dirichlet Rotated Polar Quantizer (DRPQ), although more difficult to implement than the DPQ, has lower MSE (for a possible implementation, see the Appendix).

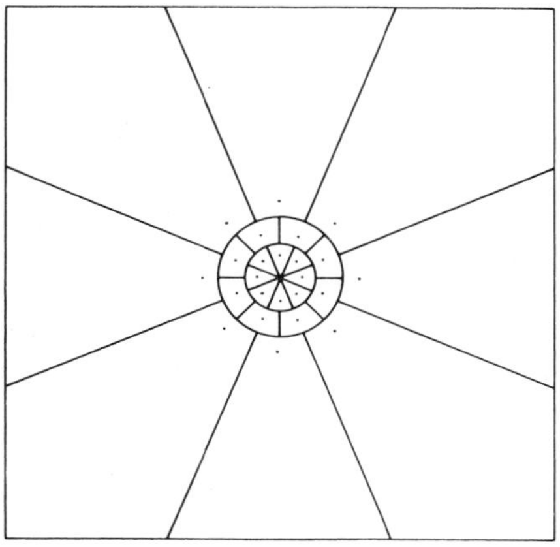

FIG. 4. Rotated polar pattern.

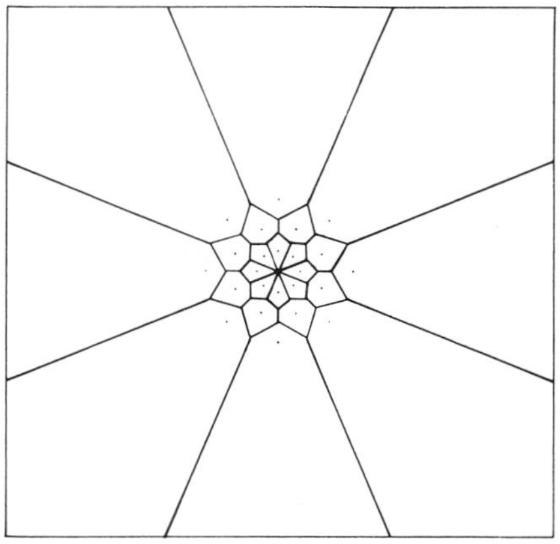

FIG. 5. Dirichlet Rotated Polar Quantizer (DRPQ) pattern.

Other rotations and permutations on the plane could be used to solve for better quantizers. However, most other patterns make (1) difficult to compute. A further extension of this rotated form is to allow a central region with N_ϕ sides and output value zero similar to Wilson's 5 through 8 patterns. The MSE savings could be dramatic, but are not considered here.

V. Examples

Equations (1) and (2) define the iterative technique. The calculations of the region probabilities and moments are as follows. For the bivariate Gaussian density, the probability of a triangular region with vertices $(0,0)$, $(h,0)$ and (h,k) has a quickly convergent series representation $V(h,k)$ (**11**). The various region probabilities are found by differencing appropriate triangular regions. As to the moments, all of the considered regions have symmetry about a ray of constant angle; hence, the region centroid is along this ray. The moment along the ray is integrated straightforwardly and the result employs an error function approximation.

For the examples, the following factorizations of N were employed:

Rectangular:	$N_x \approx N_y$
Polar:	$N_\phi \approx 2.6\ N_r$
DPQ:	$N_\phi \approx 2.6\ N_s$
DRPQ:	$N_\phi \approx N_r$.

Symmetry arguments show that the rectangular MSE is minimized if the levels are equally divided among the coordinates. Previously published results suggest the factorization for the polar scheme. As the number of levels gets large, the DPQs and the polar quantizers are equivalent, hence the asymptotic factorizations of N are the same. For the tabulated results, all factorizations for the DPQs were compared and the best result occurred concurrent with the polar factorization. For the DRPQs, all combinations were attempted for $N \leq 144$. It was seen that equal division of the levels produced optimal results. For $N > 144$, only equal factorizations were attempted. Hence, the actual error rates may be lower than the tabulated results for those values of N.

The comparison of MSE rates is in Table 1 with a plot of the results in Fig. 6. Error values for polar and rectangular quantizers are included for comparison. The number in parentheses is the actual number of levels if different from the first column. This appears due to the necessity to factor N into appropriate integers. Figures 7–13 depict the DRPQ patterns, $x, y \epsilon [-8, 8]$, for some of the values listed in Table 1.

TABLE I. *Bivariate Gaussian density quantizer's MSE values.*

N	Polar	Dirichlet Polar	Dirichlet Rotated Polar	Rectangular
16	.2396	.2391	.2224	.2350
25	.1710 (24)	.1702 (24)	.1462	.1599
36	.1176	.1174	.1052	.1159
49	.08889 (48)	.08882 (48)	.07899	.08800
64	.06973	.06967	.06134	.06908
100	.04392 (102)	.04387 (102)	.04003	.04586
144	.03244 (140)	.03241 (140)	.02816	.03268
225	.02056	.02055	$\leqslant$.01822	.02146
324	.01468 (320)	.01467 (320)	$\leqslant$.01280	.01519
529	.008904 (532)	.008899 (532)	$\leqslant$.008046	.009482
900	.005314	.005308	$\leqslant$.004684	.005668

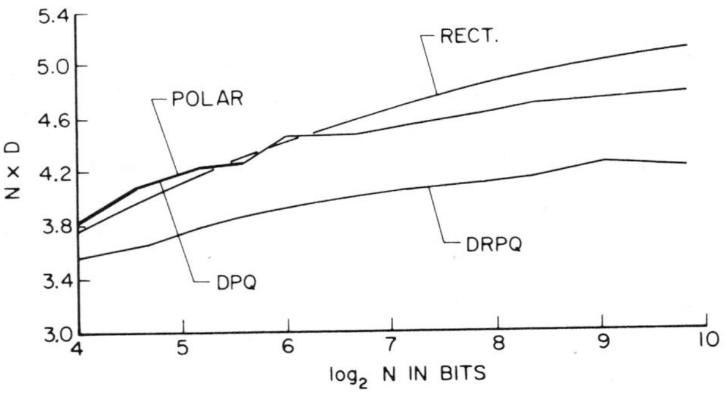

FIG. 6. Comparison of MSE rates for four bivariate quantizers (Rectangular, SPQ, DPQ and DRPQ).

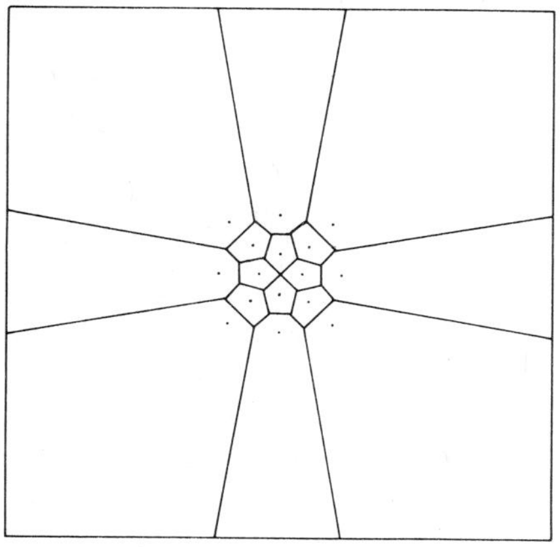

FIG. 7. $N = 16$ DRPQ pattern.

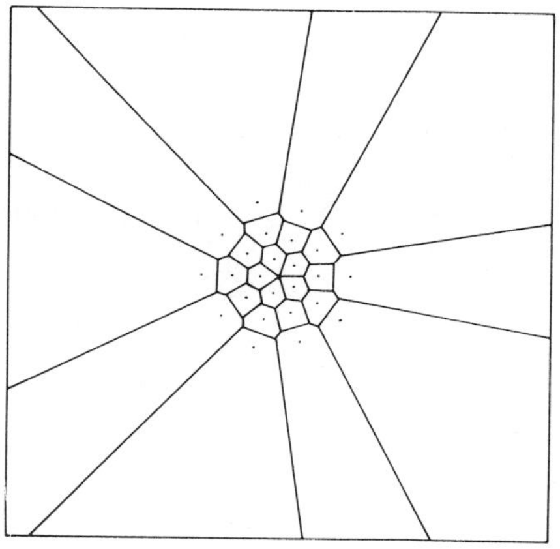

FIG. 8. $N = 25$ DRPQ pattern.

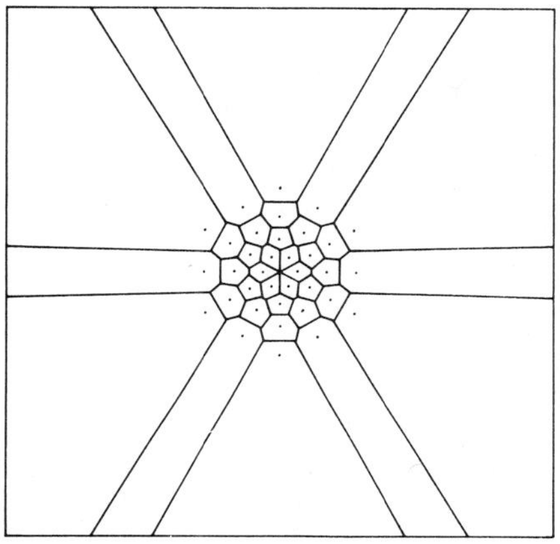

FIG. 9. $N = 36$ DRPQ pattern.

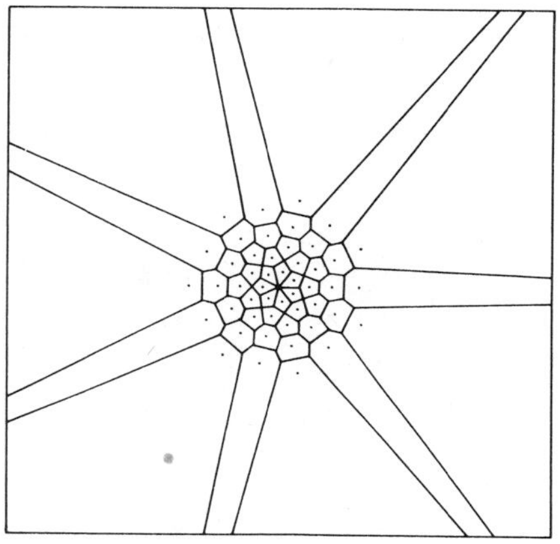

FIG. 10. $N = 49$ DRPQ pattern.

VI. Conclusions

For two reasons, the presented figures are dominated by patterns for DRPQs. The DPQ patterns are all of the same form as Fig. 3 and the DPQ's MSE is only slightly below that of the SPQ, being equal when $N \to \infty$. The DRPQ patterns are included to demonstrate the hexagonality of the quantization regions, the way in which the hexagon sizes are distributed and because the DRPQs substantially reduce MSE. At $N = 100$, the gain in SNR is 0.6 dB over rectangular and 0.4 dB over polar quantizers.

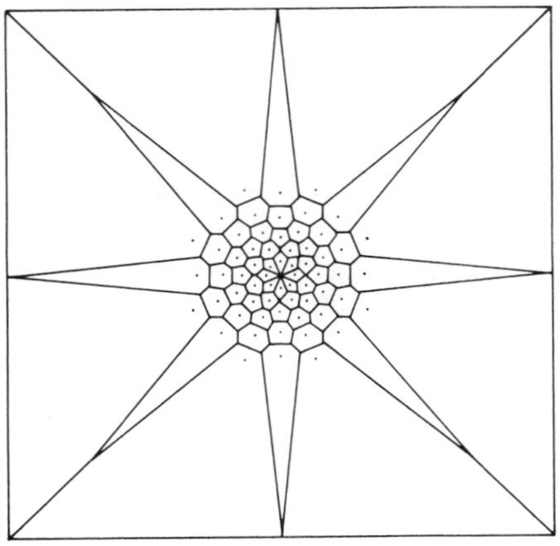

FIG. 11. $N = 64$ DRPQ pattern.

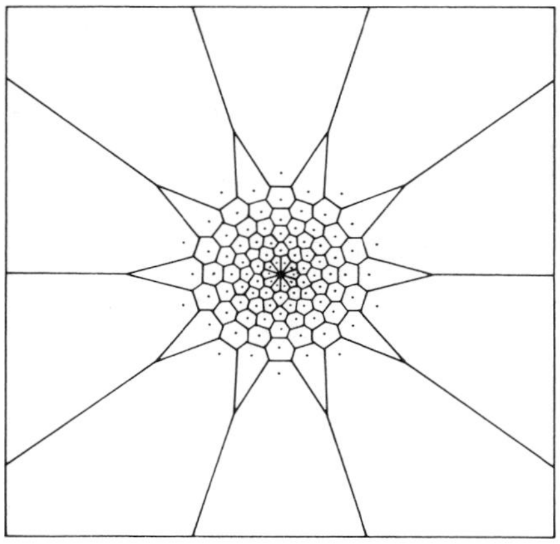

FIG. 12. $N = 100$ DRPQ pattern.

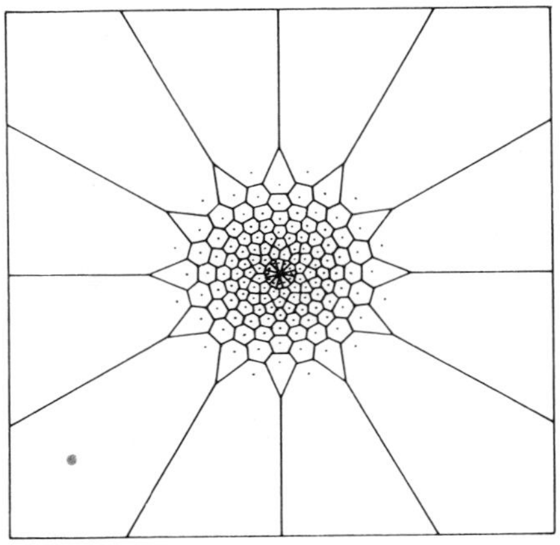

FIG. 13. $N = 144$ DRPQ pattern.

From Fig. 6, the asymptotic MSE rates can be considered. The SPQ and DPQ (equivalent as $N \to \infty$) have rate $N \times D = 4.95$ for a bivariate Gaussian source. The corresponding optimal rate is 4.03 and the rectangular rate is 5.44. From the graph, the DRPQ rate falls between the polar and optimum. Although the DRPQ is not optimal, it does always perform better than both polar and rectangular schemes.

Up to this point, this paper only considered the bivariate Gaussian case. The trapezoids of the Dirichlet Polar Quantizer and the polygons of the DRPQ become polytopes in higher dimensions. Other circularly symmetric and non-circularly symmetric densities may also be used. The difficulty in both cases is obtaining accurate probability and moment integrals.

Polar quantizers as described in the literature minimize MSE subject to independent coordinates. Loosening the coordinate selection slightly (to angle ϕ and distance s) yields DPQs, again minimizing MSE for their constraint class. Further loosening of the coordinate class yields the DRPQs with substantially reduced MSE, but increased complexity of implementation. The intuition to be gained from the work is as follows: all of the mentioned schemes (rectangular, polar, DPQ and DRPQ) minimize MSE subject to their implementation constraint. Rectangular formats retain centroids and Dirichlet partitions (necessary conditions), but lose the symmetry of the problem; polar forms preserve the problem symmetry but lose the necessary conditions; the DPQ and DRPQ schemes have both.

References

(1) J. Max, "Quantizing for minimum distortion", *IRE Trans. Inf. Theory*, Vol. IT-6, pp. 7–12, 1960.

(2) J. J. Y. Huang and P. M. Schultheiss, "Block quantization of correlated Gaussian random variables", *IEEE Trans. Comm. Sys.*, Vol. CS-11, pp. 289–296, 1963.

(3) P. Zador, "Development and evaluation of procedures for quantizing multivariate distributions," Stanford University Dissertation, Dept. of Stat. 1963.

(4) A. Gersho, "Asymptotically optimal block quantization", *IEEE Trans. Inf. Theory*, Vol. IT-25, pp. 373–380, 1979.

(5) W. A. Pearlman, "Polar quantization of a complex Gaussian random variable", *IEEE Trans. Comm.*, Vol. COM-27, pp. 892–899, 1979.

(6) J. A. Bucklew and N. C. Gallagher Jr., "Quantization schemes for bivariate Gaussian random variables", *IEEE Trans. Inf. Theory*, Vol. IT-25, pp. 537–543, 1979.

(7) S. Wilson, "Magnitude/phase quantization of independent Gaussian variates", *IEEE Trans. Comm.*, Vol. COM-28, pp. 1924–1929, 1980.

(8) P. F. Swaszek and J. B. Thomas, "*k*-Dimensional polar quantizers for Gaussian sources", *Proceedings of the* 19*th Annual Allerton Conf.*, Sept. 1981, pp. 89–97.

(9) J. A. Bucklew and N. C. Gallagher Jr., "Two-dimensional quantization of bivariate circularly symmetric densities", *IEEE Trans. Inf. Theory*, Vol. IT-25, pp. 667–671, 1979.

(10) P. E. Fleischer, "Sufficient conditions for achieving minimum distortion in a quantizer", *IEEE Int. Conv. Rec.*, part 1, pp. 104–111, 1964.

(11) N. L. Johnson and S. Kotz, "Continuous multivariate Distributions", John Wiley, New York, 1972.

Appendix

This appendix presents a possible implementation for the DRPQs. The levels factorization is $N = N_r \times N_\phi$.

1. Convert the input $\mathbf{x}$ to polar coordinates, r and ϕ.
2. Process the phase angle ϕ with a $2N_\phi$-level uniform quantizer on $[0, 2\pi)$. The output $\hat{\phi}_k$ is of the form $\pi(2k-1)/2N_\phi$ for $k \epsilon [1, 2, \ldots 2N_\phi]$.
3. Process the magnitude r with an N_r-level, lower value quantizer. The quantizer's output $\hat{r}_j$ is the magnitude value closest and less than the actual distance.

4. For a lower magnitude of level j and a phase of level k the output is either

$$\hat{r}_j\, e^{j\hat{\phi}_{k,j}} \text{ or } \hat{r}_{j+1}\, e^{j\hat{\phi}_{k,j+1}}$$

where

$$\hat{\phi}_{kj} = \frac{\pi}{N_\phi}\frac{2k-1}{2} \pm \frac{\pi}{2N_\phi}; \qquad \begin{matrix} + \text{ if } |j-k| \text{ is even} \\ - \text{ if } |j-k| \text{ is odd} \end{matrix}.$$

Compute the distances from these points to the input $re^{j\phi}$ and take the closest one as the output point.

This scheme requires no compressor functions as does the optimal scheme and can be done in real-time. Also, this implementation extends trivially to the zero-output extension of the DRPQ described at the end of Section IV.

Part VI

APPLICATIONS OF QUANTIZERS

Editor's Comments on Papers 34 Through 37

34 BUZO et al.
Speech Coding Based upon Vector Quantization

35 DALLAS
Magnitude-Coupled Phase Quantization

36 FINE
Optimum Mean-Square Quantization of a Noisy Input

37 POOR and THOMAS
Optimum Quantization for Local Decisions Based on Independent Samples

The majority of the papers in the earlier sections of this volume dealt with techniques for designing, analyzing, and implementing quantizers for signal representation without a specific application. This final section contains four papers describing typical applications of quantizers in signal processing and communications systems. The inclusion of only four papers is not meant to suggest that application-oriented articles are rare; rather, these four were chosen as representative papers demonstrating various applications of quantizers. For other examples, the reader is directed to the bibliography for this section. Papers 34, 36, and 37 also differ from most of the papers in this volume in that the performance criterion is not mean square error of the quantization noise; instead, the measure is one more commonly employed for the application.

Paper 34 by Buzo et al. develops speech coding by means of vector quantizers with the Itakura-Saito distortion measure. They discuss the aspects of computing centroids and nearest-neighbor regions for this particular measure. The topic of complexity of implementation is also considered with the design of suboptimal quantizers. Additional results on applying vector quantizers to speech coding may be found in Abut, Gray, and Rebolledo (1981), Wong, Juang, and Gray (1981), and Juang and Gray (1982). Vector quantizers have also been applied to image coding (e.g., Habibi and Wintz, 1971, and Gersho and Ramamurthi, 1982).

Paper 35 by Dallas considers bivariate magnitude/phase quantizers for computer-generated holograms using the mean squared error criterion (see also Pearlman, 1974). The cosine-coupling described here is similar to the Dirichlet polar quantization described in Paper 33. Related papers on quantization of complex valued data include Paper 28 and those of Gallagher (1976, 1978) and Pearlman (1979).

Paper 36 by Fine describes the design of scalar quantizers for use in estimating a random variable from a noisy version, employing the minimum mean squared error criterion for the difference between the original variable and the quantizer's output for the noisy input (see also Morris, 1976).

Finally, Paper 37 by Poor and Thomas considers the use of scalar quantizers in signal detection problems. For local (small-signal) detection, they maximize the slope of the power curve at the origin, yielding the locally optimum quantizer (see also Kassam, 1977, Poor and Thomas, 1977, 1978, 1980, and Poor and Alexandrou, 1980).

REFERENCES

Abut, H, R. M. Gray, and G. Rebolledo, 1981, Vector Quantization of Speech Waveforms, *IEEE Int. Conf. Acoust., Speech, and Signal Process. Proc.*, pp. 12–15.

Gallagher, N. C., Jr., 1976, Discrete Spectral Phase Coding, *IEEE Trans. Inform. Theory* **IT-22**(5):622–624.

Gallagher, N. C., Jr., 1978, Quantizing Schemes for the Discrete Fourier Transform of a Random Time Series, *IEEE Trans. Inform. Theory,* **IT-24**(2):156–163.

Gersho, A., and B. Ramamurthi, 1982, Image Coding using Vector Quantization, *IEEE Int. Conf. Acoust., Speech, and Signal Process. Proc.*, pp. 428–431.

Habibi, A., and P. A. Wintz, 1971, Image Coding by Linear Transformation and Block Quantization, *IEEE Trans. Commun. Tech.* **COM-19**(2):50–62.

Juang, B. H., and A. H. Gray Jr., 1982, Multiple Stage Vector Quantization for Speech Coding, *IEEE Int. Conf. Acoust., Speech, and Signal Process. Proc.*, pp. 597–600.

Kassam, S. A., 1977, Optimum Quantization for Signal Detection, *IEEE Trans. Commun.* **COM-25**(5):479–484.

Morris, J. M., 1976, The Performance of Quantizers for a Class of Noise-Corrupted Signal Sources, *IEEE Trans. Commun.* **COM-24**(2):184–189.

Pearlman, W. A., 1974, *Quantization Error Bounds for Computer Generated Holograms,* Stanford University Information Systems Laboratory Technical Report 6503-1.

Pearlman, W. A., 1979, Optimum Fixed Level Quantization of the DFT of Achromatic Images, *Allerton Conf. Commun. Control Comput. Proc.*, pp. 313–319.

Poor, H. V., and D. Alexandrou, 1980, The Analysis and Design of Data Quantization Schemes for Stochastic-Signal Detection Systems, *IEEE Trans. Commun.* **COM-28** (7):983–991.

Poor, H. V., and J. B. Thomas, 1977, Application of Ali-Silvey Distance Measures in the Design of Generalized Quantizers for Binary Decision Systems, *IEEE Trans. Commun.* **COM-25**(9):893–900.

Poor, H. V., and J. B. Thomas, 1978, Asymptotically Robust Quantization for Detection, *IEEE Trans. Inform. Theory* **IT-24**(2):222–229.

Poor, H. V., and J. B. Thomas, 1980, Memoryless Quantizer-Detectors for Constant Signals in m-Dependent Noise, *IEEE Trans. Inform. Theory* **IT-26**(4):423–432.

Wong, D. Y., B. H. Juang, and A. H. Gray, Jr., 1981, Recent Developments in Vector Quantization for Speech Processing, *IEEE Int. Conf. Acoust., Speech, and Signal Process. Proc.*, pp. 1–4.

BIBLIOGRAPHY

Bluestein, L., 1964, Asymptotically Optimum Quantizers and Optimum Analog to Digital Converters for Continuous Signals, *IEEE Trans. Inform. Theory* **IT-10**(4):242–246.

Broman, H., 1977, Quantization of Signals in Additive Noise: Efficiency and Linearity of the Equal Step-size Quantizer, *IEEE Trans. Acoust., Speech, and Signal Process.* **ASSP-25**(12):572–574.

Bucklew, J. A., 1984, Multidimensional Digitization of Data followed by a Mapping, *IEEE Trans. Inform. Theory,* **IT-30**(1):107–110.

Clark, B. G., 1973, The Effect of Digitization Errors on Detection of Weak Signals in Noise, *IEEE Proc.* **61**(11):1654–1655.

Goodman, L. M., 1966, Optimum Sampling and Quantizing Rates, *IEEE Proc.* **54**(1):90–92.

Goodman, L. M., and P. Drouilhet, Jr., 1966, Asymptotically Optimum Pre-emphasis and De-emphasis Networks for Sampling and Quantizing, *IEEE Proc.* **54**(5):795–796.

Kuhlmann, F., J. A. Bucklew, and G. L. Wise, 1983, Compressors for Combined Source and Channel Coding with Applications to the Generalized Gaussian Family, *IEEE Trans. Inform. Theory* **IT-29**(1):140–144.

Kurtenbach, A. J., and P. A. Wintz, 1969, Quantizing for Noisy Channels, *IEEE Trans. Commun. Tech.* **COM-17**(4):291–302.

Larson, R. E., 1967, Optimum Quantization in Dynamic Systems, *IEEE Trans. Auto. Control* **AC-12**(4):162–168.

Larson, R. E., 1972, Reply, *IEEE Trans. Auto. Control* **AC-17**(4):274–276.

Limb, J., 1969, Design of Dither Waveforms for Quantized Visual Signals, *Bell System Tech. J.* **48**(9):2555–2599.

Lippel, B., M. Kurland, and A. H. Marsh, 1971, Ordered Dither Patterns for Coarse Quantization of Pictures, *IEEE Proc.* **59**(3):429–431.

MacQueen, J., 1967, Some Methods for Classification and Analysis of Multivariate Observations, *5th Berkeley Symp. Math, Stat. & Prob.* **1:**281–297.

Mazo, J. E., 1968, Quantization Noise and Data Transmission, *Bell System Tech. J.* **47**(10):1737–1753.

Paez, M. D., and T. H. Glisson, 1972, Minimum Mean-Square-Error Quantization in Speech PCM and DPCM Systems, *IEEE Trans. Commun.* **COM-20**(4):225–230.

Peters, E. G., J. S. Boland, L. J. Pinson, and W. W. Malcolm, 1978, Quantization Effects on Signal Matching Functions, *IEEE Trans. Inform. Theory* **IT-24**(3):395–398.

Pollard, D., 1982, Quantization and the Method of k-Means, *IEEE Trans. Inform. Theory* **IT-28**(2):199–205.

Richardson, R. J., 1966, Quantization of Noisy Channels, *IEEE Trans. Aerosp. Electr. Syst.* **AES-2**(5):362–364.

Rines, K. D., and N. C. Gallagher Jr., 1979, Quantization in Spectral Phase Coding, *Johns Hopkins Conf. Info. Sci. Systems Proc.* pp. 131–135.

Roberts, L. G., 1962, Picture Coding Using Pseudo Random Noise, *IRE Trans. Inform. Theory,* **IT-8**(2):145–154.

Stroh, R. W., and M. D. Paez, 1973, A Comparison of Optimum and Logarithmic Quantization for Speech PCM and DPCM Systems, *IEEE Trans. Commun.* **COM-21** (6):752–757.

Thompson, J. E., and J. J. Sparkes, 1967, A Pseudo-Random Quantizer for Television Signals, *IEEE Proc.* **55**(3):353–355.

Varshney, P. K., 1981, Combined Quantization-Detection on Uncertain Signals, *IEEE Trans. Inform. Theory* **IT-27**(2):262–265.

Reprinted from *IEEE Trans. Acoust., Speech, Signal Process.* **ASSP-28**(5):562–574 (1980)

Speech Coding Based Upon Vector Quantization

ANDRÉS BUZO, MEMBER, IEEE, AUGUSTINE H. GRAY, JR., SENIOR MEMBER, IEEE, ROBERT M. GRAY, FELLOW, IEEE, AND JOHN D. MARKEL, SENIOR MEMBER, IEEE

Abstract—With rare exception, all presently available narrow-band speech coding systems implement *scalar quantization* (independent quantization) of the transmission parameters (such as reflection coefficients or transformed reflection coefficients in LPC systems). This paper presents a new approach called *vector quantization.*

For very low data rates, realistic experiments have shown that vector quantization can achieve a given level of average distortion with 15 to 20 fewer bits/frame than that required for the optimized scalar quantizing approaches presently in use.

The vector quantizing approach is shown to be a mathematically and computationally tractable method which builds upon knowledge obtained in linear prediction analysis studies. This paper introduces the theory in a nonrigorous form, along with practical results to date and an extensive list of research topics for this new area of speech coding.

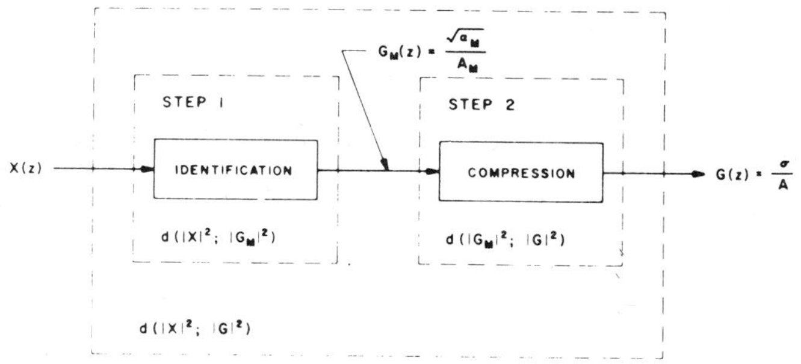

Fig. 1. Illustration showing process of LPC analysis as a two-step process.

I. Introduction

THE ubiquitous LPC technique of speech coding can be viewed as a two-step process as shown in Fig. 1. The first step is an identification process whereby an all-pole model $G_M(z)$ which best matches the input speech frame $X(z)$ (or possibly the preprocessed speech frame) is calculated. The best match is based on some predefined measure of optimality. The model $G_M(z)$ implicitly also includes a gain term so that the only remaining transmission parameter is the pitch estimate (which includes a voicing decision).

The second step is compression or quantization of the parameters from the identification step for efficient transmission or storage. A great deal has been written about the identification step (see [1], [2] and their bibliographies, for example), while substantially less has been written about the compression or quantization step [3]–[7].

Traditionally, the parameters from the identification step, such as reflection coefficients, have been individually quantized. We refer to such an approach as *scalar quantization* and it has also been called single symbol quantization. Orthogonal vector transformations have been applied with the idea of eliminating interparameter correlation. However, scalar quantization has then been applied to the transformed coefficients [8].

Other techniques, such as variable frame rate [9] or interframe coding, which can be "added on" to most speech compression systems, have been studied as a way to increase compression. To achieve highly efficient parameter compression, one must look beyond scalar quantization or the heuristics mentioned above to *rate distortion techniques* based on an appropriate fidelity criterion. Here, the fidelity criterion will include minimization of a distortion measure, not only in the identification step (as implemented in linear prediction analysis), but also in the compression or quantization step. *Vector quantization*, a design approach to this problem, has been developed during the past few years. The historical development of this area is covered in detail elsewhere [13], [14].

This paper presents a nonrigorous mathematical introduction to the new area of vector quantization in a deterministic manner along with experimental results. The presentation expands upon previously published results in linear prediction analysis. A more rigorous development of many of the theoretical concepts is presented elsewhere [10], [13], [18], [21]. In addition to the mathematical development, this paper presents the first experimental comparison between optimized scalar quantization and vector quantization. The results from both objective and subjective evaluation show dramatic bit savings for very low data rate conditions.

II. Itakura-Saito Distortion Measure

For a distortion measure to be of value in vector quantizing it must be analytically tractable, computable from sampled data, and, most important, subjectively meaningful. A distortion measure that results in the standard linear prediction

Manuscript received October 15, 1979; revised April 10, 1980. This research was supported in part by the Naval Research Laboratory under Contract NRLN00039-79-0256(S), the U.S. Department of Defense under Contract MDA904-79-C-0402, the U.S. Air Force Office of Scientific Research under Contracts F49620-78-C-0087 and F49620-79-C-0058, and by the Joint Services Electronics Program, Stanford University, Stanford, CA.

A. Buzo is with the University of Mexico, Mexico City, Mexico.

A. H. Gray, Jr. and J. D. Markel are with Signal Technology, Inc., Santa Barbara, CA 93101.

R. M. Gray is with the Information Systems Laboratory, Stanford University, Stanford, CA 94305.

analysis equations under the assumption of no distortion due to compression is also desirable.

The *Itakura-Saito distortion measure*, introduced as an error matching function, appears to satisfy all of the above requirements [11]. This section reviews several preliminary mathematical results and introduces the Itakura-Saito distortion measure along with several properties relevant to specific vector quantizer implementations.

A. Preliminaries

Let $X(z)$ represent the z transform of the windowed (and possibly preemphasized) speech data, which are to be modeled by an all-pole filter of the form

$$G(z) \triangleq \sigma/A(z) \tag{1}$$

where

$$A(z) \triangleq \sum_{k=0}^{M} a_k z^{-k}, \qquad \text{with } a_0 = 1. \tag{2}$$

In linear predictive analysis the polynomial $A(z)$ is used to minimize a residual energy. In particular, using $|X|^2$ and $|A|^2$ to denote energy density spectra

$$|X|^2 \triangleq |X(e^{j\theta})|^2 \quad \text{and} \quad |A|^2 \triangleq |A(e^{j\theta})|^2, \tag{3}$$

then the residual energy resulting from passing $X(z)$ through the inverse filter $A(z)$ is given by

$$\alpha \triangleq \int_{-\pi}^{\pi} |X|^2 \, |A|^2 \frac{d\theta}{2\pi}. \tag{4}$$

We denote the polynomial which minimizes the residual energy as $A_M(z)$ and the minimum value of the residual as α_M so that

$$\alpha \geqslant \alpha_M.$$

The model chosen in linear prediction analysis based upon the identification step in Fig. 1 is then

$$G_M(z) = \sqrt{\alpha_M}/A_M(z). \tag{5}$$

It is well known [1] that $A_M(z)$ is a minimum phase polynomial, that is, has its roots inside the unit circle so that $G_M(z)$ is stable. In addition, the minimum residual energies form a decreasing sequence as M increases, approaching a lower limit defined here as α_∞, which is given by

$$\alpha_\infty = \lim_{M\to\infty} \alpha_M = \exp\left[\int_{-\pi}^{\pi} \ln |X|^2 \frac{d\theta}{2\pi}\right]. \tag{6}$$

This lower limit α_∞ is sometimes called the one-step prediction error or gain of $|X|^2$ [1].

The actual solution process for finding $A_M(z)$ is also well known (see [1] and the references contained therein) and is not considered here.

Using arrows to denote z transform relationships, we can define the autocorrelation sequences

$$X(z)\,X(1/z) \longleftrightarrow r_x(n) = \sum_k x(k)\,x(k+n) \tag{7a}$$

$$A(z)\,A(1/z) \longleftrightarrow r_a(n) = \sum_k a_k a_{k+n} \tag{7b}$$

$$G_M(z)\,G_M(1/z) \longleftrightarrow r_M(n). \tag{7c}$$

Limits have not been placed on the summations to indicate that they are summations over all values of k. As will be pointed out, however, these will always be finite summations.

The integral which defines the residual energy in (4) can be precisely expressed for numerical evaluation as

$$\alpha = \sum_n r_x(n)\, r_a(n). \tag{8}$$

It is well known [1] that the model $G_M(z)$ matches the signal $X(z)$ in terms of the $2M+1$ term autocorrelation sequence

$$r_M(n) = r_x(n) \qquad \text{for } n = 0, \pm 1, \cdots, \pm M, \tag{9}$$

and therefore $r_M(n)$ can be used to replace $r_x(n)$ in the summation of (8).

In describing the spectral matching effects of linear prediction, Itakura and Saito introduced an "error matching function" [11] which is also referred to as the Itakura-Saito distortion measure [10]. This measure will be denoted by $d(|X|^2; |G|^2)$, where

$$d(|X|^2; |G|^2) \triangleq \int_{-\pi}^{\pi} \left[|X/G|^2 - \ln(|X/G|^2) - 1\right] \frac{d\theta}{2\pi}. \tag{10}$$

Many of its properties can be found elsewhere [10]-[12], [18]. One of the properties that interests us here is that it is a nonnegative function of the spectra, whose minimum value for a given $|X|^2$ and $G(z)$ given by (1) occurs when $G(z) = G_M(z)$. Thus,

$$d(|X|^2; |G|^2) \geqslant d(|X|^2; |G_M|^2) = \ln(\alpha_M/\alpha_\infty). \tag{11}$$

When $|X/G|$ is near unity, (10) takes on the approximate form

$$d(|X|^2; |G|^2) \cong \frac{1}{2}\int_{-\pi}^{\pi} \left[\ln(|X|^2) - \ln(|G|^2)\right]^2 \frac{d\theta}{2\pi}, \tag{12}$$

so that for small distortion the Itakura-Saito distortion measure is approximately one-half the mean-square log spectral deviation.

B. Useful Properties

A set of properties of the Itakura-Saito distortion measure to be used in the next section is now presented. A sketch of the derivation of these results is given in the Appendix. The properties are developed in detail in [10] and [12].

First, for purposes of calculation and interpretation, the Itakura-Saito distortion measure can be expressed in the form

$$d[|X|^2; |G|^2] = \alpha/\sigma^2 + \ln(\sigma^2) - \ln(\alpha_\infty) - 1 \tag{13}$$

where σ, α, and α_∞ are defined by (1), (4), and (6), respectively. Surprisingly, it can be shown to satisfy a form of "triangle equality"

$$d[|X|^2; |G|^2] = d[|X|^2; |G_M|^2] + d[|G_M|^2; |G|^2], \tag{14}$$

as shown in the Appendix. Referring to Fig. 1 we see that *the total distortion in the analysis can be viewed as precisely the sum of the distortion due to the identification step and the distortion due to the compression or quantization step.* In general, one can only hope for a triangle inequality whereby the total distortion is bounded by the sum of the individual distortions. Also, minimizing $d[|X|^2; |G|^2]$ is equivalent to minimizing $d[|G_M|^2; |G|^2]$, for $d[|X|^2; |G_M|^2]]$ is a fixed property of $|X|^2$, as indicated in (11), if M is held constant.

A second useful gain cascading property is given by

$$d[|X|^2; \sigma^2/|A|^2] = d[|X|^2; \alpha/|A|^2] + d[\alpha; \sigma^2], \tag{15}$$

which divides the distortion into two parts. The first part is independent of the gain choice σ, while the second part is apparently independent of the polynomial $A(z)$, although in fact it does depend upon $A(z)$ through the residual energy α. The first term has also been called the gain optimized Itakura-Saito distortion measure or simply the "Itakura distortion measure" [10], [12]. We will show later that this form leads to implementation of a suboptimal but storage-efficient and a practical narrow-band speech compression system.

III. Vector Quantization

The standard autocorrelation method LPC system first obtains an optimal model $G_M(z)$ for a speech frame $X(z)$, and then quantizes its parameters, leading to a quantized version $G(z)$. This system implicitly minimizes the Itakura-Saito distortion measure [1, p. 135], [10], [11] at the first step (identification of the optimal model), but does not use this distortion measure for the second step (compression or quantization of the parameters). It is therefore unknown whether any particular overall distortion from the speech spectrum $|X|^2$ to the quantized model spectrum $|G|^2$ has in fact been minimized.

An alternate approach is to define a reasonable distortion measure and attempt to choose $G(z)$ from a finite collection of vectors to minimize the overall distortion. Thus, the term vector quantization refers to the process of choosing a vector of parameters, e.g., $\{\sigma, r_a(0), r_a(1), \cdots, r_a(M)\}$, from a set which minimizes a distortion measure. As described in the previous section, the Itakura-Saito distortion measure is analytically tractable, perceptually meaningful, and readily computable (except for the term α_∞ in (13) which will be shown to be unnecessary).

One very important and useful property of the distortion measure is the triangle equality of (14). In particular, this property shows that one can minimize the overall distortion $d(|X|^2; |G|^2)$ directly in one step, or one can first obtain the ideal model $G_M(z)$ and then minimize $d(|G_M|^2; |G|^2)$, which results in a two-step process as indicated in Fig. 1 [10].

Aside from computational and storage issues, the vector quantizer being described here results in an elegant structure (shown later), once the nontrivial problem of how to determine a collection of reference or reproduction vectors or *codebook* has been resolved. The codebook is a finite set of model filters from which $G(z)$ must be found. Each code word has an index number and a stored set of parameters representing a possible $G(z)$. For example, code word number 01101001 (binary) might represent one frame of the sound /a/ as uttered by a specific test speaker, by way of its filter coefficients, reflection coefficients or autocorrelation coefficients. Then for each frame of speech, $G(z)$ is chosen from the codebook to minimize $d(|X|^2; |G|^2)$, or, equivalently, to minimize $d(|G_M|^2; |G|^2)$. This process finds the nearest neighbor to $X(z)$ in the codebook. The index to the code word is then transmitted and used at the receiver to retrieve the appropriate synthesis filter parameters.

The problems of generating an optimal codebook (in the Itakura-Saito distortion sense) and evaluating the distortion are now analyzed. Then the computational and storage issues which lead to a practical suboptimal vector quantizing system are considered.

A. Codebook Generation

The generation of a codebook to minimize distortion over a large number of test frames of speech requires an iterative process. As with many minimization problems, the procedure to be described will converge to a local minimum, but not necessarily an absolute or global minimum. The basic ideas of the procedure to be described date to Lloyd [15]. Linde *et al.* [13] and Gray *et al.* [21] contain a more detailed and general discussion.

Let $X_k(z)$ represent the z transform of the kth frame of speech, where $k = 1, 2, \cdots, K$, and K is large. These frames represent a test or learning sequence for codebook generation. We wish to model these speech frames with a finite set of models, $B = 2^b$ in number, where b is the number of bits in the codebook. Assume at this point that an initial (nonoptimum) choice has been made for a set of B code words. As a first step, each speech frame, represented by $X_k(z)$, is taken individually and assigned to a "cell" represented by a single code word. This is accomplished by finding the particular $G(z)$ out of the collection of B possible models which minimizes $d[|X_k|^2; |G|^2]$. That is, the objective is to find the particular $G(z)$ in the codebook which is the nearest neighbor to $X_k(z)$. The total overall distortion is then the sum of all the individual distortions in the database, with one from each speech frame.

The second step attempts to improve the codebook by choosing a better model for each cell. For simplicity, consider a single cell, with the frames of speech within that cell renumbered as $X_1(z), X_2(z), \cdots, X_L(z)$, so that the total distortion for that cell is given by

$$D \triangleq \sum_{k=1}^{L} d[|X_k|^2; |G|^2]. \tag{16}$$

Then we must find a single model $G(z)$ from an infinite number of possibilities to minimize the distortion in that particular cell by identifying what might be called the *centroid* of the cell. The solution to this problem is equivalent to a standard linear prediction analysis problem of minimizing $d[|\overline{X}|^2; |G|^2]$, where $|\overline{X}|^2$ is the arithmetic mean of the individual cell spectra

$$|\overline{X}|^2 \triangleq \frac{1}{L} \sum_{k=1}^{L} |X_k|^2. \tag{17}$$

Once the centroids of each cell have been found and used as the new code words (model filters), the distortion for each individual cell is minimized by the definition of a centroid. Thus the overall distortion must decrease, or at worst remain the same. One then returns to the first step to redefine the cells by assigning each speech frame to its nearest neighbor model filter. By definition of a nearest neighbor, the distortion for each individual speech frame and thus the total distortion must decrease, or at worst remain the same. Iteration of this procedure results in an overall distortion that is monotonically nonincreasing, for at each step it either decreases or remains the same. As the distortion is bounded below, it must approach a lower limit—a locally optimum value [21]. The speed of convergence and the actual value of the local minimum of the overall distortion will depend upon the initial choices for the code words at the start of the iteration.

B. Nearest Neighbor Distortion Calculation

To assign a frame of speech to a specific code word, we must find the specific $G(z)$ out of the collection of all possible models which minimizes $d[|X|^2; |G|^2]$. From (13), we note that since α_∞ depends only on the speech frame, an equivalent statement is to find the $G(z) = \sigma/A(z)$ which will minimize

$$d[|X|^2; |G|^2] + 1 + \ln(\alpha_\infty) = \alpha/\sigma^2 + \ln(\sigma^2).$$

For any single frame of speech, the residual energy α must be computed. The computation is most efficiently accomplished in the form (see the Appendix)

$$\alpha = r_a(0)\, r_x(0) + 2 \sum_{n=1}^{M} r_a(n)\, r_x(n) \tag{18}$$

where $r_a(n)$ and $r_x(n)$ are the autocorrelations of (7). The computational expression for $r_a(n)$ is given by

$$r_a(n) = \sum_{k=0}^{M-n} a_k a_{k+n} \qquad \text{for } n = 0, 1, \cdots, M. \tag{19}$$

For computational efficiency, the transmitter codebook should probably contain the normalized sequence $\{r_a(n)/\sigma^2, n = 0, 1, \cdots, M\}$, as well as the value of $\ln(\sigma^2)$, so that $\alpha/\sigma^2 + \ln(\sigma^2)$ can be evaluated with exactly $M + 1$ multiplies and $M + 2$ adds, after $r_x(n)$ is computed. The transmitter codebook then contains $M + 2$ parameters for each code word, rather than $M + 1$ parameters, which is probably a desirable tradeoff to eliminate the division by σ^2 and/or the log calculation $\ln(\sigma^2)$.

In evaluating (18), $r_x(n)$ is required for $n = 0, 1, \cdots, M$. This is the standard short-term autocorrelation sequence for the sequence $x(n)$ used in the autocorrelation method of linear prediction. For a data sequence which is truncated to $n = 0, 1, \cdots, N - 1$ samples, $r_x(n)$ is given by

$$r_x(n) = \sum_{k=0}^{N-1-n} x(k)\, x(k+n) \qquad \text{for } n = 0, 1, \cdots, M < N. \tag{20}$$

From the autocorrelation matching property, the term $r_x(n)$ can be replaced by its equivalent model autocorrelation term $r_M(n)$. This is convenient for simulation studies where test data may be stored only in the format of the model $G_M(z)$.

In summary, to find the nearest neighbor one must evaluate $\alpha/\sigma^2 + \ln(\sigma^2)$ for each entry in the codebook, and choose the code word which minimizes the result. Therefore, this procedure requires an exhaustive search for every speech frame.

C. Centroid Calculation

If the frames of speech $X_1(z), X_2(z), \cdots, X_L(z)$ are all contained within a given cell, the total distortion for that cell is given by (16), which can also be written in terms of the average spectrum of (17) as

$$D = L\, d[|\overline{X}|^2; |G|^2] + u \tag{21}$$

where u is a constant that is independent of $G(z)$, the model for the cell. Thus to find the cell centroid (the $G(z)$ that minimizes D) we have a standard linear prediction problem of modeling the average spectrum.

As a result, we can average the autocorrelation sequences for each of the speech frames to find an average autocorrelation sequence, and then solve the autocorrelation equations to give the parameters of $G(z) = \sigma/A(z)$. The constant u is not needed for these calculations. However, it represents a cost or distortion that will arise, regardless of model filter order, when dissimilar frames of speech are assigned to the same cell. See the Appendix for its exact form.

D. Speech Coder Implementation

Figs. 2 and 3 show a new speech coder structure based on the material of Section III. The system implements a *full search optimal vector quantizer*, operating directly on the speech waveform.

The analysis process consists of calculating an $M + 1$ length autocorrelation sequence and then running a full search comparison of the codebook to obtain the index $i_{\min}$ for which $\alpha/\sigma^2 + \ln(\sigma^2)$ results in minimum distortion. Using any of the standard methods of computing a pitch and voicing term with corresponding transmission index i_p, the analysis is completed.

At the synthesizer, $i_{\min}$ is inserted into the receiver codebook to obtain the synthesis parameters in a form ready for use by the synthesis structure. For example, although $M + 2$ terms $\{\ln(\sigma^2), r_a(0)/\sigma^2, \cdots, r_a(M)/\sigma^2\}$ define the analyzer code word, the corresponding synthesis code word would most likely contain the $M + 1$ terms $\{\sigma, k_1, \cdots, k_M\}$ for use with a lattice form synthesizer.

E. Discussion

An extremely interesting aspect of this development is that *no* simultaneous equations are solved for reflection coefficients or filter coefficients, even though our process is now optimal (in the Itakura-Saito sense) not only for the identification step (as in standard LPC analysis), but also for the compression or quantization step. In effect, the speech processing system based on full search optimal vector quantization becomes a one-step combined identification and compression step as shown in Fig. 4, with full optimality throughout.

Once the speech codebook has been obtained, several difficulties remain. The first problem relates to storage of a sub-

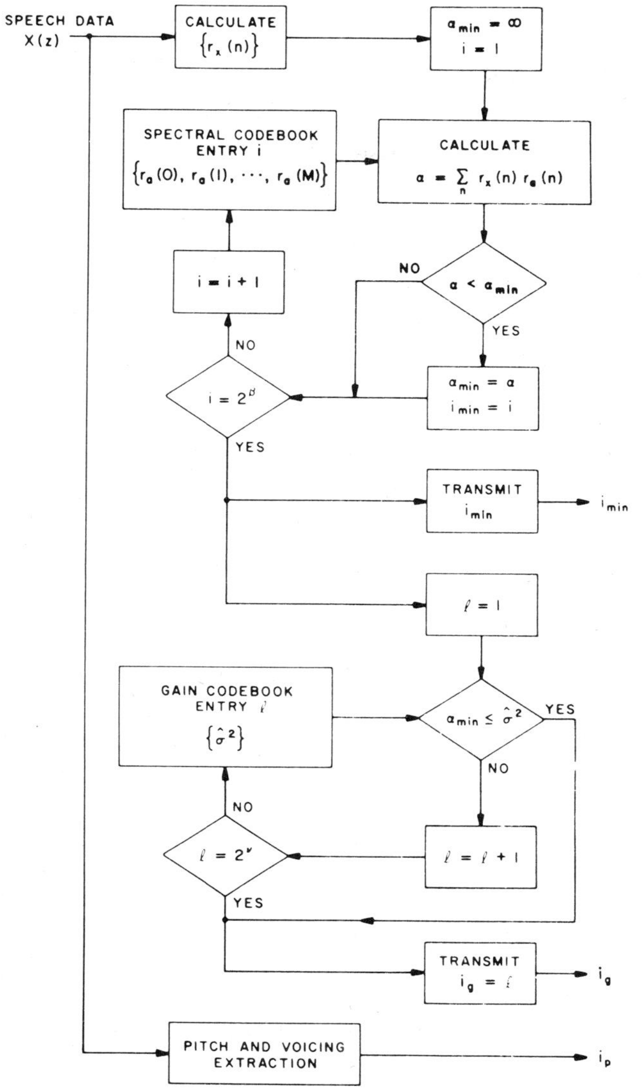

Fig. 2. Analysis structure for full search optimal vector quantized speech processor.

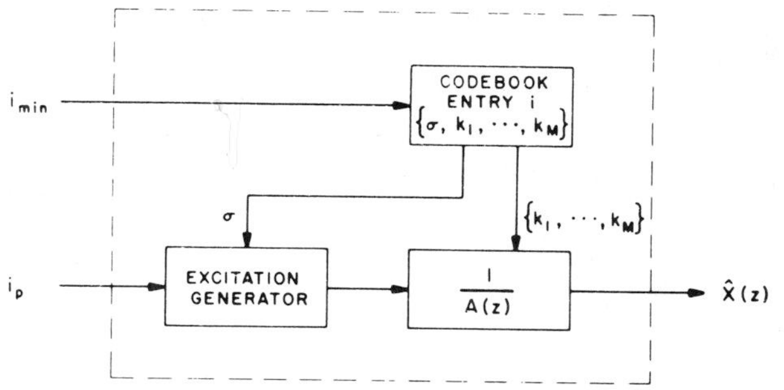

Fig. 3. Synthesis structure for full search optimal vector quantized speech processor.

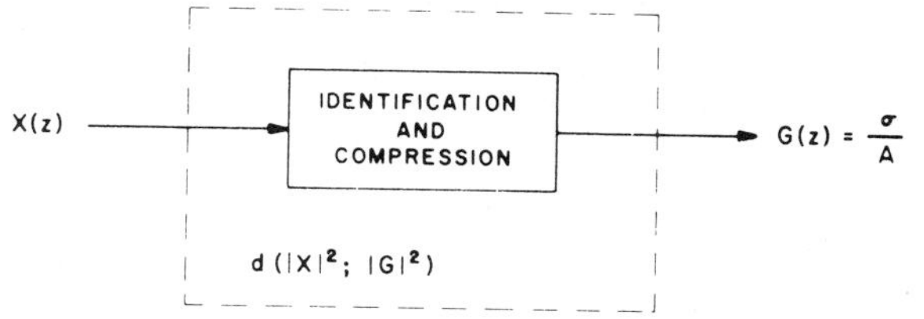

Fig. 4. A one-step identification and compression system.

stantial amount of information. Since both gain and spectral information are coded together, the codebook is larger than if they were coded independently. For example, if 15 bits are used to represent the combination of spectral and gain information for model filters of order $M = 10$, then a total of at least $2^{15} = 32\,768$ code words times 12 coefficients per code word equals 393 216 storage locations required at the transmitter and receiver. In general, $(M + 2)\,2^b$ storage locations are required with this procedure as it is presently structured. Although this amount of storage is feasible with a general-purpose computer having virtual memory, such a requirement is unrealistic for 16 bit minicomputers and special purpose real-time hardware.

A second problem relates to the amount of computation required in the determination of a code word index. Each speech frame must be compared with $B = 2^b$ possible models to choose a nearest neighbor. For example, a 15 bit codebook would require a total of 2^{15} evaluations of $\alpha/\sigma^2 + \ln(\sigma^2)$, with each evaluation requiring at least $M + 1$ multiplies. The number of these calculations grows exponentially with the number of bits.

IV. Suboptimal Vector Quantization

The rather sobering facts just described in the previous section would render the vector quantizing approach of only academic interest for codebooks greater than 10 bits were it not for two major modifications that we have developed which result in a suboptimal vector quantization system. The first modification is termed gain separation, and the second modification is termed binary tree searched coding. The latter is, to the best of our knowledge, the first suggested approach for vector searching with binary search savings. This section considers these topics. The previously described optimal system using full search and combined gain and model code words is developed in more mathematical and experimental detail in [18].

A. Gain Separation

Equation (15) illustrates a separation of the distortion into two parts. The first depends only upon the polynomial $A(z)$—not upon the gain σ—and the second depends upon the gain term (and indirectly on the polynomial through α). Rather than minimize the overall distortion, one can separately minimize the two parts of (15) by first finding $A(z)$ and then obtaining σ. The overall codebook is suboptimal since it is restricted to have a particular form. However, this approach can result in a decreased codebook size at the expense of increased overall distortion. For example, if β bits are assigned to describe $A(z)$ and ν bits are assigned to describe σ, the number of words in the codebooks for the separated parameters are reduced from $2^{\beta+\nu}$ to $2^\beta + 2^\nu$ for a full codebook containing $\beta + \nu$ bits. As will be shown, this separation leads to an approximation (further suboptimization) in the calculation of centroids needed to generate the codebook for the polynomial parameters.

1) *Nearest Neighbor Calculations*

For any given speech frame $X(z)$ we must first find the nearest neighbor in the $A(z)$ codebook which minimizes $d[|X|^2; \alpha/|A|^2]$ of (15). Substituting $\sigma^2 = \alpha$ in (1) and (13) gives an equivalent expression

$$d[|X|^2; \alpha/|A|^2] = \ln(\alpha) - \ln(\alpha_\infty). \tag{22}$$

This expression can therefore be minimized by minimizing α, the residual energy. As in Section III-A, α can be calculated by using (18). The calculation of α requires $M + 1$ multiply/add's for each entry in the codebook to determine which one gives the minimum value for a given speech frame.

Once the minimum α has been found, one then looks to the gain codebook to find which gain value will minimize the remaining part of (15), $d(\alpha; \sigma^2)$. Substituting α and σ^2 into the distortion measure (10) directly gives

$$d(\alpha; \sigma^2) = (\alpha/\sigma^2) - \ln(\alpha/\sigma^2) - 1, \tag{23}$$

which is minimized by choosing a value of σ^2 from the gain portion of the codebook. The same result can be arrived at more expeditiously by noting from (13) that for any value of gain σ^2 the overall distortion is minimized by minimizing the residual energy α, and for any value of α, the overall distortion is minimized by choosing the gain to minimize $\alpha/\sigma^2 + \ln(\sigma^2)$, or equivalently by minimizing (23). For a given product codebook, this approach gives the optimal choice of a code word. The codebook itself is suboptimal, not the assignment of the code word to a speech frame.

Since minimization of (23) is a one-dimensional problem, the codebook gains can be ordered and compared with a set of threshold values which will be stored at the transmitter, with the codebook gains stored at the receiver.

To illustrate this point, subscripts are used on the codebook gains (assumed ordered), so that the codebook consists of σ_l^2 for $l = 1, 2, \cdots, \Upsilon$, with

$$\sigma_1^2 < \sigma_2^2 < \cdots < \sigma_\Upsilon^2.$$

We shall find a set of thresholds, or cell boundaries $\hat{\sigma}_l^2$, $l = 1, 2, \cdots, \Upsilon - 1$, such that

$$\sigma_1^2 < \hat{\sigma}_1^2 < \sigma_2^2 < \hat{\sigma}_2^2 < \cdots < \hat{\sigma}_{\Upsilon-1}^2 < \sigma_\Upsilon^2.$$

At the transmitter the set of $\Upsilon - 1$ thresholds is stored, and α is compared to the thresholds to find which cell it lies in to generate an index for transmission. For example, if $\alpha \leqslant \hat{\sigma}_1^2$, the index $l = 1$ is transmitted, or if $\hat{\sigma}_4^2 < \alpha \leqslant \hat{\sigma}_5^2$, then the index $l = 5$ is transmitted. At the receiver, the index l is used to recover the stored gain term σ_l^2.

Finding the threshold values from the codebook gains is a matter of solving the equations

$$d(\hat{\sigma}_l^2; \sigma_l^2) = d(\hat{\sigma}_l^2; \sigma_{l+1}^2) \tag{24}$$

for $\hat{\sigma}_l^2$, the point "equidistant" between adjacent codebook gains. This must be done for $l = 1, 2, \cdots, \Upsilon - 1$. To solve this equation, we can use (23) to rewrite (24) in the form

$$(\hat{\sigma}_l^2/\sigma_l^2) - \ln(\hat{\sigma}_l^2/\sigma_l^2) - 1 = (\hat{\sigma}_l^2/\sigma_{l+1}^2) - \ln(\hat{\sigma}_l^2/\sigma_{l+1}^2) - 1, \tag{25}$$

which can then be solved to give

$$\hat{\sigma}_l^2 = \frac{\ln(\sigma_{l+1}^2/\sigma_l^2)}{(1/\sigma_l^2) - (1/\sigma_{l+1}^2)}. \tag{26}$$

When σ_{l+1}^2 is near σ_l^2, as it would be if there is a fine-level quantization, then (26) is not numerically efficient because of the subtraction of close numbers. In that case it is better to use a Taylor series expansion to write

$$\hat{\sigma}_l^2 = \frac{1}{2}(\sigma_l^2 + \sigma_{l+1}^2)\left[1 - \frac{2\delta^2}{3 \cdot 1} - \frac{2\delta^4}{5 \cdot 3} - \frac{2\delta^6}{7 \cdot 5} - \cdots\right] \tag{27a}$$

where

$$\delta \triangleq (\sigma_{l+1}^2 - \sigma_l^2)/(\sigma_{l+1}^2 + \sigma_l^2). \tag{27b}$$

The derivation of the Taylor series is omitted for brevity. The use of (27) rather than (26) is strongly suggested for numerical accuracy.

In summary, the threshold values can be obtained from the code word gains using (27) or (26). Finding the code index for gain requires using the residual energy α, which was obtained during the coding of the polynomial $A(z)$ in a direct comparison with the threshold values.

2) *Centroid Calculation*

In the gain separated case we are dealing with two centroids to be calculated. First, for the polynomial parameters, we would like to minimize the total cell distortion of (16). Using (15) and (22) we are attempting to minimize the sum of terms

$$D_1 = \sum_{k=1}^{L} d[|X_k|^2; \alpha^k/|A|^2] = \sum_{k=1}^{L} [\ln(\alpha^k) - \ln(\alpha_\infty^k)] \tag{28}$$

where each α^k is the "optimal" gain choice for the individual speech frames (*not* the kth power of α),

$$\alpha^k = \int_{-\pi}^{\pi} |X_k|^2 |A|^2 \frac{d\theta}{2\pi},$$

and α_∞^k is the one-step prediction error or gain of $|X_k|^2$ as in (6). Thus, the centroid problem is to choose a polynomial $A(z)$ to minimize

$$\sum_{k=1}^{L} \ln(\alpha^k) = \sum_{k=1}^{L} \ln\left[\int_{-\pi}^{\pi} |X_k|^2 |A|^2 \frac{d\theta}{2\pi}\right].$$

Clearly, finding the polynomial $A(z)$ to minimize (28) is not a trivial task. Instead of trying to solve the specific problem, we shall look to an approximate (and bounding) solution.

Each individual $X_k(z)$ has an "optimal" model whose gain or one-step prediction error term is given by α_M^k. Equation (28) is first rewritten in the form

$$D_1 = \sum_{k=1}^{L} \ln(\alpha^k/\alpha_M^k) + \sum_{k=1}^{L} \ln(\alpha_M^k/\alpha_\infty^k). \tag{29}$$

The second summation is ***independent of the parameters of the polynomial*** $A(z)$, and is simply a function of the individual speech frames. The first summation of (29) is the product of

L and the logarithm of the geometric mean of the ratios α^k/α_M^k for $k = 1, 2, \cdots, L$. An approximation to the geometric mean, and an upper bound as well, is given by the arithmetic mean so that D_1 is both approximated by and bounded above by D_2, where

$$D_2 = L \ln\left[\frac{1}{L}\sum_{k=1}^{L}(\alpha^k/\alpha_M^k)\right] + \sum_{k=1}^{L}\ln(\alpha_M^k/\alpha_\infty^k). \quad (30)$$

To minimize D_2 exactly, and thus D_1 approximately, we need to minimize the arithmetic mean of the α^k/α_M^k ratio defined by

$$\frac{1}{L}\sum_{k=1}^{L}(\alpha^k/\alpha_M^k) = \int_{-\pi}^{\pi}|\bar{\bar{X}}|^2\,|A|^2\,\frac{d\theta}{2\pi} \quad (31)$$

where $|\bar{\bar{X}}|^2$ is a normalized average spectrum given by

$$|\bar{\bar{X}}|^2 \triangleq \frac{1}{L}\sum_{k=1}^{M}|X_k|^2/\alpha_M^k. \quad (32)$$

Thus, centroid determination in the gain-separated case is similar to that in the optimal case, with the basic differences being that it is actually an approximation in the present case, and that the individual spectra (or autocorrelation sequences) are normalized before averaging.

In summary, one can average the autocorrelation sequences for all the speech frames within a cell and solve the autocorrelation equations. In the fully optimal case one does not normalize the autocorrelation sequence, but in the gain-separated case, one must normalize the autocorrelation sequences by the optimal gain coefficients, the α_M^k terms obtained from the residual energy resulting from passing the speech frame through its optimal inverse filter. In this latter case, frames of silence will count as heavily in finding a centroid as frames of high-energy voiced speech since the autocorrelation sequences are normalized. This suggests a careful screening of any reference frames used in generating the codebook.

Finding the centroid for the gain codebook is simpler, once the α^k for each frame has been found, for then one is interested in choosing a single gain term σ so as to minimize

$$D_3 = \sum_{k=1}^{L} d[\alpha^k;\sigma^2] = \sum_{k=1}^{L}[(\alpha^k/\sigma^2) - \ln(\alpha^k/\sigma^2) - 1]. \quad (33)$$

This expression can be minimized by taking σ^2 as the arithmetic mean of the individual residual energies

$$\sigma^2 = \frac{1}{L}\sum_{k=1}^{L}\alpha^k.$$

B. Binary Tree Searched Coding

To decrease the number of calculations necessary for finding a nearest neighbor code word, one can institute a binary search at the expense of doubling the storage requirements at the transmitter and increasing distortion. The systems described so far, with or without gain separation, require a large number of calculations for each speech frame to obtain a nearest neighbor growing exponentially with the number of bits. In the alternative described here and used in the experimental results, the number of calculations grows only linearly with the number of bits, but with twice the amount of storage at the transmitter and some increase in the distortion.

The codebook at the transmitter is split into levels. The first level contains only two code words and is used to split the space of speech frames into two, by one pair of residual energy evaluations and a comparison. Then each of these large cells is split into two, with a total of four code words at the second level. As we are seeking practical suboptimal systems, we will assume that only the parameters of the polynomial $A(z)$ are being treated here. Thus, if the polynomial parameters are quantized to a total of β bits, there must be a total of

$$2 + 2^2 + \cdots + 2^\beta = 2^{\beta+1} - 2$$

stored code words at the transmitter, though there are still only 2^β at the receiver. In effect, the transmitter storage is doubled for the part of the codebook holding parameters for the polynomial $A(z)$.

The number of calculations is decreased substantially in this case. Where 2^β residual energy evaluations were needed before to find a nearest neighbor, that number is now reduced to 2β.

Using the same size training sequence, calculation of the suboptimal approach is similar to that of the optimal approach except for a substantial reduction in computer time. This is because the present initialization procedures for the codebooks involve a binary splitting (see Section V), which uses all members of the training sequence at each step. For the suboptimal binary tree search procedure, one first sets up a 1 bit codebook, effectively splitting the training sequence in two. While the two halves do not have an equal number of members, they each have less than the original number, and each half is then split in two, setting up a 2 bit codebook. This procedure continues until a β bit codebook is finally established, requiring $2^{\beta+1} - 1$ code words at the transmitter, yet still only 2^β code words for the filter parameters at the receiver.

In this form of binary system with the gain separated from the model, the codebook for gain is generated as described earlier, with no change, from the resulting residual energies from the training sequence speech frames.

C. Initialization

As the codebook is at best locally optimal (or suboptimal for the gain-separated and/or binary tree searched case), the choice of initial code words can be significant in terms of the final outcome. We shall address ourselves here to the binary suboptimal coding as described in the preceding section, but many of the problems are common to the more fully optimal situation.

In the binary situation, the problem is to choose two initial code words for splitting a specific cell since this is the way the space is split.

One approach is to find the centroid of the cell, and then perturb it in some manner to give two different points. This is an ad hoc approach not guaranteed to be better than others, but it has given us reasonably satisfactory results. In particular, we find the reflection coefficients associated with the centroid, and then generate two code words by multiplying the reflection coefficients by the arbitrary factors of 1.01 and 0.99, respectively. From that point the iteration further separates the code words and "splits" the cell.

An elementary alternative to this procedure is to first find the frame in the cell farthest from the centroid (in the sense of the Itakura-Saito distortion measure) to use as an initial code word. Then one finds the frame furthest from that code word and uses that as the second initial code word. In this case the initial pair of code words are far apart rather than close together. Only experimentation will determine which approaches are more efficient in the sense of number of needed iterations and size of the local minimum that results.

It remains to be seen as to what the "best" initialization is, and how to handle a few of the problems. One problem that arises is the "empty cell." For example, if one has generated a 9 bit codebook, where only one frame of speech from the training sequence is assigned to a cell (with zero quantization distortion), there is no way to split that particular cell. This condition leads to an inefficient use of the codebook, as well as to possible error messages if one is not careful with the programming. This problem is related to the training set size.

One should have a reasonable number of training samples (speech frames) for each cell, such as at least 10 or 20, so that samples outside the training set are reasonably represented. If a reference set not exceeding $B = 2^b$ samples is chosen, zero distortion can theoretically be obtained, but only for the reference set. For a 12 bit codebook, a training sequence of 40 000–80 000 speech frames should be used. With reasonable computation speed a large number of reference frames can be obtained (in a recent study 7×10^6 disk based speech frames were obtained from conversational speech [17]). However, *substantial* amounts of computing are required to determine the codebook because of the interactive process described above (several representative times are presented in the section on experimental results).

D. Suboptimal Speech Coder Implementation

A speech processing system which implements suboptimal vector quantizing is shown in Fig. 5 (analyzer) and Fig. 6 (synthesizer). The process is quite similar to full search optimal vector quantizer implementation except that two separate codebooks are used in the analyzer and synthesizer, and the spectral parameter codebook is searched in a binary as opposed to a full search manner.

No detailed formulas have been given in this section on the suboptimal gain separated system since they have already been presented and are indicated again in Fig. 5. A few points will be emphasized, however. First, the only normalization used in the suboptimal system is in the evaluation of centroids needed to set up the codebook. No normalization is needed to use the codebook. Second, while it was computationally efficient to store the parameters $r_a(k)/\sigma^2$, for $k = 0, 1, \cdots, M$ for the fully optimal case where an evaluation of α/σ^2 was required, the suboptimal case requires only α.

For a 12 bit spectral codebook and $M = 10$ spectral coefficients, there are $2\beta = 24$ comparisons, each requiring $M + 1 = 11$ multiply/add's for a total of 264 multiply/add's. This number is essentially equivalent to the approximately $2.5M^2 = 250$ operations required to perform a single frame linear prediction analysis solution. Thus, we now have a structure that can be considered for real-time implementation with at least up to 12 bits/frame. The next section presents a set of preliminary results which compare distortion as a function of bits/frame for the new suboptimal vector quantizer with the standard scalar quantizer approach.

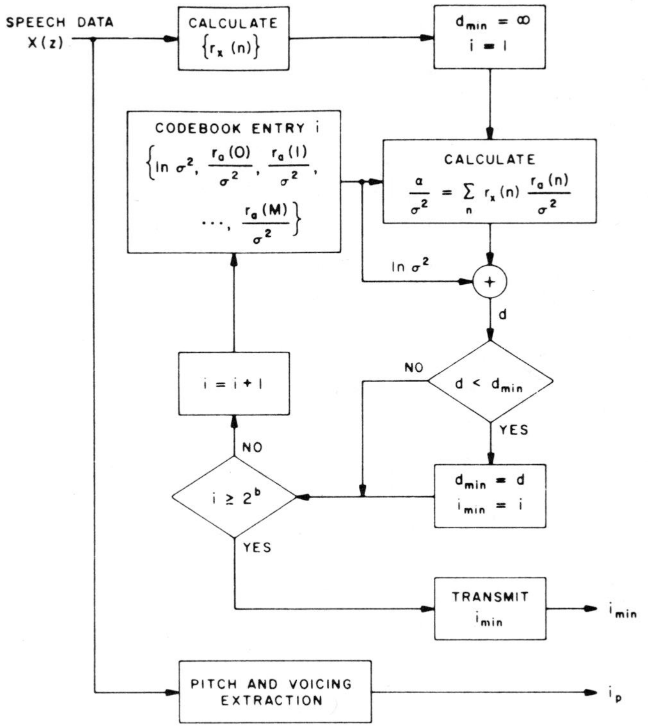

Fig. 5. Analysis structure for suboptimal vector quantized speech processor.

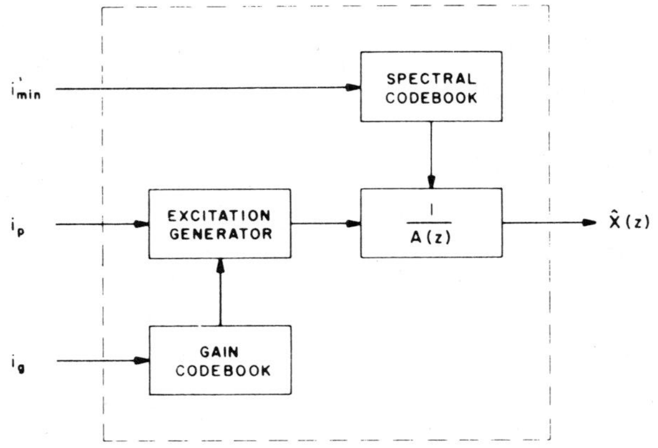

Fig. 6. Synthesis structure for suboptimal vector quantized speech processor.

V. AN EXPERIMENTAL COMPARISON BETWEEN VECTOR AND SCALAR QUANTIZATION

It is important to note that nowhere in the theory are we assured of receiving substantial bit savings when processing actual speech data. The benefits, if any, must be demonstrated by way of actual experimentation. The results presented in this section show that the benefits are substantial. These results are not intended to provide an exaustive presentation of the vector quantizer's capabilities and limitations. Rather, we describe one experiment with a substantial database which shows the dramatic benefits obtainable (in terms of bits/frame reduction for a given distortion level) to motivate additional research in this area.

A reference database of conversational speech segments from five male speakers was chosen with each speech segment consisting of approximately 20 s of speech. A 20 s test segment from a male speaker not included in the reference set was also chosen for test purposes.

The data were sampled at 6.5 kHz and digitally preemphasized. In addition, a sharp cutoff (200 Hz) high-pass filter was applied to eliminate very low-frequency energy. Previous tests have shown that if any A/D bias exists, for example, analysis coefficients will be used in representing the very low-frequency nonspeech behavior.

For this study we chose a preemphasis factor of 0.9, and filter order of 10. Analysis windows of 128 samples were chosen with a frame shift of 128 samples. A total of 5396 reference frames was obtained from analyzing 1 min, 48 s of speech. Although separate quantization tables could be obtained for voiced and unvoiced speech with possible further benefits, this experiment was based upon all frames for the reference data. Hamming windows were used.

The scalar quantizing method chosen was the uniform sensitivity quantization method proposed by Vishwanathan and Makhoul [4] and later studied by Gray and Markel [5]. The reflection coefficients are transformed into log area ratios [4] or inverse sine (theta) parameters [5] and then uniformly quantized on the basis of measured parameter sensitivities and parameter extreme values. The inverse sine transformation, which theoretically produces precisely constant sensitivity for the final coefficient, was used here. The procedure for actually measuring sensitivities and performing the bit allocation is presented elsewhere [5].

Because of computational and storage requirements for the full search optimal vector quantizer described earlier, it is of more practical interest to observe how well the suboptimal vector quantizer operates. Both the full search and binary tree searched suboptimal vector quantizer procedure were implemented to determine the tradeoff between search efficiency and distortion.

Computer programs were written in Fortran and implemented on a DEC VAX-11/780 computer system. The virtual memory capability of the VAX was found to be indispensable for this type of problem since main memory overflow occurs at about 8 bits/frame on a 16 bit computer because of the required data storage for efficient computation in the full search codebook generation.

Both the full search and binary search codebooks were obtained by the cell-splitting procedure starting with two cells, then four cells, etc., in a binary fashion to 10 bits or 1024 cells. We chose the 10 bit codebook limit for the spectral coefficients because of the database size of 5396 frames. The gain codebook was chosen to have 5 bits as this provided nearly as good quality as that obtainable at higher rates. With a standard 8 bits/frame for voicing and pitch, our highest rate

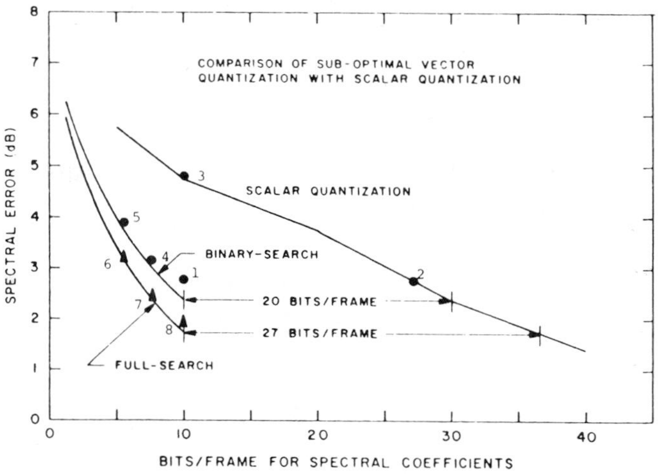

Fig. 7. Comparison of suboptimal vector quantization with scalar quantization.

system required $10 + 5 + 8 = 23$ bits per frame, or $20 \times 50 = 1150$ bits/s. The full search suboptimal vector quantizer codebook determination required 3 h, 20 min of stand-alone CPU time on the VAX system. The binary search procedure required 4 min, 47 s of stand-alone CPU time. At the completion of the binary tree searched codebook generation there were three empty cells for the 9 bit quantizer, 26 empty cells for the 10 bit codebook, and no empty cells for the full search quantizer.

For comparison with previously published results [7], the Itakura–Saito distortion measure, obtained from the codebook generation, was transformed into approximate rms log spectral error measurements using the result from (12) that for small distortion, the Itakura–Saito distortion measure is approximately one-half the mean-square log spectral deviation. The results of this experiment are shown in Fig. 7. The scalar quantization curve was obtained over the wide range of 6 bits/frame to 40 bits/frame for the spectral coefficients. The suboptimal vector quantizer was designed in increments of 1 bit/frame from 1 bit/frame to 10 bits/frame for the spectral coefficients. Both full search and binary search methods were implemented. For both the scalar and vector quantizer, the curves represent distortions measured during the reference codebook or reference quantization table design.

For a 10 bits/frame full search vector quantizer, the measured distortion is approximately 1.8 dB. The equivalent distortion point for the scalar quantizer is at approximately 37 bits/frame resulting in a difference of 27 bits/frame or a 73 percent reduction in bit rate for the equivalent distortion. When one places into perspective the fact that the field of linear prediction analysis has been concerned about bit rate reduction for a given distortion level for nearly 10 years [3], [6], [11], [12], and until recently [7] the differences between various coding schemes were 3–4 bits/frame for a given rms log spectral value, these results are remarkable.

Now, considering the suboptimal vector quantizer in more detail, we see that, whereas the full search and binary search procedures are identical at 1 bit/frame, the relative distortion of the binary search approach increases as bits/frame increase. At 10 bits/frame the distortion for the binary search is approximately 0.6 dB higher than for the full search method. Stated differently, the binary search procedure requires roughly 2 bits/frame more than the full search procedure at 8 bits/frame. This would appear to be a very reasonable tradeoff for practical implementation since $2^{10} = 1024$ residual energy calculations per frame (each requiring $M + 1$ multiply/add's) are reduced to $2 \times 10 = 20$ residual calculations. With respect to the binary search procedure, at 10 bits/frame, equal distortion for scalar quantization requires 20 additional bits/frame.

Although objective comparisons such as the rms spectral error or distortion can be very useful for comparing different speech processing systems, it is well known that subjective evaluation is ultimately most important. We therefore decided to perform at least some informal perceptual testing to compare vector and scalar quantization.

A 20 s test segment described earlier was processed resulting in 600 frames of "open-test" data. In addition, 600 frames of the reference set were processed to define "closed-test" data.

The open-test samples (indicated by closed circles) were processed with the full search vector quantizer for 6, 8, and 10 bits/frame. Scalar quantization was implemented for 28 bits/frame and 10 bits/frame. The former choice allowed us to compare the perceptual differences between the two methods for a constant distortion value. The latter choice allowed us to judge the perceptual differences between scalar and vector quantization for an identical number of bits/frame.

In addition, we obtained distortion values for the closed-test samples using the full search procedure (indicated by solid triangles) to provide a comparison with open-test samples. Numbers to the side of each dot or triangle indicate the specific synthesized speech samples.

Samples 1 and 2 are judged perceptually to be very similar, both having a distinct "warble-like" characteristic with high intelligibility.

It is important to note that the distortion level for this open-test was 2.6 dB, resulting in a scalar representation of about 28 bits/frame. In the present government ANDVT system, for example [20], 42 bits/frame are used, resulting in a distortion or average rms log spectral difference of around 1.0 dB.

When sample 1 is compared with sample 3, the perceptual differences are obvious and dramatic. For the same number of bits/frame the scalar quantizer causes an increased distortion to about 4.8 dB and a very obvious decrease in intelligibility.

It is interesting to compare the closed-test samples (6, 7, 8) with the open-test samples (1, 4, 5). In general, as bits/frame decrease, increased "warble" occurs (along with increased distortion as an objective measure). As one might expect, the closed-set samples are preferred perceptually. However, the differences are not great. We have experimentally determined that it is quite difficult to perceptually distinguish between samples whose distortion is less than about 0.5-0.7 dB even with formal *A-B* comparisons over tightly coupled headphones.

VI. Summary/Future Research

A procedure for vector quantization of speech has been developed and demonstrated to have a dramatic advantage over scalar quantization when compared on an rms log spectral measure basis. Informal listening has verified that equivalent rms log spectral measures are also, roughly speaking, perceptually equivalent. That is, at a specified dB level such as 1.8 dB, no marked preference occurs between a 35 bits/frame scalar quantized speech sample and a 10 bit full search vector quantized sample.

TABLE I
Possible Tree Structures Ranging from Binary Search to Full Search and Corresponding Number of Residual Energy Evaluations

Number of Branching Levels l	Number of Branches at Each Node n	Number of Residual Energy Evaluations nl
12	2	24 (binary search)
6	4	24
4	8	32
3	16	48
2	64	128
1	4096	4096 (full search)

Many future research areas should be considered. We will attempt to summarize here what we believe to be some of the most interesting and fruitful of these research areas. First, methods for reducing the differences in distortion between the binary search and full search method should be investigated. As the number of bits increases, the degradation of binary tree search over the exhaustive search in terms of spectral deviation was seen to increase. This is to be expected as one moves further down the tree branches.

One approach to improvement is to use trees that are not so deep as a binary tree with more branches at each node. In particular, for a case of 12 bits, one can think of the exhaustive search as a tree with only one branching node, with 2^{12} branches coming out, whereas the binary tree has a set of 12 branching levels with only two branches coming out of each node. One could easily use 6 branching levels with 4 branches at each node (as $2^{12} = 4^6$), or 4 branching levels with 8 branches at each node (as $2^{12} = 8^4$), etc.

If we let l be the number of branching levels and n be the number of branches at each level, the total number of levels will be

$$n^l = 2^b$$

where b is the number of bits in the code word. The total number of residual energy evaluations at each branching level will be n, and as there are l levels, the total number of residual energy evaluations will be nl. Table I illustrates the possibilities for a 12 bit codebook.

In this case it appears that by using $l = 6$ and $n = 4$, performance could be improved without increasing the number of residual energy evaluations.

A great deal of study needs to be done concerning the codebook generation and the local optimality of the results. In particular, there are a number of methods for initializing the codebook generation and carrying out the cell splitting. Different techniques should be compared so that both rate of convergence and size of resultant distortion can be compared.

Another topic of great concern in codebook generation is methods of treating the *empty cell* problem, where one attempts to split a cell with only one speech frame in it. In the fully optimal case, this can be handled by bypassing the split and instead using the extra code word for handling the worst speech frame in the data space (as one possibility). This approach fails for the binary tree, and at present it appears that one must either waste some of the possible tree branches by "pruning," or restart the algorithm with a different set of initial conditions. While the empty cell problem is rare, particularly when a sufficiently large database is used for a training sequence (at least 50 times the total number of cells), it can occur.

Other possible studies are listed below.

1) The use of a binary codebook as an initialization for the optimal codebook.

2) The use of a "standard" codebook, based on a variety of speakers, as an initialization for a single speaker or speaker group codebook. While this approach is straightforward for the optimal codebook, there are a number of problems in applying it to the binary tree searched codebook.

3) The use of short training sequences for initialization of the codebook.

4) The use of interframe memory to further reduce the bit rate. Simulations have shown that most code words are almost always followed by a code word from a small subset of the entire codebook. Thus, for example, a lower bit rate might be achieved by searching only a "conditional codebook" on the basis of the last code word sent and then sending the index of the code word from the smaller codebook.

5) If a low rate "side channel" is available, then the systems described here could be made adaptive by occasionally updating and improving the codebook on the basis of the data being compressed. The new, improved code words would be sent to the receiver at a much slower rate than the speech data.

6) Performance of a subjective study of the codebooks produced by the algorithm to determine whether they relate to physically perceivable "cluster" points of speech sounds such as voiced, unvoiced, male, and female. Such a connection might aid in improving design techniques for the binary search.

The suboptimal approach described here leads to a theoretical problem in that speech frames are all normalized so that frames of silence are treated as valid speech frames unless removed by preprocessing. Further study of this preprocessing and the theoretical interpretation of resultant spectral deviations is needed. The approach will probably attempt to use an elementary thresholding silence detector, which will artificially zero out the frames corresponding to silence.

While it may not be practically feasible for regular use, it is interesting to consider actually implementing the fully optimal system (including the gain term) to find a baseline for the spectral distortion and establish how much is lost in the suboptimal case where the gain is separated out. Although intuitively it seems that the effects of separating or not separating the gain term should be small, this conjecture should be verified or disproved. This topic is being actively studied at present and preliminary results are reported in [18].

Finally, a basic problem in using vector quantization for medium bit rates lies in the storage required and the size of the training sequence. For example, if we needed a 20 bit codebook to achieve a sufficiently low distortion, this would require the storage of $2^{20} = 1\,048\,576$ code words. In addition, if we required 20 frames for each code word to achieve any reasonable clustering for the reference set, this would require well over 20 million speech frames in a training sequence. The calculations required to produce this codebook would tax the capabilities of any computer and strain any researcher's computer budget.

To achieve some sort of compromise between vector and scalar quantization, some partitioning of the information in the filter description is needed. A number of possibilities exist conceptually, although many may not be practical. For example, one approach in analyzing speech would be to find the full precision model filter, use a root solver to factor it according to some rule, and finally separately code the factors for the filter. Such a procedure might be possible using two codebooks, perhaps of 12 and 11 bits, where the total number of individual code words is decreased while the total number of bits would probably have to increase. The basic shortcoming with this approach lies in the necessity of using a root finder for factoring the model filters.

A second approach, also suboptimal, is to replace the linear predictor inverse filtering operation with a cascade of two inverse filters: the first to remove the resonant information and the second for fine resolution spectral shaping. This would allow the two filters to be individually coded, again using fewer stored code words but probably more bits.

A third approach would apply a linear predictor on the basis of a cascade of second-order sections, such as described by Jackson, Rao, and Wood [19]. These filters are obtained through an iterative type of solution and would readily lead to the factoring described in the preceding section.

A forth approach would be to partition the information into frequency bands, where the input signal would be passed through a set of filters. The outputs of these filters would cover a frequency range where there could be no more than two resonances. Each of these outputs would then be applied to a linear predictor analysis of relatively low order, each model filter using a much smaller number of bits than the overall total. Some research has been published in this area under the name of "piecewise linear predictive coding" [22].

Yet another approach might be to separately evaluate the size of a codebook needed for voiced speech only and unvoiced speech only. At present all speech is mixed together and normalized so that all frames have the same weight, regardless of gain level. A previous study has shown that unvoiced frames can be represented with substantially fewer coefficients and bits/frame than voice speech frames [23].

Appendix: Derivations of Equations

We present here the derivations in abbreviated form of the equations in the main text. Earlier, more formal derivations are given in [10]. As the material of Section II-A consists only of definitions and referenced results, we start with Section II-B.

Equation (13) is identical to [1, eq. (6.10)]. Using the time domain relation for the residual energy (8), it can also be expressed in the form

$$d[|X|^2; |G|^2] = \frac{1}{\sigma^2} \sum_n r_x(n)\, r_a(n) + \ln\,[\sigma^2/\alpha_\infty] - 1. \quad \text{(A1)}$$

By direct substitution one can also write

$$d[|G_M|^2 ; |G|^2] = \frac{1}{\sigma^2} \sum_n r_M(n)\, r_a(n) + \ln [\sigma^2/\alpha_M] - 1 \tag{A2}$$

where we have used the property that

$$\int_{-\pi}^{\pi} \ln [|G_M|^2] \frac{d\theta}{2\pi} = \ln (\alpha_M) - \int_{-\pi}^{\pi} \ln [|A_M|^2] \frac{d\theta}{2\pi}$$

$$= \ln (\alpha_M) \tag{A3}$$

for the integral in the middle expression is zero [1, Sect. 6.2.1].

Equations (A1) and (A2) can be subtracted. Using the correlation matching property (9) the summations vanish, and one finds

$$d[|X|^2 ; |G|^2] - d[|G_M|^2 ; |G|^2] = \ln (\alpha_M/\alpha_\infty).$$

But from (11), $\ln (\alpha_m/\alpha_\infty)$ is the minimum distortion, $d[|X|^2 |G_M|^2]$, thus giving (14).

Equation (15) follows from a direct substitution. Starting with (13), for $G(z) = \sigma/A(z)$,

$$d[|X|^2 ; \sigma^2/|A|^2] = \alpha/\sigma^2 + \ln (\sigma^2/\alpha_\infty) - 1, \tag{A4}$$

so that by direct substitution

$$d[|X|^2 ; \alpha/|A|^2] = \ln (\alpha/\alpha_\infty). \tag{A5}$$

Then from the definition of (10), by substitution of the constants α and σ^2,

$$d[\alpha; \sigma^2] = \alpha/\sigma^2 - \ln (\alpha/\sigma^2) - 1. \tag{A6}$$

By inspection, adding the right-hand sides of (A5) and (A6), we obtain the right-hand side of (A4), hence (15) follows.

Equations (16) and (17) are definitions. Equation (18) follows from (8) and the even property of autocorrelation sequences, for

$$\alpha = \sum_{n=-\infty}^{\infty} r_x(n)\, r_a(n) = r_x(0)\, r_a(0) + 2 \sum_{n=1}^{\infty} r_x(n)\, r_a(n). \tag{A7}$$

As the filter coefficients are limited to $M + 1$ in number $\{a_0, a_1, \cdots, a_M\}$, the autocorrelation sequence $\{r_a(n)\}$ terminates at $n = M$:

$$r_a(n) = \begin{cases} \sum_{k=0}^{M-n} a_k a_{k+n} & \text{for } n = 0, 1, \cdots, M \\ 0 & \text{for } n > M, \end{cases} \tag{A8}$$

and thus (A8) is equivalent to (18).

Equation (19) follows from (A8), and (20) is the common formula for the short-term autocorrelation function for a truncated sequence.

For the centroid calculation of (21), we start with the definition of (16), rewriting the individual terms, $d[|X_k|^2, |G|^2]$ using superscripts of "k" for the "kth" frame residual energy and one-step prediction error. From (13), (4), and (6),

$$d[|X_k|^2 ; |G|^2] = \alpha^k/\sigma^2 + \ln (\sigma^2) - \ln (\alpha_\infty^k) - 1$$

$$= \frac{1}{\sigma^2} \int_{-\pi}^{\pi} |X_k|^2 |A|^2 \frac{d\theta}{2\pi} + \ln (\sigma^2)$$

$$- \int_{-\pi}^{\pi} \ln |X_k|^2 \frac{d\theta}{2\pi} - 1. \tag{A9}$$

If this expression is now summed from $k = 1$ to $k = L$, and the definition of the average $|\overline{X}|^2$ of (17) is used, then

$$D = \frac{L}{\sigma^2} \int_{-\pi}^{\pi} |\overline{X}|^2 |A|^2 \frac{d\theta}{2\pi} + L \ln (\sigma^2)$$

$$- \int_{-\pi}^{\pi} \sum_{k=1}^{L} \ln |X_k|^2 \frac{d\theta}{2\pi} - L. \tag{A10}$$

As in (A9), we also have the definition

$$d[|\overline{X}|^2 ; |G|^2] = \frac{1}{\sigma^2} \int_{-\pi}^{\pi} |\overline{X}|^2 |A|^2 \frac{d\theta}{2\pi} + \ln (\sigma^2)$$

$$- \int_{-\pi}^{\pi} \ln |\overline{X}|^2 \frac{d\theta}{2\pi} - 1,$$

which can be combined with (A10) directly to yield the result of (21), where the constant u is given by

$$u = L \int_{-\pi}^{\pi} \left[\ln |\overline{X}|^2 - \frac{1}{L} \sum_{k=1}^{L} \ln |X_k|^2 \right] \frac{d\theta}{2\pi}. \tag{A11}$$

The constant u is thus a measure of the "spread" within the cluster. The integrand in (A11) can also be looked at as the log of the ratio of the arithmetic mean of the separate spectra to their geometric mean. The logarithm is then averaged over θ.

Acknowledgment

The authors would like to express appreciation to Dr. D. Y. Wong of STI for his careful reading of the manuscript, F. Juang of STI for performing several of the experiments, G. Rebolledo of Stanford University, Stanford, CA, for helpful discussions, and Dr. D. Lee of Stanford for his assistance.

References

[1] J. D. Markel and A. G. Gray, Jr., *Linear Prediction of Speech*. New York: Springer-Verlag, 1976.

[2] J. Makhoul, "Linear prediction–A tutorial review," *Proc. IEEE*, vol. 63, pp. 561–580, Apr. 1975.

[3] F. Itakura and S. Saito, "On the optimum quantization of feature parameters in the PARCOR speech synthesizer," in *Proc. IEEE Conf. Speech Commun. Processing*, New York, 1972, pp. 434–437.

[4] R. Viswanathan and J. Makhoul, "Quantization properties of transmission parameters in linear predictive systems," *IEEE Trans. Acoust., Speech, Signal Processing*, vol. ASSP-23, pp. 309–321, June 1975.

[5] A. H. Gray, Jr. and J. D. Markel, "Quantization and bit allocation in speech processing," *IEEE Trans. Acoust., Speech, Signal Processing*, vol. ASSP-24, pp. 459–473, Dec. 1976.

[6] A. H. Gray, Jr., R. M. Gray, and J. D. Markel, "Comparison of optimal quantization of speech reflection coefficients," *IEEE Trans. Acoust., Speech, Signal Processing*, vol. ASSP-25, pp. 9–23, Feb. 1977.

[7] J. D. Markel and A. H. Gray, Jr., "Implementation and comparison of two transformed reflection coefficient scalar quantization methods," this issue, pp. 575–583.

[8] M. R. Sambur, "An efficient linear prediction vocoder," *Bell Syst. Tech. J.*, Dec. 1975.

[9] E. Blackman, R. Viswanathan, and J. Makhoul, "Variable-to-fixed rate conversion of narrow-band LPC speech," in *Conf. Rec., 1977 IEEE ICASSP Conf.*, ASSP 77CH1197-3, 1977, pp. 409–412.

[10] R. M. Gray, A. Buzo, A. H. Gray, Jr., and Y. Matsuyama, "Distortion measures for speech processing," *IEEE Trans. Acoust., Speech, Signal Processing*, vol. ASSP-28, pp. 367–376, Aug. 1980.

[11] F. Itakura and S. Saito, "Analysis synthesis telephone based upon the maximum likelihood method," in *Conf. Rec., 6th Int. Congr. Acoust.*, Y. Yonasi, Ed., Tokyo, Japan, 1968.

[12] Y. Matsuyama, A. Buzo, and R. M. Gray, "Spectral distortion measures for speech compression," Stanford University, Stanford, CA, ISL Rep. 6504-3, Apr. 1978.

[13] Y. Linde, A. Buzo, and R. M. Gray, "An algorithm for vector quantizer design," *IEEE Trans. Commun.*, vol. COM-28, pp. 84–95, Jan. 1980.

[14] L. A. Buzo de la Pena, "Optimal vector quantization for linear predictive coded speech," Ph.D. dissertation, Stanford University, Stanford, CA, Aug. 1978.

[15] S. P. Lloyd, "Least squares quantization in PCM," Bell Lab., Murray Hill, NJ, Tech. Rep., 1957.

[16] A. H. Gray, Jr. and J. D. Markel, "Distance measures for speech processing," *IEEE Trans. Acoust., Speech, Signal Processing*, vol. ASSP-24, pp. 380–391, Oct. 1976.

[17] J. D. Markel and S. B. Davis, "Text-independent speaker recognition from a large linguistically unconstrained time-spaced data base," *IEEE Trans. Acoust., Speech, Signal Processing*, vol. ASSP-27, pp. 74–82, Feb. 1979.

[18] R. M. Gray, A. H. Gray, Jr., G. Rebolledo, and J. E. Shore, "Rate distortion speech coding with a minimum discrimination information distortion measure," submitted to *IEEE Trans. Inform. Theory*.

[19] L. Jackson, R. M. Rao, and S. L. Wood, "Parameter estimation by linear prediction in cascade form," in *Conf. Rec., 1977 IEEE Conf. Acoust., Speech, Signal Processing*, ASSP 77CH1197-3, 1977, pp. 727–731.

[20] G. S. Kang, L. J. Fransen, and E. L. Kline, "Multirate processor (MRP) for digital voice communications," Naval Res. Lab., Washington, DC, NRL Rep. 8295, pp. 76–80, Mar. 1979.

[21] R. M. Gray, J. C. Kieffer, and Y. Linde, "Locally optimal block quantizer design," *Inform. Contr.*, to be published.

[22] J. E. Roberts and R. H. Wiggins, "Piecewise linear predictive coding (PLPC)," in *Proc. 1976 IEEE Int. Conf. Acoust., Speech, Signal Processing*, IEEE Cat. No. 76CH1067-8 ASSP, pp. 470–473, Apr. 1976.

[23] D. Y. Wong and J. D. Markel, "An intelligibility evaluation of several linear prediction vocoder modifications," *IEEE Trans. Acoust., Speech, Signal Processing*, vol. ASSP-26, pp. 424–435, Oct. 1978.

Andrés Buzo (S'76–M'78), for a photograph and biography, see p. 376 of the August 1980 issue of this TRANSACTIONS.

Augustine H. Gray, Jr. (S'56–M'65–SM'79), for a photograph and biography, see p. 376 of the August 1980 issue of this TRANSACTIONS.

Robert M. Gray (S'68–M'69–SM'77–F'80), for a photograph and biography, see p. 376 of the August 1980 issue of this TRANSACTIONS.

Reprinted from *Appl. Optics* **13**(10):2274–2279 (1974)

Magnitude-Coupled Phase Quantization

W. J. Dallas

A method is described for minimizing Fourier-domain phase-quantization noise in the image reconstructed from a computer-generated hologram. This method uses manipulation of the hologram magnitude to counteract the deleterious effects of phase quantization.

I. Introduction

The binary computer hologram is a device capable of producing an arbitrary optical wavefront, for a price. The more complicated and exactly implemented the wavefront, the higher the price. The complexity of the wavefront is measured by its space-bandwidth product,[1] that is, by the number of sample values necessary to specify it. For each sample the computer hologram must have one resolution cell.

The exactitude of implementation is measured by the number of complex amplitude values attainable in each resolution cell. When one uses a discreetly addressed plotter, for example, an incremental drum plotter, the number of complex amplitudes is determined by the number of addressable points in the cell. The more addressable points, the more exactly the wavefront is implemented, but also the higher the price.

We will demonstrate that the cost–fidelity trade can be made more favorable by modifying the usual relation between the discrete realizable complex amplitudes and the continuously variable desired wavefront. This modification amounts to changing the usual partitioning of the unit disk into quantization bins (Fig. 1). The modified binning procedure makes use of the magnitude (the absolute value of the complex amplitude) to counteract the effects of phase quantization. We will develop the method, assuming freedom to continuously vary the magnitude, but we will later consider the effects of magnitude quantization on the utility of the method.

For simplicity, we will consider a one-dimensional Fourier hologram. The results can be easily extended to two dimensions. The extension of the results to non-Fourier holograms can be done in a manner similar to that presented in Ref. 2.

The author is with the Physikalisches Institut der Universität Erlangen-Nürnberg, D-8520 Erlangen, Germany.
Received 12 March 1974.

II. Magnitude Coupling—An Example

We begin by looking at two different ways of quantizing the phase into two levels. We let

$$G(\nu) = A(\nu)e^{i\phi(\nu)} \quad \text{the prequantization transform,} \tag{1}$$

$$\hat{G}(\nu) = \hat{A}(\nu)e^{i\hat{\phi}(\nu)} \quad \text{the postquantization transform,} \tag{2}$$

$$g(x) = \int G(\nu)e^{2\pi i\nu x}d\nu \quad \text{the image reconstructed by } G(\nu), \tag{3}$$

$$\hat{g}(x) = \int \hat{G}(\nu)e^{2\pi i\nu x}d\nu \quad \text{the image reconstructed by } \hat{G}(\nu), \tag{4}$$

$$0 \leq \phi(\nu) < 2\pi. \tag{5}$$

Consider first conventional two-level phase quantization as illustrated in Fig. 2. We have

$$\hat{\phi}(\nu) = \begin{cases} \pi & \text{if } \pi/2 \leq \phi < 3\pi/2, \\ 0 & \text{otherwise,} \end{cases} \tag{6}$$

$$\hat{A}(\nu) = A(\nu). \tag{7}$$

Next, consider the taking of the real part, as illustrated in Fig. 2. We have here

$$\hat{\phi}(\nu) = \begin{cases} \pi & \text{if } \pi/2 \leq \phi < 3\pi/2, \\ 0 & \text{otherwise,} \end{cases} \tag{8}$$

$$\hat{A}(\nu) = A(\nu) \cos[\hat{\phi}(\nu) - \phi(\nu)]. \tag{9}$$

We have exactly the same relationship between the post- and prequantization phases. however, the postquantization magnitude now depends upon the post- and prequantization phases as well as the prequantization magnitude.

This dependence is of the form

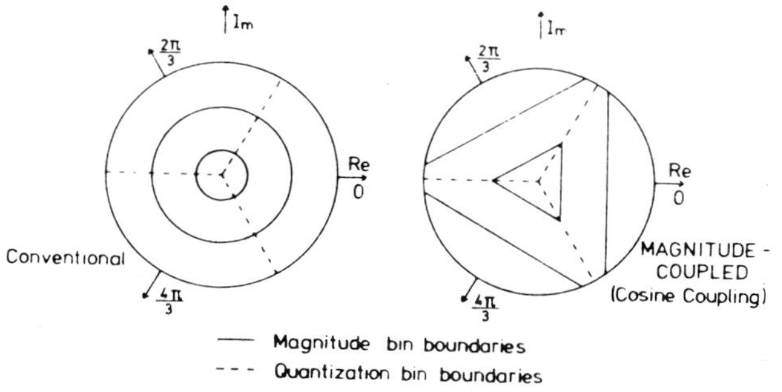

Fig. 1. Bin structure comparison.

$$\hat{A} = A \cdot W(\hat{\phi} - \phi), \qquad (10)$$

where W is a real, symmetric, positive semidefinite function we term the coupling. (Note that in Eq. (9), W is positive semidefinite because the argument $(\hat{\phi} - \phi)$ is between $(\pi/2)$ and $\pi/2$.) For conventional phase quantization, $W = 1$. For taking the real part, $W = \cos(\hat{\phi} - \phi)$.

From Fig. 2 we see that for taking the real part, the prequantization phasor is projected into the postquantization direction. This is in contrast to being pivoted into the postquantization direction as in conventional phase quantization. This projection procedure can be generalized to any number of quantization levels, and as we show in the next section, this generalization is a useful one.

III. Best Coupling

In this section we will determine the coupling W that gives the most faithful reconstruction for a given number of quantization levels. We use a mean-square measure of the reconstruction error that is appropriate when both the reconstructed magnitude and phase are of interest. Let

$$E = \int |\hat{g}(x) - g(x)|^2 dx \qquad (11)$$

be the reconstructed image error. See Ref. 3. By Parseval's theorem

$$E = \int |\hat{G}(\nu) - G(\nu)|^2 d\nu. \qquad (12)$$

We wish to minimize E, and we can do this by minimizing the integrand

$$\xi = |\hat{G}(\nu) - G(\nu)|^2. \qquad (13)$$

But

$$\xi = \hat{A}^2 + A^2 - 2\hat{A}A\cos(\hat{\phi} - \phi), \qquad (14)$$

and

$$\partial\xi/\partial\hat{A} = 2\hat{A} - 2A\cos(\hat{\phi} - \phi), \qquad (15)$$

$$\partial^2\xi/\partial\hat{A}^2 = 2 > 0,$$

so we minimize E by choosing

$$\hat{A} = A\cos(\hat{\phi} - \phi). \qquad (16)$$

Comparing Eq. (16) to Eq. (10), we see that the coupling that minimizes the reconstructed image error for a given number of quantization levels is

$$W = \cos(\hat{\phi} - \phi). \qquad (17)$$

This we term cosine coupling or cosine-coupled phase quantization. In Sec. V we will compare the reconstructed image signal to the noise ratio for holograms with conventional phase quantization, and holograms with cosine-coupled phase quantization. First, however, we must develop some mathematical tools. This we do in the next section.

IV. False Image Decomposition

In this section we will generalize the Goodman and Silvestri[4] theory of phase quantization to include magnitude couplings of the form in Eq. (10).

In general, the postquantization image $\hat{g}(x)$ is related to the prequantization image $g(x)$ by

$$\hat{g}(x) = \beta g(x) + \eta(x), \qquad (18)$$

where β is a constant and $\eta(x)$ is a function representing the quantization noise. Goodman and Silvestri demonstrated that for conventional phase quantization, the noise function could be viewed as a superposition of images, termed false images because they are not the desired or primary image $g(x)$. The authors then wrote $\hat{g}(x)$ as a single superposition. For N-level quantization

$$\hat{g}(x) = \sum_m C_m g_m(x), \qquad (19)$$

$$g_m(x) = \int A(\nu)\exp[i(mN + 1)\phi(\nu)]\exp(2\pi i\nu x)d\nu, \qquad (20)$$

$$C_m = \mathrm{sinc}[m + (1/N)]$$
$$= \mathrm{sinc}(1/N)[(-1)^m/(mN + 1)], \qquad (21)$$

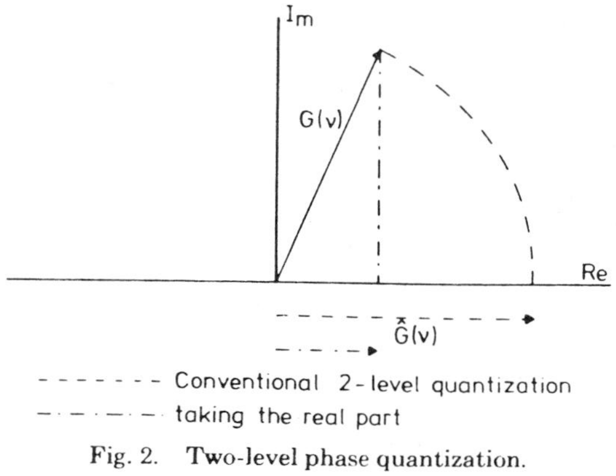

Fig. 2. Two-level phase quantization.

$$g_0(x) = g(x);\ \mathrm{sinc}(y) = \sin(\pi y)/\pi y, \tag{22}$$

$$g_n(x) = \text{false image for } n \neq 0.$$

We will show that magnitude-coupled phase quantization can be described by Eq. (19) where the images are the same as in Eq. (20); only their strengths are altered. We will then be in a position to calculate signal-to-noise ratios of various couplings, and so compare their merits. We are concerned with N-equidistant-level Fourier-domain phase quantization where

$$\hat{\phi} = 2\pi k/N, \tag{23}$$

if

$$(k - 1/2)/N \leq \phi/2\pi < (k + 1/2)/N, \tag{24}$$

$$k = 0, 1, \ldots, N - 1. \tag{25}$$

We see here that

$$-(\pi/N) \leq \hat{\phi} - \phi < \pi/N, \tag{26}$$

so that W need be defined only within this range and we may continue W in any form outside this range. We will periodically continue, at a period 2π, in order that the final results appear in a convenient form. The periodic continuation allows us to represent W by a Fourier series.

$$W(\hat{\phi} - \phi) = \sum_n B_n \exp[in(\hat{\phi} - \phi)], \tag{27}$$

$$B_n = \frac{1}{2\pi} \int_{-\pi}^{\pi} W(\zeta) \exp(-in\zeta) d\zeta, \tag{28}$$

$$B_n = B_{-n}. \tag{29}$$

Equation (29) is true because W is a real, symmetric function. Substituting Eq. (27) into Eq. (10) yields

$$\hat{A} \exp(i\hat{\phi}) = A \sum_n B_n \exp\{i[n\phi + (1 - n)\hat{\phi}]\}. \tag{30}$$

By methods similar to those used in Ref. 5, or by others detailed in Ref. 6, we can show that

$$\exp[i(1 - n)\hat{\phi}] = \sum_{m=-\infty}^{\infty} \mathrm{sinc}[m + (1 - n)/N] \exp[i(mN + 1)\phi], \tag{31}$$

so that

$$\hat{A} \exp(i\hat{\phi}) = A \sum_m \sum_n B_n \,\mathrm{sinc}[m + \{(1 - n)/N\}] \exp[i(mN + 1)\phi]. \tag{32}$$

Fourier-transforming Eq. (32) gives

$$\hat{g}(x) = \sum_m C_m g_m(x), \tag{33}$$

$$g_m(x) = \int A(\nu) \exp[i(mN + 1)\phi(\nu)] \exp(2\pi i\nu x) d\nu, \tag{34}$$

$$C_m = \sum_n B_n \,\mathrm{sinc}[m + (1 - n)/N]. \tag{35}$$

The coefficients C_m of Eq. (21), for conventional phase quantization, are seen to be special cases of Eq. (35) when

$$W(\hat{\phi} - \phi) = 1, \tag{36}$$

so that

$$B_n = \delta_{on} \tag{37}$$

V. Signal-to-Noise Ratio Comparison

Using the formalism of Sec. IV, we will in this section compare the reconstructed image signal-to-noise ratio for conventional phase quantization to that for cosine-coupled phase quantization. The signal-to-noise measure we choose is the ratio of the integrated power of the primary image to that of the false image sum; that is,

$$\mathrm{SNR} = \left[\int |c_0 g_0(x)|^2 dx\right] \Big/ \left[\int \Big| \sum_{m\neq 0} c_m g_m(x)\Big|^2 dx\right]. \tag{38}$$

In Eq. (38), if the $g_m(x)$ do not overlap (see Ref. 4), or if they have very little correlation (for instance, if a strong diffuser was imposed on the object),

$$\int g_m(x) g_n^*(x) dx \cong 0 \quad \text{for } m \neq n. \tag{39}$$

So we may neglect all cross terms in the denominator of Eq. (38). The denominator then becomes

$$\sum_{m\neq 0} |C_m|^2 \int |g_m(x)|^2 dx. \tag{40}$$

Now a careful look at Eq. (34) together with Parseval's theorem shows that

$$\int |g_n(x)|^2 dx = \int |G(\nu)|^2 d\nu \quad \text{for all } n, \tag{41}$$

and Eq. (38) becomes

$$\mathrm{SNR} = |C_0|^2 / \sum_{m\neq 0} |C_m|^2. \tag{42}$$

We now calculate the SNR first for conventional phase quantization, then for cosine-coupled phase quantization. We will take the ratio of these two SNR's as a measure of the relative merits of the two schemes.

A relation that will be useful for summing the denominator of Eq. (42) is that

$$\sum_{m=-\infty}^{\infty} \mathrm{sinc}(m - a)\,\mathrm{sinc}(m - b) = \mathrm{sinc}(a - b). \tag{43}$$

For conventional phase quantization,

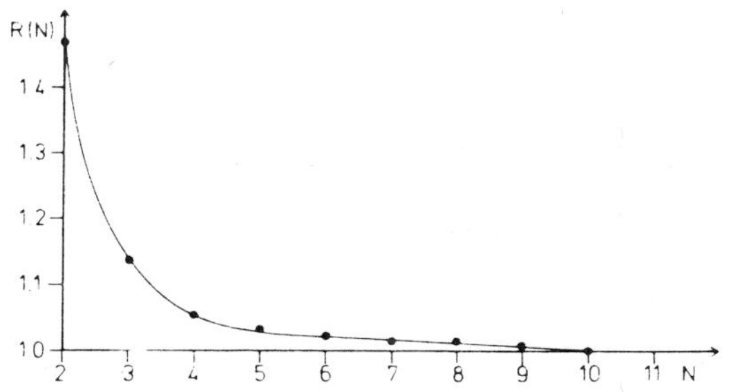

Fig. 3. Relative signal-to-noise ratio.

$$C_m = \text{sinc}(m + 1/N), \quad (44)$$

so that Eq. (43) yields

$$\sum_{m=-\infty}^{\infty} |C_m|^2 = 1, \quad (45)$$

and

$$\text{SNR1} = \text{sinc}^2(1/N)/[1 - \text{sinc}^2(1/N)]. \quad (46)$$

For cosine-coupled phase quantization, Eq. (35) gives

$$C_m = (\delta_{0m}/2) + \frac{1}{2}\,\text{sinc}[m + (2/N)]. \quad (47)$$

So here,

$$\sum_{m=-\infty}^{\infty} |C_m|^2 = \frac{1}{2}[1 + \text{sinc}(2/N)], \quad (48)$$

and

$$\text{SNR2} = [1 + \text{sinc}(2/N)]/[1 - \text{sinc}(2/N)]. \quad (49)$$

Our measure of relative merit, then, is

$$R(N) = \text{SNR2}/\text{SNR1} = [1 + \text{sinc}(2/N)]/[1 - \text{sinc}(2/N)]\cdot[1 - \text{sinc}^2(1/N)]/[\text{sinc}^2(1/N)], \quad (50)$$

where $R(N) > 1$ indicates a higher signal-to-noise ratio for cosine-coupling. In Fig. 3, $R(N)$ is plotted against the number of quantization levels N. Note that for $N = 2,3,4$, cosine-coupling shows a considerable superiority. As N increases, $R(N)$ approaches unity from above.

VI. Magnitude Coupling and Magnitude Quantization

Up to this point we have assumed a continuously variable magnitude. However, in most applications where the phase is quantized, the magnitude will also be quantized. The computer-generated binary hologram follows this rule. The question we address in this section is: How much does our choice to use magnitude-coupled phase quantization rather than conventional phase quantization influence the final quantized magnitude?

We will first consider an arbitrary magnitude coupling and an arbitrary magnitude binning scheme. We will then in Sec. VII apply our results to a particular computer-generated binary hologram with cosine-coupling.

We will develop both measures of the utility of magnitude coupling, considering magnitude quantization and a measure of the futility of applying a magnitude coupling.

By utility we mean, on the average, how much do the final quantized magnitudes differ with magnitude coupling as opposed to not using such coupling? This measure should be affected by both the magnitude of the difference and frequency, i.e., how often the difference occurs. For M magnitude-quantization levels, our measure of utility is

$$U^2(N) = \sum_{m=0}^{M-1}\sum_{q=0}^{M-1} (A_m - \hat{A}_q)^2 P_{mq}, \quad (51)$$

where P_{mq} is the joint probability that A falls in the mth quantization bin and $\hat{A}$ falls in the qth quantization bin. $A_m, \hat{A}_q$ are the corresponding quantized magnitude values.

The futility of using magnitude quantization tells how often the final quantized magnitudes are the same whether or not we use a magnitude coupling. Our futility measure is

$$F(N) = \sum_{m=0}^{M-1} P_{mm}. \quad (52)$$

To calculate the utility $U(N)$ or the futility $F(N)$, we need the joint probabilities P_{mq}. We assume

$$\hat{A} = AW(\hat{\phi} - \phi),$$

$$W(\theta) \leq 1,$$

$$W(\theta) = W^*(\theta) = W(-\theta).$$

Furthermore, let

a_m, b_m = lower and upper boundaries of mth-magnitude quantization bin,

$p(A,\phi)$ = joint probability density of A,ϕ, (53)

$$H(\zeta) = \begin{cases} 0 & \text{if } \zeta < 0, \\ 1/2 & \text{if } \zeta = 0, \text{ i.e., unit step function,} \\ 1 & \text{if } \zeta > 0. \end{cases}$$

The joint probability we seek is then

$$P_{mq} = \int_0^{2\pi} d\phi \int_{a_m}^{b_m} H(\hat{A} - a_q)H(b_q - \hat{A})p(A,\phi)dA. \quad (54)$$

To put P_{mq} in a form more amenable to direct calculation, we first note that (remembering $P_{ab} = 0$ for $b > a$)

$$P_{mq} = \sum_{\alpha=m}^{q}\sum_{\beta=m}^{q} P_{\alpha\beta} - \sum_{\alpha=m+1}^{q}\sum_{\beta=m+1}^{q} P_{\alpha\beta} + \sum_{\alpha=m+1}^{q-1}\sum_{\beta=m+1}^{q-1} P_{\alpha\beta} - \sum_{\alpha=m}^{q-1}\sum_{\beta=m}^{q-1} P_{\alpha\beta}. \quad (55)$$

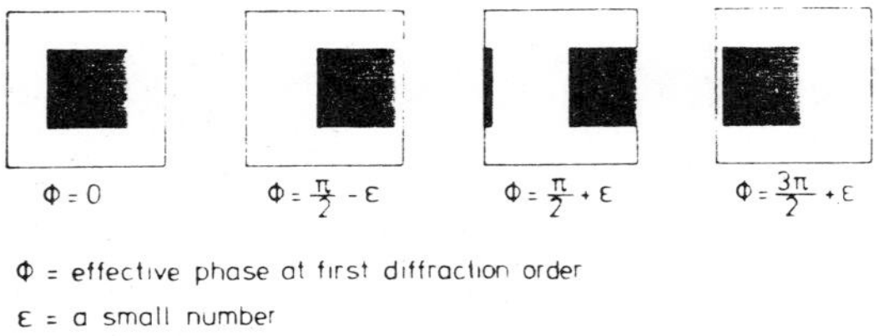

Fig. 4. Resolution cell internal structure.

Next we define

$$T(\sigma, \tau) = \sum_{k=0}^{N-1} \int_{(2\pi k/N)-t}^{(2\pi k/N)+t} d\phi \int_{\sigma/W[(2\pi k/N)-\phi]}^{\tau} p(A, \phi)dA, \quad (56)$$

where

$$t = \text{mino}[\pi/N, W_{\text{inverse}}(\sigma/\tau)], \quad (57)$$

$$\text{mino}[a, b] = aH(b - a) + bH(a - b).$$

With this definition one can show that

$$\sum_{\alpha=n}^{r} \sum_{\beta=n}^{r} P_{\alpha\beta} = T(a_n, b_r), \quad (58)$$

so that

$$P_{mq} = T(a_m, b_q) - T(a_{m+1}, b_q) + T(a_{m+1}, b_{q-1}) - T(a_m, b_{q-1}). \quad (59)$$

To proceed further, we must assume a form for the joint probability density $p(A, \phi)$. We assume A and ϕ independent and ϕ uniformly distributed. We assume a power series form for $p(A)$ so that

$$p(A, \phi) = p(A)p(\phi) = \frac{1}{2\pi} \sum_{r=0}^{\infty} D_r A^r, \sum_{r=0}^{\infty} \frac{D_r}{r+1} = 1 \quad (60)$$

for normalization. With this probability density, we have

$$T(\sigma, \tau) = \frac{N}{\pi} \sum_{r=0}^{\infty} \frac{D_r}{r+1} [t\tau^{r+1} - K_{r+1}(t)\sigma^{r+1}], \quad (61)$$

$$K_{r+1}(t) = \int_0^t \frac{d\theta}{W^{r+1}(\theta)},$$

where t is as in Eq. (57).

VII. Cosine Coupling and the Binary Hologram

We now consider the binary computer-generated hologram we use most often, a slight modification of the Lohmann type III[7] hologram. The internal resolution cell structure is shown in Fig. 4. In reality, this cell consists of $N \times N$ addressable points, so there are N possible values of the phase and M possible levels of magnitude quantization. Because we plot rectangles vertically centered, M is restricted to

$$M = 1 + N/2 \quad \text{for } N \text{ even} \quad (62)$$

$$M = (N + 3)/2 \quad \text{for } N \text{ odd}$$

We also restrict our consideration here to cosine-coupled phase quantization, so,

$$W(\theta) = \cos(\theta). \quad (63)$$

Recall that

$$F(N) = \sum_{m=0}^{M-1} P_{mm} = \sum_{m=0}^{M-1} T(a_m, b_m), \quad (64)$$

giving

$$F(N) = \frac{N}{\pi} \sum_{r=0}^{\infty} \frac{D_r}{r+1} \sum_{m=0}^{M-1} [t_m b_m{}^{r+1} - K_{r+1}(t_m) a_m{}^{r+1}], \quad (65)$$

$$K_{r+1}(t_m) = \int_0^{t_m} \frac{d\theta}{\cos^{r+1}(\theta)}; \; t_m = \text{mino}[\pi/N, \arccos(a_m/b_m)].$$

It remains only to specify the D_r, b_m, and a_m. We take two probability densities. We first consider a uniform distribution of magnitudes.

$$P_1(A) = 1; \; D_q = \delta_{q0}, \quad (66)$$

and then we consider a uniform distribution of complex amplitudes on the unit disk

$$P_2(A) = 2A; \; D_q = 2\delta_{q1}. \quad (67)$$

A comparison of the two indicates the dependence of $F(N)$ on concentration of magnitudes toward zero.

The a_m and b_m are determined by the hologram resolution cell substructure. Our substructure gives a magnitude quantization as illustrated in Fig. 5. Note especially that this quantization differs for odd and even N. For odd N the bins are not of equal size.

For N even,

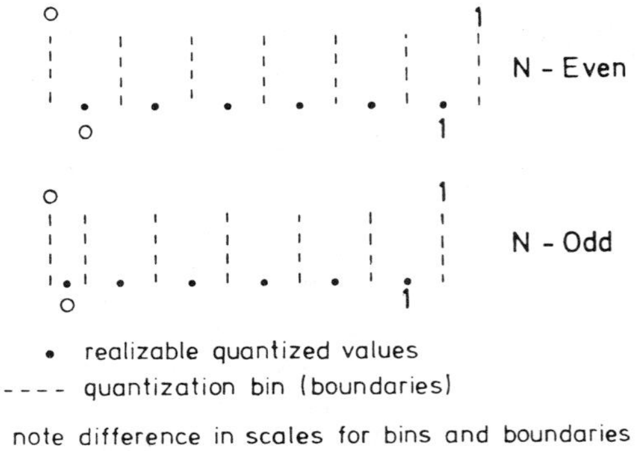

Fig. 5. Magnitude quantization.

N	F(N) for $P_1(A)$	F(N) for $P_2(A)$
2	.731	.503
3	.802	.770
4	.878	.848
5	.910	.898
6	.926	.910
7	.940	.930
8	.947	.934
9	.954	.945
10	.958	.948

Fig. 6. Futility.

$$M = (N/2) + 1,$$

$$a_m = m/M \quad 0 \le m \le M - 1, \tag{68}$$

$$b_m = (m + 1)/M \quad 0 \le m \le M - 1.$$

For N odd

$$a_0 = 0;\ a_1 = 2/(4M - 3);\ a_m = (4m - 3)/(4M - 3) \quad 2 \le m \le M - 1, \tag{69}$$

$$b_0 = 2/(4M - 3);\ b_m = (4m + 1)/(4M - 3) \quad 1 \le m \le M - 1.$$

We will not show the details of the calculations here. We list, instead, salient points.

$$\text{For } N = 2,\ t_0 = \pi/2;\ t_1 = \pi/3,$$

$$\text{for } N = 3,\ t_0 = \pi/3;\ t_1 = \pi/3;\ t_2 = \arccos(5/9), \tag{70}$$

$$\text{for } N \ge 4,\ t_m = \pi/N.$$

$$K_1(\pi/N) = \operatorname{arccosh}[\sec(\pi/N)] = \ln\left(\tan\{\pi/4[1 + (2/N)]\}\right), \tag{71}$$

$$K_2(\pi/N) = \tan(\pi/N).$$

For $N > 2$, sums over m in Eq. (65) are essentially the sums over first $M - 1$ integers for $P_1(A)$, and sums over the squares of the first $M - 1$ integers for $P_2(A)$.

Results tabulated in Fig. 6 show, for small N, a pronounced difference between cosine-coupled and conventional phase quantization. This difference is greater for $P_2(A)$ than for $P_1(A)$, showing that most of the differences between cosine couple and conventional phase quantization occur at higher amplitudes. This effect could be expected from the fact that

$$(A - \hat{A}) \propto A. \tag{72}$$

VIII. Summary

When a binary computer-generated hologram is produced, there exists a cost–fidelity trade. This trade is between the number of quantization levels and the acceptable level of quantization noise in the image. We have presented an alternative to the conventional method of quantizing phase. Our procedure uses changes in the magnitude to counteract the deleterious effects of phase quantization. We term this procedure magnitude-coupled phase quantization.

We have proved the existence of, and explicitly exhibited, a best magnitude coupling that we call cosine coupling. After generalizing the Goodman and Silvestri theory of phase quantization, we used that generalization to compare the reconstructed image signal-to-noise ratio of cosine-coupled and conventional phase-quantized Fourier holograms. For small numbers of quantization levels, cosine coupling showed a clear gain.

We then addressed the question of whether magnitude quantization negated the advantages of magnitude coupling. We developed measures of the utility of magnitude coupling in the face of magnitude quantizing and the futility of using magnitude coupling under the same circumstances. Next, we used our measure of futility (how often are the final quantized magnitudes the same whether or not we use magnitude coupling) to show that it is not futile to use cosine coupling for a binary computer-generated hologram if the number of phase quantization levels is small.

IX. Closing Remarks

We do not intend to present magnitude-coupled phase quantization as a panacea. It is a technique of refinement that is useful when one is operating near the cost–futility border. In this context, magnitude-coupled phase quantization is not in any way limited to computer holography. It should prove useful whenever quantized Fourier-domain information is stored or transmitted.

The author would like to thank Adolf Lohmann for his many helpful suggestions on the presentation of this subject.

References

1. A. W. Lohmann, IBM Research Paper RJ-438 (1967).
2. W. J. Dallas and A. W. Lohmann, Appl. Opt. **11,** 192 (1972).
3. N. C. Gallagher and B. Liu, Appl. Opt. **12,** 2328 (1973).
4. J. W. Goodman and A. M. Silvestri, IBM J. Res. Dev. **14,** 478 (1970).
5. N. M. Blachman, IEEE Trans. Info. Theory **IT-10,** 162 (1964). C. W. Helstrom, IEEE Trans. Circuit Theory **CT-12,** 489 (1965). W. J. Dallas, Appl. Opt. **10,** 674 (1971).
6. W. J. Dallas, "Computer Holograms Improving the Breed"; doctoral dissertation obtainable from University Microfilms, Ann Arbor, Michigan.
7. B. R. Brown and A. W. Lohmann, Appl. Opt. **5,** 967 (1966).

Reprinted from *IEEE Trans. Inform. Theory* **IT-11**(2):293–294 (1965)

OPTIMUM MEAN-SQUARE QUANTIZATION OF A NOISY INPUT

Terrence Fine

In this correspondence we consider the problem of the optimum mean-square quantization of a random variable Σ related to a desired random variable I through a joint probability density function $p(I, \Sigma)$; Σ might, for example, be the sum of I and independent noise N. By the quantization of the real random variable Σ we first mean the exhaustive subdivision of the real line into K sets $S_1, \cdots S_K$, not necessarily each of the form of a single interval, and second the optimum reconstruction of the value of the desired input I given only $p(I, \Sigma)$ and the information as to which one of the K possible quantization sets contains Σ. The reconstruction function that assigns a number to each of the sets $S_1, \cdots S_K$ is denoted by $R(S_j)$. The performance criterion we adopt is of the mean-square form $E[(I - R)^2]$. Our approach will provide conditions for an optimum solution and in particular will lead to a generally better solution than might be obtained by first attempting to estimate I from Σ and then quantizing and reconstructing the estimate. The proof of this superiority is based on the generality of the class of quantizers, containing as it does the possibility just mentioned, in which we seek for the optimum. A similar quantization problem has also been considered by Bluestein[1] although he adopts what appears to be a suboptimal formulation.

The procedure we employ to determine the optimum quantizer is similar to that previously discussed by the author[2] in a somewhat different connection; the earlier paper did not allow for noisy inputs, although it was more general in other respects. The design of the optimum quantizer proceeds in three steps that are analogous to those employed in extrema problems of the calculus of two variables. 1) For a given reconstruction function $R(S_j)$ the best quantization sets $S_1, \cdots, S_K$ are found. 2) For a given group of quantization sets $S_1, \cdots, S_K$ the best reconstruction function $R(S_j)$ is found. 3) The results of steps 1) and 2) are solved simultaneously for the jointly optimum quantizer. The design conditions we obtain in this manner are necessary and sufficient conditions for the solution of the problems posed in steps 1) and 2), and therefore they are necessary conditions that a solution to 3) must satisfy.

We start by assuming that the function $R(S_j)$ is given, the value of the random variable Σ has been observed, and we wish to determine to which set Σ should belong. Alternatively stated, we wish to select that set S_j such that the expected mean-square difference between the desired input I and $R(S_j)$, given the observed noisy input Σ, is less than or equal to the equivalent expression for any other reconstruction value $R(S_r)$. Thus the point Σ is assigned to the set S_j if, and only if, a comparison of conditional expectations for each $r(= 1, \cdots, K)$ yields

$$E[(I - R(S_j))^2 \mid \Sigma] \leq E[(I - R(S_r))^2 \mid \Sigma] \tag{1}$$

Observe that $R(S_i)$ is just a real number and not a random variable in (1). Expanding (1) and rearranging, we obtain

$$R^2(S_j) - R^2(S_r) \leq 2[R(S_j) - R(S_r)]E[I \mid \Sigma] \tag{2}$$

The set of K equations given by (2) can be reduced by defining S_{x_i} to be the set of values for which $R(S_j)$ is greater than $R(S_{x_i})$ and S_{y_i} to be the set of values for which $R(S_j)$ is less than $R(S_{y_i})$. It is then easy to see that the set of Σ values which are optimally quantized into the set S_j is given by

$$S_j = \left\{\Sigma : \max_{x_i} \frac{R(S_j) + R(S_{x_i})}{2} \leq E(I \mid \Sigma) \leq \min_{y_i} \frac{R(S_j) + R(S_{y_i})}{2}\right\} \tag{3}$$

Equation (3) informs us that the optimum quantization sets $S_1, \cdots, S_K$ are sets of values of inputs Σ such that the function of Σ given by $E[I|\Sigma]$ lies between two points. If, without loss of generality, we order the reconstruction outputs $R(S_j)$ in increasing order with respect to the index j then we see that (3) can be rewritten as

$$S_j = \left\{\Sigma : \frac{R(S_j) + R(S_{j-1})}{2} \leq E[I \mid \Sigma] \leq \frac{R(S_j) + R(S_{j+1})}{2}\right\} \tag{3a}$$

where we define $R(S_0)$ to be minus infinity and $R(S_{K+1})$ to be plus infinity.

The sets S_j given by (3a) are in general not single intervals, and this implies that in general the common staircase quantizer is not the optimum mean-square quantizer. However, if $E[I|\Sigma]$ is a strictly monotonic function of Σ and, therefore, has an inverse everywhere, then (3a) reduces to

$$S_j = \{\Sigma : L_j \leq \Sigma \leq L_{j+1}\} \tag{4}$$

where the endpoints L_j and L_{j+1} are obtained by applying the inverse of $E(I|\Sigma)$ to the endpoints in (3a). A particular instance in which (4) is applicable is to the case for which the actual input Σ and the desired input I are bivariate normal. In this case $E[I|\Sigma]$ is just a linear function of Σ, and staircase quantization is optimum.

The second step of the optimization procedure requires that we assume sets $S_1, \cdots, S_K$ and find the best values for the reconstruction function $R(S_j)$. We wish to select that number $R(S_j)$ that minimizes the expected quadratic difference between I and $R(S_j)$ conditional upon the information that Σ is in the known set S_j. It is well known that the number $R(S_j)$ that minimizes $E[(I - R(S_j))^2|\Sigma \varepsilon S_j]$ is given by the conditional mean of I,

$$R(S_j) = E[I \mid \Sigma \varepsilon S_j] \tag{5}$$

Equation (5) may be derived by replacing $R(S_j)$ in

$$E[(I - R(S_j))^2 \mid \Sigma \varepsilon S_j]$$

by $E[I|\Sigma \varepsilon S_j] - \delta(S_j)$, without loss of generality. Expansion of the conditional expectation directly yields that it equals

$$E[(I - E[I \mid \Sigma \varepsilon S_j])^2 \mid \Sigma \varepsilon S_j] + \delta^2(S_j)$$

and can be minimized by taking δ equal to zero. Equation (5) can be evaluated by expressing the conditional expectation in terms of integrations with respect to the joint density $p(I, \Sigma)$ as follows

Manuscript received June 1, 1964.

[1] Bluestein, L., Asymptotically optimum quantizers and optimum analog to digital converters for continuous signals, *IEEE Trans. on Information Theory (Correspondence)*, vol 10, Jul 1964, pp 242–246.

[2] Fine, T., Properties of an optimum digital system and applications, *IEEE Trans. on Information Theory*, vol 10, Oct 1964, pp 287–296.

$$R(S_i) = \frac{\int_{-\infty}^{\infty} I\, dI \int_{S_i} d\Sigma p(I, \Sigma)}{\int_{-\infty}^{\infty} dI \int_{S_i} d\Sigma p(I, \Sigma)}. \tag{6}$$

Finally the conclusions of (3a) and (6) must be solved simultaneously to find the jointly optimum solution. We observe that (3a) is a necessary and sufficient condition that S_j be optimum for a given $R(S_i)$ and (6) is a necessary and sufficient condition that $R(S_i)$ be optimum for a given S_j. However, (3a) and (6) taken simultaneously provide in general only a necessary condition for an optimum design. It is easy to see that a condition that is necessary for A to be optimum given any B must still be necessary for A to be optimum given the optimum B. The loss of sufficiency is more difficult to understand, and suffice it to say that there exist counterexamples to the assumption that sufficiency obtains.[3]

As an example illustrative of our procedure and the fact that a staircase quantizer is not always optimum we consider the problem of the binary quantization of an input Σ related to the desired input I through the joint density $p(I, \Sigma)$ given for all I and Σ by

$$p(I, \Sigma) = \frac{\sqrt{3}}{2\pi} e^{-I^2 - \Sigma^2 - I|\Sigma|}. \tag{7}$$

We first calculate $E(I|\Sigma)$ using the well-known relation

$$p(I \mid \Sigma) = \frac{p(I, \Sigma)}{p(\Sigma)} \tag{8}$$

where $p(\Sigma)$ is obtained by integrating out I in (7).

$$p(\Sigma) = \frac{1}{2}\sqrt{\frac{3}{\pi}}\, e^{-3/4\Sigma^2}. \tag{9}$$

The conditional expectation $E(I|\Sigma)$ is given by

$$E(I \mid \Sigma) = \int_{-\infty}^{\infty} Ip(I \mid \Sigma)\, dI = \frac{-|\Sigma|}{2}. \tag{10}$$

If we insert (10) in (3a) we find

$$S_1 = \left\{\Sigma : -\infty \le -\frac{|\Sigma|}{2} \le \frac{R(S_1) + R(S_2)}{2}\right\}$$

and

$$S_2 = \left\{\Sigma : \frac{R(S_1) + R(S_2)}{2} \le -\frac{|\Sigma|}{2} < \infty\right\}. \tag{11}$$

Alternatively (11) may be rewritten as

$$S_1 = \{\Sigma : \Sigma \ge -R(S_1) - R(S_2)$$

and

$$\text{or} \quad \Sigma \le R(S_1) + R(S_2)\}$$

$$S_2 = \{\Sigma : R(S_1) + R(S_2) \le \Sigma \le -R(S_1) - R(S_2)\}. \tag{11a}$$

To find the reconstruction outputs $R(S_1)$ and $R(S_2)$ we employ (11a) in (6), define A equal to the negative of the sum of $R(S_1)$ and $R(S_2)$ and obtain

$$R(S_1) = \frac{\int_{-\infty}^{-A} -\frac{|\Sigma|}{2} p(\Sigma)\, d\Sigma + \int_{A}^{\infty} -\frac{|\Sigma|}{2} p(\Sigma)\, d\Sigma}{\int_{-\infty}^{-A} p(\Sigma)\, d\Sigma + \int_{A}^{\infty} p(\Sigma)\, d\Sigma} \tag{12}$$

and

$$R(S_2) = \frac{\int_{-A}^{A} -\frac{|\Sigma|}{2} p(\Sigma)\, d\Sigma}{\int_{-A}^{A} p(\Sigma)\, d\Sigma}.$$

Employing (9) in (12) we can obtain seemingly explicit integrals for $R(S_1)$ and $R(S_2)$. However, we must still solve for A, and we do so by adding the negative of $R(S_1)$ and $R(S_2)$ to obtain

$$A = \frac{\frac{1}{3}e^{-3/4A^2}}{\int_{A}^{\infty} e^{-3/4\Sigma^2}\, d\Sigma} + \frac{\frac{1}{3}(1 - e^{-3/4A^2})}{\int_{0}^{A} e^{-3/4\Sigma^2}\, d\Sigma} \tag{13}$$

Equation (13) is a transcendental equation for A which may be solved with the aid of a table of error functions. We conclude that A is approximately 0.80. With this value of A, $R(S_1)$ equals -0.614 and $R(S_2)$ equals -0.184; note that $-R(S_1) - R(S_2)$ equals 0.798 which is close to 0.80. Therefore, the optimum binary quantizer in this case assigns the output value -0.184 to all input values Σ whose absolute value is less than 0.80 and it assigns the output value of -0.614 to the remaining possible Σ values. This is not a staircase quantizer, and a sketch of its characteristic is shown below.

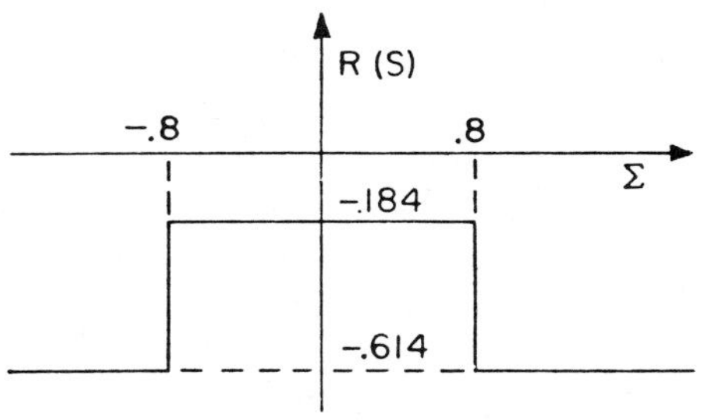

Fig. 1. Optimum quantizer for example.

[3] As a counterexample to the joint sufficiency of (3a) and (6) consider the binary quantization of the noiseless discrete random variable I which can take on either of the values 0, 0.95, 1.05, 2 each with probability 0.25. If we take $S_1 = \{I = 0\}$, $S_2 = \{I = 0.95, 1.05, 2\}$, $R(S_1) = 0$, and $R(S_2) = 1.333$ or $S_1 = \{1 = 0, 0.95\}$, $S_2 = \{I = 1.05, 2\}$, $R(S_1) = 0.475$, and $R(S_2) = 1.525$ then either set of solutions satisfies (3a) and (6) although the latter set yields a lower error.

Reprinted from *Franklin Inst. J.* **303**(6):549–561 (1977)

*Optimum Quantization for Local Decisions Based on Independent Samples**

by H. V. POOR *and* J. B. THOMAS
Department of Electrical Engineering, Princeton University, Princeton, New Jersey 08540, U.S.A.

ABSTRACT: *The problem of designing quantizers for use in decision-making systems is considered. Applying the theory of local tests, general criteria are derived for the optimal selection of quantizer parameters for the large-sample-size case. These criteria agree with previously established results based on optimization in terms of distance measures and are shown also to lead to that quantizer-decision system which is most efficient asymptotically. To illustrate the design procedure, several applications to signal detection are discussed.*

I. Introduction

A common problem in decision theory is that of deciding between two hypotheses which are, in some sense, close together. This is known as a local decision problem. An example is the detection of signals in additive noise when the signal strength is very small. For the local decision problem, the generalized Neyman–Pearson lemma gives solutions which are optimal in a statistical sense, and which often lead to physically practical systems. Frequently these designs involve non-linear operations on input data which can, at best, be implemented by approximating them with more standard devices. The purpose of this paper is to study the optimal design of decision systems when such non-linearities are replaced with quantizers.

In recent years, several authors have considered the problem of designing quantizers for use in decision systems (**1**–**5**). Various optimality criteria have been imposed including error probability (**1**, **2**), asymptotic efficiency (**3**, **5**) distance measures (**4**), and local power slope (**5**). In particular, the work of Kassam (**5**) serves as a basis for our present work. In (**5**), the design of quantizers for use in detectors for known signals in additive noise is studied. Kassam shows that the criteria of asymptotic efficiency and local power slope lead to identical quantizer designs. In this paper, we extend Kassam's work to the general problem of local decisions.

In Section II, preliminary assumptions and definitions are established, and a brief review of the local decision problem is included. The optimization of quantizers in terms of local power slope is the topic of Section III. Necessary conditions are derived for this case, and our results are seen to agree with

* This research was supported by the National Science Foundation under Grant ENG76-19808 and by the U.S. Army Research Office under Grant DAAG29-75-G-0192.

previous designs based on maximum Ali–Silvey distance criteria (**4**). Section IV considers the asymptotic efficiency of quantizer systems, and it is shown that the quantizer of Section III is most efficient asymptotically. In Section V, we consider several applications to signal detection including the detection of known and stochastic signals in white noise. A test for sufficiency of the derived necessary conditions is presented in the Appendix.

II. Preliminaries

A. *Local decisions*

Throughout this paper it is assumed that we observe a sequence $\{x_i\}_{i=1}^n \equiv \mathbf{x}$ of independent real samples, and that we have a corresponding sequence $\{P_\theta^{(i)};\ \theta \varepsilon \Theta \subseteq \mathcal{R}\}_{i=1}^n$ of indexed (i.e. parametric) classes of distributions on the real line. Singling out a particular θ_0 in a right-open set of Θ we wish to test

$$H_{\theta_0}: X_i \sim P_{\theta_0}^{(i)};\quad i = 1, \ldots, n$$

versus (1)

$$H_\theta: X_i \sim P_\theta^{(i)};\quad i = 1, \ldots, n$$

where $\theta > \theta_0$. For the local case we consider the limit as $\theta \to \theta_0$.

To find a test for (1) we impose two regularity conditions:

(i) $P_\theta^{(i)} \ll P_{\theta_0}^{(i)}$ for all $i = 1, \ldots, n$ and all $\theta > \theta_0$.

and

(ii) $\partial[\int_I \mathrm{d}P_\theta^{(i)}]/\partial\theta = \int_I (\partial L_\theta^{(i)}/\partial\theta)\, \mathrm{d}P_{\theta_0}$ for all $i = 1, \ldots, n$, all $\theta \geq \theta_0$, and all intervals I, where

$$L_\theta^{(i)} = \mathrm{d}P_\theta^{(i)}/\mathrm{d}P_{\theta_0}^{(i)};\quad i = 1, \ldots, n. \tag{2}$$

Further regularity conditions will be imposed later.

Under Conditions (i) and (ii), a size-α locally-optimum test for (1) is given by

$$\varphi_{lo}(\mathbf{x}) = \begin{cases} 1, & \mathfrak{J}(\mathbf{x}) > \tau \\ \gamma, & \mathfrak{J}(\mathbf{x}) = \tau \\ 0, & \mathfrak{J}(\mathbf{x}) < \tau \end{cases} \tag{3}$$

where

$$\mathfrak{J}(\mathbf{x}) = \partial \mathfrak{L}_\theta(\mathbf{x})/\partial\theta|_{\theta=\theta_0};$$

$\mathfrak{L}_\theta$ is the likelihood ratio between H_θ and H_{θ_0}; the threshold τ and randomization γ are chosen so that $E_{\theta_0}\{\varphi_{lo}\} = \alpha$; and we choose H_θ with probability $\varphi_{lo}(\mathbf{x})$ when $\mathbf{x}$ is observed. The generalized Neyman–Pearson lemma (**6**) asserts that φ_{lo} maximizes the slope of the power curve, $E_\theta\{\varphi\}$, at $\theta = \theta_0$, among all randomized tests φ satisfying $E_{\theta_0}\{\varphi\} \leq \alpha$.

We have the likelihood ratio

$$\mathfrak{L}_\theta(\mathbf{x}) = \prod_{i=1}^n L_\theta^{(i)}(x_i)$$

where $L_\theta^{(i)}$ is from (2). This gives the locally-optimum test statistic to be

$$\mathfrak{J}(\mathbf{x}) = \sum_{i=1}^{n} T^{(i)}(x_i) \tag{4}$$

where

$$T^{(i)} = \partial L_\theta^{(i)}/\partial\theta|_{\theta=\theta_0}; \quad i = 1, \ldots, n.$$

B. *Quantization*

Our objective is to choose a sequence $\{Q^{(i)}\}_{i=1}^n \equiv \mathbf{Q}$ of M-level quantizers to replace the non-linearities $T^{(i)}$ of Eq. (4). That is, we will consider the test

$$\varphi_{\mathbf{Q}}(\mathbf{x}) = \begin{cases} 1, \mathfrak{J}_{\mathbf{Q}}(\mathbf{x}) > \tau' \\ \gamma', \mathfrak{J}_{\mathbf{Q}}(\mathbf{x}) = \tau' \\ 0, \mathfrak{J}_{\mathbf{Q}}(\mathbf{x}) < \tau' \end{cases} \tag{5}$$

where

$$\mathfrak{J}_{\mathbf{Q}}(\mathbf{x}) = \sum_{i=1}^{n} Q^{(i)}(x_i);$$

and we wish to select the parameters of the quantizers $\mathbf{Q}$ in some optimal way.

For each i, $Q^{(i)}$ has M levels and we may represent it as a pair $(\mathbf{t}^{(i)}, \mathbf{q}^{(i)})$, where $\mathbf{q}^{(i)} \in \mathcal{R}^M$ are the levels of $Q^{(i)}$; the break-points $\mathbf{t}^{(i)}$ are such that $-\infty = t_0^{(i)} < t_1^{(i)} < \ldots < t_{M-1}^{(i)} < t_M^{(i)} = \infty$; and we take

$$Q^{(i)}(x) = q_k^{(i)} \quad \text{when} \quad x \in (t_{k-1}^{(i)}, t_k^{(i)}] \quad \text{for} \quad k = 1, \ldots, M.$$

III. Locally-Optimum Quantization

A. *Fixed breakpoints—locally-optimum choice of levels*

We first consider the case where the sequence of breakpoint vectors $\{\mathbf{t}^{(i)}\}_{i=1}^n$ is fixed, and we are free to choose only the level vectors $\{\mathbf{q}^{(i)}\}_{i=1}^n$. Here, the generalized Neyman–Pearson lemma can be used to make a locally-optimum choice of levels, as follows:

The post-quantization likelihood ratio, for fixed $\{\mathbf{t}^{(i)}\}_{i=1}^n$, is given by

$$\mathfrak{L}_\theta^{\mathbf{Q}}(\mathbf{x}) = \prod_{i=1}^{n} \{P_\theta^{(i)}(t_{k_i-1}^{(i)}, t_{k_i}^{(i)}]/P_{\theta_0}^{(i)}(t_{k_i-1}^{(i)}, t_{k_i}^{(i)}]\}$$

when

$$x_i \in (t_{k_i-1}^{(i)}, t_{k_i}^{(i)}]; \quad i = 1, \ldots, n.$$

The locally-optimum (post-quantization) test compares the statistic

$$\partial\mathfrak{L}_\theta^{\mathbf{Q}}(\mathbf{x})/\partial\theta|_{\theta=\theta_0} = \sum_{i=1}^{n} \{(\partial P_\theta^{(i)}(t_{k_i-1}^{(i)}, t_{k_i}^{(i)}]/\partial\theta|_{\theta=\theta_0})/P_{\theta_0}(t_{k_i-1}^{(i)}, t_{k_i}^{(i)}]\} \tag{6}$$

to a threshold. We note that Eq. (6) can be rewritten as

$$\partial \mathfrak{L}_\theta^{\mathbf{Q}}(\mathbf{x})/\partial\theta|_{\theta=\theta_0} = \sum_{i=1}^{n} Q^{(i)}(x_i)$$

if we make the choice

$$q_k^{(i)} = [\partial P_\theta^{(i)}(t_{k-1}^{(i)}, t_k^{(i)}]/\partial\theta|_{\theta=\theta_0}]/P_{\theta_0}^{(i)}(t_{k-1}^{(i)}, t_k^{(i)}] \quad (7)$$

for $k = 1, \ldots, M$ and $i = 1, \ldots, n$. Thus the choice Eq. (7) of levels allows the post-quantization test to be represented optimally (given $\{\mathbf{t}^{(i)}\}_{i=1}^n$) in the form of Eq. (5), and the problem of locally-optimum quantization is reduced to that of selecting the breakpoints $\{\mathbf{t}^{(i)}\}_{i=1}^n$.

Under Regularity Condition (ii), Eq. (7) becomes

$$q_k^{(i)} = \int_{t_{k-1}^{(i)}}^{t_k^{(i)}} T^{(i)}\, dP_{\theta_0}^{(i)} \Big/ \int_{t_{k-1}^{(i)}}^{t_k^{(i)}} dP_{\theta_0}^{(i)}; \quad k = 1, \ldots, M \quad \text{and} \quad i = 1, \ldots, n \quad (8)$$

where, as before, $T^{(i)} = \partial L_\theta^{(i)}/\partial\theta|_{\theta=\theta_0}$. Throughout Section III, $\{\mathbf{q}^{(i)}\}_{i=1}^n$ will denote the particular choice, given by Eq. (8), of levels and is understood to be a function of $\{\mathbf{t}^{(i)}\}_{i=1}^n$.

One further note: If we assume $T^{(i)}$ is continuous on $(t_{k-1}^{(i)}, t_k^{(i)}]$ and that $P_{\theta_0}^{(i)}$ has a continuous density (with respect to Lebesgue measure) on $(t_{k-1}^{(i)}, t_k^{(i)}]$, then, from the mean-value theorem and Eq. (8), we have that

$$q_k^{(i)} = T^{(i)}(\xi) \quad \text{for some} \quad \xi \in (t_{k-1}^{(i)}, t_k^{(i)}].$$

B. *Asymptotically-optimum choice of breakpoints*

We now wish to make a locally-optimum choice of the breakpoint vectors $\{\mathbf{t}^{(i)}\}_{i=1}^n$. To do so, we must consider the asymptotic ($n \to \infty$) case, and impose three additional regularity conditions:

(i)′ For each set $\{\mathbf{t}^{(i)}\}_{i=1}^\infty$ of non-trivial breakpoints and each $\theta \geq \theta_0$, there are numbers $e_1(\theta)$ and $e_2(\theta)$ such that $0 < e_1(\theta) \leq \text{Var}_\theta\,(Q^{(i)}) \leq e_2(\theta) < \infty$ for all i.

(ii)′ For each set of non-trivial breakpoints there is a number e_3 such that $|E_{\theta_0}\{(Q^{(i)})^3\}| \leq e_3 < \infty$ for all i.

(iii)′ For each i, $P_{\theta_0}^{(i)}$ has a continuous Lebesgue density $p_{\theta_0}^{(i)}$; and $T^{(i)}$ is continuous.

By definition, a locally-optimum choice of $\{\mathbf{t}^{(i)}\}_{i=1}^\infty$ will maximize the slope (with respect to θ) of the power curve $E_\theta\{\varphi_\mathbf{Q}\}$ at $\theta = \theta_0$, for fixed size α, over all possible sets of quantizers $\mathbf{Q}$ (note that $\mathbf{Q}$ is completely characterized by $\{\mathbf{t}^{(i)}\}_{i=1}^\infty$ since Eq. (8) gives $\{\mathbf{q}^{(i)}\}_{i=1}^\infty$ when $\{\mathbf{t}^{(i)}\}_{i=1}^\infty$ is chosen). Condition (i)′ assures that the test statistic

$$\mathfrak{J}_\mathbf{Q}(\mathbf{x}) = \sum_{i=1}^{n} Q^{(i)}(x_i)$$

will be asymptotically normally distributed under both H_{θ_0} and H_θ for any non-trivial choice of $\{\mathbf{t}^{(i)}\}_{i=1}^{\infty}$. Thus, for large n, the size of the test $\varphi_{\mathbf{Q}}$ will be α if we choose the threshold to be

$$\tau_{\alpha}^{(n)} = \Phi^{-1}(1-\alpha)[\mathrm{Var}_{\theta_0}(\mathfrak{I}_{\mathbf{Q}})]^{\frac{1}{2}} + E_{\theta_0}(\mathfrak{I}_Q) \tag{9}$$

where Φ is the standard normal distribution function. The power function of $\varphi_{\mathbf{Q}}$ is given, for large n, by

$$\beta_{\mathbf{Q}}(\theta) \triangleq E_{\theta}\{\varphi_{\mathbf{Q}}\} = 1 - \Phi([\tau_{\alpha}^{(n)} - E_{\theta}\{\mathfrak{I}_{\mathbf{Q}}\}]/[\mathrm{Var}_{\theta}(\mathfrak{I}_Q)]^{\frac{1}{2}}).$$

The slope of $\beta_{\mathbf{Q}}(\theta)$ at $\theta = \theta_0$ is now given by

$$\beta'_{\mathbf{Q}}(\theta_0) = -\eta(\Phi^{-1}(1-\alpha))\, \partial([\tau_{\alpha}^{(n)} - E_{\theta}\{\mathfrak{I}_{\mathbf{Q}}\}]/[\mathrm{Var}_{\theta}(\mathfrak{I}_Q)]^{\frac{1}{2}})/\partial\theta|_{\theta=\theta_0} \tag{10}$$

where η is the standard normal density function. Differentiating (10) gives

$$\beta'_{\mathbf{Q}}(\theta_0) = \eta(\Phi^{-1}(1-\alpha))\{\Phi^{-1}(1-\alpha)[\partial\, \mathrm{Var}_{\theta}(\mathfrak{I}_{\mathbf{Q}})/\partial\theta|_{\theta=\theta_0}]/[2\, \mathrm{Var}_{\theta_0}(\mathfrak{I}_{\mathbf{Q}})]$$
$$+ [\partial E_{\theta}\{\mathfrak{I}_{\mathbf{Q}}\}/\partial\theta|_{\theta=\theta_0}]/[\mathrm{Var}_{\theta_0}(\mathfrak{I}_{\mathbf{Q}})]^{\frac{1}{2}}\}. \tag{11}$$

We have,

$$\begin{aligned}
\partial E_{\theta}\{\mathfrak{I}_{\mathbf{Q}}\}/\partial\theta|_{\theta=\theta_0} &= \sum_{i=1}^{n}\sum_{k=1}^{M} \partial\left[\int_{t_{k-1}^{(i)}}^{t_k^{(i)}} q_k^{(i)}\, \mathrm{d}P_{\theta}^{(i)}\right]/\partial\theta|_{\theta=\theta_0} \\
&= \sum_{i=1}^{n}\sum_{k=1}^{M} q_k^{(i)} \int_{t_{k-1}^{(i)}}^{t_k^{(i)}} T^{(i)}\, \mathrm{d}P_{\theta_0}^{(i)} \\
&= \sum_{i=1}^{n}\sum_{k=1}^{M} (q_k^{(i)})^2 \int_{t_{k-1}^{(i)}}^{t_k^{(i)}} \mathrm{d}P_{\theta_0}^{(i)} = \sum_{i=1}^{n} E_{\theta_0}\{(Q^{(i)})^2\}
\end{aligned} \tag{12}$$

and, similarly,

$$\begin{aligned}
\partial\, \mathrm{Var}_{\theta}(\mathfrak{I}_{\mathbf{Q}})/\partial\theta|_{\theta=\theta_0} &= \sum_{i=1}^{n} \partial\, \mathrm{Var}_{\theta}(Q^{(i)})/\partial\theta|_{\theta=\theta_0} \\
&= \sum_{i=1}^{n}\left[E_{\theta_0}\{(Q^{(i)})^3\} - 2\left(\sum_{i=1}^{n} E_{\theta_0}\{Q^{(i)}\}\right) E_{\theta_0}\{(Q^{(i)})^2\}\right].
\end{aligned} \tag{13}$$

We note also that

$$\begin{aligned}
E_{\theta_0}\{Q^{(i)}\} &= \sum_{k=1}^{M} q_k^{(i)} \int_{t_{k-1}^{(i)}}^{t_k^{(i)}} \mathrm{d}P_{\theta_0}^{(i)} = \sum_{k=1}^{M} \int_{t_{k-1}^{(i)}}^{t_k^{(i)}} T^{(i)}\, \mathrm{d}P_{\theta_0}^{(i)} \\
&= \partial\left[\sum_{k=1}^{M} \int_{t_{k-1}^{(i)}}^{t_k^{(i)}} \mathrm{d}P_{\theta}\right]/\partial\theta|_{\theta=\theta_0} = \partial\left[\int_{-\infty}^{\infty} \mathrm{d}P_{\theta}\right]/\partial\theta = 0.
\end{aligned} \tag{14}$$

Combining Eqs. (11)–(14), we have

$$\beta'_Q(\theta_0) = \eta(\Phi^{-1}(1-\alpha))\left\{\Phi^{-1}(1-\alpha)\sum_{i=1}^{n} E_{\theta_0}\{(Q^{(i)})^3\}\Big/\left[2\sum_{i=1}^{n} \mathrm{Var}_{\theta_0}(Q^{(i)})\right]\right.$$
$$\left. + \left[\sum_{i=1}^{n} \mathrm{Var}_{\theta_0}(Q^{(i)})\right]^{\frac{1}{2}}\right\}. \tag{15}$$

Under Conditions (i)′ and (ii)′ we see that

$$\left|\sum_{i=1}^{n} E_{\theta_0}\{(Q^{(i)})^3\}/\sum_{i=1}^{n} \mathrm{Var}_{\theta_0}(Q^{(i)})\right| \le e_3/e_1(\theta_0) < \infty$$

and

$$ne_1(\theta_0) \le \sum_{i=1}^{n} \mathrm{Var}_{\theta_0}(Q^{(i)}) \le ne_2(\theta_0)$$

for all n. Thus, asymptotically, Eq. (15) becomes

$$\beta'_{\mathbf{Q}}(\theta_0) = \eta(\Phi^{-1}(1-\alpha))\left[\sum_{i=1}^{n} \mathrm{Var}_{\theta_0}(Q^{(i)})\right]^{\frac{1}{2}}.$$

From the above we see that, for the sequence $\mathbf{Q}$ to be locally optimum, we must choose $\mathbf{t}^{(i)}$ to maximize $\mathrm{Var}_{\theta_0}(Q^{(i)})$ for each i. We have

$$\mathrm{Var}_{\theta_0}(Q^{(i)}) = \sum_{k=1}^{M}\left[\int_{t_{k-1}^{(i)}}^{t_k^{(i)}} T^{(i)}\,\mathrm{d}P_{\theta_0}^{(i)}\right]^2 \Big/ \int_{t_{k-1}^{(i)}}^{t_k^{(i)}} \mathrm{d}P_{\theta_0}. \tag{16}$$

To search for maxima of Eq. (16) we set $\mathrm{grad}_{\mathbf{t}^{(i)}} \mathrm{Var}_{\theta_0}(Q^{(i)})$ equal to zero. Differentiating and using Assumption (iii)′ yields, for $k = 1, \ldots, (M-1)$,

$$\partial\, \mathrm{Var}_{\theta_0}(Q^{(i)})/\partial t_k^{(i)} = [2(q_k^{(i)} - q_{k+1}^{(i)})T^{(i)}(t_k^{(i)}) + (q_{k+1}^{(i)})^2 - (q_k^{(i)})^2]p_{\theta_0}^{(i)}(t_k^{(i)}). \tag{17}$$

Setting Eq. (17) equal to zero, we arrive at necessary conditions for the optimum $\mathbf{t}^{(i)}$ to satisfy; namely,

$$\text{(a)}\quad T^{(i)}(t_k^{(i)}) = (q_k^{(i)} + q_{k+1}^{(i)})/2; \quad k = 1, \ldots, (M-1)$$

where (from IIIA) (18)

$$\text{(b)}\quad q_k^{(i)} = \int_{t_{k-1}^{(i)}}^{t_k^{(i)}} T^{(i)}\,\mathrm{d}P_{\theta_0}^{(i)} \Big/ \int_{t_{k-1}^{(i)}}^{t_k^{(i)}} \mathrm{d}P_{\theta_0}^{(i)}; \quad k = 1, \ldots, M.$$

We note that Eq. (18a) is an asymptotic result while Eq. (18b) is valid for any sample size. Equations (18) agree with necessary conditions derived in Ref. (**4**) for a quantizer to maximize a class of distance measures in the local case. The maximum-distance conditions are valid for any sample size, a fact which provides some reinforcement for Eq. (18a) in the small-sample-size case.

The Condition (18b) is sufficient for a maximum (given $\{\mathbf{t}^{(i)}\}_{i=1}^{\infty}$) since its validity rests only on the generalized Neyman–Pearson lemma and not on the existence of a stationary point; however, the sufficiency of Eq. (18a) must be checked by examining the definiteness of the matrix of second partial derivatives. There is a general test for the sufficiency of Eq. (18a) and a discussion of this topic is included in the Appendix.

IV. Asymptotically-Most-Efficient Quantization

A. *Asymptotic relative efficiency*

We now wish to consider the optimization of the test $\varphi_{\mathbf{Q}}$ in terms of the alternate criterion of asymptotic relative efficiency (ARE). The ARE is frequently used to measure the relative performance of two tests under local, large-sample-size conditions. Recall that, for the situation of Eq. (1), the asymptotic efficiency of a test φ_1 relative to another test φ_2 is defined as (**7**),

$$\text{ARE}_{\varphi_1,\varphi_2} = \lim_{\substack{n\to\infty \\ \theta\to\theta_0}} e(\alpha, \theta, n)$$

where $e(\alpha, \theta, n)$ is the relative number of samples φ_2 requires to achieve the same power that φ_1 achieves for sample size n when both φ_1 and φ_2 are operating at level α and the distribution parameter is θ. The parameter θ approaches θ_0 in a way so that the power converges to a number $\beta \in (\alpha, 1)$. For many important cases the ARE is independent of α.

If φ_1 and φ_2 are of the form

$$\varphi_j(\mathbf{x}) = \begin{Bmatrix} 1, \mathfrak{T}_j(\mathbf{x}) > \tau_j \\ \gamma_j, \mathfrak{T}_j(\mathbf{x}) = \tau_j \\ 0, \mathfrak{T}_j(\mathbf{x}) < \tau_j \end{Bmatrix}; \quad j = 1, 2 \tag{20}$$

where

$$\mathfrak{T}_j(\mathbf{x}) = \sum_{i=1}^{n} \mu_j^i(x_i)$$

and the $\{\mu_j^i\}$ are deterministic functions, then $\text{ARE}_{\varphi_1,\varphi_2}$ can be calculated readily. Under various groups of assumptions on $\mathfrak{T}_j(\mathbf{x})$ (e.g. asymptotic normality) we have (**8, 9**)

$$\text{ARE}_{\varphi_1,\varphi_2} = \eta_1/\eta_2 \tag{21}$$

where

$$\eta_j = \lim_{n\to\infty} \{[\partial E_\theta\{\mathfrak{T}_j\}/\partial\theta|_{\theta=\theta_0}]^2/[n \text{ Var}_{\theta_0}(\mathfrak{T}_j)]\} \tag{22}$$

is the efficacy of detector φ_j. Note that, for a given sequence $\{P_\theta^{(i)}\}_{i=1}^{\infty}$ of classes of distributions, the test with the higher efficacy is the most efficient asymptotically.

B. *The maximum-efficacy quantizer sequence*

From Eq. (5), we see that the test φ_Q is in the form of Eq. (20). If we again impose Regularity Conditions (i)′ and (iii)′ of Section IIIB, and consider only those quantizer sequences for which

$$\sum_{i=1}^{n} \sum_{k=1}^{M} q_k^{(i)} \int_{t_{k-1}^{(i)}}^{t_k^{(i)}} T^{(i)}\, dP_{\theta_0}^{(i)} \neq 0 \quad \text{for} \quad n = 1, 2, \ldots, \tag{23}$$

then we can apply the method of efficacies [Eqs. (21) and (22)] to compare various quantizer systems. [Note: we are no longer restricting the choice of levels to be given by Eq. (8).] Thus, the asymptotically-most-efficient quantizer sequence (under our restrictions) will be that which maximizes efficacy over all possible sequences.

The efficacy of the test $\varphi_{\mathbf{Q}}$ based on the sequence $\mathbf{Q}$ of quantizers can be calculated from Eq. (22) to be

$$\eta_{\mathbf{Q}} = \lim_{n\to\infty} \eta_n(\mathbf{Q})$$

where

$$\eta_n(\mathbf{Q}) = \left[\sum_{i=1}^{n}\sum_{k=1}^{M} q_k^{(i)} \int_{t_{k-1}^{(i)}}^{t_k^{(i)}} T^{(i)}\,\mathrm{d}P_{\theta_0}^{(i)}\right]^2 \Big/ n\left[\sum_{i=1}^{n}\sum_{k=1}^{M} (q_k^{(i)})^2 \int_{t_{k-1}^{(i)}}^{t_k^{(i)}} \mathrm{d}P_{\theta_0}^{(i)} - \left(\sum_{i=1}^{n}\sum_{k=1}^{M} q_k^{(i)} \int_{t_{k-1}^{(i)}}^{t_k^{(i)}} \mathrm{d}P_{\theta_0}^{(i)}\right)^2\right]. \tag{24}$$

As in Section III, it is convenient to determine first the optimum level vectors $\{\mathbf{q}^{(i)}\}_{i=1}^{\infty}$ in terms of the breakpoints $\{\mathbf{t}^{(i)}\}_{i=1}^{\infty}$. Define, for $n = 1, 2, \ldots,$

$$m(\mathbf{x}; \{\mathbf{t}^{(i)}\}_{i=1}^{n}) = \sum_{i=1}^{n} M^{(i)}(x_i) \tag{25}$$

where

$$M^{(i)}(x) = \int_{t_{k-1}^{(i)}}^{t_k^{(i)}} T^{(i)}\,\mathrm{d}P_{\theta_0}^{(i)} \Big/ \int_{t_{k-1}^{(i)}}^{t_k^{(i)}} \mathrm{d}P_{\theta_0}^{(i)} \quad \text{when} \quad x \in (t_{k-1}^{(i)}, t_k^{(i)}], \quad k = 1, \ldots, M.$$

We have

$$E_{\theta_0}\{m(\mathbf{x};\ \{\mathbf{t}^{(i)}\}_{i=1}^{n})\} = 0$$

for all n and all $\{\mathbf{t}^{(i)}\}_{i=1}^{n}$. Equation (24) can be rewritten as:

$$\eta_n(\mathbf{Q}) = \mathrm{Cov}_{\theta_0}^2(\mathfrak{F}_{\mathbf{Q}}, m(\{\mathbf{t}^{(i)}\}_{i=1}^{n}))/n\ \mathrm{Var}_{\theta_0}(\mathfrak{F}_{\mathbf{Q}}). \tag{26}$$

Equation (26) and the Schwarz inequality imply that

$$\eta_n(\mathbf{Q}) \le \mathrm{Var}_{\theta_0}(m(\{\mathbf{t}^{(i)}\}_{i=1}^{n}))/n$$

with equality if and only if

$$\mathfrak{F}_{\mathbf{Q}}(\mathbf{x}) = a\,m(\mathbf{x}; \{\mathbf{t}^{(i)}\}_{i=1}^{n}) + b$$

for all $\mathbf{x} \in \mathcal{R}^n$ and some numbers $a \neq 0$ and b. Thus, given $\{\mathbf{t}^{(i)}\}_{i=1}^{\infty}$, the maximum possible efficacy will be achieved if we choose

$$\mathfrak{F}_{Q}(\mathbf{x}) = m(\mathbf{x}; \{\mathbf{t}^{(i)}\}_{i=1}^{n}) \quad \text{for} \quad n = 1, 2, \ldots$$

i.e. if we take [from Eq. (25)]

$$q_k^{(i)} = \int_{t_{k-1}^{(i)}}^{t_k^{(i)}} T^{(i)}\,\mathrm{d}P_{\theta_0}^{(i)} \Big/ \int_{t_{k-1}^{(i)}}^{t_k^{(i)}} \mathrm{d}P_{\theta_0}^{(i)} \quad k = 1, \ldots, M; \quad i = 1, 2, \ldots \tag{27}$$

Comparing Eqs. (8) and (27) we see that, for fixed $\{\mathbf{t}^{(i)}\}_{i=1}^{\infty}$, this choice of levels is both locally-optimum and asymptotically-most-efficient.

Applying Eq. (27) to $\eta_n(\mathbf{Q})$ of Eq. (24) we have

$$\eta_n(\mathbf{Q}) = \sum_{i=1}^{n} \mathrm{Var}_{\theta_0}(Q^{(i)})/n$$

where $Q^{(i)}$ is determined by $\{\mathbf{t}^{(i)}\}_{i=1}^{\infty}$ through Eq. (27). Thus, as in Section III, the best quantizer $Q^{(i)}$ maximizes $\mathrm{Var}_{\theta_0}(Q^{(i)})$, and Eq. (18a) gives the necessary conditions. Equation (27) comes from the Schwarz inequality and therefore is sufficient; however, as before, the sufficiency of the necessary conditions in Eq. (18a) must be checked.

V. Applications to Signal Detection

To illustrate the results of the preceding sections it is interesting to consider some applications to signal detection. The first of these is Kassam's quantization results (**5**) for a known signal in additive noise. The other two examples are models for stochastic signals in noise. A treatment of the locally-optimum detection problem (without quantization) for these examples can be found in (**10**) for the first case (A), and in (**11**) for the final two (B and C).

A. *Known signals in additive noise (Kassam's result)*

Here, the situation of Eq. (1) is

$$H_{\theta_0}: X_i \sim f(x); \quad i = 1, \ldots, n$$

versus (28)

$$H_\theta: X_i \sim f(x - \theta s_i); \quad i = 1, \ldots, n$$

where f is a differentiable Lebesgue density function; the signal sequence $\{s_i\}_{i=1}^{n}$ is known; and the parameter θ is positive. For the local case we consider $\theta \to 0^+$.

The ith likelihood ratio for (28) is

$$L_\theta^{(i)}(x) = f(x - \theta s_i)/f(x)$$

and differentiating yields

$$T^{(i)}(x) = -s_i f'(x)/f(x).$$

From Eq. (18), the optimum quantizer sequence is given by

$$Q^{(i)}(x) = s_i Q(x)$$

where $Q = (\mathbf{t}, \mathbf{q})$ is the solution to

$$-f'(t_k)/f(t_k) = (q_k + q_{k+1})/2; \quad k = 1, \ldots, (M-1)$$

and

$$q_k = [f(t_{k-1}) - f(t_k)] \Big/ \int_{t_{k-1}}^{t_k} f(x)\,dx; \quad k = 1, \ldots, M.$$

The regularity conditions of Section III are satisfied if $0 < c_1 \le |s_i| \le c_2 < \infty$ for all i and for some c_1 and c_2.

B. *Stochastic signals in noise* (*additive-noise model*)

For this case, the situation in Eq. (1) becomes

$$H_{\theta_0}: X_i \sim f(x); \quad i = 1, \ldots, n$$

versus (29)

$$H_\theta: X_i \sim \int_{-\infty}^{\infty} f(x - \theta^{\frac{1}{2}} s)\, \mathrm{d}G_i(s); \quad i = 1, \ldots, n$$

where f is a twice-differentiable Lebesgue density function; $\{G_i\}_{i=1}^{n}$ is a sequence of zero-mean distribution functions corresponding to a sequence of independent samples from a stochastic signal; and θ is positive. Again, the local case corresponds to the limit as $\theta \to 0^+$.

The ith likelihood ratio for (29) is

$$L_\theta^{(i)}(x) = \int_{-\infty}^{\infty} f(x - \theta^{\frac{1}{2}} s)\, \mathrm{d}G_i(s)/f(x)$$

and the ith locally-optimum non-linearity is

$$T^{(i)}(x) = \sigma_i^2 f''(x)/2f(x)$$

where

$$\sigma_i^2 = \int_{-\infty}^{\infty} s^2\, \mathrm{d}G_i(s).$$

The locally-optimum quantizer sequence is given by

$$Q^{(i)}(x) = \sigma_i^2 Q(x)/2$$

where $Q = (\mathbf{t}, \mathbf{q})$ is the solution to

$$f''(t_k)/f(t_k) = (q_k + q_{k+1})/2; \quad k = 1, \ldots, (M-1)$$

and

$$q_k = [f'(t_k) - f'(t_{k-1})] / \int_{t_{k-1}}^{t_k} f(x)\, \mathrm{d}x; \quad k = 1, \ldots, M.$$

The regularity conditions of Section III are satisfied if $0 < c_3 \leq \sigma_i^2 \leq c_4 < \infty$ for all i and some c_3 and c_4.

C. *Stochastic signals in noise* (*scale-change model*)

For this model, Eq. (1) becomes

$$H_{\theta^0}: X_i \sim f(x); \quad i = 1, \ldots, n$$

versus

$$H_\theta: X_i \sim f(x/\nu_i)/\nu_i; \quad i = 1, \ldots, n$$

where

$$\nu_i = [1 + \theta \sigma_i^2/\sigma^2]^{\frac{1}{2}};$$

f is a differentiable Lebesgue density with variance σ^2; the sequence $\{\sigma_i^2\}_{i=1}^n$ is a sequence of signal variances; and θ is positive. For the local case we consider the limit as $\theta \to 0^+$.

We have

$$L_\theta^{(i)}(x) = f(x/\nu_i)/\nu_i f(x)$$

and

$$T_\theta^{(i)}(x) = (\sigma_i^2/2\sigma^2)(-xf'(x)/f(x) - 1).$$

The locally-optimum quantizer sequence is found from

$$Q^{(i)}(x) = (\sigma_i^2/2\sigma^2)Q(x)$$

where $Q = (\mathbf{t}, \mathbf{q})$ is the solution to

$$(-t_k f'(t_k)/f(t_k) - 1) = (q_k + q_{k+1})/2; \quad k = 1, \ldots, (M-1)$$

and

$$q_k = [t_{k-1}f(t_{k-1}) - t_k f(t_k)] \Big/ \int_{t_{k-1}}^{t_k} f(x)\,dx.$$

As in Subsection *VB*, the regularity conditions of Section III are satisfied if $0 < c_3 \le \sigma_i^2 \le c_4 < \infty$ for all i and for some c_3 and c_4.

Appendix

A test for sufficiency of the necessary conditions

From Sections III and IV we see that the optimum choice of the quantizer $Q^{(i)}$ is made by maximizing $\mathrm{Var}_{\theta_0}(Q^{(i)})$ over $\mathbf{t}^{(i)}$ while making the choice given by Eq. (8) for $\mathbf{q}^{(i)}$. Equation (18a) gives necessary conditions for a maximum. A sufficient condition for a set $\mathbf{t}_0^{(i)}$ satisfying (18a) to give a local maximum of $\mathrm{Var}_{\theta_0}(Q^{(i)})$ is that the $(M-1)\times(M-1)$ matrix $\underset{\sim}{\mathcal{H}}^{(i)}$ whose j-kth element is given by

$$h_{jk}^{(i)} = \partial^2 \mathrm{Var}_{\theta_0}(Q^{(i)})/\partial t_k^{(i)}\,\partial t_j^{(i)}\big|_{\mathbf{t}^{(i)}=\mathbf{t}_0^{(i)}} \tag{A1}$$

be negative definite.

For each i, $\underset{\sim}{\mathcal{H}}^{(i)}$ will be a symmetric tridiagonal band matrix; i.e.

$$h_{kj}^{(i)} = h_{jk}^{(i)} = \begin{cases} b_k^{(i)}, & \text{if } j = k+1 \\ a_k^{(i)}, & \text{if } j = k \\ 0, & \text{if } |j-k| > 1. \end{cases} \tag{A2}$$

The definiteness of this type of matrix may be checked by a simple test (**12**);

Lemma: An $(M-1)\times(M-1)$ matrix $\underset{\sim}{\mathcal{H}}^{(i)}$ whose elements are as in (A2) is negative definite if, and only if, $l_k^{(i)} < 0$ for $k = 1, \ldots, (M-1)$, where $l_1^{(i)} = a_1^{(i)}$ and $l_{k+1}^{(i)} = (a_{k+1}^{(i)} - (b_k^{(i)})^2/l_k^{(i)})$.

Proof: The matrix $\underset{\sim}{\mathcal{H}}^{(i)}$ is negative definite if and only if $(-\underset{\sim}{\mathcal{H}}^{(i)})$ is positive definite; and a symmetric matrix is positive definite if and only if all of its principal minors have positive determinants. Thus, $\underset{\sim}{\mathcal{H}}^{(i)}$ is negative definite if and only if

$$|-\underset{\sim}{\mathcal{H}}_k^{(i)}| > 0 \quad \text{for} \quad k = 1, \ldots, (M-1) \tag{A3}$$

where $\underset{\sim}{\mathcal{H}}_k^{(i)}$ is the kth principal minor of $\underset{\sim}{\mathcal{H}}^{(i)}$. It can be shown easily that

$$l_k^{(i)} = |\underset{\sim}{\mathcal{H}}_k^{(i)}|/|\underset{\sim}{\mathcal{H}}_{k-1}^{(i)}| \quad \text{for} \quad k = 1, \ldots, (M-1)$$

(we take $|\underset{\sim}{\mathcal{H}}_0^{(i)}| \triangleq 1$). We have

$$|-\underset{\sim}{\mathcal{H}}_k^{(i)}| = (-1)^k\, |\underset{\sim}{\mathcal{H}}_k^{(i)}|$$

which implies

$$l_k^{(i)} = -|-\underset{\sim}{\mathcal{H}}_k^{(i)}|/|-\underset{\sim}{\mathcal{H}}_{k-1}^{(i)}|; \quad k = 1, \ldots, M.$$

Thus, from (A3), $\mathcal{H}^{(i)}$ is negative definite if and only if $l_k^{(i)} < 0$ for $k = 1, \ldots, (M-1)$, which completes the proof.

Returning to the particular matrix $\underset{\sim}{\mathcal{H}}^{(i)}$ of Eq. (A1) we have, deleting superscripts,

$$b_k = (q_{k+2} - q_{k+1})(q_{k+1} - q_k)p_{\theta_0}(t_k)p_{\theta_0}(t_{k+1})/2 \int_{t_k}^{t_{k+1}} \mathrm{d}P_{\theta_0}; \quad k = 1, \ldots, (M-2)$$

and

$$a_k = 2p_{\theta_0}(t_k)(q_{k+1} - q_k)[\partial[(q_k + q_{k+1})/2]/\partial t_k - T'(t_k)]; \quad k = 1, \ldots, (M-1)$$

where

$$p_{\theta_0} = \mathrm{d}P_{\theta_0}/\mathrm{d}\lambda.$$

This test may or may not simplify for a particular problem; however, most of the quantities involved must be calculated to solve the necessary conditions [Eq. (18)] and thus will be readily available for a quick computational test for sufficiency.

References

(1) Y-C. Ching and L. Kurz, "Nonparametric detectors based on m-interval partitioning", *IEEE Trans. Inf. Theory,* Vol. IT—18, No. 2, pp. 251–257, March 1972.

(2) W. J. Bushnell and L. Kurz, "The optimization and performance of detectors based on partition tests", *Proc. 12th Allerton Conf. Circuit and System Theory,* pp. 1016–1023, Oct. 1974.

(3) S. A. Kassam and J. B. Thomas, "Generalizations of the sign detector based on conditional tests", *IEEE Trans. Commun.,* Vol. COM-24, No. 5, pp. 481–487, May 1976.

(4) H. V. Poor and J. B. Thomas, "Maximum-distance quantization for detection", *Proc. 14th Allerton Conf. Circuit and System Theory,* pp. 925–934, Sept.–Oct. 1976.

(5) S. A. Kassam, "Optimum quantization for signal detection", *IEEE Trans. Commun.,* Vol. COM-25, No. 5, pp. 479–484, May 1977.

(6) T. S. Ferguson, "Mathematical Statistics", Academic Press Inc., New York, 1967.

(7) G. E. Noether, "Elements of Nonparametric Statistics", John Wiley, New York, 1967.

(8) J. Capon, "On the asymptotic efficiency of locally optimum detectors", *IRE Trans. Inf. Theory,* Vol. IT—7, No. 2, pp. 67–71, April 1961.

(9) C. W. Helstrom, "Statistical Theory of Signal Detection", Pergamon, Oxford, 1968.

(**10**) J. H. Miller and J. B. Thomas, "Detectors for discrete-time signals in non-Gaussian noise", *IEEE Trans. Inf. Theory*, Vol. IT—18, No. 2, pp. 241–250, March 1972.

(**11**) H. V. Poor and J. B. Thomas, "Local detection of stochastic signals in additive noise", *Proc. 14th Allerton Conf. Circuit and System Theory*, pp. 432–440, Sept.–Oct. 1976.

(**12**) P. E. Fleischer, "Sufficient conditions for achieving minimum distortion in a quantizer", *IEEE Int. Conv. Rec.*, pp. 104–111, 1964.

AUTHOR CITATION INDEX

SUBJECT INDEX

About the Editor

PETER F. SWASZEK received the B.S.E.E. degree from New Jersey Institute of Technology in 1978 and the M.S.E.E. and Ph.D. degrees from Princeton University in 1980 and 1982, respectively. From 1982 to 1984 he was an assistant professor of electrical engineering at the University of Washington. Since 1984 he has been an assistant professor of electrical engineering at the University of Rhode Island. Professor Swaszek became interested in quantization problems during his studies at Princeton University, devoted a portion of his doctoral thesis to several aspects of the problem, and has published several papers in the area.